Trace Element Analytical Chemistry in Medicine and Biology

Trace Element Analytical Chemistry in Medicine and Biology

Volume 4

Proceedings of the Fourth International Workshop
Neuherberg, Federal Republic of Germany,
April 1986

Editors
Peter Brätter · Peter Schramel

Walter de Gruyter · Berlin · New York 1987

Editors
Peter Brätter, Professor, Dr. rer. nat.
Hahn-Meitner-Institut Berlin GmbH
Research Group – Trace Elements in Health and Nutrition
Glienicker Straße 100
D-1000 Berlin 39
Germany

Peter Schramel, Dr. phil.
Gesellschaft für Strahlen- und Umweltforschung mbH
Institut für ökologische Chemie
Ingolstädter Landstraße 1
D-8402 Neuherberg
Federal Republic of Germany

CIP-Kurztitelaufnahme der Deutschen Bibliothek

Trace element analytical chemistry in medicine and biology : proceedings of the . . . internat. workshop / . . . Internat. Workshop on Trace Element Analyt. Chemistry in Medicine and Biology. – Berlin ; New York : de Gruyter

NE: International Workshop on Trace Element Analytical Chemistry in Medicine and Biology

4. Neuherberg, Federal Republic of Germany, April 1986. – 1987.

ISBN 3-11-010905-0 (Berlin)
ISBN 0-89925-296-6 (New York)

 Printing: Druckerei Gerike GmbH, Berlin. – Binding: Dieter Mikolai, Berlin. – Printed in Germany.

PREFACE

This year the fourth workshop on "Trace Element Analytical Chemistry in Medicine and Biology" was organized again by the GSF (Gesellschaft für Strahlen- und Umweltforschung mbH) and the working group "Trace Elements in the Life Sciences" of the AGF (Arbeitsgemeinschaft der Großforschungseinrichtungen) at Neuherberg.

Six years ago we started this series of meetings with international and multi-disciplinary character in order to intensify the mutual transfer of scientific knowledge in this field. The general theme of the first workshop was the analysis of essential trace elements. Two years later the aim of the second meeting was to stimulate discussion on the deficiency problems of essential trace elements and their analysis. At the 3rd workshop a lot of time was devoted to the increasing importance of the trace element selenium and recent developments in the field of some "newer trace elements".

It was the aim of the present workshop to consider modern nutritional aspects of trace elements as well as their role in clincal states. The invited experts dealt with the important aspects of bioavailibility (A.E. Harmuth-Hoene), dietary requirements (L.M. Klevay) and recommendations (R. Schelenz), interactions (R.J. Shamberger) and toxic action (M. Anke) of trace elements as well as their role in metabolic processes (M. Kirchgessner, C.F. Mills), infant nutrition (H.B. von Stockhausen), parenteral nutrition (K. Ladefoget, W. Prellwitz) diagnosis and therapy of cancer (G.N. Schrauzer) and psychiatric disorders (C.J.M.van Tiggelen). Each of the short contributions was presented in the auditorium and extensively discussed.

It is clear that our knowledge of trace element distribution in the body, the delineation of deficiency situations of the organism, and the identification of trace element-containing biologically active molecules depend on the contemporary state of the art in trace element analytical chemistry. Although at this workshop the analysis of samples did not feature prominently, this aspect will be a major feature at the next meeting. Now as before it is our hope that developments in analytical chemistry will help to extend the research frontiers in biochemistry, medicine, biology and nutrition related to trace elements.

This book contains the 56 papers accepted by the scientific committee. As it was again our aim to produce the book fairly rapidly, the manuscripts were not extensively edited. We are indebted to all the participants, for their presentations and for their valuable support of interdisciplinary research.

Our thanks are also due the chairmen, the referees of the poster session and all who contributed to the discussions.

October 1986

P. Brätter

P. Schramel

CONTENTS

Session V: SIGNIFICANCE OF TRACE ELEMENTS IN MEDICINE
Plenary Lectures given by: G.N. Schrauzer
C.J.M. van Tiggelen

Session I

MODELS FOR TRACE ELEMENT METABOLISM

Plenary Lectures given by:

M. Kirchgeßner

C.F. Mills

Chairman:

L.M. Klevay

TRACE ELEMENT REQUIREMENT MODELS BASED ON PHYSIOLOGICAL AND METABOLIC PROCESSES

M. Kirchgeßner, A.M. Reichlmayr-Lais, E. Weigand

Institut für Ernährungsphysiologie der Technischen Universität München D-8050 Freising-Weihenstephan

Introduction

Because of essential functions of trace elements in metabolism, mainly as integral constituents or as effectors of enzymes and hormones, their nutritional supply must be sufficiently high to meet the metabolic requirements. The supply status of a trace element may be divided into five ranges (see Kirchgeßner and Reichlmayr-Lais 1985):

1. deficient: characterized by clinical symptoms

2. suboptimal: there are no conspicuous clinical symptoms, however biochemical responses are measurable

3. optimal: optimal functioning of metabolism, promoting normal health and performance

4. subtoxic: intake in excess of the optimal supply characterized by biochemical changes without manifest clinical symptoms

5. toxic: characterized by clinical symptoms.

Trace Element – Analytical Chemistry in Medicine and Biology, Vol. 4

Over a certain range of supply, which differs from element to element, the body is capable by regulating absorption, turnover and excretion of maintaining stable trace element concentrations that are functionally necessary and physiologically tolerable for cells and their subcellular compartments. This drive for homeostasis may also be interpreted as the measures for matching the nutritional supply with the current metabolic needs. The analytically determined amount of a trace element in the food cannot be accounted as fully available for meeting the functional requirements, because "losses" occur during absorption and intermediary metabolism. Dietary allowance must, therefore, consider the efficiency with which a trace element is utilized for its functional roles (Kirchgeßner et al. 1974). For this very reason, a distinction is made between the gross and net requirement. The gross requirement refers to the dietary quantity of intake of the trace element that is adequately

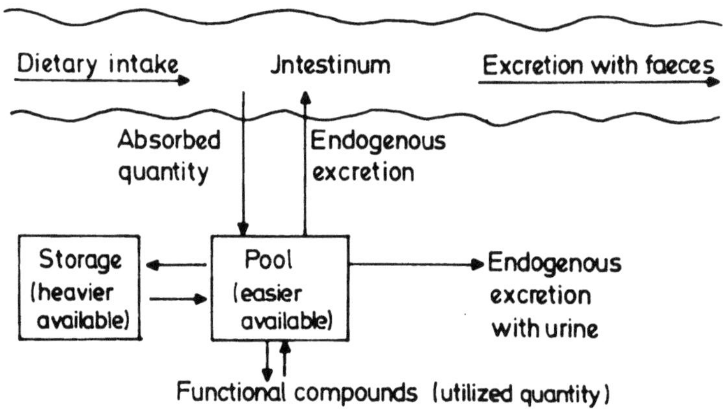

Fig. 1: Model for metabolism of trace elements

high, accounting for losses, to meet the net requirement, i. e. the functional needs at the molecular level. In Fig. 1 a simplified model of the metabolism of trace elements is shown. It elucidates that losses may be encountered not only during absorption from the digestive tract but also during intermediary metabolism. This makes evident the complexity of trace element utilization and hence of assessing requirements.

In the following, three concepts of assessing trace element requirements are presented and evaluated.

1. Balance Studies

In balance studies, intake is set against excretion. In the assessment of trace element excretion, only the amounts eliminated in feces and urine are generally considered. Losses via sweat, epithelial slouyhing or hair are largely neglected, in part, because their separate collection demands additional efforts. Neglecting these surface losses seems justified only if they are very small compared with the amounts excreted in feces and urine.

In the adult human or animal, the balance is zero under steadystate conditions, that is intake equals excretion. In the state of production, in particular body growth or gravidity, trace element balance is positive provided that the dietary supply is adequate, this is to say that retention occurs. Of the adult body the requirements are surely not met if the balance is zero or even negative in productive states of the body or if the balance is negative in the state of maintenance. The task of spezifying the required intake on the basis of balance data is aggravated by the homeostatic regulation of trace element metabolism. Within certain bounds of supply, the body attempts to compensate for a reduction in dietary intake by increasing the efficiency of absorption and/or lowering the

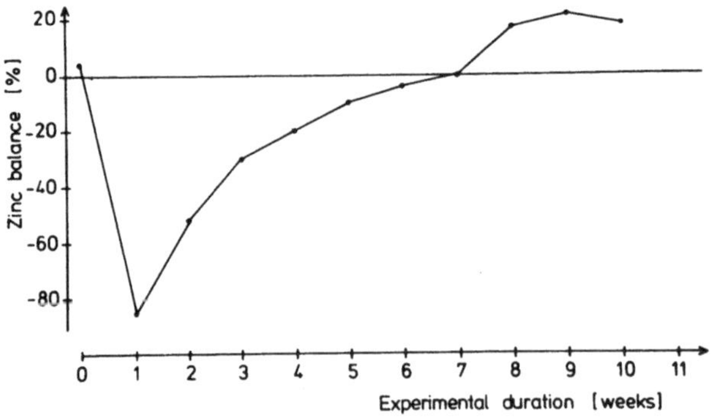

Fig. 2: Effect of depletion of Zinc on Zinc balance

excretion rate, and, conversely, attempts to counterbalance an increased supply by reducing the absorption rate and/or raising the excretion rate. In studies with dairy cows, Kirchgeßner und Schwarz (1976) observed that Zn balance (including the output in milk) approached the zero level after an initial period of negative balance in response to dietary Zn depletion even though the cows ultimately showed clinical signs of Zn deficiency (Fig. 2). Evidently, the body is capable of attaining a zero balance at different levels.

Also, time plays a role in reaching a steady balance. Interpretation of balance data, therefore, must take into account the duration of the study, in particular the length of the preliminary and the collection periods. If, for examples the length of the experiment is too short, balance may not get be fully adapted to the particular supply level and the findings may be misleading. Therefore, aspects of the length of the preliminary and collection periods are of particular importance in situations were trace element supply was inadequate before the balance study. Consequently, the supply status of the trace element under study must be evaluated critically when results of balance studies are interpreted.

Additional difficulties arise in the determination of the requirements for growth and reproduction. The optimal extent of positive balance in these stages of production is difficult to decide upon. Closely associated with the positive retention is the question of the desirable extent of trace element storage. In this context, superretention during gravidity is also of concern. This extra retention in maternal tissues outside the reproductive organs leads to reserves that are depleted again in the course of lactation. Superretention has been found to be particularly pronounced, provided the trace element supply was adequate, in the case of copper more than in the case of other trace elements investigated (Kirchgeßner et al. 1978).

Further limitations to the specification of trace element requirements by balance studies are imposed by analytical aspects. Since balance data are computed as the difference between intake and excretion and its values are often very small compared with its determining quantities, it is frequently impossible to assess its magnitude with satisfactory precision. The large variation mainly results from the summation of random, and perhaps, even systematic error sources encountered in the determination of intake and excretion. A change in matrix is also involved. In assessing intake and excretion, utmost care is needed to accurately determine these quantities and to avoid contaminations in processing the samples. Accordingly, it is commonly nessessary to use materials low in trace elements and to pretreat them as an additional precaution. Of particular relevance are these analytical aspects in the case of the so-called ultra trace elements which are present in foods, the body and excreta only in minute concentrations or are particularly difficult to analyze.

2. Dose-response Relationships

In the derivation of requirements from dose-response relationships, the response of a suitable criterion is investigated in relation to the dose over a relatively wide range of supply. As a rule, the relationships between criterion and dose are exponential and may be analyzed by regression analysis. Fig. 3 shows ideal dose-response relationships. An optimal level of supply of the trace element is reached if the differential response to the further increase in the supply of the element is relatively small or even absent. The response

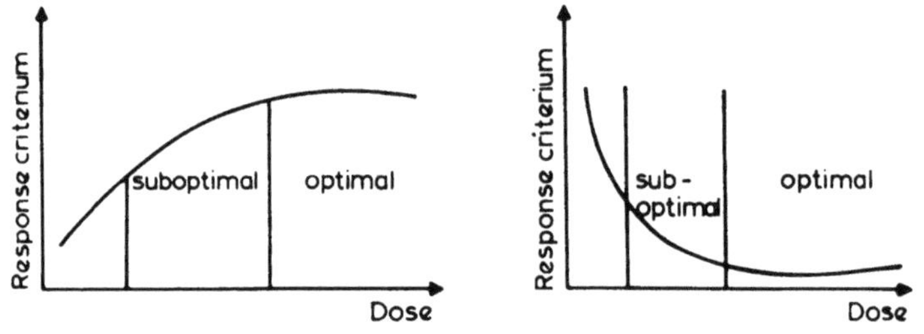

Fig. 3: Dose-response relationships

criteria to be chosen have to be specific for and sensitive to the particular trace element under study. A criterion is sensitive if it reveals latent states of deficient supply and thus responds to even slight changes in supply. Suitable criteria are for the most part biochemical parameters showing adverse changes in response to suboptimal supply as compared to optimal intake. The intensity of suboptimal supply of a trace element may be subdivided into two stages:

Degree 1: characterized by

a) an increase in fractional retention of the trace element as the result of an improved efficiency of true absorption and/or decreased rate of excretion from endogenous sources

b) a mobilization of the trace element from depot compounds, leading to a depletion of the analytical content of these depot compounds,

c) an increase in the binding capacity of transport proteins.

Degree 2: characterized by

a decrease in functional compounds, for example, a reduction in the activity of enzymes.

While the biochemical changes that indicate latent states of an adequate trace element supply are also prevalent in the case of a conspicuous deficiency, there are no manifest clinical symptoms in the situation of suboptimal supply.

There are several advantages to studies of dose-response relationships in which the dose level starts with a suboptimal supply level. First, it is not necessary to compound diets severely deficient of the element concerned, which task may be particularly arduous in the case of ultra trace elements. Secondly, this approach, avoiding severe depletions may also be acceptable for studies with humans.

Proceeding from these boundaries of the suboptimal supply range, a variety of response criteria may be chosen such as retention, absorption, excretion of the trace element, mobilization of storage compounds, the binding capacity of transport proteins and the activity of functional compounds. In the following a few examples shall be presented for these parameters.

In the case of nickel, a specific functional compound is not yet known for the animal or human body. It has, therefore, been attempted to estimate the requirement of this trace metal on the basis of its fractional retention in relation to intake

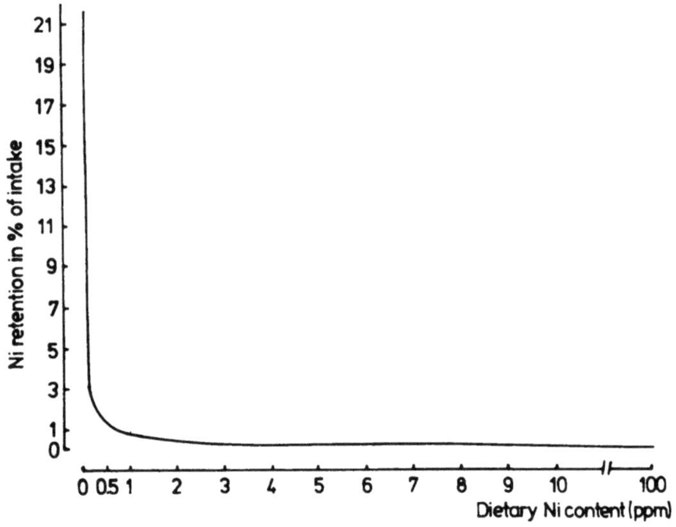

Fig. 4: Relative Ni-retention in relation to dietary Ni-content

(Kirchgeßner et al. 1984). Fig. 4 shows the relative retention of nickel by groups of growing rats fed a diet of different Ni concentrations. On the basis of this response curve, it is proposed that the growing rat requires between 1 and 10 ppm Ni in the diet.

An example of deriving the dietary requirement of a trace

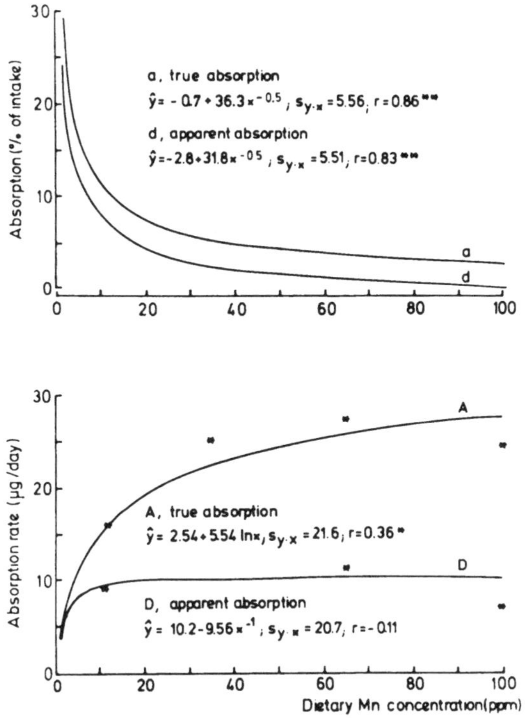

Fig. 5: Relative and absolute true and apparent absorption of Mn dependent on dietary Mn supply

element from absorption data determined in response to the dietary supply of the element is presented in the case of the Mn requirement of the growing rat. Fig. 5 shows both the apparent and true absorption of manganese as daily rate and as percentage of intake in relation to the dietary Mn concentration (Weigand et al. 1986). The response curve of the percentage absorption is similar to that of the relative Ni retention. Fig. 5 illustrates that the true and apparent Mn absorption, expressed as daily rate, have a steep incline over the low range of dietary Mn supply. Above a Mn concentration of about 15 ppm in the diet, the response of the true absorption rate rapidly falls and the apparent absorption rate varies around a plateau value of somewhat less than 10 ug Mn/day. The difference between the curves of true and apparent absorption reflects the fecal excretion of manganese from endogenous sources. The increasing rate of endogenous excretion is clear evidence that the Mn supply status is controlled not only by a pronounced change in the absorptive efficiency but also by a responsive endogenous excretion on the metal. The minimum dietary Mn requirement of the fast growing rat is approximately at the dietary Mn concentration where the apparent absorption rate approaches its plateau value. This indicates that the requirement is already met at a dietary level where percentage absorption has not yet reached the right section of the response curve where it assumes an asymptotic course. Accordingly, the asymptote approach of the relative absorption curve would estimate the requirement to be higher. This situation also has to be clarified for other elements. As an example of a functional compound serving as a response criterion, the activity of the glutathione peroxidase of the liver in relation to different dietary Se supply is presented in Fig. 6 according to a study by Sunde et al. (1981). Apart from the activity of the glutathione peroxidase in liver, the response of the activity of this enzyme may also be analyzed in whole blood, plasma, erythrocytes or platelets in relation to Se nutrition.

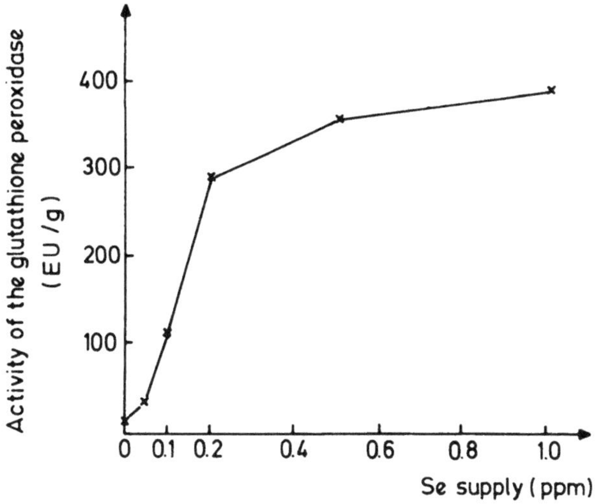

Fig. 6: Activity of the glutathione peroxidase in liver of rats given different Se supply

Generally, model studies must be conducted for each element to find response criteria that are most suitable for assessing the dietary requirement. It must be considered that certain criteria, for example, enzyme activities may respond differently depending upon organ or tissue. In all cases, however, it may be advisable to investigate several criteria for the derivation of requirements by dose-response relationships.

Two different approaches may be followed in the experimental design of studies for deriving dose-response relationships:

1. The response criterion is analyzed after the dietary supply of the trace element has been varied over a wide range, including suboptimal to optimal levels.

2. Following a period of depletion, perhaps to the point of clinical deficiency symptoms, the effect of repletion at graded dose levels on the response criterion is investigated, or the element is supplied in successively higher doses until a level is reached at which the criterion no longer responds.

In the case of either method, care must be taken that the response criterion is analyzed at the time when it has adjusted completely to the particular dose level in order to avoid misinterpretations of the data. This aspect of proper timing is especially important when the dose-response curve is assessed after a deficiency may exert on absorption, metabolism and excretion of the trace element.

Some major points have to be taken into account when requirements are derived by the dose-response concept:

1. The method leads to estimates of the gross requirement. The response criterion may reach its plateau value at different levels of dietary intake depending upon the efficiency with which the element may become available for absorption and metabolism. Exogenous factors influencing trace element utilization include the type of chemical compound and interactions with other dietary constituents. Strictly speaking, estimates of the requirement are valid only for the situation under which they have been determined.

2. The net requirement depends upon the intensity of production of the body, for example, the rate of live weight gain. Accordingly, the response criterion may plateau at a different level of intensity of the response criterion depending upon the height of the net requirement. Therefore, dose-response relationship have to be reassessed whenever productivity materially affects the net requirement.

3. Different response criteria may reach their response optimum at different dose levels. The requirement is assumed to be met when even the most sensitive criterion has reached its optimum.

3. Factorial Model for Deriving Requirements

According to the factorial model for deriving trace element requirements, the total net requirement (N) is comprised of a maintenance component (N_m) and a production component (N_p) :

$$N = N_m + N_p.$$

The net requirement for maintenance accounts for the inevitable losses of the trace element which occur during the continuous turnover of functional compounds and renewal of epithelia. Experimentally, the net requirement for maintenance is most often determined under the nutritional situation of suboptimal or even deficient supply of the trace element of concern in order to avoid the additional excretion of the element from endogenous sources which occurs in response to higher intakes for homeostatic reasons and is regarded as a quantity affecting the efficiency of metabolic utilization. The inevitable losses are usually considered as constant for a given situation. Certainly, these are not really constant, but are most likely to vary with the supply status. It may be assumed that the mean turnover rate of functional compounds is slowed with the advance of a nutritional depletion of the trace element and that a higher portion of the element is reutilized in metabolism. Otherwise, the inevitable endogenous losses may be related to metabolic body size or body mass as has been demonstrated, for example, in the case of zinc (Weigand and Kirchgeßner 1977).

The net requirement for production is determined by the type and intensity of productive functions beyond body maintenance. It is frequently estimated by analyzing the retention of the trace element in the products formed. There are restrictions to the applicability of this approach to human studies. For example, the net requirement for pregnancy could be assessed only by analyses of stillbirths or aborted fetuses. Estimation of the net requirement for production by analyzing the retention of the trace element in the products is not without problems, because the retention per unit of product formed is not necessarily constant, but may vary in response to a number of factors including, for example, the supply status or interactions with other elements.

The efficiency of utilization of the particular trace element is needed to compute the gross requirement (G), i.e. the necessary daily intake to meet the net requirement (N). The computational formula for the gross requirement is $G = N/c$, where c defines the efficiency of total utilization, which may vary within the boundaries of $0 \leq c \leq 1$. Total utilization may be influenced by numerous factors not only affecting absorption, but also intermediary metabolism and excretion of the trace metal. Such factors are, for example, level of intake, supply status, chemical nature of the element, dietary composition, interactions with other dietary constituents, pH and ionic conditions in the gastrointestinal tract etc. In this factorial concept of deriving the gross requirement, the overall efficiency of utilization or availability is not equated with absorbability as it has been done in other models (e.g., NRC). The concept presented here and previously (Kirchgeßner et al. 1974, Weigand and Kirchgeßner 1977/78) is baced on the fact that the absorbed amount of a trace element is not a priori fully available for the biosynthesis of functional compounds. Accordingly, the model concept of total utilization (c) entails its partitioning into two partial efficiencies, (1) the absorbability (a) and

(2) the metabolic efficiency (q), whereby the relationship $c = a \; q$ holds. This concept thus also takes factors into account which exert their effect on trace element utilization at the site of intermediary metabolism. Besides, trace elements are often absorbed in amounts exceeding the net requirement and hence the excessive quantity is then either stored or excreted because of homeostatic regulations and is consequently not used for the synthesis of functional compounds. Overall, this factorial model of total utilization represents a concept of a broader and more general applicability.

The partial efficiency termed absorbability of fractional true absorption is defined as the ration of the truly absorbed amount (A) over intake (I):

$$\text{absorbability} = \frac{A}{I}$$

The metabolic efficiency is defined as the ratio of the amount of the element utilized for metabolic functions (Q) over the amount truly absorbed (A):

$$\text{metabolic efficiency} = \frac{Q}{A}$$

The total utilization defining the overall efficiency represents the product of absorbability and metabolic efficiency or, correspondingly, the ratio between the amount utilized in metabolism and the amount of intake:

$$\text{total utilization} = \frac{A}{I} \cdot \frac{Q}{A}$$

and hence

$$\text{total utilization} = \frac{Q}{I}$$

A major obstacle of deriving gross requirements by the factorial concept lies in the practical difficulties of determining total utilization. Theoretically, it would be desirable to obtain a direct integral measurement of the quantity of the trace element used for functional roles per unit of time. This approach however, proves impracticable for the simple reasons that, first, there is still a lack of information on the biochemical role for a number of elements and that secondly, trace elements frequently have functions in numerous compounds which cannot be integrated, at the present, to an overall quantitative value.

In certain cases, total utilization may be approximated from determinations of apparent absorption or retention (Weigand and Kirchgeßner 1979, 1980). This shall be demonstrated by the following sequence of equations, in which capital letters express absolute quantities with the dimension of mass permit of time:

$$D = I - F$$

$$d = \frac{D}{I}$$

$$R = I - F - U$$

$$r = \frac{R}{I}$$

$$Q = I - F - U - S + F_m + U_m$$

$$Q = D - U - S + F_m + U_m$$

$$c = d - \frac{U+R}{I} + \frac{F_m + U_m}{I}$$

D = apparent absorption

r = apparent retention

I = intake

F = fecal excretion

U = urinary excretion

d = fractional apparent absorption

r = fractional apparent retention

F_m = inevitable endogenous fecal losses

or, alternatively,

$$Q = R-S+F_m+U_m$$

U_m = inevitable endogenous urinary loss

$$c = r-\frac{S}{I} + \frac{F_m+U_m}{I}$$

S = retention without being used for functional roles, i. e. storage

If it holds, for example that $R \geq S$, F_m and U_m , it follows that $Q \sim R$ and hence $c \sim r$. This is most likely the case in situations of high productivity, for instance rapid growth rate. The values of R and r can be estimated by conventional balance studies or, in the case of animal studies, also by carcass analysis.

This way illustrate the practical application of the factorial concept for deriving requirements. At the present stage of trace element research, however, the application of this model concept may be limited for the most part to situations of high productivity. The approximation of total utilization in these cases seems acceptable from the standpoint that recommended dietary allowances should contain safety margins. It should be ascertained, however, that such safety margins do not approach subtoxic levels of supply. Also, safety margins should not be so ample as to visk imbalances.

Conclusion

The net requirement of trace element is not of constant magnitude, but depends upon the height of the maintenance requirement and the type and intensity of production.

Accordingly, the determination of requirements must take into account the particular situation of life. The gross requirement, which specifies the necessary dietary intake, must be sufficiently high so that the net requirement is adequately net. Quantitative information on trace element utilization is a major prerequisite for computing gross requirements. The efficiency with which trace elements are utilized depends upon numerous factors such as, for example, chemical nature of the trace element, dietary composition, interactions with other dietary constituents, especially with other mineral elements, supply status, pH conditions within the gastrointestinal tract. Because of the multiplicity of influencing factors and the complexity and dynamics of trace element metabolism, the experimental determination of trace element utilization represents a challenging, multifaceted task that can be solved only approximately even under strictly standardized nutritional situations. For reliable recommendations of dietary allowance it is, never the less, necessary to invest special research efforts into the metabolism and utilization of trace elements. Under practical situations, the specification of the efficiency of utilization is thwarted especially because of a continual change in dietary composition. It is therefore necessary to rely on the approximation of assuming average values of utilization.

In the derivation of requirements, the theoretical concept of the factorial method fits best the fundamental concepts of trace element metabolism. The limits in the determination of the efficiency of utilization set at the same time the limits to the applicability of this method. Estimating requirements by the factorial method may be an especially sensible approach whenever the requirement for production is very much higher than that for maintenance, so that total utilization or may be approximately equated with apparent absorption apparent retention.

Considering the present state of knowledge of trace element metabolism and the methodological prerequisites, the dose-response relationship may currently appear as the method of choice for determining requirements. It is, however, recommended to analyze several response criteria. Model studies must be conducted to find response criteria suitable for estimating requirement, suitable even in studies with humans. However, continual changes in dietary composition again impose a very serious obstacle. It is therefore necessary that experimental estimates of the requirement are provided with a certain margin of safety for the specification of a recommended dietary allowance. The margin of safety must however, not be so high so as to impose the situation of trace element imbalances or subtoxic effects. Safety margins are needed not only because of intrapersonal variability, but also because of interpersonal variability.

In view of the fact that, within certain boundaries of trace element intake, the body is capable of responding to changes in the dietary supply with equivalent changes in the excretion, balance data are very difficult to interpret and, therefore, should no longer be applied for estimating requirements.

References

Kirchgeßner, M., Müller, H.L., Weigand, E., Graßmann, E., Schwarz, F.J., Pallauf, J., Roth, H.-P.: Z. Tierphysiol., Tierernährg. u. Futtermittelkde. 34, 3 (1974)

Kirchgeßner, M., Schwarz, W.: Arch. Tierernährg. 26, 3 (1976)

Kirchgeßner, M., Spoerl, R., Schneider, U.A.: In: Trace element metabolism in man and animals -3 (ed. M. Kirchgeßner), ATW, Freising, p. 440 (1978)

Kirchgeßner, M., Reichlmayr-Lais, A.M., Maier, R.: Z. Tierphysiol., Tierernährg. u. Futtermittelkde. 52, 27 (1984)

Kirchgeßner, M., Reichlmayr-Lais, A.M.: In: Trace elements in man and animals -5 (eds. O.F. Mills, I. Bremner, J.K. Chesters), Commonwealth Agricultural Bureaux, p. 884 (1985)

Sunde, R.A., Sonnenburg, W.K., Gutzke,G.E., Hoekstra, W.G.: In: Trace element metabolism in man and animals -4 (ed. J.McC Howell, J.M. Gawthrone, C.L. White) Australian Academy of Science, Canberra, p. 165 (1981)

Weigand, E., Kirchgeßner, M.: Z. Tierphysiol., Tierernährg. u. Futtermittelkde. 39, 84 (1977)

Weigand, E., Kirchgeßner, M.: Z. Tierphysiol., Tierernährg. u. Futtermittelkde. 39, 325 (1977/78)

Weigand, E., Kirchgeßner, M.: Biol. Trace Elem. Res. 1, 347 (1979)

Weigand, E., Kirchgeßner, M.: J. Nutr. 110, 469 (1980)

Weigand, E., Kirchgeßner, M., Helbig, U.: Biol. Trace Elem. Res. (in press) (1986)

MODELS FOR TRACE ELEMENT METABOLISM IN HUMANS

Colin F. Mills
Biochemistry Division, Rowett Research Institute, Bucksburn, Aberdeen, Scotland

Introduction

Appreciation of the significance of the essential and toxic trace elements for health has grown rapidly during the last 20 years. During that time, studies with experimental animals have gradually changed in emphasis from demonstrations of the gross clinical effects of deficiencies and toxicities to investigations of early pathological responses. From these we have come to appreciate that the absence of diagnostically specific clinical signs is no guarantee that tissue damage is not occurring and, furthermore, that when external evidence of a pathological response eventually appears, its causes are often difficult to identify from the clinical picture alone. In such situations, the ability to resolve the problem depends upon the availability of information from experimental studies describing relationships between trace element intake and animal response, with the latter defined either from evidence of <u>physiologically significant</u> anomalies in tissue trace element patterns or biochemical evidence of characteristic metabolic defects.

The experimental approaches needed to define the tolerable extremes of trace element exposure and to devise more appropriate procedures for detecting pathological responses involve deliberate depletion or excessive loading with the element under investigation. Ethical considerations will frequently

Trace Element – Analytical Chemistry in Medicine and Biology, Vol. 4

preclude such studies with human subjects, particularly during pregnancy or early postnatal development when the risks of incurring permanent damage are particularly high. Since high risk situations are of particular interest it is inevitable that much of the information upon which risk assessment is based will continue to be derived from studies with experimental animal models, the advantages and limitations of which must be kept continually under review.

At the molecular or functional level there is a remarkable consistency in the evidence derived from a wide range of species that the essential trace elements serve identical and often highly specific functional roles. Typical examples are the respective roles of copper, zinc and selenium in the enzymes cytochrome oxidase, alkaline phosphatase and glutathione peroxidase, all of which irrespective of species, decline in activity as tissue deficiency of their inorganic component develops. However, much more relevant to the problem of selecting an appropriate animal model is that both the sensitivity to and the pathological expression of such changes can differ both between species and within a single species at differing stages of development (1). Such factors, together with species differences in sensitivity to interactions involving dietary constituents influencing trace element absorption and tissue redistribution, are all relevant to the selection of an appropriate model intended to simulate the response of man to changes in element supply.

Objectives of modelling

Our objectives in attempting to develop animal models of trace element metabolism in man are frequently ambitious. They can include:-

(i) attempts to focus attention on the principal pre-clinical and clinical pathological consequences of deficiency or excess in human subjects,

(ii) to assist with the development of more appropriate direct or indirect biochemical criteria from which to achieve early diagnosis of such problems and,

(iii) to explore, tentatively, the likely limits of tolerable exposure to element deficiency or excess and to identify the principal dietary and other variables likely to modify minimal requirements or tolerance.

This paper will illustrate some of the successes achieved and some of the problems encountered in attempts to meet such objectives by animal modelling of the effects caused by zinc and copper deficiencies and lead excess.

Comparative aspects of trace element availability

Evidence that trace element absorption and tissue utilisation can be influenced sufficiently by a range of common dietary constituents to induce pathological responses is much more firmly established for non-human species than for man (2). Whether the data derived from animal studies have implications for the trace element nutrition of man depends upon adequate consideration of both physiological and dietary differences as indicated by the following examples.

For most monogastric animals, the supply of physiologically available Zn is dictated both by Zn intake and by dietary content of antagonists to Zn absorption such as the phytic acid associated with components of the fibre fraction of cereals and the protein fraction of some legume seeds. Studies with pigs, chicks and rats indicated more than 20 years ago that Zn requirement was directly proportional to phytate intake. It was also recognised that the antagonistic effects of phytate were potentiated by increases in dietary calcium and possibly ameliorated by increased dietary protein. That the antagonistic potency of phytate can thus vary has not only been ignored in many attempts to define dietary zinc availability to animals but has contributed to the controversy currently surrounding the significance of phytate-containing diets in the aetiology of zinc deficiency in man. In this context, the rat may be a reasonable model for man only if efforts are made to equate the normally differing dietary contents of the principal potentiators of the phytate-Zn antagonism, i.e. calcium and probably magnesium. Neglect of this precaution by the use of dietary regimes which, for the rat, normally contain more than twice the level of Ca in human diets quite understandably yields data which grossly overestimate of significance of phytate in the Zn nutrition of man. For the rat model, the ratio [phytate][Ca] (moles/kg diet) in excess of 3.5 realistically describes diets in which available Zn is insufficient to maintain optimum growth (3). While the validity of this model for man has yet to be established retrospective analysis of data from a range of Zn-balance trials suggests that a ratio exceeding 0.5 may provoke negative zinc balance. This discrepancy may be more apparent than real in view of the differing criteria used in rat and human trials to detect adverse effects on Zn utilisation.

The above simple example of the need for rigorous control and simulation of dietary conditions when attempting to "model" responses to variations in trace element supply and availability is not unique. Compositional variables, also involving dietary phytate, Ca (4) and possibly protein, certainly modify lead uptake by the rat and, according to recent evidence, probably affect Pb retention by man (Rose, H.E. personal communication). Additional evidence that the physical characteristics of protein precipitates formed during intragastric and intraduodenal digestion (5) influence the uptake of zinc, cadmium and possibly other metals, emphasises the necessity of rigorous control of dietary protein sources in model studies intended to characterise availability phenomena. This requirement extends even to the selection of homologous or heterologous milks when investigating the uptake of adventitious contaminants from a milk diet.

Failure to take account of such variables may well have contributed to the difficulty in devising reliable animal models of Pb toxicity and, particularly, in attempts to compare dose/response relationships with man. That differences in dietary composition are highly relevant is indicated by evidence that the remarkable differences in the efficiency of Pb absorption exhibited by the rodent and the human subject (approx. 0.1-1% efficiency, rodent; 10% efficiency, man) are eliminated if the rodent is given a low fibre semi-synthetic diet. Rodent "modelling" with adequate dietary control has highlighted the significance of low dietary Ca, P (6) and Fe (7) in promoting Pb uptake (all of which effects have been confirmed in human subjects (8,9)). Additionally such models indicate that Pb uptake is enhanced strongly by a liquid milk diet; a possibility which has yet to be investigated in human subjects.

Major differences in gut physiology which, through their influence on digestive processes, influence the forms in which elements are presented to the gut mucosa, can have a major influence on model validity. Typical of such are the microbial events within the rumen or caecum which modify the forms of dietary Cu, Mo, Se and S (2) species ultimately presented to the gut mucosa. Such chemical transformations impose severe limitations upon the human relevance of absorptive studies based upon the lamb, guinea pig or rabbit. In these species, microbiological processes within the gut not only modify the valency state of transition elements but also by reductive generation of sulphide or selenide, radically increase the opportunities for formation of insoluble metal derivatives of these particular anions and thus tend to reduce metal bioavailability, particularly of Cu, Mo and Pb. An interesting exception to this generalisation is that for diets containing Hg or Pb, there are indications that microbial methylation within the caecum can enhance bioavailability and the tissue incorporation of methyl mercury (10) or tetramethyl lead (11).

Remarkably few guidelines exist from which to develop appropriate animal models to predict the changes in trace element absorption which may occur postnatally in the human infant. Studies with rats indicate that prior to gut "closure" at approximately 16 days of age, absorption of the essential elements Fe and Cu is probably pinocytic and, in contrast to the older animal is not homeostatically regulated (12). Highly efficient pinocytic absorption of lead and of cadmium also occurs at this stage (for review, 13). If to this we add the finding that absorption of Zn by both unweaned rat and the human infant is more effective from homologous than from heterologous milk and the evidence that absorption of many trace elements from a milk diet is more efficient than from a solid diet, it will be apparent that the modelling of events at this stage of development is a particularly complex task. Since it is suspected that adventitious contamination of liquid diets offered before weaning has a disproportionally large influence on the uptake, the tissue retention and the risks of toxicity from several trace metals (e.g. Pb, Cd and Cu) it is imperative that greater attention should be devoted to the development of models appropriate both to the premature and normal human infant.

In concluding this brief account of the modelling of trace element absorptive phenomena it is important to caution against experimental approaches based solely upon administration of test doses of elements or their antagonists in drinking water or aqueous gavages except in instances where these vehicles are the specific objects of investigation. For example, evidence is emerging that the mutually competitive interactions between Fe and Zn that influence absorption of these elements are expressed much more strongly when imbalance is induced by administration of aqueous solutions rather than by modifying the Fe and Zn content of the diet (15). Similar considerations may influence the antagonistic effects of high intakes of Zn, Fe and Cd on Cu absorption. Furthermore the absence of food from the proximal gut of the rat enhances between 2 and 10 fold the carcass uptake of Zn, Cu, Fe, Pb and Hg and enhance the proportion of the retained dose present in gut tissue (14). That similar variables influence the fate of elements in man is typified by data from one study indicating that the half-time for faecal excretion of a dose of ^{65}Zn given in solution to starved subjects was approximately 13 days whereas, when given with food was only 3.5 days. Rigid compatibility of experimental protocols is thus essential both for comparative studies and to define the intrinsic properties of absorptive processes within a single species.

Modelling of pathological responses

Effective animal modelling of human responses to anomalies in trace element supply serves two objectives. It can alert investigators to subtle pathological changes that are not immediately evident from a clinical examination of human subjects at risk and can provide guidelines for discriminating effects attributable to short or long-term exposure to anomalies in trace element supply. Furthermore, if accompanied by biochemical studies, such modelling can frequently indicate the likely value and limitations of diagnostic or monitoring procedures based on measurements of tissue trace element content or the presence of abnormal enzyme or metabolite concentrations in plasma or urine. Typical aspects of the value of such modelling are considered below.

Zinc deficiency or excess

Under the closely controlled protocols usually applied in studies with experimental animals, measurements of changes in plasma or hair Zn or of plasma alkaline phosphatase activity can provide a reasonably reliable indication of Zn status. However, such studies have also revealed limitations to these approaches. Thus, plasma Zn, even in Zn-deficient animals increases rapidly if tissue growth is limited by other nutritional insults and is depressed markedly by concurrent infection or other stress. Under extreme situations these effects may well be paralleled by corresponding changes in the pathological manifestations of deficiency. Thus, the skin lesions (parakeratosis) of Zn deficiency in the animal model are provoked by physical abrasion or ectoparasite activity; in contrast, such lesions may never appear if the concurrent intake of protein is low. Animal modelling has also shown that the teratogenic consequences of Zn deficiency in the rat are reduced or eliminated if skeletal resorption is promoted by a low Ca diet (16) or if tissue catabolism is provoked by a low food intake (17).

Virtually all these situations influencing the interpretation of data for plasma and hair Zn have been found to have their parallels in studies of Zn-responsive conditions in man. Nevertheless, species differences in response do arise as typified by the work of Hurley and her colleagues (18,19) on zinc deficiency in rodents, ruminants and laboratory primates.

These and other comparative studies have particularly highlighted marked species differences in susceptibility to teratogenic effects of Zn deficiency during pregnancy which, superficially at least, appears to be related directly to total maternal anabolic activity during pregnancy. Thus susceptibility is high in the rat but, on current evidence, virtually non existent in the primate although late foetal growth is affected in all species.

Comparative studies suggest that the critical threshold values for plasma Zn may be higher for the laboratory primate or human subject than for other species. However, studies with all primates, including man, illustrate a complexity of relationships between plasma Zn concentrations and the pathological expression of Zn deficiency that is not always revealed in studies with other animal species maintained under closely controlled experimental conditions. Among such findings are an initial decline in plasma Zn when the growth of deficient infants resumes after Zn supply is increased and evidence of an inverse relationship between plasma Zn and growth of the human foetus, infant and laboratory primate judged to have a "marginal" Zn status as defined by other biochemical and pathological criteria (18-21).

Such ostensibly paradoxical findings indicate, firstly, the need for better understanding of factors governing the equilibria between Zn in plasma and other tissues and, secondly, the need for better criteria than plasma Zn for assessing Zn status. Work with rats suggests that plasma concentrations of the Zn-induced protein, metallothionein, may offer a more suitable and physiologically relevant alternative (22,23). Its validity as a marker of the Zn status of man has yet to be assessed. Other inadequately explored aspects of the potential relevance of data from animal models also exist. Thus there are clear indications from studies with rats that the efficiency of Zn absorption is enhanced markedly during the last third of pregnancy and during early lactation (24). This finding may well be relevant to growing evidence that past predictions of Zn requirements for human pregnancy and lactation are gross overestimates when derived by a factorial approach assuming no adaptive increase in the efficiency of Zn absorption. Also widely ignored is evidence from work with rats and ruminants that while Zn is normally well tolerated, susceptibility to Zn intoxication is particularly high during the terminal stages of pregnancy when the element is rapidly translocated to the developing foetus (25,26). Indications of the possible relevance of this finding are unfortunately provided by case histories of abortions in women given 75 mg Zn orally/day - a dose rate which although high would normally be well tolerated by non-pregnant subjects (25).

Copper deficiency

The growth of studies of copper deficiency in man provides clear examples of the value, the limitations and, sometimes, the misinterpretation of data derived from work with experimental animals. Initial studies with animal models concentrated on the origins and characteristics of the anaemia of Cu deficiency, paid little attention to adverse effects on growth and virtually ignored other aspects. The failure of one group of investigators to produce anaemia in infants subjected to a period of Cu depletion (27), coupled with a wider appreciation of the extensive precautions to exclude contamination that are needed to induce Cu deficiency anaemia in rats and swine, established the belief that Cu deficiency was unlikely to arise in human subjects.

Although one case report later described the development of a Cu-responsive anaemia in socially deprived infants offered diluted cow's milk, the wider pathological implications of Cu deficiency in man were only appreciated after more extensive studies with laboratory animals. Such reports of the effects of Cu deficiency on skeletal and connective tissue development, cardiovascular and CNS integrity, hair structure and pigmentation and foetal survival all pointed very clearly to the fact that anaemia is a relatively late consequence and is thus unlikely to be the principal pathological feature of human Cu deficiency.

Features common to nutritional Cu deficiency in human infants, adolescents and many experimental animal models include defects in bone connective tissue (elastin and collagen) biosynthesis and in neutrophil development or function (for review see 28). Data from animal models indicate that the metabolic defects responsible for all these lesions can develop before overt signs of deficiency appear.

The significance for man of many other features of Cu deficiency in experimental animals has not yet been explored adequately. These include cardiomegaly and its associated connective tissue changes, decreased energetic efficiency of the heart, defects in catecholamine metabolism in the heart (29), intestine and other tissues and metabolic lesions influencing mitochondrial energy metabolism in a variety of tissues. Some lesions, such as defects in aortic elastin morphology (30) and those responsible for foetal death and resorption which develop in young rats given diets only "marginally" deficient in Cu (e.g. 2 mg Cu/kg diet), merit closer investigation with respect to the effects of Cu deficiency in other species, including man.

In remarkable contrast, virtually all the Cu deficiency lesions demonstrated in experimental animals have now been detected in infants with Menkes' disease (28). Indeed, the identity of this disease as a genetic defect in Cu metabolism was first suggested from the similarity of structural defects in the hair of affected subjects with those of the wool of Cu deficient sheep. The other major characteristic of Menkes' disease, its clinical manifestations of neuropathological defects, appears superficially similar to that caused by degenerative lesions of the CNS in neonatal Cu-deficient lambs. However, there are suspicions that the pathogenesis of these two disorders may differ with cerebral vascular lesions restricting oxygen supply playing a much more prominent role in Menkes' disease.

The cardiac lesions and cardiovascular functional abnormalities of Menkes's disease also resemble those found in Cu deficient rats. The extent to which such lesions also develop as a consequence of nutritional Cu deficiency in man is a topic currently exciting attention (31,32). The heart of the Cu deficient rat is enlarged, fragile, flaccid, becomes oedematous under workload and exhibits arrhythmia and other ECG abnormalities (29,34). Biochemical studies indicate changes in collagen type, decreased concentrations of noradrenaline and nucleotide triphosphates and a reduced energetic efficiency. For reasons unknown as yet, susceptibility to cardiac damage appears to be exacerbated by diets high in fructose (33). It is also claimed that a concurrent deficiency of selenium or a high dietary Zn/Cu ratio increases mortality from myocardial rupture in Cu deficient rats.

For obvious reasons, much less is known of any cardiopathological changes in human subjects with a low Cu status. However such consequences must be considered more closely in the light of recent evidence of the development of tachycardia and other ECG abnormalities in 5 of 25 human subjects offered diets providing approx. 0.36 mg Cu/megacal for periods between 7 and 15 weeks (31,32). Appraisal of the validity of the rat or pig model of such possibly Cu-related cardiovascular phenomena is probably contingent upon closer *in vivo* monitoring of cardiac output and related electrophysiological changes in human volunteers undergoing Cu depletion.

Pathological aspects of the modelling of lead intoxication

A wide range of neurological lesions has been reported from studies of lead toxicity conducted with laboratory animals exposed to relatively high doses of Pb in food or drinking water. Biochemical findings in such models include depression of acetyl choline release, changes in catecholamine metabolism and depressed oxidative metabolism of the brain cortex. However, the doses of Pb required to provoke such changes or to produce consistent behavioural abnormalities are substantially higher than those claimed to provoke adverse effects in man. The possible relevance of differences in dietary composition influencing the dynamics of Pb absorption in studies with man and laboratory animals has been commented upon earlier in this paper. However it is also evident that the typical blood Pb concentrations associated with such pathological responses in laboratory rodents are frequently 2-3 fold higher than in human subjects believed to be at risk.

These and other limitations of rodent models of Pb toxicity have been considered _in extenso_ elsewhere. The possibly greater relevance of laboratory primate studies is suggested by work with rhesus monkeys with blood Pb in the range 0.4-0.6 mg/l thus simulating blood Pb levels in children believed to be at risk. Young monkeys so treated developed cognitive defects which persisted upto 3 years after Pb withdrawal (35). It is still not clear whether the limitations of the rodent model reflect uncharacterised physiological or metabolic differences in the response to Pb or limitations in our ability to detect in the rodent the subtle behavioural and cognitive changes that are considered significant in Pb-exposed children

Conclusion

Although necessarily restricted in scope, this paper will have indicated aspects of both the potential and the limitations of animal modelling of human trace element metabolism. For reasons outlined in the introduction it is clear that reliance on such approaches must continue. Thus it is essential to emphasise that one of the most significant limitations to progress arises from the frequently inadequate incremental approaches to the definition of dose/response relationships and the failure to define changes in these relationships at differing stages of physiological development of possibly useful model species. Tedious though such systematic studies may appear, they are essential if the potential of models is to be explored adequately. They provide the only means of defining the change in sensitivities to trace element-dependent processes that occur as synthetic and metabolic activities change during growth and development.

Obviously, it is unrealistic to expect that any animal model can reflect all the relevant aspects of human trace element metabolism for which information is required to define tolerance or pathological risks. Nevertheless, with careful interpretation of adequate data derived from models, our ability to predict and anticipate human responses is increasing rapidly.

Animal models have contributed substantially to understanding of the genetic origins of defective Cu and Zn metabolism in Menkes' disease and acrodermatitis enteropathica in man as well as suggesting approaches to clarifying the pathologies arising from dietary deficiencies of Cu, Zn and Mo deficiencies in man (36). Less satisfactory is the status of animal modelling of the claimed effects of selenium deficiency as the sole relevant variable in the aetiology of Keshan disease in man - a situation which should perhaps be interpreted as suggesting an inadequate understanding of the causes of this "nutritional" disease rather than of inadequate modelling.

The future development and application of animal modelling depends greatly upon a wider appreciation of those physiological and dietary variables that modify the response of individual species to changes in trace element intake. Included among these is genetic variability, operating even within species, that influences both the kinetics of element absorption and retention and the pathological response to deficiency or excess. Apart from situations such as the dramatic clinical consequences of acrodermatitis enteropathica and Menkes' disease in the infant, little is yet known of the significance of such genetic variability in trace element metabolism in man. Nevertheless, evidence of its relevance to animals is growing rapidly from studies such as those demonstrating marked genetic differences in susceptibility to both deficiency and toxicity of copper in sheep and corresponding evidence of variability in responses to copper and manganese exposure in rats and mice.

Although much remains to be done, the achievements of modelling have already been substantial. Furthermore, there is clear evidence of a failure to explore the possible significance of some findings that may have relevance to man. Thus, epidemiological studies of Pb and Cd "toxicities" have almost totally disregarded the weight of evidence from animal models that dietary composition modifies toxicity. Possibly of much wider significance is the inadequate exploration of findings that a marginal Fe status, totally irrelevant from the standpoint of its minor haematological consequences, has a marked influence on the voluntary activity of rats (37).

Such considerations and the progressively increasing demand for methods for biological monitoring of the effects of changing dietary patterns or the growing availability of "novel" foods on trace element metabolism in man indicate the opportunities that exist for the development of animal modelling techniques whose applicability and limitations are responsibly considered and clearly defined.

References

1. Mills, C.F.: Phil. Trans. R. Soc. Lond. B. 294, 199-213 (1981).
2. Mills, C.F.: Ann. Rev. Nutr. 5, 173-193 (1985).
3. Davies, N.T., Carswell, A.J.P, Mills, C.F.: In Trace Elements in Man and Animals - TEMA 5 (eds. Mills, C.F., Bremner, I., Chesters, J.K.), pp. 456-458 (1985) CAB Press, Farnham Royal
4. Rose, H.E., Quarterman, J.: In Trace Elements in Man and Animals - TEMA 5 (eds. Mills, C.F., Bremner, I., Chesters, J.K.), pp. 524-526 (1985) CAB Press, Farnham Royal.
5. Blakeborough, P., Gurr, M.I., Salter, D.N.: Br. J. Nutr. 55 (in press) (1986).
6. Quarterman, J., Morrison, J.N., Humphries, W.R.: Environ Res. 17, 60-67 (1978).
7. Six, K.M., Goyer, R.A.: J. Lab. Clin. Med. 76, 933-942 (1970).
8. Blake, K.C.H., Mann, M.: Environ. Res. 30, 188-194 (1983).
9. Watson, W.S., Moore, M.R., Hume, R.: Lancet ii, 236-237 (1980).
10. Horsburgh, H.: Studies on the Methylation of Mercury by Gut Organisms, M.Sc. Thesis, Univ. Aberdeen, 1978
11. Schmidt, V., Huber, F.: Nature, Lond. 259, 157-158 (1976).
12. Linder, M.C., Munro, H.N.: Enzyme 15, 111-138 (1973).

13. Mills, C.F., Davies, N.T.: In Development of Mammalian Absorptive Processes (eds. Elliott, K., Whelan, K.), pp. 247-266 (1979) Excerpta Med, Amsterdam

14. Quarterman, J., Morrison, E.: Br. J. Nutr. 46, 277-287 (1981).

15. Fairweather-Tait, S.J., Payne, V., Williams, C.M.: Br. J. Nutr. 52, 79-86 (1984).

16. Hurley, L.S., Tao, S.H.: Am. J. Physiol. 222, 322-325 (1972).

17. Record, I.R., Dreosti, I.E., Manuel, S.J., Buckley, R.A., Tulsi, R.S.: In Trace Elements in Man and Animals - TEMA 5, (eds. Mills, C.F., Bremner, I., Chesters, J.K.), pp. 210-213 (1985), CAB Press, Farnham Royal

18. Golub, M.S., Gershwin, M.E., Hurley, L.S., Baly, D.L., Hendrix, A.G.: Am. J. Clin. Nutr. 39, 879-887 (1984).

19. Golub, M.S., Gershwin, M.E., Hurley, L.S., Saito, W.Y., Hendrix, A.G.: Am. J. Clin. Nutr. 40, 1192-1202 (1984).

20. Golden, B.E., Golden, M.H.N.: Am. J. Clin. Nutr. 34, 892-899 (1981).

21. Golden, M.H.N., Golden, B.E.: Am. J. Clin. Nutr. 34, 900-908 (1981).

22. Bremner, I., Mehra, R.K., Sato, M.: In Trace Elements in Man and Animals - TEMA 5, (eds. Mills, C.F., Bremner, I., Chesters, J.K.), pp. 581-587 (1985), CAB Press, Farnham Royal

23. Sato, M., Mehra, R.K., Bremner, I.: J. Nutr. 114, 1683-1689 (1984).

24. Davies, N.T., Williams, R.B.: Br. J. Nutr. 38, 417-423 (1977).

25. Kumar, S.: Nutr. Rept. Int. 13, 33-36 (1976).

26. Campbell, J.K., Mills, C.F.: Environ. Res. 20, 1-13 (1979).

27. Wintrobe, M.M., Cartwright, G.E., Gubler, C.J.: J. Nutr. 50, 395-419 (1953).

28. Danks, D.M.: In Biological Roles of Copper (eds. Evered, D., Lawrenson, G.), pp. 209-226 (1980) Excerpta Med. Amsterdam

29. Prohaska, J.R., Heller, L.J.: J. Nutr. 112, 2142-2150 (1982).

30. Hunsaker, H.A., Morita, M., Allen, K.G.D.: Atherosclerosis, 51, 1-9 (1984).

31. Reiser, S., Smith, C.J., Mertz, W., Holbrook, J.T., Scholfield, D.J., Powell, A.S., Canfield, W.K., Canary, J.J.: Am. J. Clin. Nutr. 42, 242-251 (1985).

32. Klevay, L.M., Inman, L., Johnson, L.K.: Metabolism, 33, 1112-1118 (1984).

33. Fields, M., Ferretti, R.J., Smith, C.J., Reiser, S.: Am. J. Clin. Nutr. 39, 289-295 (1984).

34. Klevay, L.M., Viestenz, K.E.: Am. J. Physiol., 240, H185-H189.

35. Bushnell, P.J., Bowman, R.E.: J. Toxicol. Env. Hlth. 5, 1015-1020 (1979).

36. Aggett, P.J., Davies, N.T.: Proc. Nutr. Soc. 39, 241-248 (1980).

37. Finch, C.A., Miller, L.R., Inamdar, A.R., Person, R., Seiler, K., Mackler, B.: J. Clin. Invest. 58, 447-453 (1976).

Session II

TRACE ELEMENTS IN NUTRITION

Plenary Lectures given by:

L.M. Klevay

H.B. von Stockhausen

R. Schelenz

A.E. Harmuth-Hoene

Chairman:

M. Kirchgeßner

M. Anke

Poster Reports:

M. Anke

R. Schelenz

DIETARY REQUIREMENTS FOR TRACE ELEMENTS IN HUMANS

Leslie M. Klevay

United States Deptartment of Agriculture, Agricultural Research Service, Grand Forks Human Nutrition Research Center, Grand Forks, North Dakota, USA.

Introduction

Study of the essentiality of nutrients is largely a phenomenon of the twentieth century. After Aristotle recorded the toxicity of arsenicals (1) and Holmes (2) and Semmelweis (3) showed how to avoid puerperal fever, physicians were well acquainted with the concept that diseases could be caused by external agents. Acceptance of the concept that diseases could occur because an agent was absent required general reevaluation of medical thought.

The origin of the deficiency concept is widely associated with Lind (4) who cured scurvy with citrus fruit. However, Boyle concluded approximately a century earlier that animals dying in a vacuum perished from the lack of an essential substance (5). Funk (6) promoted the concept of "vitamines". Vitamins were renamed by the excision of the terminal e after it was realized that not all of them were amines.

Although there may be thousands of chemicals in foods, the number of essential nutrients probably is not much greater than fifty if one counts the amino acids, the minerals, the vitamins, water and a few trace elements currently under study. Often the experiments that prove essentiality provide the first estimates of requirements.

Trace Element – Analytical Chemistry in Medicine and Biology, Vol. 4

Nutritional Essentiality

A substance is considered a nutritional essential for an organism if the organism can neither grow nor complete its life cycle in the absence of the substance. This concept was best outlined by Bowen (7) who extended Arnon's (8) criteria for plants to animals. It is difficult, perhaps impossible, to obtain evidence in conformity with this rigid definition, even for plants or animals. Essentiality for humans sometimes is based on inference. It is unethical to satisfy this definition by subjecting people to severe experimental conditions.

Surprisingly little has been written on the definition of nutritional essentiality (9-25). Table 1 contains a summary of criteria suggested by several authors. Criteria are classified as definitive or inferential; criteria for which there was little agreement among authors are omitted.

TABLE 1. CRITERIA OF NUTRIENT ESSENTIALITY

Definitive	
Restricted Diets:	Supplemented Diets:
interrupt the life cycle	prevent pathology
impair function	provide relief
Inferential	
Atomic number is low	
Biological activity	
Homeostatic control	
Toxicity is low	
Ubiquitous	

Definitive criteria

The criteria (Table 1) illustrate the concepts of the authors but do not necessarily reflect the words used by the authors. For example, Bowen and Arnon are quoted above regarding the definitive criteria. Hegsted (16) referred to a requirement "...for normal growth and development...". Mertz (19) wrote "absence...is incompatible with life" and Frieden (25) defined essential elements as those "required for the maintainance of life" and quoted Bertrand's work of 1912 regarding less severe deficiency states.

Several of the authors noted that function is impaired if too little of an essential nutrient is ingested. Cotzias (15) mentioned "structural and physiological abnormalities". Some authors noted that more than one function could be impaired by deficiency of one nutrient.

McCollum (9) referred to a "diet which causes malnutrition" and the identification of "protective foods". Mertz (18) referred to "supplementation" and "therapeutic trial" as means of improving an impaired function. Later he mentioned prevention or cure by an essential element (20). Underwood (13) mentioned "responses in growth" and Cotzias (15) wrote that an essential element "either prevents or reverses" abnormalities and included biochemical improvement as appropriate evidence of essentiality. Others (24) refer to "normal function and health".

In summary, anatomical, biochemical and physiological abnormalities produced by restricted diets and relieved by supplemented diets are acceptable as evidences of essentiality. A nutrient can have more than one function. Death and improved growth are mentioned specifically; other impairments are permissible but are not specified.

Inferential criteria

The definition of essentiality is simple in theory and severe in practice. Many experiments have been done with nutrients that satisfy the definitive criteria. Some characteristics of essential nutrients have been inferred from these experiments. The inferential criteria are not very helpful in predicting which elements will be found essential after appropriate experiments.

Schwartz (23) noted that no elements with atomic number greater than that of iodine has been proved to be essential. Schroeder (11) suggested lanthanum as an upper limit. Underwood (22) suggested that gallium and germanium may be worthy of study for essentiality because a substantial number of essential elements have atomic numbers between 23 and 34. Consideration of atomic number is peculiar to trace elements; however, other nutritional essentials usually have low molecular weights.

Mildvan (26) has estimated that more than 27% of known enzymes "have metals built into their structures, require added metals for activity, or are further activated by metal ions". Underwood (13) quoted Green (27) as saying that "enzymatic catalysis... is the only rational explanation of how a trace substance can produce profound biological effects". Frieden (25) however, cautioned that stimulatory effects are not always evidence of essentiality and that a decreased amount of an essential nutrient may be beneficial. This latter situation can arise if the dose of the nutrient is too large or if the dose is sufficient to interfere with the utilization of another nutrient.

Several authors have emphasized that essential elements are under homeostatic control. Perry et al. (12) noted that the value for the 90th percentile for metal concentrations in organs divided by that of the 10th percentile (80% range) was greater for nonessential than for essential elements. A quotient of 7.0 separated the two groups. Liebscher and Smith (17) proposed several distinctions, the simplest of which was that essential elements have normal distributions and nonessential elements have log-normal distributions.

Schroeder (11) suggested that the dose of an essential element equal to that of the body burden is readily tolerated. Schwarz (23) believed that toxicity does not mitigate against essentiality.

Several authors have suggested that essential elements are ubiquitous. Cotzias (15) emphasized presence in healthy tissues. However, contemporary analysts can measure small amounts and low concentrations accurately and with such ease (28),that general detectibility is not very helpful in distinguishing between essential and nonessential elements. Vallee (10) has warned that the "concentration of an element in biologic matter is not the sole criterion...of importance... in health and disease...". Underwood (13) and Mertz (19) warned against uncritical reliance on analysis in establishing deficiency. One can infer that similar caution is desirable in establishing essentiality.

In addition to the inferential criteria listed in Table 1, numerous other criteria have been suggested (9,17,19,21). Although some of these may be helpful, these criteria have received less attention. Careful reading permits the conclusion that exceptions are numerous.

After essentiality has been established, measurement of nutritional requirements becomes of prime importance. Experiments with animals are more likely to yield rapid progress than those with humans.

Zones of nutriture

Dann and Darby (29) provided instruction in the search for impaired function. Tracing the history of nutritional appraisal to earlier study of protein, they emphasized vitamins. Their five zones of nutriture, which are equally applicable to trace elements, are summarized in Table 2. Saturation implies maximal storage of the nutrient without early toxicity. Potential deficiency exists when departure from saturation is sufficiently great that a new stress on the organism will induce illness or when a suitable test reveals reduced functional capacity. Latent deficiency disease is the mildest clinically detectable form of deficiency disease. This zone is characterized by vague, nonspecific symptoms that do not permit a definite diagnosis. Improvement occurs on nutritional replacement. Clinically manifest disease may be mild or severe. Severe illness can be diagnosed without the assistance of the laboratory or therapeutic trial. Some examples of recent research that illustrate these principles are provided.

TABLE 2. ZONES OF NUTRITURE*

1. Saturation
2. Unsaturated, but functionally unimpaired
3. Potential deficiency disease
4. Latent deficiency disease
5. Clinically manifest deficiency disease

*After Dann and Darby (29).

Guiding Principles

Animal studies

Many of characteristics are useful in assessing vitamin nutriture and establishing nutritional requirements. Methods of assessing trace element nutriture and establishing requirements are less well developed. Experiments with vitamins can be useful guides.

The principles which guide the measurement of the nutritional requirements of humans are similar to those used for animals. A few years ago I determined the biotin requirement of rats fed egg white (30). Many students of trace element metabolism use egg white as a dietary protein because it is of high nutritive quality and it is low in copper and zinc. Unfortunately one of the proteins in it interferes with the metabolic utilization of biotin. Weanling rats fed 1.0 or 1.5 mg of biotin per kg of diet weighed 45 and 72%, respectively of those fed 2.0 mg/kg after sixty days (30). This impairment corresponds to the fifth zone of nutriture (Table 2). Growth in response to the lower amounts of biotin was clearly inferior; greater amounts of biotin gave no benefit. The requirement was set at 2 mg/kg.

We also assessed several characteristics of copper deficiency in a group of rats (31). Table 3 summarizes the data. The rats were equally malnourished as a group; variation between rats sometimes was considerable. Some animals were ill (zone 5, Table 2); others had functional impairment (zone 3, Table 2). Variation among the characteristics was even greater. Liver iron was nearly three times normal; body weight was decreased by one fourth. Many of these characteristics cannot be measured on people. They illustrate some types of useful measurements, however. In principle functional tests based on cardiac catecholamines can be devised. There is no theoretical obstacle to those tests being validated.

TABLE 3. INDICES OF COPPER DEFICIENCY IN RATS

Greatest Change	Other Findings:
Liver iron	Abnormal electrocardiograms
Heart dopamine	Achromotrichia
Liver copper*	No anemia
Plasma cholesterol	Ventricular aneurysms
Heart weight	
Heart norepinephrine*	
Body weight*	
Least change	

*Decreased, others are increased

After Klevay, et al. (31)

The chemical reactivity of transition elements makes probable the existance of several chemical forms of these elements in both diets and in organs. Vallee (10) suggested that if it had been hypothesized --- prior to the discovery of vitamin B_{12} --- that cobalt is a significant factor in pernicious anemia, pharmacologic experiments with cobalt would have been negative. Measurement of cobalt in organs would have been similarly uninformative.

Underwood (13) noted that it often is difficult to correlate the deficiency state with "subnormal levels of the elements in the blood or organs". Mertz (20) expanded on this view. Copper deficiency produced decreases in cardiac copper ranging from 26 to 42% in three replications of an experiment (32) with rats. However, the highest control value for cardiac copper was 2 1/2 times the lowest value.

Funtional impairment as a result of dietary deficiency usually is demonstrated by the methods of biochemistry. In principle, the techniques of anatomy or physiology are equally useful.

Human studies

Studies of vitamin nutriture are more complete than those on trace elements and can be a useful guide. Theone, Mock and Kien, et al. (33-36) have studied the biochemical characteristics of human biotin deficiency. Table 4 summarizes some of their measurements. Underwood (22) and others (24) have emphasized that judgement is necessary in interpreting data on requirements. Data for some nutrients often are incomplete. In principle there could be ten requirements for biotin based on the characteristics in Table 4. It is unlikely that they would differ by an order of magnitude, however.

TABLE 4. ASSESSMENT OF BIOTIN NUTRITURE*

Biotin Enzymes	Biotin Concentration In:
Proprionyl-CoA carboxylase	Plasma
3-Methylcrotonyl-CoA carboxylase	Erythrocytes
Pyruvate carboxylase	Whole blood
	Urine

Urinary Metabolites
Methylcitrate
3-Methylcrotonylglycine
3-Hydroxyisovalerate

*After Theone, Mock, Kien, et al. (33-36)

Requirements for Trace Elements

Balance studies

Although balance studies usually are not mentioned in discussion of criteria for measuring nutritional requirements, these studies have been prominent in nutrition. Comparisons between dietary intake and excretory loss are made. Individuals (or groups) who are neither accumulating nor losing the nutrient being studied are said to be in balance. Loss of nutrients via the surface of the body generally is ignored. Occasionally urinary loss is ignored if the daily loss in urine is small in comparison to that in feces.

Claude Bernard (37) thought that balance studies are "...like trying to tell what happens inside a house by watching what goes in by the door and what comes out by the chimney." His disenchantment was related to his discovery "...that sugar is continually produced in animals...". The law of conservation of mass will provide some protection to those who study trace elements by the balance method. Nevertheless balance studies are not perfect (38,39).

We measured the copper balance (40) of 13 men. Each was fed diets made from conventional foods for two to seven periods lasting 30 days. Observations were made during 47 periods. Dietary intake was balanced by urinary and fecal loss at a mean daily intake of 1.30 mg. This estimate of the requirement was within the scatter of points around the graph drawn by Cartwright (41) from which the daily requirement of 2 mg is derived (42). We also considered surface loss of copper; this loss will increase the daily amount to obtain balance to 1.55 mg.

Levander and Morris (43) concluded that approximately 1 µg of dietary selenium per kg of body weight was needed to maintain balance in North America. This amount was considerably higher than that required in New Zealand where low selenium status is common. Apparently historical selenium intake affects the selenium requirement as measured by balance. Mertz (20) noted that balance studies are difficult to interpret.

Part of this difficulty is related to the fact that measurements done on dietary samples reduced to relatively simple solutions by the action of acids, oxidizing agents and heat reveal little about chemical form. Chromium and selenium are two nutrients in which chemical forms may be particularly important.

Trace Element Requirements of Humans

Physiological change

Studies with humans, which are not free of risk, must be done cautiously to avoid injury. Goldberger (44) depleted men of protein and niacin until they got pellagra and then cured them with a good diet in pioneering depletion studies. Restoration of an impaired function is both ethically desirable and scientifically rewarding.

Severe experiments with young people similar to those on the biotin (30) or copper (31) requirements of rats should not be done. However, keen observers can learn from accidents or experiments of nature. The virtual disappearance of Keshan disease, a cardiomyopathy, after Chinese diets were supplemented with selenium (45) has led to the conclusion that the unsupplemented diets were deficient (46). McCollum (9) emphasized the role of chance in the discovery of new nutrients. Chance may be of assistance in the establishment of requirements.

Levander (46) reviewed methods used in the assessment of selenium nutriture including dietary intakes, tissue levels, urinary levels and functional measurements. Interacting substances were considered. He concluded that no single measurement is satisfactory. Maximal activity of glutathione peroxidase, a selenoenzyme, occurs when selenium in whole blood is less than 0.10 µg/ml. Dietary amounts of selenium that produce this enzyme activity may correspond to the first zone of nutriture (Table 2).

Measurement of serum ferritin is thought to be useful in assessing iron nutriture as anemia seems to occur late in iron depletion. Among nonanemic, phlebotomized men with initial serum ferritin less than 25 ng/ml, the amount of iron loss necessary to decrease serum ferritin by 1 ng/ml was five times as great (25.3 mg vs 4.5 mg) as that of similar men with serum ferritin greater than 25 ng/ml (47). Thus the definition of normal ferritin is uncertain. We think it is greater than 25. These data correspond to the third zone of nutriture (29).

Cholesterol in plasma of a man (48) increased by 16% during 105 days of dietary copper depletion. Repletion with copper diminished the concentration to slightly below the control value in 39 days. The copper requirement exceeded 0.8 mg/day; it may have exceeded 1.5 mg/day as urinary and fecal losses exceeded this amount during the initial, control period. Two similar men (49) responded to copper depletion with a decrease in glucose tolerance that returned to normal following repletion. These data, which may illustrate the fourth zone of nutriture (Table 2), are consonant with epidemiology of ischemic heart disease and extensive experiments with animals (50,51). Substantial numbers of diets in the industrialized world contain less copper than these amounts.

The efficiency with which dietary copper can be utilized is affected by at least a dozen chemicals. These chemicals affect both copper metabolism and cholesterolemia. This duality of action prompted the suggestion that a new class of chemicals, those both cholesterotropic and cuprotropic, has been identified (52,53). The origin of the suffix -tropic is from the Greek τροπ-, which indicates a turning in direction in response to a stimulus. In this sense the chemicals are the stimuli that change the directions of cholesterol and copper.

Some of these chemicals are hypercholesterolemic and copper inhibiting: ascorbic acid, cadmium, cholesterol plus cholic acid, fructose, glucose, histidine, sucrose, and zinc. Others are hypocholesterolemic and copper enhancing: calcium, clofibrate, guaran (the principal polysaccharide from guar seeds), and sodium phytate. The identification of these chemicals and the probability that others remain to be discovered complicate the measurement of copper requirements. Zinc was the first of these chemicals to be discovered (54); its influence on copper requirements was mentioned by Underwood (13).

Men fed a diet containing 2.7 to 5.0 mg of zinc daily for 24 to 40 weeks responded with declines in plasma zinc and serum testosterone (55). Plasma zinc decreased earlier than serum testosterone and never was abnormally low (>70 µg/dl). Serum testosterone was low after repletion with zinc had been begun and plasma zinc had begun to increase. These data may correspond to the fourth zone of nutriture (Table 2).

A diet low in zinc (2.6 mg Zn per day) fed to postmenospausal women for 125 days resulted in a decreased ability to clear a standard dose of ethanol (56). The maximal concentration in blood after ingestion was doubled, and after 4 hours the concentration was increased four-fold. Alcohol clearance returned to normal after repletion with zinc. No differences in plasma zinc were found. It is assumed that the activity of the zinc metalloenzyme alcohol dihydrogenase (E.C.) was impaired. These data may correspond to the third zone of nutriture (Table 2).

Summary

Criteria for the identification of essential nutrients were enumerated and classified. Emphasis was placed on definitive criteria which can be simplified to the induction and relief of deficiency. Inferential criteria supported by several authors were collected. These and similar criteria with less support are not very helpful when putative nutrients are considered *a priori*, but they seem to be characteristic of known essentials.

Experiments with animals illustrate the principles by which human nutriture can be assessed and by which human requirements can established. Depletion experiments with human volunteers were described briefly to illustrate some functional impairments that have been attributed to marginal diets. Diets that impair function are considered inadequate. Evidence in support of essentiality can be used to establish requirements for nutrients generally recognized as essential. As a nutritional requirement is somewhat dependent on the test that defines it, requirements are ideally based on several experiments using different methods.

References

1. Klevay, L.M.: Pharmacology and toxicology of heavy metals: Arsenic. Pharmac. Ther. A, 1: 189-210, 1976.

2. Holmes, O.W.: N. Engl. Q. J. Med. Surg. 1, 503-530 (1843).

3. Semmelweis, I.F.: The Etiology, the Concept and the Prophylaxis of Childbed Fever. C.A. Hartleben's Verlags Expedition, 1861. Pest Hungary. Special edition: Classics of Medicine Library. Birmingham, 1981.

4. Lind, J.: A Treatise on the Scurvy, 3rd Edition. S. Crowder, et al. London, 1772. Special Edition: The Classics of Medicine Library. Birmingham, 1980.

5. Mettler, C.C., Mettler, F.A.: History of Medicine. Blakiston, Philadelphia, 1947, p. 129.

6. Funk, C.: The etiology of the deficiency diseases. J. State Med. 20,341-368 (1912).

7. Bowen, H.J.M.: Trace Elements in Biochemistry. Academic Press, London, 1966, p. 102.

8. Arnon, D.J.: Lotsya 3, 31-40 (1950).

9. McCollum, E.V.: A History of Nutrition. Houghton Mifflin, Boston, 1957, p. 376.

10. Vallee, B.L.: J. Chron. Dis. 9, 74-79 (1959).

11. Schroeder, H.A., Balassa, J.J.: J. Chron. Dis. 14, 408-425 (1961).

12. Perry, H.M., Jr., Tipton, I.H., Schroeder, H.A., Cook, M.D.: J. Lab. Clin. Med. 60, 245-253 (1962).

13. Underwood, E.J.: Trace Elements in Human and Animal Nutrition, 2nd edition. Academic Press, New York, 1962, pp. 1-9.

14. Tipton, I.H., Cook, M.J.: Health Phys. 9, 103-145 (1963).

15. Cotzias, G.C.: Importance of trace substances in environmental health as exemplified by manganese. D.D. Hemphill, ed. Proceddings of the First Conference on Trace Substances in Environmental Health. University of Missouri Press, Columbia, 1967, pp. 5-19.

16. Hegsted, D.M.: The beneficial and detrimental effects of fluoride in the environment. D.D. Hemphill, ed. Proceedings of the First Conference on Trace Substances in Environmental Health. Columbia, 1967, p. 105-113.

17. Liebscher, K., Smith H.: Arch. Environ. Health 17, 881-890 (1968).

18. Mertz, W.: Physiol. Rev. 49, 163-239 (1969).

19. Mertz, W.: Fed. Proc. 29, 1482-1488 (1970).

20. Mertz, W.: Ann. N.Y. Acad. Sci. 199, 191-201 (1972).

21. Nielsen, F.H.: Essentiality and function of nickel. W.G. Hoekstra, J.W. Suttie, H.E. Ganther, W. Mertz, eds. Trace Element Metabolism in Animals - 2. University Park Press, Baltimore, 1974, pp. 381-395.

22. Underwood, E.J.: Trace Elements in Human and Animal Nutrition, 4th edition. Academic Press, New York, 1977, pp. 1-12.

23. Schwarz, K.: Essentiality versus toxicity of metals. S.S. Brown, ed. Clinical Chemistry and Chemical Toxicology of Metals. Elsevier/North-Holland, Biochemical Press, 1977, pp. 3-22.

24. Anon.: Recommended Dietary Allowances. Food and Nutrition Board, National Research Council, National Academy of Sciences, Washington, DC. 1980, p. 3.

25. Frieden E.: A survey of essential biochemical elements. E. Frieden, ed. Biochemistry of the Essential Ultratrace Elements. Plenum, New York, 1984, pp. 1-15.

26. Mildvan, A.S.: Metals in enzyme catalysis. P.D. Boyer, ed. The Enzymes, Vol. 2. Academic Press, New York, 1970, pp. 445-536.

27. Green,D.E.: Adv. Enzymol. 1, 177-198 (1941).

28. McLafferty, F.W.: Science 226, 251-253 (1985).

29. Dann, W.J., Darby, W.J.: Physiol. Rev. 25, 326-346 (1945).

30. Klevay, L.M.: J. Nutr. 106, 1643-1646 (1976).

31. Klevay, L.M., Milne, D.B., Wallwork, J.C.: Nutr. Rep. Int. 31, 963-971 (1985).

32. Lau, B.W.C., Klevay, L.M.: J. Nutr. 112, 928-933 (1982).

33. Thoene, J., Baker, H., Yoshino, M., Sweetman, L.: N. Engl. J. Med. 304, 817-820 (1981).

34. Mock, D.M, de Lorimer, A.A., Leibman, W.M., Sweetman, L., Baker, H.: N. Eng. J. Med. 304, 820-823 (1981).

35. Kien, C.L., Kohler, E., Goodman, S.I., Berlow, S., Hong, R., Horowitz, S.P., Baker, H.: J. Pediatr. 99, 546-550 (1981).

36. Thoene, J.G., Lemons, R., Baker, H.: N. Engl. J. Med. 308, 639-642 (1983).

37. Bernard, C. An Introduction to the Study of Experimental Medicine. Translated by H.C. Greene. The MacMillan Company, New York, 1927. Special edition: The Classics of Medicine Library, Birmingham, 1980, p.130.

38. Hegsted, D.M.: Nutr. Rev. 10, 257-259 (1952).

39. Isaksson, B., Sjögren, B.: Proc. Nutr. Soc. 26, 106-116 (1967).

40. Klevay, L.M., Reck, S.J., Jacob, R.A., Logan, G.M., Jr., Munoz, J.M., Sandstead, H.H. Am. J. Clin. Nutr. 33, 45-50 (1980).

41. Cartwright, G.E. Copper metabolism in human subjects. W.D. McElroy, B. Glass, eds. Copper Metabolism. Johns Hopkins University Press, Baltimore, 1950, pp. 274-314.

42. Anon.: Recommended Dietary Allowances. Food and Nutrition Board, National Research Council, National Academy of Sciences, Washington, DC. 1974, pp. 95-96.

43. Levander, O.A., Morris, V.C.: Am. J. Clin. Nutr. 39 809-815 (1984).

44. Goldberger, J.: Goldberger on Pellagra. M. Terris, ed. Baton Rouge, Louisiana State University Press, 1964.

45. Anon.: Chinese Med. J. 92, 471-476 (1979).

46. Levander, O.A.: Fed. Proc. 44, 2579-2583 (1985).

47. Jacob, R.A., Sandstead, H.H., Klevay, L.M., Johnson, L.K.: Blood 56,786-791 (1980).

48. Klevay, L.M., Inman, L., Johnson, L.K., Lawler, M., Mahalko, J.R., Milne, D.B., Lukaski, H.C., Bolonchuk, W., Sandstead, H.H.: Metabolism 33, 1112-1118 (1984).

49. Klevay, L.M., Canfield, W.K., Gallagher, S.K., Henriksen, L.K., Lukaski, H.C., Bolonchuk, W, Johnson, L.K., Milne, D.B., Sandstead, H.H.: Nutr. Rept. Internat. 33:371-382 (1986).

50. Klevay, L.M.: Biol. Trace Element Res. 5, 245-255 (1983).

51. Klevay, L.M.: The role of copper, zinc and other chemical elements in ischemic heart disease. O.M. Rennert, W.-Y. Chan, eds. Metabolism of Trace Metals in Man: Developmental Biology and Genetic Implications. CRC Press, Boca Raton, FL., 1984, pp. 129-157.

52. Klevay, L.M.: Drug-Nutr. Interactions 2, 131-137 (1983).

53. Klevay, L.M.: Cholesterotropic and cuprotropic chemicals. Proceedings of the 5th Conference on Trace Element Metabolism in Man and Animals, Aberdeen, Scotland, in press.

54. Klevay, L.M.: Am. J. Clin. Nutr. 26, 1060-1068 (1973).

55. Abbasi, A.A., Prasad, A.S., Rabbani, P., DuMouchelle, E. J. Lab. Clin. Med. 96, 544-550 (1980).

56. Milne, D.B., Canfield, W.K., Gallagher, S., Johnson, L.K., Klevay, L.M., Mahalko, J. Am. J. Clin. Nutr., in press.

PRACTICAL ASPECTS OF TRACE ELEMENT SUPPLEMENTATION IN EARLY INFANCY

Hans-Burckhard v.Stockhausen
Universitäts-Kinderklinik Würzburg, D-8700 Würzburg

Introduction

There is no doubt that pregnancy and early infancy are periods of life when special consideration must be given to nutrition and efficient supply of additional substances, such as vitamines and trace elements. Under normal conditions the supply of the mother, the fetus, and the newborn with nutritive substances may be considered efficient.

It has only been in the course of increasingly successful neonatal intensive care that pediatricians have had to encounter the features of severe trace element deficiency. Today, the rate of survival in premature infants of birthweights between 500 and 1000 grs already exceeds 50 percent. Subsequent to complicated surgical treatment for malformations of the gastro-intestinal-tract also premature and young infants may be fed totally parenterally, for months. In the beginning, most of the infusion solutions, especially amino acid mixtures, were contaminated with various kinds of trace elements. Thus there was no immediate need for taking trace element deficiency into account. Today, infusion solutions are subject to demands of highest purity, consequently most solutions are practically free of trace elements.

Trace element deficiency may also occur in enteral nutrition provided the storage of trace elements in patients, such as premature infants, has been reduced primarily or foods do not contain sufficient amounts of trace elements. Today, every pediatrician recognizes the problem area of trace element supplementation in early infancy. Nevertheless, emphasis must be laid on the fact that no appropriate solution has been found yet. Some major aspects that are important in practice are discussed in the following report.

Trace Element – Analytical Chemistry in Medicine and Biology, Vol. 4

1. Which supply of trace elements of the young newborn is considered to be normal ?

Before considering trace element substitution a definite agreement must be reached on to what degree the presence of trace elements is to be regarded as normal and at what degree their absence is to be regarded as deficient. This is the very issue that seems to have been causing enormous difficulties. Physiologically the fetus has stored nutritive substances as well as a number of trace elements, by the end of gestation. At birth, the concentration of many trace elements present in the infant's liver is much higher than at any other future time of his life (23). The content of copper is sixteenfold, that of zinc and iron almost twofold. Through the composition of the mother's milk, at the beginning of lactation, the newborn will enjoy this abundance of trace elements for a short time, post partum. Normally the term of gravidity anabolism, as introduced by Kirchgessner, applies to both the animal and the human. This is why one is not surprised to note that an increase in serum levels that often only takes place for a short time, during the first days of life, will be followed by a significant decrease and a reduction in stored reserves, during the first months of life. The supply of trace elements will only improve increasingly in the course of the second half of the first year of life.

The reactions of iron, zinc, and selenium serum levels are most characteristic, during the first year of life. Between the third and sixth month there is a strong decline in the serum levels of these elements (4,14,17). As reported in this volume already by Brätter et al. the decline is stronger in formula-fed infants than in breast-fed infants. Yet, nobody would expect these formula-fed infants who all took a normal development and were clinically healthy to have suffered any harm. With the range of physiologic variations obviously being very wide trace element deficiency is a rareness in young infants.

The decision to define what a normal status is and what is not becomes even more difficult in cases where pregnancy ended pre-term with the fetus being unable to participate in intra-uterine accumulation of trace elements, during the last weeks. In those circumstances various authors no longer use intake by breast-milk, post partum, to measure the requirements of a pre-term infant but have chosen intra-uterine storage of trace elements as a measurement, instead. If acceptable at all, this suggestion can at least not be accepted for all trace elements as will be discussed in the following examples of iron and copper.

There is a stronger decline in hemoglobin in small premature infants than in full-term newborns. It is a known fact that the content of iron, at birth, in small premature infants is only of a small proportion as compared to the full-term infants. Were those small infants administered iron for prevention, immediately post partum, the decline in hemoglobin would be even more evident because any supply of iron, at this early stage of life, will increase hemolysis. Furthermore, these infants that are already more susceptible to infections than others would become even more vulnerable. In a small premature infant of 1000 grs physiologic iron storage will not meet the requirements of the first six months of life, however, it will last till the end of the second month, at the least. To supplement iron before the stores have been exhausted is of no use, moreover, it may even be dangerous to do so. After the second month of life iron supplementation cannot be avoided in small premature infants (27).

The serum copper level of cord blood amounting to 20-30 µg/dl, at birth, only reaches 30 % of normal concentration in adults despite the large copper store of the liver. This results from low ceruloplasmin present in the serum of premature and full-term infants binding 96 % of serum copper. Although premature infants have a negative copper balance (4), post partum, until the beginning of the second month, serum copper increases, post partum, even in small premature infants and it does not make any difference whether foods are supplemented by copper or not (10). The premature and full-term infant first live on their copper stores. There is no need for supplementing like in utero, postpartum, it might even be dangerous to do so.

As to the zinc balance it is negative in premature infants even up to the sixtieth day of life according to investigations carried out by Dauncey (4). Yet, Gibson and DeWolfe did not find a difference in serum-zinc-levels between small premature and full-term infants, during the first half of the first year of life (6). Despite their initially worse position and, in most cases, reduced supply there is generally no evidence of trace element deficiency in premature infants. Levels of various elements are obviously below the limits of full-term infants. However, if one dealt with the matter purely mathematically on the basis of lacking intra-uterine accumulation, post partal supply, and rapid extra-uterine growth the results would show a more inefficient supply.

2. When is trace element deficiency in early infancy due to occur ?

Table 1 shows the main causes of significant trace element deficiency. The smaller and the more premature pre-term infants are the more supply of trace elements will be within the lowest range of normal standards. If the infant needs parenteral infusion therapy for a longer period, in addition, or if there are even nutritional disturbances accompanied by diarrhoea supply of trace elements may become insufficient, altogether. In this respect infants with short-bowel-syndrome and enteral fistulae are especially vulnerable.

Table 1: Causes of relevant trace element deficiency

1. Very small and immature pre-term infants (1000 g)
2. Long-term parenteral nutrition
3. Recurrent diarrhoea and intestinal malabsorption
4. Long-term catabolism
5. Short-bowel-syndrome
6. Very low concentration of some trace elements in human milk
7. Inborn errors of metabolism

3. Which are the trace elements that are likely to develop clinically relevant deficiencies ?

Subsequent to necrotising enterocolitis, during the second week of life, a premature infant with short-bowel-syndrome and anus praeter received parenteral nutrition, from his birth to the eightth month of life. Only from his seventh month of life he was successfully given enteral nutrition. From the fifth month of life the glutathione peroxidase activity in the erythrocytes was below the standards we found in healthy infants (fig.1). Glutathione peroxydase in erythrocytes represents the selenium status fairly well. However, its activity will only slow down if there has been a selenium deficiency, for a period of at least two months.

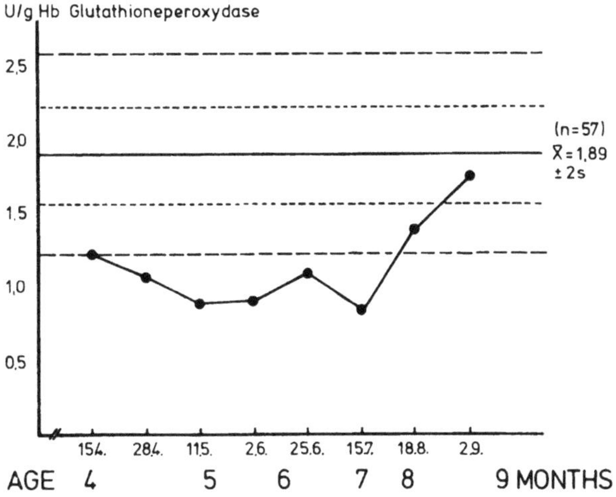

Fig. 1: Glutathione peroxidase activity of erythrocytes in a small premature infant with short-bowel-syndrome recieving parenteral nutrition from birth to eightth month in relation to normal standards of 57 healthy infants.

The selenium concentration in the serum, however, decreases much more quickly. Our own studies in sick pre-term and full-term infants have shown selenium in the serum to drop to 25 ng/ml after 2-3 weeks of parenteral

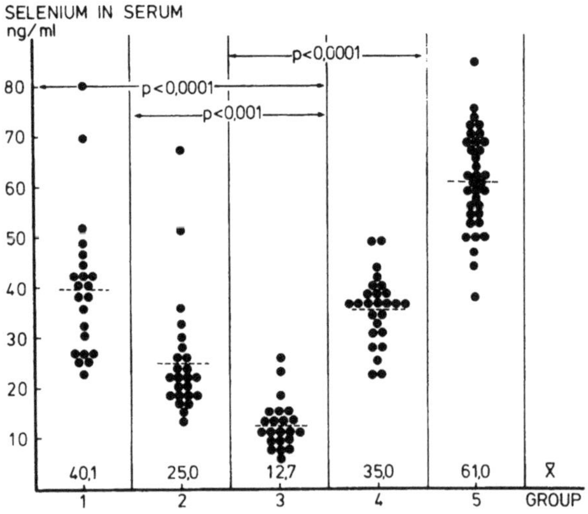

Fig. 2: Selenium concentration in serum of sick premature infants with long term parenteral nutrition on 1-3 day (1), 2-3 week (2), and 5-8 week (3) in comparison to healthy 3-month - old infants formula-fed (4) or breast-fed (5).

nutrition and to 12.7 ng/ml after 5-8 weeks (fig.2). Our data are lower than those of Amin that have been mentioned time and again (3). Ours are similar to those observed by Mrs. Lombeck in children with phenyketonuria receiving a dietary food poor in selenium, for several months (18). However, neither Mrs. Lombeck nor Amin nor our group could provide any evidence for clinical symptoms to result from selenium deficiency. It seems probable that increased transaminases which are occasionally observed in the course of parenteral nutrition are to be attributed to selenium deficiency. Furthermore, it also seems that ventilated premature infants have a higher risk to develop bronchopulmonary dysplasia because of a rise in oxygen toxicity resulting from selenium deficiency. However, cardiomyopathia as observed in adult patients has not been found so far in selenium deficiency, in early infancy. Selenium substitution has not been generally recommended, it has only been applied sporadically. To our knowledge, there is no study to prove that selenium substitution has a positive effect on ventilation and long-term intensive care of premature infants.

However, there is no doubt that zinc deficiency if supplementation is inadequate, in early infancy, it will manifest itself clinically. In our own study in sick pre-term and full-term infants we observed serum zinc level to go down to 0.55 µg/ml subsequent to 5-8 weeks of parenteral nutrition (fig.3). This value is usually suggested as the lower limit of standard. At no time, did any of the infants show any clinical features of zinc deficiency. Tyrala has also found that serum zinc levels between 40-50 µg/dl have no negative effects on the development of premature infants (30).

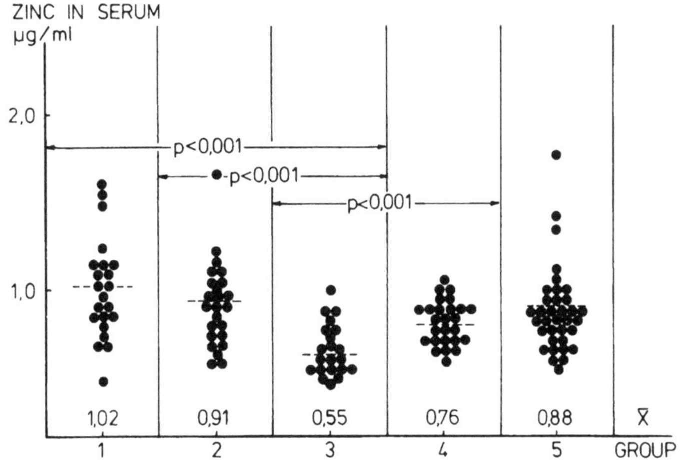

Fig. 3: Zinc concentration in serum of sick pre-term infants with parenteral nutrition on 1-3 day (1), 2-3 week (2), and 5-8 week of life (3) in comparison to healthy 3-month-old infants formula-fed (4) or breast-fed (5).

Mild or moderate zinc deficiency may not be excluded in individual cases of our study groups (tab. 2). According to the literature the earliest date of obvious zinc deficiency in premature infants is 5-7 weeks post partum (28). As a rule, however, zinc deficiency will occur after 3-4 months when insufficiently supplied.

There are also reports on copper deficiency with severe clinical symptoms in premature infants following very long periods of parenteral nutrition (9,13), obviously it is a rare event. Moreover, the condition will manifest itself only after a longer latency of 7-9 months. Features of copper deficiency can be fairly characteristic (tab.3). Altogether, there is no doubt that apart from iron there is also the danger of trace element deficiency in zinc and copper (tab.4). Premature infants tend to having decreased serum levels in some other elements, too. However, no such deficiencies in most of the trace elements have been described, so far.

Tab.2: Clinical features of zinc deficiency

Feature	mild	moderate	severe
Skin: Alopecia	-	-	+
Bullous-pustular dermatitis	-	-	+
Delayed wound healing	-	+	+
Gastrointestinal: Anorexia	-	+	+
Taste dysfunction	-	+	+
Diarrhoea	-	-	+
Endocrine: Growth retardation	-	+	+
Oligospermia	+	+	+
Hypogonadism	-	+	+
Insulin hypersensitivity	-	+	+
Metabolic: Weight loss	+	+	+
Hyperammonemia	+	+	+
Neuropsychiatric: Depression	-	+	+
Irritability	-	+	+
Emotional dysfunction	-	-	+
Ophthalmologic: Impaired dark adaption	-	+	+
Immunologic: Immune dysfunction	-	+	+
Recurrent infections	-	-	+

Tab. 3: Features of copper deficiency in infants

1. Hepatosplenomegaly
2. Roentgenographic bone changes
   - Osteoporosis, fractures
   - Blurring and cupping of metaphysis
   - Subperiosteal new bone formation
3. Sideroblastic anemia (resistant to iron)
4. Neutropenia
5. Hypotonia
6. Pallor and hypopigmentation

Tab. 4: Essential trace elements ranged by necessity of supplementation

Trace element	Necessity of supplementation
Iron	+ + +
Zinc	+ +
Copper	+
Iodine	(+)
Cobalt	(+)
Selenium	?
Manganese	? ?
Chromium	? ?
Molybdenum	? ? ?

4. Which is the appropriate dosage in trace element supplementation?

Trace element substitution is largely discussed. However, nobody knows exactly how much is needed. For preventive supplementation of formula-milk and infusion solutions the average amounts in human milk of the first and second month of lactation might serve as a guide (tab.5). The recommendations of the "Committee of Dietary Allowance" only partially correspond with the average intake by human milk, whereas the recommendations of A.M.A. on parenteral supplementation of trace elements appear rather arbitrary (2). At present, only recommendations for iron, zinc, copper, and selenium are based on balance studies (3,16,32). To establish a positive zinc and copper balance in premature infants 438 µg/kg and day of zinc and 63 µg/kg and day of copper were found to be the appropriate amounts that correspond approximately with those assumed in intra-uterine accumulation (32). To date no agreement has been reached whether this is the goal in parenteral nutrition. In my view this is not what we are to aim at (tab.5).

Tab.5: Trace element intake by human milk and formula milk in coparison to some recommendations for trace element supplementation

	Daily intake of trace elements (µg/kg)							
	Fe	Zn	Cu	J	Se	Mn	Cr	Mo
Human milk (early lactation)	100	300	60	6	4	2	0.05	0.1
Formula milk	700	800	150	8	1	30	1.5	6
Committee on Dietary Allowance	2000	600	120	8	5	100	5	8
A.M.A. for TPN (2)	-	100-300	20	-	-	2-10	o.2	-
Hörnchen 1980 (12)	110	156	31	-	3.2	9.1	10.4	-
Inzolen pro infantibus (Dr. Köhler, Germany)	89,3	98,1	31,8	0.38	-	27,4	10,4	-
Our suggestion for trace element supplementation in TPN in early infancy	(100)	150	15	(1-2)	2	(2)	(0.05)	-

5. Aspects of trace element administration.

Enterally, all trace elements that are present in food in the form of cations are not so well absorbed as those present as anions. As a rule, absorption improves considerably in the form of complex compounds. Despite higher concentrations in formula-milk the supply of trace elements of the newborn by breast-milk is more efficient because of its greater bio-availability. This applies especially to zinc and iron. The good absorption results mainly from the large content of lactalbumin in breast-milk which is in the range of 85 %. A zinc absorption study of different types of milk supplemented with zinc 65 and then fed to a group of adolescent volunteers showed that absorption is highest from human milk while additives of citrate have not affected the zinc absorption (26). According to findings by Voyer et al. the absorption of zinc, especially in premature infants, can be considerably improved if MTC is used as an additive, that will result in a rise in fat absorption up to 90 % (28). It is also interesting to note that the content of zinc in human milk can be improved by administration of zinc to the mother (15).

Remarkably, the absorption of copper and manganese is poor, even with high concentrations in foods (11,29). Even with manganese concentration in formula-milk being fiftyfold higher Stastny did not find any difference in serum manganese levels when compared to breast-fed infants (26).

When considering selenium substitution selen-methionin has been found to be excelently absorbed (20). In general, there is no trace element deficiency in young infants any longer provided they are given more and more solid foods composed of cereals and vegetables with added meat and liver starting at the beginning of the 6th month of life.

There are far more problems in long-term parenteral nutrition. These problems could easily be solved if the substitution of erythrocytes, plasm, and albumin always needed by premature infants also met the babies' requirements for trace elements. In our unit pre-term and full-term infants that are parenterally fed are administered from the first day of life 10 ml/kg of serum, every other day and approximately 15 ml/kg of packed red cells usually every 2-3 weeks. Thus their iron and copper intakes are likely to be sufficient for a rather long period of time. Even after 6 months of parenteral nutrition we found copper levels within normal standards and ceruloplasmin in the serum according to age. Infused serum contains on the average four to fivefold amounts of ceruloplasmin as compared to young infants the half-life of which is five days. When a group of premature infants had been given parenteral nutrition accompanied by intakes of blood and serum for a period of several months, Manser observed higher ceruloplasmin and higher serum copper levels in his study group as compared to an orally-fed control group (21). Erythrocytes only transfer small amounts of copper to the infant, but they transfer relatively large amounts of zinc. The content of selenium in total blood increases subsequent to a transfusion of erythrocytes. However, it has to be remembered that selenium as well as other trace elements are only available to the patient when erythrocytes are destroyed which happens at a rate of 1-2 % per day. If only serum and erythrocytes are administered parenterally for several weeks, a general decline in serum zinc and selenium cannot be prevented, even if serum contains zinc bound to albumin and $alpha_1$-macro-globulin, that is in a physiological form. In comparison to other trace elements, parenteral zinc supply is relatively non-toxic, and may be used in premature infants even in doses up to 800 µg/kg and day (32). Unfortunately accelerated intestinal excretion must be taken into account if such high doses are given rather rapidly. Obviously parenteral zinc intake does not stimulate the synthesis of metallothionein to the same extent as oral intake does. When parenterally administered zinc as well as other cationic trace elements should be given as organic complex compounds, such as acetates, aspartates, and glucogenates.

When a combination of zinc and copper is administered an important role must be attributed to the quantitative relationship of the two elements as demonstrated by Hörnchen (12). A ratio of zinc and copper of 1:1.5 will result in a further decline in these elements, in particular where zinc is concerned. Changing the ratio to 6:1 with only decreasing the supply of copper both elements will rise in the infant's serum. Finally it should be remembered that zinc is also absorbed by skin when applied in the form of zinc-oils and zinc-pastes.

6. Which are the side effects and complications in trace element substitution in early infancy?

Basically there is a vital difference between preventive supplementation and directed therapy in clinically relevant deficiency of a specific trace element. Mixtures of different trace elements are not recommended for directed therapy because when the dose is increased too many accompanying trace elements might be administered and have a toxic effect. Especially, premature infants suffer most frequently from deficiencies and they are most easily rendered susceptible to trace element intoxication. Premature infants always have a deficiency in carrier proteins. When avoiding the mucosal barrier by parenteral administration of trace elements the immature liver might be overstrained. Copper and manganese intoxications are most threatening despite the relatively wide therapeutical range of the latter element. In premature infants there is practically always an iron deficiency which will increase manganese toxicity and also intensify the absorption of lead, cadmium, and cobalt (24).

Excretion of manganese, mainly in stools, in the range of 1-2 µg/kg and day will always be the same regardless of how large a dose the infant is administered. As half-life after parenteral supplementation amounts to 37 days as opposed to 10.5 days after oral supply parenteral supplementation can easily lead to toxic manganese accumulation (33). Similar to copper manganese is excreted into the bowels mainly in bile. Especially during parenteral nutrition, the biliary acid flow is reduced in premature infants. In cases of cholestasis or intestinal obstruction manganese and copper in serum may increase. Young premature infants are most vulnerable to copper intoxication for copper stores are large and the synthesis of ceruloplasmin cannot be augmented even if copper is supplied (16). We therefore agree with Hanschke that copper supplies above 20 µg/kg and day should be avoided in premature infants as long as they do not show normal growth and development (8).

Zinc is the trace element that has the largest therapeutical safety limits as compared to other trace elements. Furthermore, in special situation such as acute catabolism, diarrhoea, parenteral supply of amino acids as well as the administration of diuretics, corticoids, and chelating agents additional zinc will temporarily be needed. Nevertheless we must remember that too high an intake of zinc alone may have indirect toxic effects as it results in an increased decline in copper and iron. Supply of zinc should be reduced in renal insufficiency and persistent obstipation.

All pediatricians are extremely reluctant in parenteral administration of iron, even in small doses. Small premature infants in intensive care units, treated by long-term parenteral nutrtion and ventilation, are extremely vulnerable to infections due to deficient humoral and cellular immune functions. Iron ions may stimulate acute bacterial infections, especially by gram-negative bacteria.

Since no clear clinical evidence of selenium deficiency in early infancy, has been provided, so far, the well-know toxicity of selenium has caused great reluctance in substituting selenium. However, it has been interesting to observe that newborn rats are considerably growth retarded if the dam will receive increasing amounts of selenium in drinking water (25). Furthermore, one should remember that selenium deficiency might cause an increased pulmonary oxygen toxicity. On the other hand, an increased toxicity in lung tissue may occur due to demethyl-selenite that is breathed off by the lung as a final product of selenium metabolism when selenium is excessively supplied (5).

Finally I should like to mention a topic the importance of which has not yet been fully evaluated. In almost all units for premature infants and neonates end-sterilised formula-milk is used in one-way bottles unless human milk is available. In those bottles Maillard-products will develop by irreversible complexes of protein and carbohydrates, a process that is dependent on temperature and duration of storage. These products are not toxic. Nevertheless, they are well absorbed and will subsequently be excreted without any metabolic process. These substances may bind trace elements to an unknown extent which are then irreversibly lost (1).

7. What kind of monitoring does trace element supplementation require?

In long-term parenteral nutrition it is a normal rule to control macroelements, such as sodium, potassium, calcium, and chlorine, at regular intervals. The intervals are dependent on the severity of the basic illness or accident or immaturity. As a rule, acute deviations of essential elements develop the more slowly the smaller the concentrations of serum are. Nevertheless therapeutic trace element supplementation should not be started unless the exact trace element status has been determined. Some authors postulate that long-term parenteral nutrition is only to be administered when at any time determinations of trace elements are possible. As the control parameters suitable for monitoring trace elements show this postulate can practically not be fullfilled (tab.6). The best method for measuring trace elements is neutron activation analysis. Unfortunately the results will at its earliest only be available three months later. Thus, this method is of scientific interest only.

Tab. 6: Monitoring of trace element supplementation.

Zinc:	Serum zinc level Zinc in the erythrocyte Zinc in the leucocyte Hair zinc concentration alcaline phosphatase
Copper:	Serum coper level Ceruloplasmin
Iron:	Hemoglobin MCV (mean cell volume of erythrocytes) Serum iron level Transferrin Ferritin
Selenium:	Serum selenium level Selenium in the erythrocyte Glutathione peroxidase in the erythrocyte Glutathione peroxidase in serum Selenium excretion in urine

Emission spectrophotometry may be quicker. However this method is only available in a small number of places. Another method is atomic absorption spectroscopy which can be performed in a number of larger hospitals. In older infants ceruloplasmin is considered as a useful parameter, but neither ceruloplasmin levels nor copper levels can provide any information about the copper status in early infancy, especially in premature infants.

Zinc-serum levels are reliable as long as there are no infections or any other acute diseases accompanied by stress and shock. Then zinc levels are incorrectly low because of an increased uptake of zinc by metallotheoneins. Unfortunately, alcaline phosphatase which has been mentioned frequently will only decline at a profound zinc deficiency. Today, iron substitution is best monitored by a number of parameters. Determination of hemoglobin and examination of erythrocytes in a blood smear can be easily done.

The search for appropriate parameters for monitoring selenium substitution is a special field of interest. Determination of selenium by using neutron activation analysis is too time-consuming. Glutathione peroxidase in erythrocytes will only show increases and decreases 1-2 months after the amount of selenium has changed. According to Mrs. Lombeck a useful recent method has been established in measuring the activity of glutathione peroxidase in plasm (19). This is a relatively rapid way of documenting the effects of selenium substitution. Apart from determining serum levels which is always too complicated and unreliable because of technical errors there are no additional parameters for monitoring substitution of all other trace elements.

Summary and conclusions

1. In contrast to the breeding of animals and the cultivation of plants trace element deficiency of clinical relevance in early infancy is rare.
2. It is of vital importance to know the conditions under which trace element deficiency is due to occur.
3. Patients with short-bowel-syndrome and extremely premature infants, both needing parenteral nutrition for months, are especially vulnerable to developing trace element deficiencies.
4. Whenever trace elements are substituted this must be done with a clear view of the goal that is to be achieved. Even in premature infants imitating intra-uterine accumulation of trace elements need not necessarily be the only goal.
5. To treat trace element deficiency pure separate preparations of all those trace elements for which clinically relevant deficiencies are

known available on the market. There should be no therapy without monitoring.

6. To prevent trace element deficiency it is suggested to supplement infants either with human serum or infusions, e.g. amino acid solutions containing physiological doses of trace elements.

7. Special foods for premature infants and patients suffering from chronic intestinal diseases or inborn errors of metabolism are to be supplemented appropriately.

References

1. Agget, P.J.: Clin. Endocrinol. Metab. 14, 513-543 (1985).
2. A.M.A.: Department of foods and nutrtion: J.A.M.A. 241, 2051-2054 (1979).
3. Amin, S., Chen, S.Y., Collip, P.J., Castro-Magana, M., Maddaiah, V.T., Klein, S.W.: Nutr. Metab. 24, 331-340 (1980).
4. Dauncey, M.J., Shaw, J.C.L., Urman, J.: Pediatr. Res. 11, 991-997(1977)
5. Diskin, C.J., Tomasso, C.L., Alper, J.C., Glaser, M.L., Fliegel, S.E.: Arch. Intern. Med. 139, 824-826 (1979).
6. Gibson, R.S., DeWolfe, M.S.: Acta Paediatr. Scand. 70, 497-500 (1981).
7. Halliday, H.L., Lappin, T.R.J., McClure, G.: Biol. Neonate 45, 228-235 (1984).
8. Hanschke, F.: In: Spurenelemente (Eds. Gladtke, E., Heimann, G., Lombeck, I., Eckert, I.) pp 172-182, Thieme, Stuttgart ·New York 1985
9. Heller, R.M., Kirchner, S.G., O'Neill, J.A.Jr., Hough, A.J.Jr., Howard, L., Kramer, S.S., Green, H.L.: J. Pediatr. 92, 947-949 (1978).
10. Higashi, A., Matsuda,I., Masumoto, T., Saikusa, H., Yabuso, M., Oka,Y.: Tohoku. J. exp. Med. 146, 253-263 (1985).
11. Hillman, L.S., Martin,L., Fiore, B.: J. Pediatr. 98, 311-313 (1981).
12. Hörnchen, H.: In: Klinische Ernährung 2 (Ed. Ibe, K.) pp 33-43, Zuckscherdt, München 1980
13. Karpel, J.T., Peden,V.H.: J. Pediatr. 80, 32-36 (1972)
14. Kasparek, K., Feinendegen, L.E., Lombeck,I.: Eur. J. Pediatr. 126, 199-202 (1977).
15. Krebs, N.F., Hambidge, K.M., Jacob, M.A., Rasbach, J.O.: Am. J. Clin. Nutr. 41, 560-570 (1985).
16. Lockitch, G., Godolphin, W., Pendray, M.R., Riddel, D., Quigley, G.:

J. Pediatr. 102, 304-308 (1983).

17. Lombeck, I., Kasparek, K., Harbisch, H.D., Feinendegen, L.E., Bremer, H.J.: Eur. J. Pediatr. 125, 81-88 (1977).

18. Lombeck, I., Kasparek, K., Feinendegen, L.E., Bremer, H.J.: In: Spurenelemente (Eds. Gladtke, E., Heimann, G., Eckert,I.) pp 119-128, Thieme, Stuttgart · New York 1979

19. Lombeck, I., Menzel, H., Steiner,G., Kasparek, K.: Klin. Pädiatr. 194, 303-305 (1982).

20. Luo, X., Wei, H., Yang, C., Xing, J., Liu, X., Qiao, C., Feng, Y., Liu, J., Liu, Y., Wu, Q., Liu, X., Guo, J., Stoecker, B.J., Spallholz, J.E., Yang,S.P.: Am. J. Clin. Nutr. 42, 439-448 (1985).

21. Manser, J.I., Crawford, C.S., Tyrala, E.E., Brodsky, N.I., Grover, W. D.: J. Pediatr. 97, 795-799 (1980).

22. Matseshe, J.W., Philips, S.F., Malagelada, J.R., McCall, J.T.: Am. J. Clin. Nutr. 33, 1946-1953 (1980).

23. Meinel, B., Bode, J.C., Koenig, W. Richter, F.W.: Biol Neonate 36, 225-232 (1979).

24. Pleban, P.A., Numerof, B.S., Wirth, F.H.: Clin. Endocrinol. Metab. 14, 545-566 (1985).

25. Ponce, E., Cacciabaudo, J., Chen, S.Y., Collip, P.J., Dincsoy, M.Y.: Pediatr. Res. 19, 230A (1985).

26. Sandström, B., Cederblad, A., Lönnerdal, B.: Am. J. Dis. Child. 137, 726-729 (1983).

27. Siimes, M.A., Järvenpää,A.L.: J.Pediatr. 101, 277-280 (1982).

28. Sivasubramanian, K.N., Hoy, G., Davitt,M.K., Henkin, R.I.: Lancet 1, 508 (1978).

29. Stastny, D., Vogel,R.S., Picciano, M.F.: Am.J. Clin. Nutr. 39, 872-878 (1984).

30. Tyrala, E.E., Manser, J.I., Brodsky, N.L., Tran, N.: Acta Paediatr. Scand. 72, 695-698 (1983).

31. Voyer, M., Davakis, M., Antener,I., Valleur, D.: Biol. Neonate 42, 87-92 (1982).

32. Zlotkin, S.H., Buchanan, B.E.: J. Pediatr. 103, 441-446 (1983).

33. Zlotkin, S.H.: In: Trace elements in nutrition of children. (Ed. Chandra, R.K.) pp 175-180, Raven Press, New York 1985

DIETARY RECOMMENDATIONS FOR TRACE ELEMENT INTAKE

An Analyst's Point of View

R. Schelenz *)

INTERNATIONAL ATOMIC ENERGY AGENCY
Department of Research and Isotopes
Wagramer Strasse 5, P.O.Box 100
A-1400 Vienna, Austria

This contribution is intended to comment on human dietary recommendations for trace element intake from an analytical and not from a toxicological point of view. It is not the intention of the author to argue about the reliability of the established Recommended Dietary Allowances (RDA).

Recommended Dietary Allowances (RDA)

According to the National Research Council of the USA the "Recommended Dietary Allowances (RDA) are the level of intake of essential nutrients considered, in the judgement of the Committee on Dietary Allowances of

*) The views expressed in this article do not necessarily represent those of the International Atomic Energy Agency and the Agency accepts no responsibility for them.

Trace Element – Analytical Chemistry in Medicine and Biology, Vol. 4

the Food and Nutrition Board on the basis of available scientific knowledge, to be adequate to meet the known nutritional needs of practically all healthy persons" (1).

The RDAs are set up for average daily intake by population groups and not for specific individuals over a certain time. They are estimated to meet the requirements of most of the individuals in the population. Lower intakes than recommended do not necessarily mean inadequate intake, but lower intakes increase the risk that the intake level is no longer to be considered as safe. As a convention some people understand an intake less than 2/3 of the recommendation to be critical for an individual.

Mostly RDAs are used for

a. assessing food intake by groups or individuals, planning therapeutic diets and institutional meals, planning national food supplies
b. nutrition labelling
c. expressing nutritional quality of foods using the nutrient density index

Fig. 1: Average dietary daily intake of copper
by each of 26 females (25 - 45 years)
for 4 individual days (2)

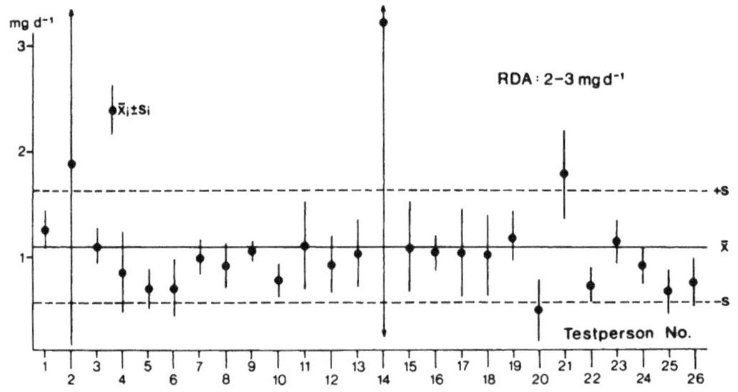

ad a) Figure 1 shows the average dietary daily copper intake by 26 females between 25-45 years for 4 individual days (2). The filled in circles indicate the mean intake over a period of 4 days whereas the vertical lines represent the standard deviation of this average intake which should not be confused with the analytical standard deviation of the mean. The RDA for copper is 2 -3 mg per day, and the average dietary copper intake of all the probands was determined to be 1.1 mg per day which is only 37 - 55 % of the recommended value. The high intake of more than 3 mg per day by one woman is probably due to the consumption of beef liver. To explain such unexpected values the food records of the probands in dietary studies should be kept as precise as possible.

ad b) RDAs are also used as denominators in food industry to express the contribution of a special food to the dietary intake of different nutrients, the so called food labelling.

Fig. 2: Cornflakes with milk
Contribution to diet of 8 year old boy

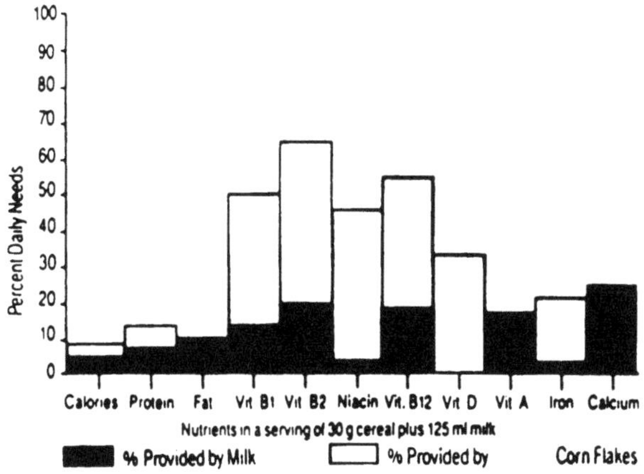

Figure 2 demonstrates the contribution provided by corn flakes with milk to the diet of an eight year old boy. The actual intake of Fe and Ca provided by this meal is divided by the recommended intake of these elements (Fe: 10 mg per day; Ca: 800 mg per day) (1). As can be seen from figure 2, approximately twenty percent of Fe and twenty five percent of Ca of the daily intake is provided by this breakfast constituent.

ad c) Besides the comparison of the daily intake of essential nutrients with the recommended values, the "Quality Index" also represents a suitable standard. The Quality Index (QI) is the quotient of the actual nutrient density related to 1 MJ energy of the diet and the energy related nutrient density which can be calculated from the recommended nutrient intakes. QI < 1 indicates marginal, QI > 1 luxurious intake of an individual nutrient (3).

The right end of the dashed bar means (Figure 3): lower limit of the recommended intake as reference for QI. The left end of the dashed bar means: upper limit of the recommended intake as reference for QI. Referring to energy and analyzing freeze-dried material it should be noted that (because of evaporation during the drying process) the energy contribution due to the alcohol consumption is often neglected. With respect to the nutrient density, the values for Fe, Zn, Cu and Se are marginal whereas the values for Na, Cl, Ca, Mg and P are well above those recommended. The results shown in figure 3 originate from the Study performed at Karlsruhe, 1981 (2).

Essential Food Constituents

In table 1 the RDAs for some elements are shown for females aged 23-50 years with 55 kg body weight. For zinc, besides the RDA of 15 mg per day (i.e. regarding zinc as an essential element) the provisional maximum acceptable daily load of zinc is given as proposed by World Health Organization (WHO)/Food and Agriculture Organization (FAO) of the United Nations (i.e. regarding zinc as a toxic metal) and is calculated on the basis of a body weight of 55kg (4 - 9).

Fig. 3: Nutrient Density Profile:
Average nutrient consumption of 26 women
(104 total daily diets) per 1 MJ

Nutrient	Dimension	Recommended Intake per day*	Recommended Intake per MJ and day	QI	Quality-Index 0 0.5 1.0 1.5 2.0
Energy	MJ	8.8	1.00	1.00	----------|
Protein	g	52	5.91	1.32	----------|---
Fat	g	68	7.73	1.42	----------|-----
Linoleic Acid	g	10	1.14	1.23	----------|--
Sodium	g	2-3	0.24-0.34	1.05-1.56	----------|- - - -
Chlorine	g	3-5	0.34-0.57	0.93-1.57	--------- |- - - -
Potassium	g	2-3	0.23-0.34	0.91-1.35	-------- -|- -
Calcium	mg	700	79.5	1.23	----------|--
Magnesium	mg	220	25.0	1.36	----------|-----
Iron	mg	18	2.04	0.59	-------
Zinc	mg	15	1.70	0.65	--------
Copper	mg	2-3	0.23-0.34	0.35-0.52	------
Phosphorous	mg	700	79.5	1.69	----------|--------
Manganese	mg	2.5-5.0	0.28-0.57	0.67-1.36	--------- |- - -
Selenium	µg	50-200	5.7-22.7	0.27-1.09	------ - -|- - - -
Thiamine	µg	1400	159.1	0.89	--------

* Mainly according to RDA and "Empfehlungen für die Nährstoffzufuhr. Deutsche Gesellschaft für Ernährung" Umschau-Verlag, Frankfurt/Main 3. Auflage, 1975 (see also reference 2)

Tab. 1: Recommended daily dietary allowances
Population group: females
Age: 23 - 50 years; body weight: 55 kg

	mg/day	Max. Acceptable Daily Load mg/day
Calcium	800	
Phosphorous	800	
Magnesium	300	
Iron	18	
Zinc	15	16.5 - 55 (provisional)
Iodine	0.15	

Obviously the recommendations for Zn nearly overlap, and are inconsistent with each other.

When calculated on the basis of 6kg body weight for babies, the ranges of desirable intakes recommended (3.1 - 12.5 mg per day (9); 3 mg per day (1)) actually overlap the acceptable daily load of zinc (4). This is valid also for the dietary iron intake by babies where the recommended intake (10 mg per day) (1)(9) and the maximum acceptable daily loads (4.8 mg per day) (4) are overlapping and are inconsistent.

Safe and Adequate Daily Dietary Intakes (SADDI)

Where insufficient information is available on which to base allowances, these figures are given as ranges of recommended intakes called Estimated Safe and Adequate Daily Dietary Intakes (SADDI) (1) (Table 2).

It has to be emphasized that the provisional maximum acceptable daily load of copper proposed by WHO/FAO (i.e. regarding copper as a toxic metal) (4), when calculated on the basis of a 55kg body weight, actually overlaps the range of desirable intakes of SADDI (i.e. regarding copper as an essential trace element)(1,5,9).

Tab. 2: Estimated safe and adequate daily dietary intakes (SADDI) for adult females and 55 kg body weight (ranges of recommended intakes)

	SADDI mg/day	Max. Acceptable Daily Load mg/day
Copper	2.0 - 3.0	2.75 - 27.5 (provisional)
Manganese	2.5 - 5.0	
Fluoride	1.5 - 4.0	
Chromium	0.05 - 0.2	
Selenium	0.05 - 0.2	
Molybdenum	0.15 - 0.5	
Sodium	1100 - 3300	
Potassium	1875 - 5625	
Chloride	1700 - 5100	

Obviously, these recommendations are inconsistent with each other, and some reconsideration is suggested. This is also to be noted for the daily intake of copper of babies via mother's milk where the recommendations (0.5 - 0.7 mg per day (1); 0.48 mg per day (9)) overlap the maximum acceptable daily loads (0.3 - 3 mg per day (4)).

Food Contaminants

For the purpose of the Codex Alimentarius under the Joint FAO/WHO Food Standards Programme "contaminant" means (4): "Any substance not intentionally added to food, which is present in such food as a result of the production (including operations carried out in crop husbandry, animal husbandry and veterinary medicine), manufacture, processing, preparation, treatment, packing, packaging, transport or holding of such food or as a result of environmental contamination. The term does not include insect fragments, rodent hairs and other extraneous matter."

According to these definitions and considering toxicological evaluations the following "limits" have been recommended (4, 6) (Table 3):

Tab. 3: Maximum acceptable daily load/ provisional tolerable daily intake based on mg element/kg body weight calculated from weekly intakes

	MAXIMUM ACCEPTABLE DAILY LOAD MG/KG BODY WEIGHT (4)		PROVISIONAL TOLERABLE DAILY INTAKE MG/KG BODY WEIGHT (6)
As	0.002 (7)		
Cd			0.0067 - 0.0083
Cu	0.05 - 0.5	(provisional)	
Fe	0.8		
Pb			0.05
Hg (total)			0.005
Hg (Methylmercury)			0.0033*)
Sn	20	(provisional)	
Zn	0.3 - 1.0	(provisional)	

*) expressed as Hg

The answer to the question whether the occurrence of a substance in the diet may be harmful for human health requires a comprehensive knowledge of the dietary intake of that substance. Therefore, both the concentration of the contaminant in the diet and the amount of the dietary consumption need to be known. Such information is often very incomplete or only available as rough estimates.

Dietary Intake

The "Guidelines for the Study of Dietary Intakes of Chemical Contaminants" prepared under the joint sponsorship of FAO, WHO and the United Nations Environmental Programme (UNEP) provide detailed instructions for conducting three basic and practical procedures for estimating actual dietary exposures of consumers to chemical contaminants (10).

Three types of intake studies are recommended:
- market basket studies
- selective studies of individual foodstuffs (composite diet)
- duplicate diet studies (duplicate portion technique)

The choice of which of these techniques are to be used will depend on the objectives and the available resources in assessing the consumer's dietary intake. Each approach has its own strengths as well as limitations for obtaining information on actual dietary exposures.

- Market basket studies

  These studies are conducted analyzing a pattern of individual foodstuffs and using official food consumption statistics. The average intake per capita of the population is estimated from the analytical results taking into account the average consumption of the specific food item. This procedure suffers from several disadvantages, because mainly raw foodstuffs are analyzed. The changes in elemental composition caused by kitchen-like preparation of individual foodstuffs (e.g. washing, cleaning, cooking) are often neglected or only considered as a rough estimate. It was found, e.g. that the arsenic content of potatoes is removed almost quantitatively after peeling and washing (11). The food consumption statistics do not consider the fact of non-eaten portions of a meal as well as throw away losses in the household, which were estimated to be 10% of the consumption. Further this technique gives no idea of the extent to which the dietary intake by an individual with his particular eating and drinking habits is affected.

- Composite Diet

  This method, as applied by the Food and Drug Administration (FDA) of the United States seems to be in between market basket studies and duplicate portion techniques. According to the official food consumption statistics the individual food items which mainly contribute to the average food intake by the population are bought from the market. After composing meals out of these foodstuffs and kitchen-like preparation using standard procedures, aliquots are analyzed. The actual dietary daily intake by a particular element is calculated from these results. No information can be obtained with respect to the intake by an individual.

- Duplicate Diet Studies

  Although the duplicate portion technique is time consuming and more expensive compared to the other methods, it appears to have some benefits, whereby identical aliquots or totals of the daily consumed and prepared food are analyzed. Nearly all effects influencing the actual dietary intake of an individual from harvesting or production of a foodstuff up to eating a prepared meal are considered.

According to the Austrian Nutrition Report 1982 (12), the average dietary weekly intake of cadmium using the market basket technique is 470µg and is calculated to be 170µg per capita of the population when using the duplicate diet technique. According to the recommendation of the Joint FAO/WHO Expert Committee, the provisional acceptable weekly intake for cadmium for an adult of 60kg body weight is 400-500µg (6). Assuming that the average of 470µg cadmium intake according to the market basket study is correct, the limit of 500µg cadmium is nearly met and with respect to individual eating and drinking habits this limit will surely be exceeded by some individuals. If the intake of 170µg Cd according to the duplicate diet study is correct, only approximately one third of the upper limit will be obtained.

The dietary daily intake of Zn and Fe for females using the duplicate portion technique is reported in the figures 4 and 5 (2).

Fig. 4: Average dietary daily intake of Zinc by each of 26 females (25 - 45 years) for 4 individual days (2)

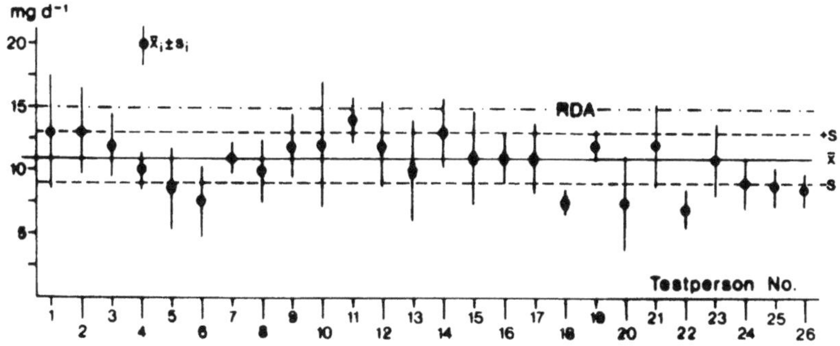

Fig. 5: Average dietary daily intake of iron by each of 26 females (25 - 45 years) for 4 individual days (2)

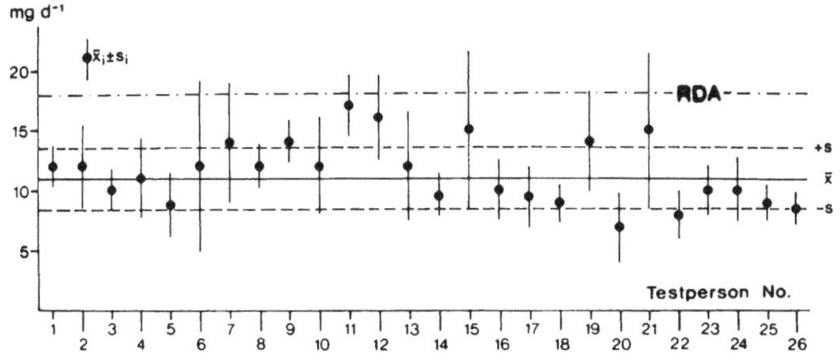

A marginal intake for both elements was found. To compare these findings, studying a relatively small group of persons, some literature values for the dietary daily intake of Fe and Zn by females are shown in the tables 4 and 5 (13 - 24).

Excluding the results of a study from New Zealand for Zn (19), the dietary intake of both Fe and Zn has been found lower than the respective RDAs.

Tab. 4: Average dietary daily intake of iron
by females (some literature values)
RDA: 18 mg per day

Average daily intake in mg	Country	Year	Diet	Literature
16.0 *)	USA	1977	College meals	(14)
10.6 *)	USA	1978	Self selected	(15)
10.9	FRG	1978	Food records	(16)
12.3	UK	1981	Food tables	(13)
13.1 *)	FRG	1981	Food record (24 hrs.)	(17)
11 *)	FRG	1981	Self selected	(2)

* = analyzed

Tab. 5: Average dietary daily intake of zinc
by females (some literature values)
RDA: 15 mg per day

Average daily intake in mg	No. of probands	Country	Year	Diet	Literature
9.8-14.4 *)	13	USA	1954	Self selected	(18)
18.4 *)	4	New Zealand	1973	Metabolism study	(19)
13.3 *)	36	USA	1976	Self selected	(20)
11.6 *)	8	New Zealand	1977	Catering	(21)
8.9 *)	7	New Zealand	1977	Self selected	(21)
7.6 *)	1	Scotland(UK)	1979	Household	(22)
10.1 *)	100	Canada	1981	Food tables	(23)
9	?	Australia	1981	Market basket	(24)
11 *)	26	FRG	1981	Self selected	(2)

* = analyzed

Limit of Detection

According to Keith et al. (25) the limit of detection (LoD) is defined as the lowest concentration level that can be determined to be statistically different from a blank. The LoD is defined as three times the standard deviation of the analytical blank (26). The ideal analytical blank is suggested to be determined in an identical matrix in the absence of this particular element. In practice this cannot be achieved, due to minor or trace concentrations of the element to be determined, e.g. in reagents and acids, even of high purity quality, a significant amount will probably be detected. Figure 6 schematically gives a picture of a possible way to use the expression of LoD in a more practical way. The limit of quantitation (LoQ) is defined as the level above which quantitative results may be obtained with a specific degree of confidence. This is ten times the standard deviation (s.d.) of the analytical blank, corresponding to an uncertainty of ± 30 % in the measured value (10 s.d. ± 3 s.d.) at the 99 % confidence level.

There are two main problems which should be considered for analytical results at or near the limit of detection (LoD). The uncertainty of the result can approach or even equal the reported value. In addition, a confirmation of the reported value is impossible because the identification is mainly related to the sensitivity and selectivity of the analytical method used. These influences decrease with increasing concentration of the element of concern. Therefore Keith concluded "quantitative interpretation, decision-making, and regulatory actions should be limited to data at or above the LoQ", which is ten times the standard deviation of the analytical blank (Figure 6) for the respective method (25). This actually means that RDAs or tolerable maximum intakes as well as legally established maximum permissible concentrations should be considered using the LoQ as a decision criterion.

Fig. 6: Significance Levels

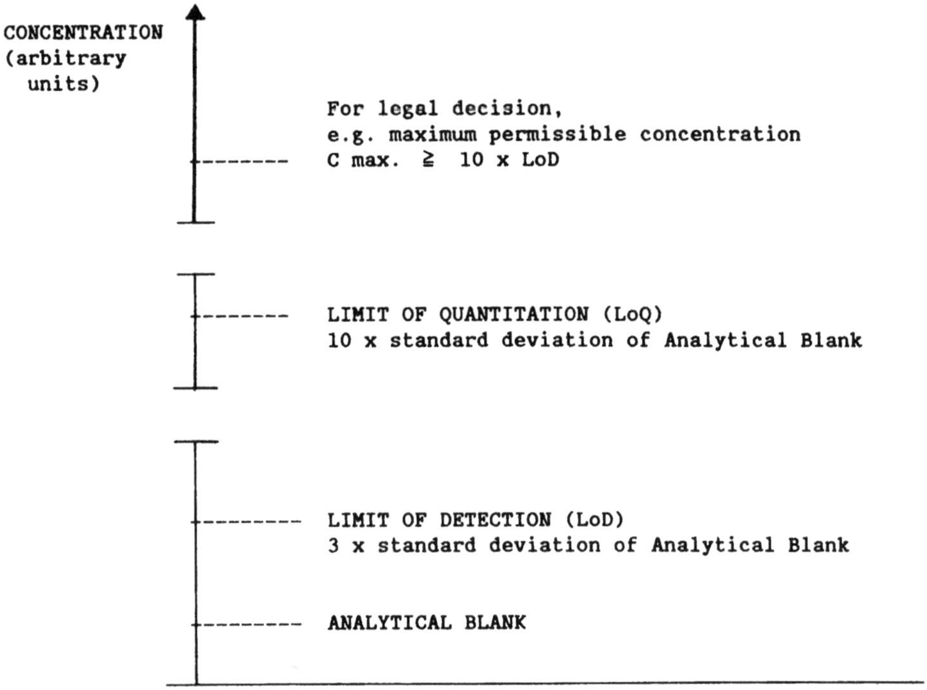

In recent Co-ordinated Research Programmes initiated by the International Atomic Energy Agency (IAEA), a panel of Consultants recommended the use of ten times the LoD as the limit for legal decisions or to confirm whether a recommended or tolerable intake of a particular element is achieved. This probably has some educational effect in causing the participating laboratories to pay more attention to the use of a suitable analytical method and presumably in reducing the analytical background by improvements in the laboratory environment.

Sensitivity of Analytical Methods

In table 6 an attempt has been made to compare LoDs of three different analytical methods (AAS, NAA, ICP-AES) with LoDs required to meet the respective RDAs or provisional tolerable weekly intakes. The RDAs, SADDIs (data given as ranges) and WHO values (data within brackets) refer to females, aged 23-50. They are converted to 55kg body weight, assuming an average food consumption of 2080g per day, which has been found in the Karlsruhe Study 1981 (2). The LoDs are not theoretical ones but have been determined in routine analytical work in the laboratory, analyzing freeze-dried material without any further preconcentration. Therefore, in most cases the LoDs are higher than those reported in the literature because they are related to practical trace element analysis considering also dilution factors using AAS, ICP-AES as well as RNAA. Only such elements are listed for which nutritional standards or limits exist.

To explain this for chromium as an example, comparing the calculated laboratory LoDs for different analytical techniques with the concentrations to be met according to RDA or SADDI and WHO recommendations, no problems should arise in meeting the required LoD range of 24-96 µg/kg if adequate analytical methods are used, which in this case are AAS-GF, INAA and RNAA. According to the figures given in table 6 all elements for which food standards exist can be determined with sufficient reliability in using freeze-dried material, without any preconcentration, the appropriate analytical methods.

Tab. 6: Limit of detection (LoD) µg/kg
of different analytical methods and techniques

Element	AAS			NAA		ICP-AES		Concentration to be met related to food standards
	Flame	Others	GF	INAA	RNAA	SM	SQ	
As		50(HT)					1750	(53)
Ca	4.5×10^{3}	-				25	6.5	3.85×10^{5}
Cd	3000		15		3	170	85	(25-31)
Cr	23000		15	20	1.5	650	205	24-96
Cu	6500		8		120	365	180	962-1442
Fe	8.8×10^{4}		15	300		380	155	8.65×10^{3}
Hg		20(CVT)		5	0.5		850	(19 total)
K	1500			1.5×10^{6}			2.1×10^{6}	$9.0\times10^{5}-2.7\times10^{6}$
Mg	1000					25	5	1.44×10^{5}
Mn	300		4	430	10	60	55	$5.2-10.4\times10^{3}$
Mo	37500				2		265	72-246
Na	1.0×10^{3}			4.0×10^{4}			0.95×10^{3}	$5.29\times10^{5}-1.59\times10^{6}$
Pb	37500		6				1400	(189)
Se		20(HT)		30	1		2500	24-96
Zn	1000			20		285	60	7200

AAS = Atomic Absorption Spectrometry, dilution 1 : 100
CVT = Cold Vapour Technique
GF = Graphite Furnace
HT = Hydride Technique
NAA = Neutron Activation Analysis
I = Instrumental R = Radiochemical separation
ICP-AES = Inductively Coupled Plasma - Atomic Emission Spectrometry; dilution 1 : 50
SM = Simultaneous determination SQ = Sequential determination

Analytical Quality Control

In 1985 the Analytical Quality Control Services (AQCS) of the IAEA organized a worldwide intercomparison for the determination of trace elements in a freeze-dried human diet with 66 participants using different analytical methods. The composition of the diet represents average eating and drinking habits of the Finnish population according to the food consumption statistics. Some preliminary results are given in the figures 7 and 8.

As expected (Table 6) most of the results for Hg have been reported for CVT (A4) besides that for NAA (N2) (Figure 7). The three CVT values at higher mercury concentration, which are nearly one order of magnitude higher than most of the other reported values, may lead to the assumption that possibly contamination has occured. The drawn line represents the element concentration to be met related to food standards (last column, Table 6).

Fig. 7: Hg in mixed human diet H-9, (in µg/kg)

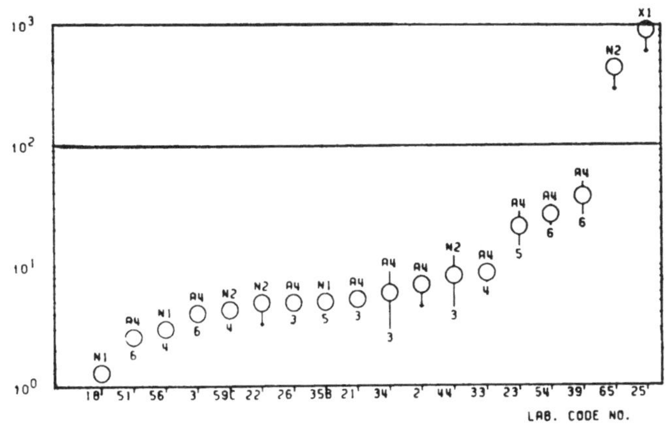

According to the LoDs given in table 6 the most sensitive methods for the determination of chromium should be in the order: **RNAA (N2)**, **AAS-GF (A3)** and sequential ICP-AES (E5). In fact some sort of clustering may be observed in figure 8. At lower concentrations (left side) the frequency of RNAA determinations seems to be higher. Whereas in the central concentration range results of the AAS-GF technique appear more frequently compared with other methods. Going to higher concentrations above the RDA value (right side) the first results for ICP-AES appear. The four highest values have been reported for Flame-AAS (A-1) and XRF (X1). Figure 8 actually reflects the increasing order of the limits of detection for RNAA, AAS-GF and ICP-AES as well. According to the reported results Flame-AAS and XRF do not seem to be adequate methods for the determination of chromium in human diets at that concentration level. The drawn line represents the element concentration to be met related to food standards (last column, Table 6).

Fig. 8: Cr in mixed human diet H-9 (mg/kg)

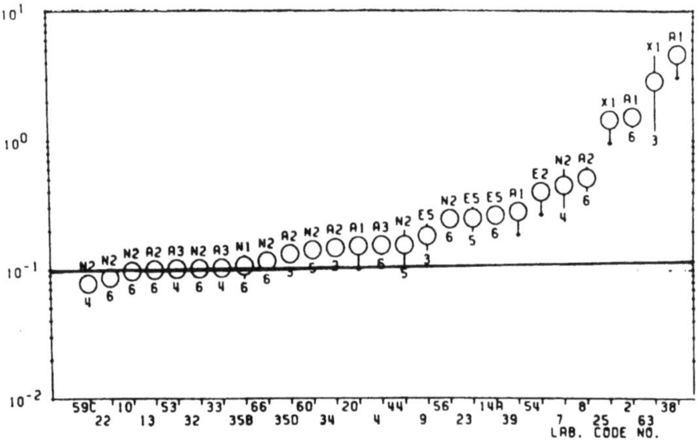

Contamination

As to contamination some feedback can be given from an enquiry asking analytical laboratories where contamination possibly occurred (Table 7). Most frequently the decomposition and separation step of the sample followed by collection and storage of the samples have been reported.

Tab. 7: Stages where contamination occurred

Stages	No. of Laboratories (Total = 96)	%
Collection of sample	44	46
Storage of sample	44	46
Comminution (pulverization)	17	18
Sieving	10	10
Filtration	17	18
Desiccation	11	11
Decomposition of sample	62	65
Separation	53	55
Instrumental measurement	31	32

Another source of contamination may be the laboratory atmosphere or environment (Table 8). In trace analysis unfiltered air may be the reason for possible contamination by most of the elements. Whereas smoking in the laboratory may interfere, especially in the determination of bromide.

Biological Reference Materials

The choice of the biological reference materials may also affect the analytical result (Table 9). A certified value for, e.g. chromium in citrus leaves and oyster tissue associated with an error of 25 and 39 %, respectively at a level of 0.7 - 0.8 ppm is confusing. Whereas an error of 27 % for milk powder at a level 300 times lower seems to be comparatively excellent. The IAEA is not excluded in this respect as can be seen for cotton cellulose with an error of 27 %. A similar picture appears for selenium in biological reference materials (Table 10). Tables 9 and 10 are taken from a compilation which contains trace element data for 60 Biological Reference Materials from 9 producers and has recently been published by the IAEA (27). It is more or less confusing to see in the T-column of table 10 that several different expressions are used to quantify the reliability of the given selenium concentrations. An approach to this problem has been made by the International Organization of Standardization (ISO) giving some definitions in the ISO Guide 35 (28).

Elemental standard solutions used for calibration purposes are more accurate. However, their use suffers from the disadvantage that, in analyzing foodstuffs or diets, the effect of the decomposition of the biological material is often neglected. Whereas elements in standard solutions occur mostly as ions, the binding of elements in biological matrices is a more complex one. Therefore, e.g. incomplete decomposition of the sample or volatilization of particular elements may lead to inaccurate results if elemental standard solutions are used.

A comprehensive monograph describing the problems arising in the analysis of trace elements in biological materials also considering the use of different analytical techniques has been published by the IAEA (29).

Tab. 8: Elemental levels
in laboratory atmosphere and environment

Element	Non-filtered air	Filtered air	Tobacco smoke	Cosmetics	Sweat	Skin	Hair
	µg/g dust	µg/g dust	µg/g	µg/g	µg/ml	µg/g wet	µg/g
Al	3000	6				1-2	4-29
As	55	<0.01	2.85			0.06-0.10	0.2-3.7
Br	23	<0.02	71.50	4000	0.2-0.5	4-10	<1-53
Ca	2690	<0.004		$6x10^4$	4-10	250	200-3190
Cd	2.8	0.1				0.032	0.24-2.7
Cl	1.5	<0.005		630	1054-2041	2654	950-4805
Co	9	0.1	0.034	1.12	0.02	0.05	
Cr	39	<0.006	0.35			0.10	0.1-3.6
Cs	1.3	<0.01	0.01			0.03	0.4-1.1
Cu	213	<0.02			0.1-1.5	10-20	11-32
F	1.0	0.10			0.2-2.0		
Fe	3230	<0.006	7.30	1100	0.5-1.5	10	5-68
Hg						0.10	1.3-7.6
I	2.9	<0.01			0.01	0.10	<1-15
K	7920	<0.004		250	176-350	3067	150-863
Mg	2.4	<2390			1-48	58	19-163
Mn	116	<0.006			0.06	0.1	0.3-5.7
Mo						0.05	0.06-0.21
Na	2950	134			1017-3370	2000	18-1720
Ni	70	0.50			0.05-0.2	0.05	0.6-6.5
P	1150	1.50		1400	0.2-1.5	327	83-165
Pb	2150	<0.04			0.05-2.7	0.10	3-70
Rb	23	<0.01				2	0.2-0.50
S	20000	<0.003		400		1500	47700
Sb	14.8	<0.03	0.15			0.04	0.1-3
Se	0.6	<0.02	0.22			0.25	0.6-2.53
Si			$9x10^4$			100-200	20-1950
Sn	9.6	<0.05					?
Sr	13.5	<0.01				0.10	0.05-0.92
Th							?
Ti	258	3		6300		0.50	0.05-14
Tl						0.03	0.012
U	0.1	<0.01					0.00013
V	259	0.005				0.15	0.005-0.53
Zn	1640	<0.02	10	$3.5x10^4$	1	6-20	99-450

Tab. 9: Standardized chromium in biological material

Material	Code	Concentration MG/KG OR MG/L	T	Error %
Hay powder	IAEA-V-10	6.5	certified	12
Tomato leaves	NBS-SRM-1573	4.5	certified	11
Pine needles	NBS-SRM-1575	2.6	certified	7.7
Lobster	NRCC-TORT-1	2.4	certified	25
Brewers yeast	NBS-SRM-1569	2.12	certified	2.4
Human hair	NIES-CRM-5	1.4	certified	14
Fish flesh	IAEA-MA-A-2	1.3	certified	7.7
Pepperbush	NIES-CRM-1	1.3	information value	
Mussel tissue	IAEA-MA-M-2	1.25	recommended conc.	27
Copepod	IAEA-MA-A-1	1.1	certified	18
Citrus leaves	NBS-SRM-1572	0.8	certified	25
Oyster tissue	NBS-SRM-1566	0.69	certified	39
Mussel	NIES-CRM-6	0.63	certified	11
Pine needles	NBS-SRM-1575	<0.5	information value	
Kale	Bowen's kale	0.369	indicated conc.	27
Milk powder	IAEA-A-11	0.26	information value	
Cotton cellulose	IAEA-V-9	0.11	certified	27
Urine (spiked)	NBS-SRM-2670	0.005	certified	7.1
Urine	BI-CUM-3	0.0556	certified	14
Urine	BI-CUM-2	0.0290	certified	17
Urine (normal)	NBS-SRM-2670	0.013	information value	
Animal muscle	IAEA-H-4	0.010	information value	
Blood	BI-CBM-1	0.0099	certified	12
Urine	BI-CUM-1	0.0028	certified	25
Milk powder	NBS-SRM-1549	0.0026	certified	27
Bovine serum	NBS-RM-8419	0.00030	recommended value	17

Tab. 10: Standardized selenium in biological material

Material	Code	Concentration MG/KG OR MG/L	T	Error %
Lobster	NRCC-TORT-1	6.88	certified	6.8
Horse kidney	IAEA-H-8	4.67	certified	6.4
Albacore tuna	NBS-RM-50	3.6	recommended value	11
Copepod	IAEA-MA-A-1	3.0	certified	6.7
Mussel tissue	IAEA-MA-M-2	2.27	recommended value	19
Oyster tissue	NBS-SRM-1566	2.1	certified	24
Fish flesh	IAEA-MA-A-2	1.7	certified	18
Mussel	NIES-CRM-6	1.5	information value	
Human hair	NIES-CRM-5	1.4	information value	
Wheat flour	NBS-SRM-1567	1.1	certified	18
Bovine liver	NBS-SRM-1577a	0.71	certified	9.9
Urine (spiked)	NBS-SRM-2670	0.46	certified	6.5
Rice flour	NBS-SRM-1458	0.40	certified	25
Animal muscle	IAEA-N-4	0.28	certified	12
Animal blood	IAEA-A-13	0.24	certified	33
Kale	Bowen's kale	0.134	indicated conc.	15
Milk powder (spiked)	BCR-CRM-150	0.127	information value	
Milk powder (spiked)	BCR-CRM-151	0.125	information value	
Milk powder	NBS-SRM-1549	0.11	certified	9.0
Serum	NYE-105	0.0907	certified	
Milk powder	BCR-CRM-063	0.088	information value	
Urine	NYE-108	0.0494	certified	
Milk powder	IAEA-A-11	0.034	recommended conc.	21
Urine (normal)	NBS-SRM-2670	0.030	certified	27
Citrus leaves	NBS-SRM-1572	0.025	information value	
Hay powder	IAEA-V-10	0.022	information value	
Bovine serum	NBS-RM-8419	0.016	recommended value	13
Cotton cellulose	IAEA-V-9	0.015	information value	

Conclusion

1. Trace element concentration levels meeting RDAs and SADDIs for essential elements as well as provisional tolerable intakes for toxic elements can be determined in total diets (using the adequate analytical method), without preconcentration procedures after freeze-drying.
2. Some recommendations for daily dietary intakes, e.g. Fe, Zn, Cu, compared to maximum daily loads of these elements are inconsistent with each other and reconsideration is proposed.
3. The choice of an adequate analytical method increases the probability in obtaining reliable results within the limits of detection in most cases.
4. Ameliorating the LoD by cleaner laboratory environment will also increase the quality of the analytical results.
5. The choice of a suitable biological multielement reference material is inevitable to improve the accuracy of trace element results and to reduce matrix effects.
6. A total diet multielement reference material with improved confidence intervals is urgently needed to get more reliable trace element nutritional data.

References

1. Recommended Dietary Allowances, Food and Nutrition Board, National Academy of Sciences - National Research Council; 9th revised edition. Washington, D.C., USA, (1980).
2. Essential and toxic food constituents in the daily total diet; Editor: R. Schelenz; Reports of the Federal Research Centre of Nutrition, Karlsruhe, Federal Republic of Germany, Report: BFE-R--83-02 (in german, engl. abstract), (December 1983).
3. Schelenz, R.: Intake of Zn, Mn and Se by Adult Females; Trace Element - Analytical Chemistry in Medicine and and Biology, Vol. 3, 73 - 89, 1984; Editors: P. Brätter, P. Schramel; Walter de Gruyter & Co., Berlin - New York 1984
4. Contaminants Codex Alimentarius, Volume XVII, 31, First Edition; Joint FAO/WHO Food Standards Programme (CAC, Vol. XIVV - Ed. 1) Rome, 1984
5. World Health Organization (WHO): Evaluation of Certain Food Additives and the Contaminants; Techn. Rep. Ser. No. 683, Geneva, 1982
6. World Health Organization (WHO): Evaluation of Certain Food Additives and the Contaminants Mercury, Lead and Cadmium; Techn. Rep. Ser. No. 505, Geneva, 1972
7. World Health Organization (WHO): Evaluation of Certain Food Activities and Contaminants; Techn. Rep. Ser. No. 696, Geneva, 1983
8. World Health Organization (WHO): Trace Elements in Human Nutrition; Techn. Rep. Ser. No. 532, Geneva, 1973
9. Handbook on Human Nutritional Requirements; World Health Organization (WHO), Geneva, 1974
10. Guidelines for the Study of Dietary Intakes of Chemical Contaminants; WHO-EFP/83.53, Geneva, 1983
11. Schelenz, R., Boppel, B.: Veränderungen der Gehalte von Arsen, Blei, Cadmium und Quecksilber bei der haushaltsüblichen Zubereitung; Landwirtsch. Forsch., Kongressband 1982, Sonderheft 39, 342-351 (1983).

12. Oesterreichischer Ernährungsbericht 1982; Herausgeber: Bundesminister für Gesundheit und Umweltschutz; Wien, 1982
13. Bingham, S., Mc. Neil, N.I., Cummings, J.H.: The Diet of Individuals: A Study of a Randomly-Chosen Cross Section of British Adults in a Cambridgeshire Village; Br. J. Nutr. 45, 23 - 35 (1981).
14. Walker, M.A., Page, L.: Nutritive Content of College Meals; J. Am. Diet. Ass. 70, 260 - 266 (1977).
15. Wolf, W.R.: Assessment of Inorganic Nutrient Intake from Self-Selected Diets; Human Nutrition Research, BARC Symposium, Allanheld, Osmun, Totawer, ed. G.R. Bucher
16. Einkommens- und Verbrauchsstichproben 1978 (EVS 78); cited in: "Material zum Ernährungsbericht 1980"; Deutsche Gesellschaft für Ernährung, 9 - 49, Frankfurt/Main, 1980
17. Arab, L., Schellenberg, B., Schlierf, G.: Nutrition and Health. A Survey of Young Men and Women in Heidelberg; Am. Nutr. Metab. 26 (1) 224 p (1982).
18. Tribble, H., Scoular, F.I.: Zinc Metabolism of Young College Women on Self-Selected Diets, J. Nutr. 52, 209 (1954).
19. Robinson, M.F., McKenzie, J.M., Thomson, C.D., VanRij, A.I.: Metabolic balance of zinc, copper, cadmium, iron, molybdenum and selenium in young New Zealand women; Br. J. Nutr. 30 (1973) 195
20. White, H.S.: Zinc content and zinc-to-calorie ratio of weighed diets; J. Am. Diet. Ass. 68, 243 (1976).
21. Guthrie, B.E., Robinson, M.F.: Daily Intake of Manganese, Copper, Zinc and Cadmium by New Zealand Women, Br. J. Nutr. 38, 260 - 266 (1977).
22. Lyon, T.D.B., Smith, H., Smith, L.B.: Zinc Deficiency in the West of Scotland? A Dietary Intake Study; Br. J. Nutr. 42, 413 - 416 (1979).
23. Gibson, R.S., Scythes, C.A.: Trace Element Intakes by Women; Br. J. Nutr. 48, 241 - 248 (1982).
24. English, R.: The Market Basket (Noxious Substances) Surveys: Zinc Content of Australian Diets; J. Food and Nutrition 38, 63 - 65 (1981).

25. Keith, L.H., Crummett, W., Deagan Jr., J., Libby, R.A., Taylor, J.K., Wentler, G.: Principles of Environmental Analysis, Anal. Chem. 55, 2210-2218 (1983).
26. Currie, L.A.: Limits for Qualitative Detection and Quantitative Determination, Application to Radiochemistry, Anal. Chem. 40, 586-593 (1968)
27. Muramatsu, Y., Parr, R.M.: Survey of currently available reference materials for use in connection with the determination of trace elements in biological and environmental materials. Report IAEA/RL/128, (December 1985).
28. International Organization for Standardization (ISO): Certification of Reference Materials - General and Statistical Principles, ISO Guide 35 (E), (1985).
29. Elemental Analysis of Biological Materials; Current Problems and Techniques with Special Reference to Trace Elements, Techn. Rep. Ser. No. 197, International Atomic Energy Agency (IAEA), Vienna, 1980

DIETARY FIBER AND THE BIOAVAILABILITY OF ESSENTIAL TRACE ELEMENTS, A CONTROVERSIAL TOPIC

Anna-Elisabeth Harmuth-Hoene
Federal Research Center for Nutrition
D-7500 Karlsruhe

Introduction

Based on the observation that many of the so-called "Western diseases" prevalent in industrial societies are rarely known in countries of the Third World, it has been suggested that a liberal intake of indigestible dietary fiber has a protective effect against such diseases. This hypothesis has generated innumerable studies broadening our present-day knowledge of the various functions of dietary fiber in the organism. Today, dietary fiber is considered an essential nutrient and a daily intake of at least 30 g is recommended. In view of such strong evidence of the beneficial action of dietary fiber, possible negative side-effects like the interaction with the absorption of essential minerals and trace elements must be considered.

Definition of Dietary Fiber

Today, the generally accepted definition of dietary fiber refers to the residue derived from plant foods, which is resistant to human digestive enzymes, although partial degradation by human colonic bacteria may take place. The components of dietary fiber are cellulose, hemicellulose, pectins, gums, mucilages, algal polypeptides and lignin. They occur in the form of structural materials of the cell wall and as storage polysaccharides, which include some water-soluble constitu-

Trace Element – Analytical Chemistry in Medicine and Biology, Vol. 4

ents. Crude fiber, a definition found in the older literature, refers to acid- and alkali-insoluble plant residue, which only includes cellulose and lignin but not many of the other polysaccharides not digested by man.

Mechanisms by which Dietary Fiber Affects the Bioavailability of Trace Elements

1) Cation-exchange capacity

The bioavailability of elements is usually determined by balance studies. A positive balance means intake is higher than excretion; a negative balance means excretion is higher than intake. Many observations have shown increased fecal excretion of minerals and trace elements when elevated amounts of certain dietary fiber materials were consumed, sometimes even resulting in negative balance. This has prompted a number of *in vitro* studies of the cation-exchange capacity of various fiber components. Cellulose, hemicellulose and lignin contain functional groups such as carboxyl and amino groups which become negatively charged as the pH increases during the passage of the ingesta from the stomach to the intestinal tract. Since the chemical composition and the number of the functional groups vary considerably for each fiber preparation it is not surprising that the results of *in vitro* binding studies show little agreement.

The binding of Cu, Zn and Fe to isolated alkali-soluble and water-soluble hemicellulose from rice bran and the release of these elements after treatment with three enzymes is illustrated by an investigation of MOD and coworkers (26) in Table 1. The predominant sugar components in the alkali-soluble fraction are arabinose and xylose and in the water-soluble fraction arabinose and galactose. The protein content in the water-soluble fraction is about 10 times that of the

Table 1: Binding of copper, zinc and iron by rice bran hemicellulose and enzymatic release, adapted from (26)

	Cu	Zn	Fe
bound to a.s.[1)] hemicellulose ppm	9949	9399	5088
released by hemicellulase, %, pH = 4.0	26	64	0
released by pepsin, %, pH = 5.5	37	8	0
released by trypsin, %, pH = 8.1	59	0	7
bound to w.s.[2)] hemicellulose ppm	7179	3000	5047
released by hemicellulase, %, pH = 4.0	21	92	36
released by pepsin, %, pH = 5.5	45	73	60
released by trypsin, %, pH = 8.1	41	21	37

[1)] alkali-soluble [2)] water-soluble

alkali-soluble fraction. These data indicate that Cu is predominantly bound to the protein or glycoprotein of hemicellulose rather than to the carbohydrate moiety and thus should be available for absorption to a large extent. In alkali-soluble hemicellulose Zn is primarily bound to polysaccharide, whereas in the water-soluble fraction carbohydrate as well as protein or glycoprotein are involved in the binding of Zn. Iron seems to be tightly bound by alkali-soluble hemicellulose in contrast to water-soluble hemicellulose. Hemicellulase occurs only as bacterial enzyme in the lower intestinal tract of man and monogastric animals. Minerals released in this part of the intestine are hardly available for absorption.

Since 1971 more than 20 *in vitro* studies on the binding of minerals to native or isolated dietary fiber have been published. Some investigators find that the various components of cereal fiber have a strong binding capacity for Ca (3, 11, 18, 30), Mg (3, 11), Zn (16, 21, 28, 29, 30, 33), Fe (3, 16, 21,

26, 27, 28, 30), and Cu (21, 26, 28).

The results of other studies indicate that phytic acid, the hexakis dihydrogenphosphate of myoinositol, is mainly responsible for the low bioavailability of Ca (39), Zn (9, 14), and Fe (9, 33) in cereal products. Metal binding to fiber components was found to increase as the pH goes up from 4 to 7 (28, 30, 42).

2) Reduced intestinal transit time

Human radiological studies suggested that transit time of the ingesta through the small intestine (24) as well as total transit time (5, 41) are shortened by a liberal intake of dietary fiber. In contrast, when breath hydrogen following the ingestion of dietary fiber was measured, mouth-to-cecum transit time was prolonged (22). It is conceivable, however, that delayed gastric emptying was responsible. Rapid passage of the ingesta through the absorptively active area of the small intestine could reduce absorption of trace elements as well as other nutrients. Support for this assumption comes from measurements of zinc-65-absorption and -turnover in rats maintained on diets with a cellulose content increasing from 0 to 10 % (2). At a cellulose content of 6 % and higher a significant decrease of biological half-time as well as zinc absorption was evident (Table 2). In contrast to these results MERCURIO and BEHM (25) could not find any effect of the level of neutral detergent fiber from various fiber sources on mineral excretion in rats despite a significant decrease of transit time with increasing fiber level. In a review of the effects of dietary fiber intake in man, KELSAY (19) found that the intestinal transit time in healthy subjects and patients with diverticular disease was decreased in 8 studies, whereas no effect was observed in 5 other investigations. These contradictory results may have been caused by widely differing

Table 2: Effect of dietary cellulose on absorption and turnover of zinc-65 administered to rats, adapted from (2)

dietary cellulose %	t 1/2 (first component) hours	^{65}Zn-absorption %
0	17.6 ± 1.2	33.8 ± 4.8
2	19.3 ± 2.0	28.5 ± 2.3
4	20.8 ± 1.0	32.7 ± 1.4
6	12.9 ± 0.4	22.7 ± 1.8
8	12.5 ± 0.8	17.5 ± 2.2
10	11.2 ± 0.4	18.9 ± 1.8

amounts of fiber ingested as well as by large variations of transit time between subjects at the outset of the study.

3) Increased endogenous secretion of trace elements

Some components of dietary fiber seem to increase bile and pancreatic secretions into the small intestine (38). Considerable amounts of zinc and copper are secreted with bile and of zinc with pancreatic juice into the intestinal lumen (37, 40), from where they are reabsorbed and returned to the liver. This could possibly apply to other trace elements as well. This enterohepatic circulation can be interrupted by binding of the elements to undigested fiber components, resulting in an increased fecal excretion. Another source of endogenous trace elements in feces originates from increased desquamation of intestinal mucosal cells caused by high fiber intake (4, 17, 36).

Factors influencing the Bioavailability of Trace Elements

1) Source and composition of dietary fiber

The composition of various naturally occurring as well as of isolated fiber shows marked differences, and the interaction of the various components with trace element absorption differs widely. In adolescent boys increases in fecal excretion of zinc, copper and magnesium were more pronounced on a diet supplying 21 g of hemicellulose than on comparable diets with equal amounts of cellulose or pectin (8). Pectin showed the least interference with mineral utilization and retention. Our own studies with rats also indicated considerable differences in mineral absorption when five different isolated fiber preparations were fed in a semisynthetic diet (15). Any statement concerning the effect of dietary fiber on trace element availability should, therefore, include a specification of source and composition of the material investigated.

2) Other food constituents

One has to be cautious in extrapolating from the binding properties of isolated fiber components *in vitro* to the ability of the body to absorb trace elements when ingested with the binding agent. During gastric digestion at a low pH minerals originally bound to fiber are largely liberated and, as the pH rises during intestinal transit, will be bound by various soluble ligands present in the ingested food or coming from gastric or intestinal secretions. *In vitro* studies showed citric acid to be most effective in solubilizing Fe > Cu > Zn > Mg > Ca, in descending order (23). Histidine was quite effective in binding Zn, Fe, Mg and Cu. Ascorbic acid solubilized more than 50 % of Mg and 30 % of Fe, lactose bound more than 50 % of Mg. Copper was completely solubilized by all ligands investigated (Table 3). These data illustrate that a major factor controlling the bioavailability of trace elements and minerals are small organic ligands. By their action some of the elements originally bound to dietary fiber will be transformed to soluble compounds, which are readily absorbed.

In addition to dietary fiber various other food constituents have an inhibitory effect on the absorption of trace elements and will thus enhance the effect of fiber. In a human study the daily intake of 300 - 400 mg of oxalic acid from spinach caused negative Ca-, Mg- and Zn-balances in adult males on a high fiber diet (20). In rats, however, oxalic acid did not affect Zn-availability (45). Inhibitory effects of specific minerals have also been suggested for phenols, caffeine and tannic acid.

Table 3: Solubilizing effect of various low molecular weight chelating agents, adapted from (23)

	% of added minerals remaining in solution on raising pH from 1 to 7				
	Ca	Mg	Zn	Fe	Cu
EDTA	100	81	100	100	100
Citric acid	63	84	93	100	100
Histidine	13	51	49	38	100
Ascorbic acid	13	55	2	29	100
Lactose	13	56	1	22	100

Table 4: Content of dietary fiber and phytate in cereal products and legumes

	Dietary fiber[1] g/100 g	Phytate[2] g/100 g
Cereals		
Wheat bran	44.0	3.001
Wheat, shredded	12.3	1.481
Wheat flour, wholemeal	9.6	0.845
Bread, whole wheat	8.5	0.390
Bread, white	2.7	0.069
Corn flakes	11.0	0.048
Corn, canned kernels	5.7	0.031
Rice crispies	4.5	0.232
Beans		
Lima beans, raw	21.6	1.010
French beans, cooked	3.2	0.097
Peas		
Chick peas, boiled	6.0	0.207
Green peas, boiled	5.2	0.028

1) McCance and Widdowson's: The composition of Foods, 4th Edition, Elsevier/North-Holland Biomedical Press, Amsterdam, New York, Oxford, 1978

2) Oberleas, D., Harland, B.F.: J. Am. Diet. Ass. 79, 433-436 (1981)

Phytic acid is present in all cereals, in many legumes and nuts. At the pH encountered in the intestine it is a powerful chelator of most of the essential minerals and trace elements. Phytic acid is deposited as the soluble Na- and K-salt in the pericarp and aleuron layer of cereals, e.g. at the same sites where minerals and trace elements are found. Foods rich in fiber also contain fairly high amounts of phytate (Table 4), and the question whether fiber or phytate largely determines the bioavailability of minerals in cereal products has been the subject of numerous investigations. In a series of studies in an Iranian village population a high incidence of zinc and iron deficiency was found and related to the consumption of tanok, an unleavened bread made from high-extraction wheat flour with a high content of phytate as well as dietary fiber (11, 12). Initially it was thought that phytate was responsible for the interference with mineral absorption (32). Based on the results of more recent *in vitro* and *in vivo* studies, however, REINHOLD and coworkers (31) concluded that fiber rather than phytate was the determinant factor in causing mineral deficiency.

In contrast to these findings it has been shown in human (1, 43) and rat studies (6, 10) that phytate rather than fiber limited the bioavailability of iron and zinc. These are only a few of many contradictory results, which demonstrate the difficulty to distinguish between the effect of fiber and phytate on trace element absorption. Some of these conflicting results can be explained by differences in the amounts of fiber ingested. The diet of the Iranian village population studied by REINHOLD and his group contained 15 - 17 g/1000 kcal., whereas the fiber content of a Western-type vegetarian diet lies around 10 - 11 g of fiber/1000 kcal. Studies of SANDSTEAD and his group involving a large number of human volunteers indicated that moderate intakes of dietary fiber do not adversely affect the absorption of trace elements (34).

The amount and source of dietary protein affects the bioavailability of trace elements. The beneficial effect of increased animal protein on the bioavailability of zinc from a diet high in fiber and phytate was observed in rats (7) and human subjects (35). These results support the above discussed assumption that an ample supply of proteins or peptides may facilitate mineral complexation, thus overriding the binding by dietary fiber.

3) Age and mineral status of the experimental animal or human subject

The absorptive capacity of the intestinal mucosa for iron, zinc and possibly some other trace elements is controlled by a very sensitive homeostatic mechanism, which responds to the need of the organism for these elements. Depletion or increased requirements caused by growth, pregnancy, lactation or disease will therefore overcome to some extent the inhibitory effect of dietary fiber on trace element absorption, resulting in a relatively high bioavailability. Optimal mineral stores on the other hand will have the opposite effect.

4) Adaptation to a high fiber intake

The immediate effect of an increased fiber intake, as observed in many studies, is to put the organism into apparent negative balance, indicated by an increased fecal excretion of trace elements. Results from long-time studies on animals as well as human subjects indicate, however, that an adaptation of mineral absorption will compensate for initial losses when high fiber diets are fed over an extended period of time. We monitored fecal excretion of several trace elements by growing rats, fed diets containing 10 % of gum guar or agar-agar continuously over a period of 21 weeks (15). Despite increased fecal losses of iron, copper and zinc during the initial phase, when compared to a low fiber control diet, the retention of these elements, as measured by carcass analysis, did not differ significantly from those observed in the control group (Tables 5, 6).

In adult men consuming wheat bran muffins which supplied them with 16 g of neutral detergent fiber/day for 15 consecutive days, a distinct increase in apparent absorption of Zn, Fe, Mn, and Ca was observed with time (13). The effect of time was

Table 5: Apparent absorption of trace elements by rats maintained on diets containing 10 % of gum guar or agar-agar for 21 weeks, mean values ± s.d., n = 8. Dietary content of Fe = 134-144 mg/kg; Zn = 7.11-7.46 mg/kg; Cu = 1.25 mg/kg

week of balance	Diet	Fe %	Zn %	Cu %
	Control	27.2 ± 4.4	55.1 ± 6.7	5.4 ± 2.8
1	Gum guar	15.8 ± 3.9*	50.4 ± 5.4	-22.0 ± 9.8*
	Agar-agar	13.5 ± 7.9*	25.8 ± 5.4*	-42.1 ± 3.2*
	Control	9.6 ± 3.0	66.7 ± 3.8	0.5 ± 11.7
5	Gum guar	14.4 ± 3.0*	60.7 ± 4.1*	1.9 ± 14.1
	Agar-agar	4.6 ± 4.7	45.0 ± 3.3*	-7.8 ± 12.7
	Control	7.4 ± 3.6	54.2 ± 3.4	4.1 ± 5.5
21	Gum guar	8.0 ± 6.7	45.2 ± 5.4*	-0.6 ± 11.8
	Agar-agar	0.4 ± 2.7*	38.8 ± 2.7*	5.0 ± 15.9

*: significant differences to control values, $p < 0.05$, t-Test
from: Harmuth-Hoene, A.E. and Schelenz, R. (15)

Table 6: Carcass trace element content in rats maintained on diets containing 10 % of gum guar or agar-agar for 21 weeks. Mean values ± s.d., n = 8

Diet	mg/100 g of carcass[1)]		
	Fe	Zn	Cu
Control	6.65 ± 0.87	3.10 ± 0.58	0.20 ± 0.01
Gum guar	6.58 ± 1.11	3.14 ± 0.72	0.19 ± 0.03
Agar-agar	6.95 ± 1.11	3.26 ± 0.20	0.19 ± 0.02

1) cecum and intestinal contents removed
from: Harmuth-Hoene, A.E. and Schelenz, R. (15)

somewhat less pronounced on absorption of Cu and Mg. Further supporting evidence for the existence of an adaptive mechanism was recently presented by a Norwegian group of workers (44). They observed the influence of 33 g of bran and 29 g of gum guar daily consumed in bread by 28 diabetic patients over a period of 3 months. Serum concentrations of Ca, Mg, Fe, Zn, Cu and Se, measured at the end of each supplementation period, did not differ from the values obtained after consumption of a low fiber bread.

Conclusion

Evaluation of numerous human, animal and *in vitro* studies of the effect of dietary fiber on the bioavailability of essential trace elements published during the last 15 years reveals indeed more controversy than consensus. There is hardly a single aspect on which general agreement has been reached so far, indicating that further research effort is needed to clarify a number of open questions. Generally an increased intake of dietary fiber, which is frequently associated with phytate, is followed by an immediate increase of fecal excretion of trace elements. Whether this is due to the action of fiber or phytate or perhaps to both is debatable. High fiber foods, especially cereals are also rich sources of trace elements, which can offset a reduced fractional absorption (availability) by a higher total absorption. The ability to adapt to a high fiber intake with time will, within certain limits, protect the organism against continuous high fecal losses of essential trace elements. On extremely high fiber and phytate diets deficiency of zinc and iron are likely to occur as has been shown in Iran. There is, however, little evidence that the fiber content of a so-called fiber-rich Western diet will sufficiently interfere with trace element absorption to cause a state of deficiency in a healthy population.

References

1. Andersson, H., Nävert, B., Bingham, S.A., Englyst, H.N., Cummings, J.H.: Br. J. Nutr. *50*, 503-510 (1983).
2. Becker, W.M., Hoekstra, W.G.: Intestinal Absorption of Metal Ions, Trace Elements and Radionuclides, eds. S.C. Skoryna, D. Waldron-Edward, Pergamon Press, Oxford 1970
3. Camire, A.L., Clydesdale, F.M.: J. Food Sci. *46*, 548-551 (1981).
4. Cassidy, M.M., Lightfoot, F.G., Grau, L.E., Story, J.A. Kritschewsky, D., Vahouny, G.V.: Am. J. clin. Nutr. *34*, 218-228 (1981).
5. Cummings, J.H.: Dietary Fiber, ed. K.W. Heaton, J. Libbey Co. Ltd. London 1979
6. Davies, N.T., Hristic, V., Flett, A.A.: Nutr. Rep. int. *15*, 207-214 (1977).
7. Davies, N.T.: Dietary Fiber in Health and Disease, eds. G.V. Vahouny, D. Kritschewsky, Plenum Press New York • London 1982
8. Drews, L.M., Kies, C., Fox, H.M.: Am. J. clin. Nutr. *32*, 1893-1897 (1979).
9. Ellis, R., Morris, E.: Cereal Chem. *58*, 367-370 (1981).
10. Fairweather-Tait, S.J.: Brit. J. Nutr. *47*, 243-249 (1982).
11. Haghsenass, M., Mahloudji, M., Reinhold, J.G., Mohammadi, N.: Am. J. clin. Nutr. *25*, 1143-1146 (1972).
12. Halsted, J.H., Ronaghy, H.A., Haghsenass, M., Amirhakimi, R.M., Barakat, R.M., Reinhold, J.G.: Am. J. Med. *53*, 277-284 (1972).
13. Harland, B.F., Morris, E.R.: Dietary Fiber Prospectives, ed. A.R. Leeds, J. Libbey Co. Ltd. London 1985
14. Harland, B.F., Harland, J.: Cereal Chem. *57*, 226-229 (1980).
15. Harmuth-Hoene, A.E., Schelenz, R.: J. Nutr. *110*, 1774-1784 (1980).
16. Ismail-Beigi, F., Faradji, B., Reinhold, J.G.: Am. J. clin. Nutr. *30*, 1721-1725 (1977).
17. Jacobs, L.: Am. J. clin. Nutr. *37*, 954-960 (1983).
18. James, W.P.T., Branch, W.J., Southgate, D.A.T.: Lancet *I*, 638-639 (1978).
19. Kelsay, J.L.: Am. J. clin. Nutr. *31*, 142-159 (1978).
20. Kelsay, J.L., Prather, E.S.: Fed. Proceeds. *40*, 854 (1981).

21. Lee, K., Garcia-Lopez, J.S.: J. Food Sci. *50*, 651-653 (1985).

22. Leeds, A.R., Gassul, M.A., Metz, G.L., Jenkins, D.J.A.: Lancet *II*, 1213 (1975).

23. Lyon, D.B.: Am. J. clin. Nutr. *39*, 190-195 (1984).

24. McCance, R.A., Prior, K.M., Widdowson, E.M.: Brit. J. Nutr. *7*, 98-104 (1953).

25. Mercurio, K.C., Behm, P.A.: J. Food Sci. *46*, 1462-1463 (1981).

26. Mod, R.R., Ory, R.L., Morris, N.M., Normand, F.L.: J. Agric. Food Chem. *29*, 449-454 (1981).

27. Prabhavathi, T., Narasinga, R.: J. Sci. Food Agric. *30*, 597-600 (1979).

28. Reilly, C.: Biochem. Soc. Transact. *7*, 202-204 (1979).

29. Reinhold, J.G., Faradji, B., Ismail-Beigi, F.: Trace Elements in Human Health and Disease, eds. A.S. Prasad, D. Oberleas, Acad. Press, New York 1976

30. Reinhold, J.G., Ismail-Beigi, F., Faradji, B.: Nutr. Rep. int. *12*, 75-85 (1975).

31. Reinhold, J.G., Faradji, B., Abadi, P., Ismail-Beigi, F.: J. Nutr. *106*, 493-503 (1976).

32. Reinhold, J.G., Nasr, K., Lahimgarzadeh, A., Hedayati, H.: Lancet *I*, 238-288 (1973).

33. Rendelman, J.A., Grobe, C.A.: Cereal Chem. *59*, 310-317 (1982).

34. Sandsted, H.H., Klevay, L.M., Jacob, R.A., Munoz, J.M., Logan, J.M., Reck, S.J., Dintzis, F.R., Inglett, G.E., Shuey, W.C.: Dietary Fibers: Chemistry and Nutrition, eds. G.E. Inglett, S.I. Falkenhage, Acad. Press, New York 1979

35. Sandström, B., Arvidsson, B., Lederblad, A., Björn-Rasmussen, E.: Am. J. clin. Nutr. *33*, 739-745 (1980).

36. Scheibel, M.S., Metha, T.: Nutr. Res. *5*, 81-93 (1985).

37. Schneeman, B.O., Lönnerdal, B., Keen, C.L., Hurley, L.C.: J. Nutr. *113*, 1165-1168 (1983).

38. Schneeman, B.O.: Dietary Fiber in Health and Disease, eds. G.V. Vahouny, D. Kritschewsky, Plenum Press, New York · London 1982

39. Schweizer, T.F., Frølich, W., Del Vedovo, S., Besson, R.: Cereal Chem. *61*, 116-119 (1984).

40. Spencer, H., Rosoff, B., Feldstein, A., Cohn, S.H., Gusmano, E.: Radiat. Res. *24*, 432-445 (1965).

41. Spiller, G.A., Wong, L.G., Whittam, J.H., Scala, J.: Nutr. Rep. int. *25*, 23-30 (1982).

ESTIMATION OF TRACE ELEMENT BIOAVAILABILITY: INTESTINAL ABSORPTION FROM IRON ENRICHED DIETETIC FOODS

Ch. Hansen, E. Werner, P. Roth

Ges. für Strahlen- und Umweltforschung, Institut für Biophysikalische Strahlenforschung, Frankfurt am Main

J.P. Kaltwasser

Zentrum der Inneren Medizin, Abt. für Hämatologie, J.W. Goethe-Universität, Frankfurt am Main

Introduction

Iron is an essential trace element for human beings because of its function in oxygen transport and in catalytic processes. Normal iron content of human subjects amounts to about 4 g. Although iron is one of the most abundant elements on earth, iron deficiency is a very frequent deficiency disease not only in developing but also in industrialized countries. This is due to the poor solubility of ferric compounds, which results in a low availability of food iron for intestinal absorption (1). Since body iron status is regulated by the adaptation of uptake in the upper small intestine, an adequate intake of bioavailable iron with the diet is necessary to avoid a negative iron balance. Therefore, in many countries fortification of dietary constituents with iron is performed in order to increase the iron intake of the total population (1,2,3). In Germany no such fortification program is carried out. For prophylactic iron supplementation of groups at risk of iron deficiency iron enriched dietetic foods and beverages are distributed commercially.

In the present study we have investigated the fractional intestinal absorption of iron from three iron enriched dietetic foods, two grape juices and one malt extract.

Trace Element – Analytical Chemistry in Medicine and Biology, Vol. 4

Subjects and methods

A total of 120 healthy volunteers who gave informed consent were included in the study, which was approved by the governmental authorities. For each product intestinal iron absorption was measured in 20 volunteers (10 males and 10 females) with normal body iron status and in 20 subjects with depleted iron stores. Differentation of subjects into the two groups was based upon the measurements of serum iron concentration, saturation of total iron binding capacity, and serum ferritin concentration (Table 1).

Table 1: Parameters of Body Iron_Status of the 120 Volunteers Included in the Study ($\bar{x} \pm SD$)

Subjects	Age years	Haemoglobin Conc. g/l	Serum Iron Conc. µg/dl	Saturation of TIBC %	Serum Ferritin Conc. µg/l
Product A					
normal ♂	32 ± 7	155 ± 9	112 ± 29	42 ± 12	103 ± 42
normal ♀	43 ± 18	138 ± 8	123 ± 36	46 ± 13	76 ± 56
iron depl. ♂	43 ± 13	144 ± 6	70 ± 21	24 ± 8	18 ± 8
iron depl. ♀	33 ± 10	138 ± 10	104 ± 66	33 ± 23	12 ± 5
Product B					
normal ♂	34 ± 5	153 ± 9	126 ± 55	49 ± 20	89 ± 40
normal ♀	40 ± 9	133 ± 6	88 ± 30	30 ± 11	57 ± 31
iron depl. ♂	46 ± 7	147 ± 12	83 ± 36	26 ± 10	12 ± 6
iron depl. ♀	35 ± 10	135 ± 9	86 ± 48	28 ± 14	11 ± 6
Product C					
normal ♂	35 ± 6	154 ± 7	100 ± 28	36 ± 8	104 ± 47
normal ♀	38 ± 17	141 ± 11	103 ± 17	37 ± 9	53 ± 28
iron depl. ♂	44 ± 9	150 ± 9	115 ± 34	37 ± 12	17 ± 4
iron depl. ♀	37 ± 13	132 ± 6	83 ± 18	28 ± 7	14 ± 3

The features of the three products investigated are given in Table 2. In order to compensate for individual variations of intestinal iron absorption, in each subject a second absorption test was performed with a reference dose of the same iron content as the dietetic product. As reference dose an aqueous solution of ferrous sulphate and 500 mg of ascorbic acid was used which is known to show an optimal bioavailability.

Table 2: Iron Enriched Dietetic Products Tested in the Study

Product	Iron compound	Iron dosage mg/l	Iron content of the test dose mg
A: Grape juice	ferrous gluconate	30	5
B: Grape juice	ferric saccharate	150	22
C: Malt extract	ferrous gluconate	840	20

For the absorption measurements the products were labelled by adding the corresponding ^{59}Fe tagged compounds. For the assessment of intestinal iron absorption the subjects were given the test dose after an overnight fast in the morning of day 0. No food or drink was allowed for the following 2 hours. Whole body counter measurements were performed before and after application and on day 14 thereafter. The second dose was administered on day 14 and retention of the ingested iron was measured after another fortnight (4).

Percentage intestinal iron absorption A_{Fe} was calculated from

$$A_{Fe} = \frac{a_{ret} \cdot k_{Fe}}{a_{oral}} \cdot 100\%$$

a_{ret}: activity retained in the body on day 14
a_{oral}: administered activity
k_{Fe}: correction factor for radioactive decay of ^{59}Fe

Results and discussion

Mean values of iron absorption from the three dietetics as well as from the reference doses in normal and iron depleted subjects are given in Table 3. The data show that iron absorption from all three tested dietetics is significantly lower than the absorption from the

Table 3: Intestinal Iron Absorption ($\bar{x} \pm$ SEM) from Three Iron Enriched Dietetic Foods and from a Reference Dose of Equal Iron Content as Ferrous Sulphate + 500 mg of Ascorbic Acid in Aqueous Solution in Subjects with Normal Body Iron Status and in Subjects with Depleted Iron Stores

Test dose	normal subjects		iron depleted subjects	
	men	women	men	women
Grape juice A (5 mg Fe^{++} glucon.)	3.2 $\pm$ 0.4	2.4 $\pm$ 0.3	17 $\pm$ 5	15 $\pm$ 3
Aqueous solution (5 mg Fe^{++} sulph. + 500 mg asc. acid)	34 $\pm$ 4	26 $\pm$ 4	44 $\pm$ 8	55 $\pm$ 6
Grape juice B (22 mg Fe^{+++} sacch.)	1.6 $\pm$ 0.4	1.9 $\pm$ 0.5	15 $\pm$ 3	8 $\pm$ 1
Aqueous solution (22 mg Fe^{++} sulph.+ 500 mg asc. acid)	20 $\pm$ 3	21 $\pm$ 2	50 $\pm$ 3	37 $\pm$ 3
Malt extract (20 mg Fe^{++} gluc.)	1.6 $\pm$ 0.2	1.7 $\pm$ 0.3	13 $\pm$ 3	8 $\pm$ 3
Aqueous solution (20 mg Fe^{++} sulph.+ 500 mg asc. acid)	26 $\pm$ 2	25 $\pm$ 3	41 $\pm$ 5	40 $\pm$ 5

the aqueous solution of ferrous sulphate. For normal subjects the absorption values from the reference dose were found to fall into the expected range, but normally females show higher absorption values than males. As expected intestinal iron absorption from the reference dose was higher in subjects with body iron depletion. For the doses administered in this study (5 to 22 mg of iron), there is no clear dependence of the fractional absorption from the iron content. For the dietetic products the differences in iron absorption between normal and iron depleted subjects is more pronounced than for the reference dose. In Table 4 the mean values of the individual ratios between the absorption values of the dietetic product and the absorption values of the reference dose are given. In subjects without need for additional iron supply about ten times less iron is absorbed from the product as compared to the ferrous sulphate standard. For subjects in need of iron supplementation this ratio is increased up to one third. This result shows that the regulatory mechanisms for iron uptake may more adequately respond to changes in food iron content than to administration of a pure chemical compound.

Table 4: Ratio (R) of Fractional Iron Absorption ($\bar{x} \pm$ SEM) from Iron Enriched Dietetic Products and Aqueous Solution of Ferrous Sulphate in Subjects with Normal Body Iron Status (R_N) and in Subjects with Depleted Iron Stores (R_D)

Iron enriched dietetic food	normal subjects R_N		iron depleted subjects R_D	
	men	women	men	women
Grape juice A (5 mg Fe^{++}gluc.)	0.11 ± 0.02	0.12 ± 0.03	0.35 ± 0.07	0.29 ± 0.05
Grape juice B (22 mg Fe^{+++}sacch.)	0.07 ± 0.02	0.09 ± 0.02	0.30 ± 0.05	0.23 ± 0.04
Malt extract (20 mg Fe^{++}gluc.)	0.07 ± 0.01	0.07 ± 0.01	0.33 ± 0.04	0.16 ± 0.04

From these data one may calculate that consumption of the recommended amounts of the dietetics would improve the average iron balance in subjects with iron depletion by 2 to 5 mg of iron per day. This could result at least in a steady state balance in most subjects with iron depletion. On the other hand there seems to be no inappropriate high increase in the body iron status of normal subjects during long term use of the iron enriched dietetics investigated.

Although the costs are rather high, the iron enriched grape juices as well as the malt extract can be useful sources of iron supplementation in subjects with increased demands, since they are advantageous over pure chemical compounds with respect to taste and form of intake. Colour and taste may have been the reason for choosing grape juice as a vehicle for iron enrichment of the diet. Such dietetic foods are on the market since a long time. In the meantime it has been demonstrated, that tannins which are contained in the shell of red grapes in considerable amounts are potent inhibitors of intestinal iron absorption (1). Therefore, other beverages like orange juice (5) may be better vehicles for iron enrichment.

Acknowledgement

The authors wish to thank Mrs. I. Ehmann, Mrs. H. Hahn, Mrs. A. Huck, and Mrs. U. Tacke for excellent technical assistance.

References

1. Bothwell, T.H., R.W. Charlton, J.D. Cook and C.A. Finch. 1979 Iron metabolism in man. Blackwell Scientific Publ., Oxford, London, Edinburgh, Melbourne.

2. Bread and Flour Regulations.1953.Statutory Instruments 1435, H.M.S.O.

3. Hallberg, L. 1982. Iron nutrition and food iron fortification. Seminars in Haematology 19, 31-41.

4. Werner, E., J.P. Kaltwasser and P.B. Bechstein.1978. In: Nuklearmedizin: Nuklearmedizin und Biokybernetik, Band 2 (K. Oeff and H.A.E. Schmidt, eds.) Medico Informationsdienste, Berlin. pp.122-125.

5. Layrisse, M.. 1975. In: Iron metabolism and its disorders (Kief, H., ed.) Exerpta Medica, Amsterdam, Oxford, American Elsevier Publishing Co Inc. New York. pp. 25-33.

DIURNAL CHANGES OF HUMAN SERUM IRON, ZINC, COPPER, SELENIUM, AND MAGNESIUM CONCENTRATIONS

J.J. Lehto, H. Mussalo-Rauhamaa

Department of Public Health, University of Helsinki
SF-00290 Helsinki, Finland

Introduction

Concentrations of many elements in human serum are subject to considerable diurnal variation. Zinc concentrations, for example, seem to be higher after a period of fasting than one hour after a meal (1). In addition, it is also well known that humans exhibit a circadian variation in the urinary excretion of several electrolytes (2). Unfortunately all clinical laboratories do not take these phenomena into account in their sampling procedures, which may therefore complicate comparisons between analytical results and reference values.

In order to determine whether a shorter period of fasting, e.g. four hours, is enough to bring the concentrations of elements in serum to the same levels as those following an overnight fast, studies were conducted of the iron, zinc, copper, selenium, and magnesium concentrations in the serum of 17 subjects. Samples were taken following an overnight fast, one hour after a midday meal, and four hours after a midday meal. To evaluate whether fluctuations in element concentrations were caused by food intake or merely by increased physiological activity, samples were also taken from a smaller group which fasted during the entire sampling period.

Trace Element – Analytical Chemistry in Medicine and Biology, Vol. 4

Materials and methods

Sample donors. Fourteen females and three males were recruited for the study. The subjects were healthy, except one who was epileptic, and aged 18-55 years (mean $\pm$SD:34 $\pm$9 years). The subjects were interviwed regarding their consumption of mineral and vitamin supplements during the five months preceeding the experiment.

Sample collection. The first blood samples were obtained at 9 a.m. following an overnight fast. The second samples were taken one hour after the midday meal (11 a.m. to 0.30 p.m.) and the last samples were collected three hours later. Subjects maintained their standard activity during the course of the experiment, and were allowed to have a breakfast of their own selection after the first samples were taken. Near midday subjects received a meal of meat and a meal of fish an the consecutive days of the experiment, and following the midday meal the subjects abstained from food until the last sample was taken. Additional samples were taken on a third day, when five subjects fasted over the entire sample collection period.

Serum samples were taken from venous blood by stainless steel needles (Neolus^R, Terumo, Belgium). The serum was separated by centrifuging and then stored at 20 ^{o}C until analysis.

Sample analysis. Zinc, copper, and magnesium concentrations were determined by flame atomic absorption spectrometry. The details of the method are given elsewhere (3). The selenium concentrations were determined by electrothermal atomic absorption spectrometry using a method described by Alfthan and Kumpulainen (4) which consists of a 1 to 9 dilution with water and the utilization of nickel as a matrix modifier. For the atomic absorption determinations a Perkin-Elmer 5000 spectrometer was used with an HGA-400 graphite furnace in electrothermal mode. Iron concentrations were determined by a colorimetric method (5) which involved Hitachi Perkin-Elmer Model 135 UV-VIS spectrophotometer. Commercial standard reference materials Seronorm^R (Nyegaard & Co, Oslo) and Moni-Trol^R (Dade, Miami, Florida) were used to control analytical quality.

Results

The results for the two days, even with different midday meals, were essentially the same and the pooled results are presented in Table 1. Alterations in the mean concentrations of the elements and their statistical significance are calculated with respect to the mean values after an overnight fast.

Table 1. Concentrations of elements in serum after fasting and midday meal. Mean values (SD) are given.[a)]

Element	Unit	Concentration in serum		
		After fasting	1 h after meal	4 h after meal
Cu	µmol/l	19.3 (6.7)	19.2 (6.5)	19.8 (6.7)**
Zn	µmol/l	12.4 (1.7)	11.4 (1.1)***	10.2 (1.0)***
Se	µg/l	73.6 (10.6)	76.4 (10.5)	74.1 (9.8)
Fe	µmol/l	20.9 (6.5)	19.6 (5.8)	16.8 (4.3)***
Mg	mmol/l	0.78 (0.05)	0.82 (0.08)**	0.84 (0.07)***

a) The means after eating are compared to the means after fasting. The significance of the differences (paired 't' test) are shown as follows: * = $0.01<p<0.05$, ** = $0.001<p<0.01$, *** = $p<0.001$

Even though the copper concentration in the serum was only 2.6 % higher four hours after lunch than it was after prolonged fasting, the difference was found to be statistically significant. The decrease in zinc concentration with respect to values following fasting was higly significant ($p<0.001$) for samples both one and four hours after a midday meal. Selenium concentrations were found to be highest at one hour following a meal, however, the increase was not statistically significant. Magnesium concentrations in serum increased steadily from the value following fasting. Magnesium levels were 5 % higher one hour after a meal and 7 % higher four hours after a midday meal. Both increses were found to be statistically significant. Serum iron concentration decreased only slightly from morning to noon, but was 20 % lower in the last sampling period than in the first period.

The results for the elemental determinations in the sera of the five subjects which abstained from food during the eight hours after the overnight fast are presented in Table 2.

Table 2. Elemental serum concentrations during prolonged fasting. Number of subjects N = 5, mean (SD) values are given.

Element	Unit	Concentration at different sampling times		
		at 9.00	12.00	15.00
Cu	µmol/l	14.6 (2.8)	14.2 (2.3)	13.9 (2.2)
Zn	µmol/l	13.5 (1.4)	13.6 (0.6)	12.9 (0.9)
Se	µg/l	74.4 (8.0)	80.0 (6.8)	78.2 (7.3)
Mg	mmol/l	0.76 (0.06)	0.73 (0.07)	0.71 (0.06)
Fe	µmol/l	16.1 (7.6)	16.1 (6.1)	16.1 (4.5)

For the subjects which underwent prolonged fasting, copper and zinc levels remained constant or decreased slightly during the day. The maximum mean selenium concentration was observed to occur during the second sampling period, in this instance as well. The decrease in the magnesium concentration in serum samples over the afternoon was obvious, whereas iron concentration remained very stable over all three sampling periods.

Discussion

According to Jacob (6), copper levels in plasma exhibit diurnal variation, with the highest values occuring in the morning. In agreement with the present study, Lahey et al. (7) observed that the concentration of copper in serum remains nearly unchanged or increases slightly during the course of the day. The observed change in copper concentration in our experiment was not as pronounced as the one observed by Lifschitz and Henkin (2), for, in their experiment, serum copper at 2 p.m. was 15 % higher than at 6 a.m. after six hours' fast.

The reports concerning zinc fluctuations in serum are not completely in agreement. Most studies suggest a significant decrease in serum zinc concentrations during the course of the day. The strongest change was obseved by Persiegehl et al. (8), who found that three hours after subjects received 0.5 g of protein per kg of body weight zinc levels had decreased by 40 - 50 %. Oral glucose doses decreased serum zinc by a maximum of 20 %. Dawson and Walker (9,10), Hetland and Brubakk (1), and Hellwege (11) have observed concentration decreases in order of 10 - 20 %, which is in close agreement with our results.

The fluctuations in zinc, magnesium, and iron concentrations during the period of prolonged fasting were clearly different from those days when food intake was normal. This suggests that the reason to diurnal variation of concentrations of these elements in serum is not caused by increased physical activity or stress in daytime.

The diurnal changes in copper, zinc, magnesium, and iron concentrations in human serum observed in this study were significant, which makes it important to standardize the time of day at which specimens are taken and the food intake beforehand. This is absolutely necessary for scientific research laboratories, while for practical diagnostic purposes, results obtained from samples drawn without a preceeding period of fasting are still of use. Comparison to reference values will only be valid when the influence of food intake prior to sampling will be discovered.

Acknowledgement

The authors acknowledge the excellent technical assistance of Ms Laila Saarvo.

References

1. Hetland, Ö., E. Brubakk. 1973. J.Clin.Lab.Invest. 32, 225.

2. Lifschitz, M.D., R.I. Henkin. 1971. J.Appl.Physiol. 31, 88.

3. Salmela, S., E. Vuori. 1984. At. Spectrosc. 5, 146.

4. Alfthan, G., J. Kumpulainen. 1982. Anal.Chim.Acta 140, 221.

5. Ruutu, R. 1975. Clin.Chim.Acta 61, 229.

6. Jacob, R.A. 1986. In: Textbook of Clinical Chemistry (N.W. Tietz, ed.). W.B. Saunders Company, New York, p. 972.

7. Lahey, M.E., C.J. Gubker, G.E. Cartwright, M.M. Wintrobe. 1953. J.Clin.Invest. 32, 322.

8. Persigehl, M., A. Höck, K. Kasparek, E. Land, L. Feinendegen. 1974. Z.Klin.Chem.Klin.Biochem. 12, 171.

9. Dawson, J.B., B.E. Walker. 1969. Clin.Chim.Acta 26, 465.

10. Walker, B.E., J.B. Dawson, J. Kelleher, M.S. Losowsky. 1973. Gut 14, 943.

11. Hellwege, H.H. 1970. Klin.Wschr. 48, 1063.

TRACE ELEMENT CONCENTRATION IN SERUM OF INFANTS IN RELATION TO DIETARY SOURCES

P. Brätter, V.E. Negretti de Brätter, U. Rösick and H.B. v. Stockhausen*

Hahn-Meitner-Institut Berlin, FRG
*Kinderklinik und Kinderpoliklinik der Universität Würzburg,FRG

Introduction

The feeding practice adopted by the mother of a newborn will directly influence its trace element status. After delivery the child has to build up its own regulation of the metabolic processes. Especially at this stage of life a sufficient supply of essential nutrients is required in order to ensure normal development. Therefore the feeding practice adopted may be critical to its well-being.

Since the bioavailability of an element determines the extent of utility, it is essential not only to determine its total concentration in the nutrient fluid, but also the chemical form in which it occurs. The extent of adsorption and utilization by the infant may be reflected in the levels of the trace elements in the blood or blood constituents. Few studies have investigated the influence of trace element intake on the blood levels. Therefore one aim of this work was to study serum levels of the elements Cu, Co, Cr, Cs, Fe, Mn, Rb, Se and Zn in healthy infants aged 12 to 14 weeks receiving a controlled intake of these elements solely through breast-feeding or formula-feeding. Another aspect of this study was to obtain physiological reference ranges for serum trace elements. These values are needed as reference values for trace element serum concentration of hospitalized infants in the course of long-term total parental nutrition.

Copper, iron and zinc were selected because they have received plenty of attention from clinicians in the diagnosis of trace element abnormalities. Despite its vital function manganese belongs

Trace Element – Analytical Chemistry in Medicine and Biology, Vol. 4

to the problem elements. As shown by Cornelis and Versieck (1) the literature values cover some order of magnitudes. On the basis of our present knowledge it is fair to say that high serum values in healthy subjects can be attributed to contamination of the sample before analysis. Rubidium was of interest because its beneficial action is still under discussion. Cobalt and chromium belong to the vital elements. Although cesium doesn't belong to this group we included its values since it might be helpful in the estimation of possible body burden by the radioactive species Cs-137. Selenium is of special interest with regard to the formation of the selenium containing enzyme glutathion peroxidase. Because the child has to build up and maintain his own antioxidant defense mechanism a sufficient supply with selenium from nutrient fluids is important.

Subject and Methods

140 full-term healthy infants aged 12-14 weeks were examined. All infants were born in the children's clinic of the medical school of Lübeck, FRG. They were divided into two groups solely breast-fed (n=45) and exclusively formula-fed (n=95) subjects. The formula-fed group was further divided into 4 subgroups according to different brands of adapted cow milk formulas (Humana, Multival, Pre-Aptamil and PreBeba) with which they were fed, respectively. None of the infants received mixed feeding.
A further group consisted of preterm and fullterm infants receiving total parental nutrition. In addition to the parental regimen they received serum infusion 3 times a week and partially albumin and erytrocyte-concentrate. Trace element supplementation was not included in the infusion program. The serum selenium and zinc concentration of 24 hospitalized children were examined in the course of intravenous nutrition for up to 8 weeks.
All materials used in the sampling of blood or breast milk were found to be free of contamination with the elements of interest. Blood samples were collected in acid cleaned test tubes without additives. Serum was obtained after centrifugation of the blood samples at 3000 rpm for 10 min. The sera (0.5 to 1 ml) for trace element determination were stored at -30°C until analysis. Mature

breast milk was obtained by palpation of the breast and directly collected into precleaned polyethylene tubes.
The concentration of the trace elements Co, Cr, Cs, Fe, Rb, Se and Zn in serum and milk samples was determined by instrumental neutron activation analysis. 100 - 150 µl of the fluid was transferred into high purity quartz vials and ovendried at 50°C. The vials were sealed using a specially designed quartz burner. All treatments were carried out in a dust free laboratory. The samples were irradiated together with standard materials for 10 days at a thermal neutron flux density of $5 \cdot 10^{13}\, n \cdot cm^{-2} \cdot s^{-1}$ in the rotating device of the Berlin Research Reactor BER II. After a cooling time of about 8 weeks the outer walls of the quartz vials were carefully cleaned (HF, EDTA, Aceton) to avoid background interferences and subsequently measured by aid of high resolution spectrometry using a Ge(Li) well detector. For the determination of Cu and Mn graphite furnace atomic absorption spectrometry was used (Perkin Elmer 5000 with HGA 400).

Results and Discussion

Table 1 lists the mean trace element content of mature breast milk of mothers involved in the study. The corresponding trace element concentrations in serum of the exclusively breast-fed healthy infants given in table 2 might be used as references. Assuming that the breast milk may satisfy the growing demands of healthy infants of this age, the four subgroups, each of them receiving another commercially available infant formula, can be compared with the fully breast-fed group in order to examine the trace element supply via that nutrient fluids. In table 3 the serum trace element concentrations are given on the dry weight as well as on a wet weight basis for all groups investigated.
Applying Wilcoxon's rank test to the results no statistically significant differences in the serum levels of Cu, Cr, Fe and Mn were found. On the other hand significantly lower serum concentrations were found for Se in all formula-fed groups (tab. 5). From the comparison of the Se content of mature breast milk and infant milk formulas (tab. 1) it can be seen, that the lower Se-intake via the formulas is reflected in the infants serum. This finding

Tab. 1: Trace Element Content (ng/g wet weight) of Mature Human Milk and Infant Milk Formulas

Element	Human Milk	Multival	Pre-Aptamil	Humana
Cu	440 ± 130	455	< 0.5	< 0.5
Fe	780 ± 440	770	1000	2890
Rb	650 ± 130	950	730	550
Zn	2900 ± 990	6960	1110	1020
Co	0.58 ± 0.41	1.6	1.5	1.9
Cr	4.4 ± 4.2	5.7	3.9	5.3
Cs	3.8 ± 1.8	2.5	1.3	1.8
Se	15.3 ± 5.7	4.6	5.3	8.1
Dry weight (%)	12.6 ± 1.2	11.9	10.8	14.4

Tab. 2: Serum Trace Element Content (wet weight) of Healthy Breast.Fed Infants (Age 12 - 14 w)

Element	N	$\bar{x} \pm s$	Dimension	Method	Range
Mn	19	0.58 ± 0.30	ng/ml	AAS	0.10 - 1.10
Cu	27	0.89 ± 0.23	µg/ml	AAS	0.59 - 1.39
Se	45	58.8 ± 11.0	ng/g	NAA	38.0 -85.0
Co	35	1.05 ± 0.89	ng/g	NAA	0.20 - 4.90
Cr	19	3.9 ± 2.2	ng/g	NAA	1.1 - 8.4
Zn	44	0.80 ± 0.13	µg/g	NAA	0.56 - 1.16
Fe	31	1.35 ± 0.38	µg/g	NAA	0.72 - 2.09
Rb	45	0.27 ± 0.05	µg/g	NAA	0.18 - 0.43
Cs	12	1.7 ± 1.0	ng/g	NAA	0.8 - 4.2
Dry Weight	35	7.71 ± 0.78	%	---	5.05 - 9.32

TAB. 3: TRACE ELEMENTS (SE, ZN, FE, RB, CO, CR, CS) IN SERUM SAMPLES OF HEALTHY INFANTS (AGE 12 - 14 W). COMPARISON OF BREAST-FED INFANTS WITH FORMULA FED INFANTS.

SELENIUM (NG/G)

		DRY WEIGHT			WET WEIGHT		
FEEDING	N	x̄ ± s	RANGE	MEDIAN	x̄ ± s	RANGE	MEDIAN
HUMAN MILK	45	778 ± 137	490 - 1150	780	58.8 ± 11.0	38 - 85	58.0
PRE-APTAMIL	29	412 ± 90	180 - 570	440	31.4 ± 5.4	21 - 45	31.0
HUMANA	33	408 ± 76	190 - 620	410	30.1 ± 5.9	13 - 48	30.0
MULTIVAL	27	480 ± 116	150 - 680	490	34.4 ± 7.4	10 - 49	36.0
PRE-BEBA	6	248 ± 39	200 - 310	245	19.8 ± 2.8	16 - 23	18.5

ZINC (/UG/G)

HUMAN MILK	44	10.67± 1.59	7.7 - 14.7	10.55	0.803 ± 0.134	0.56 - 1.16	0.805
PRE-APTAMIL	30	9.68± 3.00	5.2 - 21.7	9.19	0.743 ± 0.219	0.52 - 1.72	0.705
HUMANA	33	9.61± 1.48	6.3 - 12.1	9.62	0.706 ± 0.119	0.43 - 0.96	0.720
MULTIVAL	28	11.10± 1.40	9.1 - 14.7	10.77	0.799 ± 0.124	0.62 - 1.20	0.790
PRE-BEBA	6	10.41± 1.72	8.1 - 12.6	10.60	0.813 ± 0.151	0.59 - 1.01	0.845

IRON (/UG/G)

HUMAN MILK	31	18.0± 5.1	10.0 - 27.7	18.3	1.35 ± 0.38	0.72 - 2.1	1.37
PRE-APTAMIL	19	16.3± 6.2	9.9 - 30.1	13.8	1.24 ± 0.45	0.74 - 2.2	1.14
HUMANA	27	17.8± 6.0	8.1 - 30.1	17.1	1.29 ± 0.42	0.63 - 2.1	1.26
MULTIVAL	8	15.0± 5.5	8.4 - 26.4	15.2	1.06 ± 0.34	0.64 - 1.7	1.13
PRE-BEBA	5	16.5± 4.9	11.0 - 23.1	17.6	1.32 ± 0.43	0.80 - 1.8	1.48

TABL.3: CONT.

		DRY WEIGHT			WET WEIGHT		
FEEDING	N	x̄ ± s	RANGE	MEDIAN	x̄ ± s	RANGE	MEDIAN
RUBIDIUM (μg/g)							
HUMAN MILK	45	3.58± 0.62	2.5 - 5.7	3.47	0.268±0.047	0.18-0.43	0.270
PRE-APTAMIL	27	4.03± 0.64	2.6 - 5.1	4.08	0.307±0.039	0.23-0.40	0.310
HUMANA	35	2.82± 0.39	2.1 - 3.5	2.87	0.209±0.031	0.15-0.28	0.210
MULTIVAL	23	4.52± 1.10	2.6 - 7.6	4.35	0.344±0.068	0.22-0.45	0.320
PRE-BEBA	6	2.64± 0.48	1.9 - 3.1	2.67	0.207±0.038	0.15-0.25	0.220
COBALT (ng/g)							
HUMAN MILK	35	13.6±11.7	3 - 63	10.0	1.05 ± 0.89	0.2 - 4.9	0.80
PRE-APTAMIL	29	12.6± 7.7	3 - 38	10.0	0.96 ± 0.80	0.2 - 3.3	0.70
HUMANA	35	8.7± 4.9	3 - 20	10.0	0.51 ± 0.35	0.2 - 1.5	0.40
MULTIVAL	27	20.9±24.6	3 - 120	11.0	1.62 ± 1.84	0.3 - 6.7	0.70
PRE-BEBA	6	8.7± 1.5	7 - 10	9.0	0.52 ± 0.10	0.4 - 0.7	0.50
CHROMIUM (ng/g)							
HUMAN MILK	19	49 ± 27	14 - 118	43	3.9 ± 2.2	1.1 - 8.4	3.4
PRE-APTAMIL	11	37 ± 24	9 - 88	30	2.8 ± 1.9	0.7 - 7.0	2.3
HUMANA	4	50 ± 32	12 - 90	50	3.7 ± 2.3	1.0 - 6.6	3.7
MULTIVAL	10	69 ± 38	20 - 120	61	5.6 ± 3.7	1.5 - 13.0	4.6
PRE-BEBA	1	20			1.7		
CESIUM (ng/g)							
HUMAN MILK	12	21.3±12.1	10 - 45	17.0	1.66 ± 1.01	0.8 - 4.2	1.15
PRE-APTAMIL	9	13.3±12.1	10 - 20	10.0	0.92 ± 0.27	0.6 - 1.5	0.80
HUMANA	3	14.0± 6.9	10 - 22	10.0	0.97 ± 0.67	0.4 - 1.7	0.80
MULTIVAL	9	18.3±13.7	10 - 50	10.0	1.31 ± 1.08	0.5 - 3.8	1.00
PRE-BEBA	4	8.7± 5.5	4 - 16	7.5	0.62 ± 0.46	0.3 - 1.3	0.45

TAB. 4: DRY MASS (%) OF SERUM SAMPLES OF HEALTHY INFANTS (AGE 12-14W). COMPARISON OF BREST-FED AND FORMULA FED INFANTS.

FEEDING	N	$\bar{x} \pm s$	RANGE	MEDIAN
HUMAN MILK	35	7.71 ± 0.78	5.1 - 9.3	7.75
PRE APTAMIL	28	7.74 ± 0.75	6.8 - 9.7	7.60
HUMANA	35	7.41 ± 0.57	6.5 - 9.5	7.36
MULTIVAL	27	7.75 ± 1.15	5.8 - 11.2	7.56
PRE-BEBA	6	7.83 ± 0.48	7.3 - 8.6	7.90

is in agreement with data from the U.S.A. (2,3) and from Japan (4). In addition to the lower intake there is probably a difference in the bioavailability of selenium compounds. According to Hatano et al. (4) selenium compounds in breast milk seem to be more biologically available. As regards Se it can be concluded that breast feeding is probably the more favored method of nutrition.

With respect to Zn, its content in two formulas was lower than that of human breast milk (tab. 1) and correspondingly lower concentrations of Zn in serum were observed (tab. 3, 5). Although the Zn-content of Multival was about twice as high as the breast milk, the serum Zn was not found to be elevated. In fact, the levels are similar. That might be explained in two ways. Firstly, above the certain level of dietary Zn increased intake will not be reflected in serum. Secondly, the bioavailability of Zn is lower in the formula.

Despite a 3-fold higher Fe content in one formula no statistically significant differences in the infants' sera levels were found. That is comparable to the findings for Zn mentioned above.

The Co concentration of all formulas investigated was measured to be about 3 times higher than that of breast milk. Despite this similarity we observed in one formula-fed group (Humana) significantly lower serum Co levels, when compared to the breast-fed

TAB. 5: Trace Element in Serum of Healthy Infants (Age 12 - 14 w)
Comparison of Breast- and Formula-feeding
Differences vs. Human Milk - tested by Wilcoxon's Rank Test

Element	Pre-Aptamil	Humana	Multival	Pre-Beba
Se	↓↓↓	↓↓↓	↓↓↓	↓↓↓
Zn	↓↓	↓↓	N.S.	N.S.
Fe	N.S.	N.S.	N.S.	N.S.
Rb	↑↑↑	↓↓↓	↑↑↑	↓↓
Co	N.S.	↓↓↓	N.S.	N.S.
Cr	N.S.	N.S.	N.S.	only 1 sample
Cs	↓	N.S.	N.S.	↓
Cu	N.S.	N.D.	N.S.	N.D.
Mn	N.S.	N.D.	N.S.	N.D.
Dry Mass	N.S.	↓↓	N.S.	N.S.

↑↑↑, ↑↑, ↑ Increase
↓↓↓, ↓↓, ↓ Reduction

Statistical Significance ($P < 0.001$; < 0.01; < 0.05)

N.S. statistically not significant

N.D. not determined

GROUP	SUBJECT	STATUS	TYPE OF NUTRITION	AGE	N
1	PREMATURE AND NEW-BORN	HEALTHY	ORAL MILK FORMULA	3 D	33
2	INFANTS	HEALTHY	BREAST-MILK	12 - 14 W	35
3			MILK FORMULA		26
4	PREMATURE AND	HOSPITALIZED	INFUSION	1 - 3 D	22
5	NEW-BORN,	(OPERATION,		2 - 3 W	24
6	INFANTS	ARTIFICIAL RESPIRATION, ETC.)		5 - 8 W	22

TAB. 6: EXAMINATION OF INFANTS IN RELATION TO AGE AND TO DIETARY SOURCES.

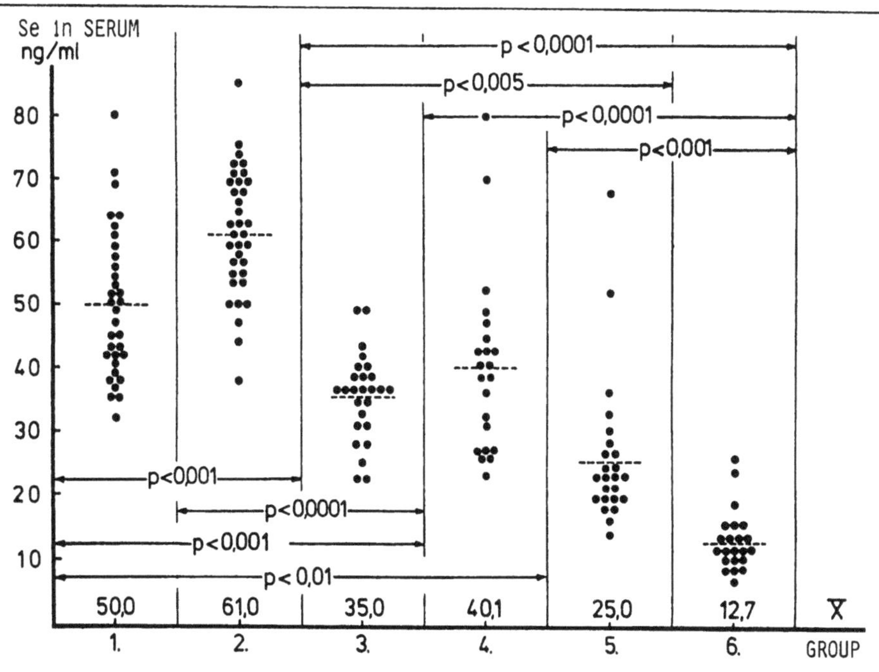

Fig.: 1 SELENIUM IN SERUM OF INFANTS

Comparison of Breast-feeding, Formula-feeding and Total Parenteral Nutrition

group. This could be due to differences in the form of Co. In contrast to Zn the serum rubidium levels of the subgroups were found to be significantly different and directly correlated with the Rb content of the respective dietary source.
Regarding the dry mass of the serum samples (tab. 4) statistically significant lower values have been found in the Humana group (tab. 5).
The trace element serum levels obtained from the healthy infants have been used as a basis for comparison in our studies on the nutrition of infants suffering from different diseases. Table 6 gives detailed information on the groups examined. Group 1 consisted of 33 healthy premature and newborn infants who have been under clinical observation and were all oral-fed. The groups 2 and 3 represent the healthy breast-fed and formula-fed infants aged 12 - 14 weeks, respectively. The hospitalized infants on long-term intravenous nutrition were classified according to different periods (group 4, 5 and 6).
Among the elements considered in this study, Se showed dramatic changes. After 2 - 3 weeks of total parental nutrition we observed a significant decrease of the serum levels. After 5 - 8 weeks of infusion therapy the Se values dropped to about 12 ng/ml serum. That is about 20% of the serum levels of Se in healthy breast-fed infants of this age [fig. 1].
These findings indicate the development of a Se deficiency in the course of total parental nutrition. With respect to supplementation strategies, it must be emphasized that intravenous Se administration is still a problem. The small margin between beneficial and toxic action of this element makes the choice of the suitable chemical form and the dose difficult, particularly in the neonatal period.

Summary

The serum concentration of the trace elements Cu, Cs, Co, Cr, Fe, Mn, Rb, Se and Zn of 140 healthy infants aged 2 to 3 months receiving the fed solely through breast-feeding or formula-feeding (4 different brands) was determined by means of instrumental neutron activation analysis and graphite furnace AAS (Cu, Mn). In

addition the variation of the trace element serum level of infants in the course of long-term total parental nutrition (TPN) was examined. No statistically significant differences between breast-fed and formula-fed healthy infants were found with regard to Cu, Cr, Fe and Mn in serum. The serum Se concentration was lower in the all formula-fed subgroups compared with breast-fed infants. The serum Zn data suggest that the bioavailability of Zn for infants nutrition is higher in breast milk than in formulas. The serum Rb level corresponds directly to the variation in the Rb content of the dietary sources.
After two weeks of intravenous feeding the serum Se concentration becomes lower as compared to the healthy control group. After six weeks of TPN the variation of Se in serum tends towards a lower limit of about 12 ng Se/ml. This suggests the need for supplementation during prolonged TPN.

References

1. Cornelis, R., Versieck, J. Anal.Chim.Acta 116 (1980) 217

2. Picciano, M.F., Trace Elements in Nutrition of Children, (ed. R.K. Chandra) Nestle Nutrition Vevey, Raven Press New York 157-169 (1985).

3. Smith, A.M., Picciano, M.F., Milner, J.A., Am. J. Clin. Nutr. 35, 521-526 (1982).

4. Shuichi Hatano, Katsuaki Aihara, Yoshikazu Nishi and Tomofusa Usui, J. Pedriatr. Gastroenterol Nutr., Vol.4, 1 (1985) 87-92

INGESTION/EXCRETION EXPERIMENTS FOR ARSENIC IN FISH

M. Burow and M. Stoeppler+)
Institute of Applied Physical Chemistry (ICH-4), Nuclear Research Centre (KFA) Juelich, Federal Republic of Germany

Introduction

Arsenic occurs in the food chain in two significantly different groups of compounds. In most terrestric materials, inland and sea water inorganic As(III) and As(V) compounds together with simple methylated metabolites like methylarsonic acid (MMA) and dimethyl arsinic acid (DMA) predominate. In aquatic species from fresh water and sea water, however, chemically very stable and thus probably far less toxic organoarsenic compounds, are present (e.g. arsenocholine, arsenobetaine, arsenolecithines) in some crustaceae and teleost fish identified to be nearly exclusively arsenobetaine (1,2).

From own investigations and those of others we knew that marine and inland water fish contain mainly these chemically stable forms of arsenic and only spurious amounts of inorganic arsenic (3,4,6). Thus it was of particular interest to investigate, if ingestion of fish would result in any deterioration of organo-arsenic compounds in the human body or already during preparation (thermal treatment) of the fish meal.

In the first study, performed with trout (Salmo spec.), a fresh water fish with comparatively low arsenic content (see table 1), the data for about 50 % of the volunteers indicated a possible additional formation of inorganic arsenic (6). This could be due to a partly decomposition of arsenobetaine or another organo-arsenic compound in the human body or during preparation (heating) of the fish. Additionally we found that only a part of the ingested

+) Author to whom correspondence should be addressed

Trace Element – Analytical Chemistry in Medicine and Biology, Vol. 4

arsenic was rapidly excreted via urine, the predominant excretion route. This showed a distinct storage capacity within body compartments also for organoarsenicals that might depend on the abolute amount given.

In order to obtain more information about these facts and also if there might be some differences in physicological behaviour between fresh water fish and marine fish we performed two additional experiments with a marine platfish with much higher arsenic content (Microstromus kitt) and another (Solea solea) with an arsenic content similar to that of the initially investigated trout.

Materials and Methods

All fish consumption studies have been performed in mixed groups of male and female volunteers by analyzing 24 h urine samples from all participants prior to fish meals (background levels) and continuously in distinct intervals for at least 96 h after fish consumption. The arsenic values found are given as μg (or mg) per g creatinine. The data further have been computed as absolute excretion compared to the ingested amount (obtained by analysis of fish fillets taken for the experiments) and expressed as % excreted arsenic. In order to start with low endogenous arsenic levels the participants were not permitted to consume fish for one week prior to the experiment.

Total arsenic determinations in fish and urine samples were performed after complete wet digestion by nitric, perchloric and sulfuric acid using a programmable automated digestion system. If only inorganic arsenic, MMA and DMA ("reactive arsenic") have to be determined, fish samples are dissolved at ambient temperature in 5 % Tetramethylammoniumhydroxide (TMAH) prior to the hydride evolution step (7,9), while this can be done directly in urine without pretreatment (8,9). Details of the digestion and hydride AAS determination conditions are given elsewhere in detail (7,9).

Results and Discussion

In table 1 an overview is given on all ingestion/excretion experiments hitherto performed with fresh water fish and marine fish also including the previous study with trout (Salmo spec.) (6).

Table 1: Overview on hitherto performed ingestion/excretion experiments

Species	As Content (mg/kg)	Average Ingested amount per person (mg)	Number of volunteers	Excretion maximum after consumption (h)	half-time (h)
Salmo Spec (fresh water)	2.3	0.45	10	3.6	8.8
Solea solea	5.6	0.7	4	-	-
Microstomus kitt	46.3	16.1	4	4.5	7.6

Note: For Solea solea only 24 h urine samples were analyzed.

The results clearly indicate, that the half time for arsenic excretion remains nearly unchanged irrespective of the total amount ingested. A typical example for the time dependent excretion of organic (non reactive) arsenic is shown for Microstomus kitt in figure 1.

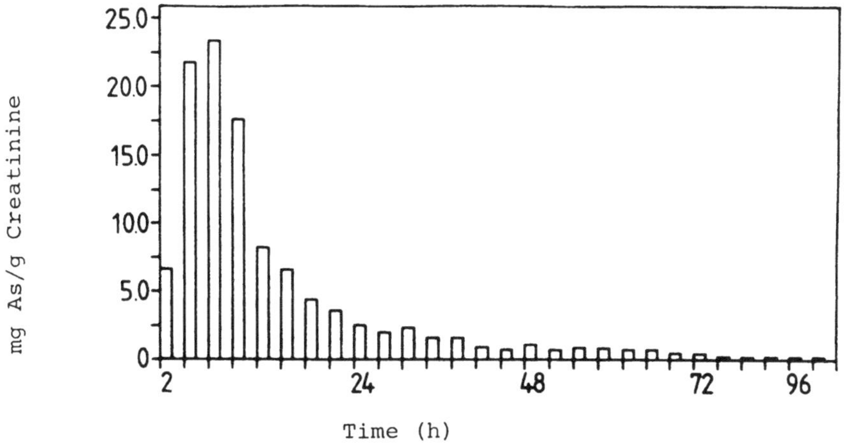

figure 1: Time dependent excretion of organic arsenic after consumption of Microstomus kitt

A semi-lorgarithmic excretion pattern is given for Microstomus kitt with high arsenic content in figure 2.

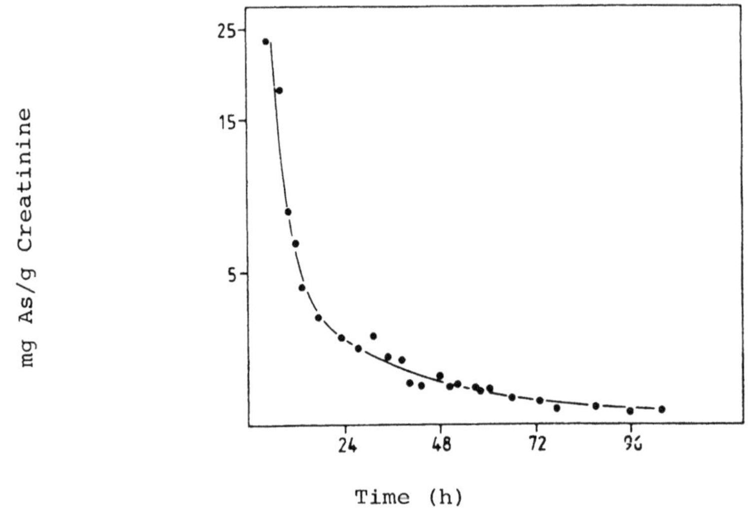

figure 2: Semi logarithmic excretion pattern for organic arsenic in human urine after consumption of Microstomus kitt

However, as was to be expected, there could be seen a clear relation between the integrated time dependent percentage of excreted arsenic and arsenic content in fish in all experiments, see figure 3. This confirms the existence of distinct body stores for organoarsenic compounds like arsenobetaine and that obviously these stores have a limited capacity. From figure 3 it is also clear that there is no significant difference in arsenic excretion pattern between fresh water fish and seawater fish which is an indirect confirmation that organoarsenic compounds are most probably the same for fresh water fish and marine fish.

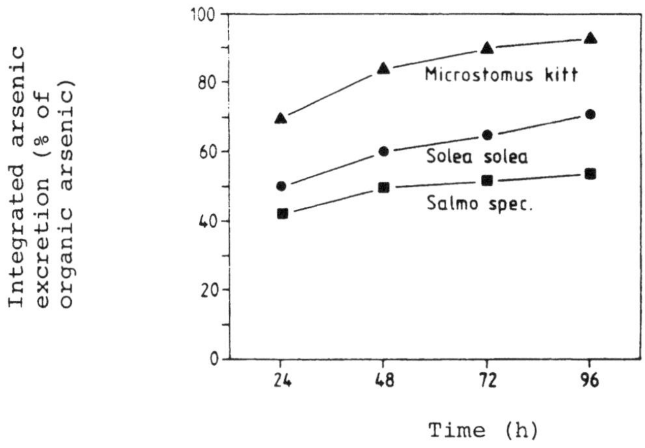

figure 3: Integrated arsenic excretion (organic arsenic) in urine for 96 h

As far as an increase of inorganic arsenic in urine is concerned, the experiments performed with Microstomus kitt could also answer this question. Analysis of inorganic arsenic in fillet prior and after preparation (heating) showed within the error limits of the TMAH-method the same, very low "reactive arsenic" content i.e. $\leq$0.2 % of the total arsenic in fillet. The increase of "reactive arsenic" in urine after fish consumption showed the same ratio, i.e. only $\leq$0.2 % of the total arsenic excreted could be identified as reactive arsenic.

These findings thus allow the following conclusions that are in good agreement with recent findings of others too (e.g. 10):

- The percentage of excretion of organo (non reactive) arsenic compounds via urine depends on the absolute amount ingested, however, with approximately the same half time irrespective of the total amount taken.
- Excretion pattern appears not to differ significantly between fresh water fish and marine fish. The biological half time for these compounds thus is significantly lower than the half time for inorganic arsenic and metabolites as found in our first series of experiments (6).

- There is no indication that stable organoarsenic compounds (arsenobetaine) can be deteriorated and transformed into less stable and possibly toxic arsenic compounds neither during preparation (thermal treatment) nor in the human body after ingestion.

References

1. World Health Organization. 1981. Environmental Health Criteria 18, Arsenic, Geneva
2. Philips, D.J.H. and M.H. Depledge. 1985. Marine Environ. Res. 17, 1-12
3. Burow, M. and M. Stoeppler to be published
4. Maher, W.A. 1983. Mar. Poll. Bull. 14, 308-310
5. Luten, J.B., G. Riekwel-Booy, J.v.d. Greef and M.C. ten Noever de Brauw. 1983. Chemosphere, 12, 131-141
6. Apel, M. and M. Stoeppler. 1983. In: Proc. Int. Conf. Heavy Metals in the Environment, Heidelberg, CEP Consultants, Edinburgh. pp. 517-520
7. Stoeppler, M., M. Burow, F. Backhaus, W. Schramm and H.W. Nürnberg. 1986. Marine Chemistry, 18, 321-334
8. Norin, H. and M. Vahter. 1981. Scand. J. Work Environ. Health, 7, 38-44
9. Stoeppler, M. and M. Apel. 1984. Fresenius Z. Anal. Chem. 317, 226-227
10. Luten, J.B. and G. Riekwel-Booy. 1983. In: Trace Element-Analytical Chemistry in Medicine and Biology, Vol. 2 (P. Brätter and P. Schramel, eds.) Walter de Gruyter & Co., Berlin-New York, pp. 276-286

INGESTION OF MERCURY DURING EARLY INFANCY

V.E. Negretti de Brätter, P. Brätter
Hahn-Meitner Institute Berlin for Nuclear Research, F.R.G.
Trace Elements in Health and Nutrition

J. Müller, P. Weigert
Federal Health Office Berlin , F.R.G.

St. Götze, G. Karkut
Klinikum Steglitz Berlin , F.R.G.

Introduction

Human milk is the most complete feed during early infancy, consequently, it is necessary to know its composition and trace element content in order to guarantee healthy baby growth.
However, the health of the baby might be put at risk if the mother is exposed to environmental pollutants. Therefore, it is very important to control the variation of the content of hazardous metals like mercury in the breast milk during the lactation period.
For adults with a 70 kg body weight the estimated upperlimit of mercury intake is 0.35 mg/week (1,2). In contrast, the tolerable mercury level for the growing infant is not known.
In order to estimate the daily ingestion of this metal via breast feeding a collaborative study was undertaken. The total mercury content in blood serum and breast milk of 149 mothers in West Berlin were analysed.
For the accurate determination of mercury in these biological fluids special care is required. Handling procedures were developed in order to minimize the risks of contamination or losses of mercury during the sample collection, storage, pre-treatment and digestion.

Experimental

Sample collection. All materials were previously washed with nitric acid and rinsed in quartz re-distilled water. The hospital personal involved in the sampling were instructed with regard to the contamination problems in the course of collection. After collection, all samples were immediately deep frozen (-25°C) until analysis. Breast milk: The samples were collected on the first day and after ten days of breast feeding. The milk samples were obtained by manual pressing of the breast and directly filled into the containers avoiding any contact with the breast nipple. A good aid for the stimulation of the milk secretion was the simultaneous feeding of the child with the other breast.
Blood serum: The blood samples were collected on the first day of lactation, after overnight fast. All mothers were advised not to take any drugs prior to the blood collection. Contamination free syringes were used. After coagulation (about 40 minutes waiting time) the serum was separated by centrifugation.
Procedures. The samples were pre-treated and digested as follows: 2 ml sample aliquots were oven dried at 50°C. After drying 1 ml nitric acid was added and the covered sample was left overnight. The walls of the above mentioned sample container were thoroughly rinsed with 1 ml HNO_3 and subsequently transferred into a quartz tube. 50 µl perchloric acid was added and covered with a quartz lid. The sample was decomposed under pressure at 160°C for 4 hours. After digestion, the sample volume was made up to 10 ml with deionised water. For this procedure only quartz containers which had been washed with nitric acid and rinsed thoroughly with quartz re-distilled water were used.
Reagents. A solution of 3% (v/w) $NaBH_4$ in 1% (v/w) NaOH was prepared daily and filtered before use. A custom made stock standard solution of 1000 µg Hg/ml in 1.8% nitric acid (Alfa Products Thiokol/Ventron. Div.) was used. A silicon anti-foaming agent was added to the sample mixture. The standard curve was made by adding 5, 7.5 and 10 ng Hg to a solution of 1.5% (v/w) HNO_3. All reagents were of ultra pure grade.
Apparatus. A Perkin-Elmer Model 503 atomic-absorption spectrophotometer equipped with a hollow cathode lamp, a Perkin-Elmer Model MHS-20 hydride system and a Perkin-Elmer Model 56 recorder was

used for the mercury determination. In order to increase the mercury output a gold/platin net accessory was included. Instrument settings and instrumental condition are shown in Table 1.

Tab.1: Instrument Settings and instrumental Conditions for the Mercury Determination with hydride AAS cold vapor Technique.

Instrument settings		Instrumental conditions	
Purge I	100 s	Wavelenght	253.6 nm
Reaction	15 s	Spectral band width	0.7 nm
Purge II	30 s	Reaction volume	10 ml
Cell Temp.	200 ^{o}C		

Accuracy and precision of the methodology. The accuracy of the method was tested using as control the following standard reference materials: BCR 151 milk powder (European Community Bureau of Reference) and NBS 1571 orchard leaves (National Bureau of Standards). The standards and samples were similarly pre-treated and digested. Table 2 shows the results of the accuracy control.

Tab.2: Accuracy Control for the Mercury Determination with hydride AAS cold vapor Technique.

Standard Reference Material	This Work (ng Hg/g $\pm$ % RSD)	Certified Values (ng Hg/g $\pm$ % RSD)
BCR 151	111.2 $\pm$ 26.8	101.0 $\pm$ 10.0
NBS 1571	145.0 $\pm$ 11.5	155.0 $\pm$ 9.7

The precision of the method was established by repeated analysis of a selected sample collective. The relative standard deviation of the day to day precision of the mercury determination is given in Table 3.

Tab.3: Precision Control for the Mercury Determination with hydride AAS cold vapor Technique.

		Day to day precision (9 days) % RSD
Human milk	colostrum	12 % - 35 %
	mature	12 %
Blood serum		6 %

The reproducibility of the signal for one measurement was 3% RSD. An overall relative standard deviation of 6% (human milk) and 4% (blood serum) was found. The detection limit of the developed methodology for the mercury determination was 0.5 ng. This value was determined by the blank levels in the digestion step.
Dry matter of blood serum. In this study the dry matter percentage shows a variation range between 4.0% and 7.8%. This is a consequence of the hemo-dilution effect associated with pregnancy and parturition.

Results and Discussion

The mercury values in colostrum, mature milk and blood serum are given in Table 4.

Tab.4: Mercury Content of human Milk and Blood Serum (µg/kg).

	weight	n	Arithm.Mean ± SD	Median	Range	98 % Percentile
Colostrum	dry	142	21.0 ± 16.2	17.4	0.8 - 155.8	60.4
	wet	142	2.6 ± 2.0	2.2	1.0 - 18.0	8.1
Mature Milk	dry	112	11.2 ± 7.0	9.6	2.0 - 47.6	46.1
	wet	112	1.5 ± 1.2	1.3	0.3 - 9.7	8.9
Blood Serum	dry	136	19.4 ± 11.0	17.8	0.05 - 81.4	58.4
	wet	136	1.0 ± 0.5	1.0	0.05 - 4.0	2.0

The mercury content in blood serum of the mothers was found to be 1.0 µg/kg (range: 0.05-4.0 µg/kg). This is in very good agreement with the results obtained by Schierling and Schaller (3), Stoeppler et al. (4) and Schaller (5), who found the mean 0.6-0.7 ng/ml (range: 0.2-1.9 ng/ml) and correspond to levels of German people living in non-polluted areas.

A correlation coefficient of 0.6 was found between the mercury content in blood serum and colostrum.

Furthermore, this study demonstrated that the mercury level of the breast milk decreases from 2.6 µg/kg (colostrum) to 1.5 µg/kg (mature milk).

On the basis of these results, assuming a daily breast milk consumption of 600 g and a mean weight of 3.9 kg, the estimated daily mercury ingestion of a 1 month old infant via mature milk is 0.23 μg/kg body weight.

Acknowledgement

We thank the valuable collaboration of Ms. D. Melchert in the performing of this work.

References

1. 16th Report of the Joint FAO/WHO Expert Committee on Food Additives, Evaluation of Certain Food Additives and the Contaminants Mercury, Lead, and Cadmium.
WHO Technical Report Series No. 505, Geneva, 1972.

2. 22nd Report of the Joint FAO/WHO Expert Committee on Food Additives, Evaluation of Certain Food Additives and the Contaminants.
WHO Technical Report Series No. 631, Geneva, 1978.

3. Schierling, P. und Schaller, K.H. 1981. Arbeitsmedizin 16, Nr. 3, 57-61.

4. Stoeppler, M., Dürbeck, H.W. und Nürnberg, H.W. 1982. Talanta 29, 963-972.

5. Schaller, K.H. 1982. Staub-Reinh. Luft 42, Nr. 4, 142-144.

AN INTERNATIONAL COLLABORATIVE RESEARCH PROGRAMME ON MINOR AND TRACE ELEMENTS IN TOTAL DIETS

R.M. Parr
Division of Life Sciences, International Atomic Energy Agency,
A-1400 Vienna, Austria

Introduction

A research programme was initiated by the IAEA in March 1985 to obtain reliable comparative data on the average daily intakes of 24 nutritionally important minor and trace elements (including some common toxic trace elements) via diets consumed in a variety of developed and developing countries. At present, 16 institutes in 12 countries, including the IAEA's own laboratory, are contributing to the programme either as collection centres or reference analytical laboratories. The concept of this programme foresees (1) a core programme to which all participants are contributing, and (2) a supplementary programme, the details of which differ from one participant to another.

Core & Supplementary Programmes

Under the core programme, representative total diets are being collected in 11 countries (up to 60 specimens per country). These are analysed in selected reference analytical laboratories according to a protocol designed to provide adequate analytical quality assurance. The elements of interest are: Al, As, Ca, Cd, Cl, Cr, Cu, F, Fe, Hg, I, K, Mg, Mn, Mo, Na, Ni, P, Pb, Se, Si, Sn, V and Zn as well as total energy, fibre and phytate. Under the supplementary programme, each collection centre is following its own interests, e.g. the determination of selected elements in staple food

Trace Element – Analytical Chemistry in Medicine and Biology, Vol. 4

Table 1 Project Summaries

COLLECTION CENTRE	CORE PROGRAMME Study groups	SUPPLEMENTARY PROGRAMME
INPA-CNPq, Manaus, BRAZIL (H.A. Marinho)	1. Low social class, Manaus, The Amazon	Effects of Zn supplementation on infant growth, prevalence and duration of diarrhoeal disease and appetite
Dalhousie University, Halifax, CANADA (A. Chatt)	1. General Canadian population (4 regions)	Development of analytical methods (e.g. NAA via short-lived activation products; speciation studies)
Institute of Health Beijing, CHINA (Liu Shengjie)	1. Urban 2. Rural, northern province (wheat eating) 3. Rural, southern province (rice eating)	Ca, Cu, Fe, Mg, Mn, Zn (by AAS); Mo (by electrochem); Se (by fluorimetry) in individual food items
Nuclear Research Centre, Teheran, IRAN (A.G. Gharib)	1. Urban (higher literacy) 2. Urban (lower literacy) 3. Rural	Cu, Fe, Zn etc. (by NAA, AAS, ICP, PIXE etc.) in individual food items
ENEA-CRE Casaccia, Rome, ITALY (G. Ingrao)	1. General population, industrial area 2. General population, mountainous area (traditional diet) 3. General population, area with high intake of seafood	Cd, Co, Cr, Fe, Pb, Sb Se, Zn (by NAA and AAS) in main food items; Hg and methyl-Hg in seafood and human tissues
Universidad Complutense, Madrid, SPAIN (C. Camara-Rica)	1.Urban, higher middle class 2.Urban, lower middle class	As, Cu, Se, Pb, Zn (by AAS), I (by ASV), F (by ISE) in individual food items

products. A brief description of the core and supplementary programmes at each collection centre is given in Table 1.

The data collected will be used:

(1) to compare in the selected population groups, actual intakes with present estimates of requirements (see Table 2) in order to detect possible deficiencies and/or imbalances in trace element intakes, thereby opening the way for intervention efforts; in the case of toxic trace elements the comparison will be made with provisional tolerable intakes (see Table 2); and

(2) to test and, if possible, confirm the validity of presently estimated requirements (i.e. if a healthy population group is found with intakes lower than the "requirement" then this would be evidence that the requirement should be reviewed).

A variety of analytical techniques is being used, in both the core and supplementary programmes. They include neutron activation analysis, atomic absorption spectrometry, and inductively-coupled plasma atomic emission spectrometry.

Reference Analytical Laboratories

Samples collected under the core programme will all be analysed by selected reference analytical laboratories. Generally, all the analyses for any one analyte (an element or other component such as fibre) will be made by only one reference laboratory. However, for each analyte, a backup laboratory will be appointed for cross checking of some analyses, and possibly also for other forms of assistance.

Reference analytical laboratories have been tentatively identified as follows (see list of participants):
Vienna, Austria (R. Schelenz): Ca Cl Cu K Mg Mn Na (by NAA and ICP)
Gaithersburg, USA (G.V. Iyengar): Cd Cr Fe Hg Mo Se Zn (NAA)

Project Summaries (continuation)

COLLECTION CENTRE	CORE PROGRAMME Study groups	SUPPLEMENTARY PROGRAMME
Atomic Energy Commission Khartoum, SUDAN (A.S. Hassan)	1. Urban, lower middle class	Ca, Mg, Na, K, Cu, Fe, Zn (by AAS, XRF and NAA) in individual food items and snuff
University of Lund, SWEDEN (M. Abdulla)	1. General Swedish population	Dietary intakes of four risk groups (pregnant women, 7-year old children, patients with diabetes mellitus, patients with food allergy)
Institute of Nutrition Bangkok, THAILAND (Songsak Srianujata)	1. Active working adults and pregnant women, NE region 2. Ditto for another region with other nutritional habits	As, Ca, Cd, Cu, Fe, K, Mg, Na, P, Pb, Zn (by AAS), plus possibly Cr and Se; also energy, fibre, phytate; in individual food items
Middle East Technical University Ankara, TURKEY (N.K. Aras)	1. Rural 2. Urban, upper middle class	Multi-element studies (by AAS, NAA) of individual food items
FDA, NBS, USDA, Washington, U.S.A. (G.V. Iyengar)	1. General US population (composite total diets)	(Reference analysts for several minor and trace elements, and other components)

Toronto, Canada (S.S. Krishnan): Al (AAS)
Madrid, Spain (C. Camara-Rica): Pb (AAS)
Beltsville, USA (W. Wolf): phytate
Reference analytical laboratories are still being sought for the following analytes: As F I Ni P Si Sn V fibre and energy.

Analytical Quality Assurance

In the first phase of the programme, participants are required to validate their analytical methods using mainly the following reference materials: (1) IAEA mixed human diet, H-9, and (2) mixed diet reference material, NBS RM 8431. Various other mixed diet samples will be prepared and used for quality control purposes during the course of the programme.

For H-9, an intercomparison has recently been organized by IAEA. Sixty-six laboratories (including most of the participants in this research programme) have so far submitted their data. On the basis of the results reported, recommended values can now be proposed for 23 elements (see Table 3).

For reasons connected with quality assurance, the IAEA is also providing all collection centres with certain standard items of equipment, e.g. pre-cleaned specimen jars, and a food blender with a plastic bowl and titanium blades (mainly so as to reduce contamination with elements such as chromium).

Present status and future plans

Four collection centres have already started to provide total diet specimens for analysis; the remaining collection centres are expected to reach this stage during the next few months. About 200 diet samples will be provided during the next year, shipped to the IAEA's laboratory, homogenized, aliquotted, and distributed to the reference analytical laboratories. The programme as a whole is expected to continue until 1989,

Table 2. Recommended Daily Dietary Intakes of Essential and Toxic Elements. (The values quoted refer to adults. For infants children and adolescents, different intake values of essential elements are recommended. Data from US National Academy of Sciences for essential elements and from FAO for toxic elements.)

Element	Quantity	Intake value per day
As	maximum acceptable daily load	2 µg/kg body-weight
Ca	recommended dietary allowance	800 mg
Cd	provisional tolerable intake	1.0-1.2 µg/kg body-weight
Cl	estimated safe and adequate intake	1700-5100 mg
Co	recommended dietary allowance	3 µg vitamin B_{12} [a]
Cr	estimated safe and adequate intake	50-200 µg
Cu	estimated safe and adequate intake maximum acceptable daily load	2-3 mg 0.05-0.5 mg/kg body-weight[c]
F	estimated safe and adequate intake	1.5-4.0 mg
Fe	recommended dietary allowance maximum acceptable daily load	10 or 18 mg (age/sex dependent) 0.8 mg/kg body-weight
Hg	provisional tolerable intake	0.7 µg/kg body-weight total Hg 0.5 µg/kg body-weight methyl-Hg[b]
I	recommended dietary allowance	150 µg
K	estimated safe and adequate intake	1875-5625 mg
Mg	recommended dietary allowance	300 mg (female); 350 mg (male)
Mn	estimated safe and adequate intake	2.5-5.0 mg
Mo	estimated safe and adequate intake	0.15-0.5 mg
Na	estimated safe and adequate intake	1100-3300 mg
P	recommended dietary allowance	800 mg
Pb	provisional tolerable intake	7.1 µg/kg body-weight
Se	estimated safe and adequate intake	50-200 µg
Sn	maximum acceptable daily load	20 mg/kg body-weight[c]
Zn	recommended dietary allowance maximum acceptable daily load	15 mg 0.3-1.0 mg/kg body-weight[c]

a. expressed as vitamin B_{12}, which contains about 4% by weight of cobalt

b. expressed as weight of mercury

c provisional

Table 3. Elemental Composition of IAEA Mixed Human Diet, H-9[a)]
(all results expressed on a dry weight basis)

Element	No of accepted lab. means	Overall mean ± SE[b]	Unit
Al	11	17.4 ± 3.3	mg/kg
As	13	90 ± 14	μg/kg
Br	12	7.5 ± 0.3	mg/kg
Ca	41	2.31 ± 0.04	g/kg
Cd	23	31.9 ± 1.9	μg/kg
Cl	18	12.4 ± 0.3	g/kg
Co	16	44 ± 2	μg/kg
Cr	18	0.14 ± 0.01	mg/kg
Cu	44	2.8 ± 0.1	mg/kg
Fe	50	33.5 ± 0.5	mg/kg
Hg	11	4.9 ± 0.7	μg/kg
K	35	8.4 ± 0.2	g/kg
Mg	26	782 ± 11	mg/kg
Mn	50	11.9 ± 0.2	mg/kg
Mo	10	0.23 ± 0.02	mg/kg
Na	31	8.1 ± 0.1	g/kg
Ni	10	0.26 ± 0.02	mg/kg
P	15	3.44 ± 0.06	g/kg
Pb	19	0.15 ± 0.02	mg/kg
Rb	19	8.0 ± 0.17	mg/kg
Se	19	0.11 ± 0.01	mg/kg
Sr	15	3.0 ± 0.2	mg/kg
Zn	58	27.6 ± 0.3	mg/kg

a) The elements listed are those for which the quoted values can be tentatively considered to be reliable

b) Overall mean of accepted laboratory means ± standard error

during which time approximately 500 samples will be collected and analysed. Under the supplementary programme, participants will at the same time be doing related research as outlined in Table 1.

Participants

The IAEA is grateful to the following principal investigators whose institutes are contributing to the programme either as collection centres or reference analytical laboratories.

Dr. M. Abdulla, University of Lund, Sweden
Prof. N.K. Aras, Middle East Technical University, Ankara, Turkey
Dr. C. Camara-Rica, Ciudad Universitaria, Madrid, Spain
Dr. A. Chatt, Dalhousie University, Halifax, Nova Scotia, Canada
Dr. A.G. Gharib, Nuclear Research Centre, Teheran, Iran
Dr. A.S. Hassan, Sudan Atomic Energy Commission, Khartoum, Sudan
Dr. G. Ingrao, ENEA, Rome, Italy
Dr. G.V. Iyengar, National Bureau of Standards, Gaithersburg, U.S.A.
Dr. S.S. Krishnan, Toronto General Hospital, Toronto, Ontario, Canada
Dr. Liu Shengjie, Institute of Health, Beijing, China
Dr. H.A. Marinho, Instituto Nacional de Pesquisas de Amazonia, Manaus, Brazil
Dr. Songsak Srianujata, Institute of Nutrition, Bangkok, Thailand
Dr. J.J. Tanner, Food and Drug Administration, Washington, D.C., U.S.A.
Dr. W. Wolf, US Department of Agriculture, Beltsville, U.S.A.
Dr. R. Zeisler, National Bureau of Standards, Gaithersburg, U.S.A.

At the IAEA in Vienna, Dr. R.M. Parr is acting as programme manager and Dr. R. Schelenz as responsible analyst.

THE INFLUENCE OF HOME PREPARATION AND EATING HABITS ON DAILY CHROMIUM INTAKE

O. Van Schoor and H. Deelstra

Dept. Pharmaceutical Sciences, University of Antwerp, U.I.A.
Universiteitsplein 1, B-2610 Wilrijk, Belgium

Introduction

Chromium has been recognised to be an important trace element in human nutrition. An accurate analytical methodology is indispensable to establish the dietary intake of chromium with foods and diets. Because chromium is widely used in metal coatings and alloys, strict contamination control is needed during chromium analysis.

The daily chromium intake can be calculated from the chromium content of different foodstuffs and the mean daily intake of these foods by different populations. There are, however, few reliable data on the chromium content of individual foodstuffs. Daily chromium intake can also be assessed by determination of the chromium content of laboratory-prepared typical diets (1). None of these two methods takes into account the possible influence of home preparation and eating habits on the chromium content of these meals. In this study, this influence was studied using prepared and non-prepared duplicate meals.

Materials and Methods

Sample collection

During three consecutive days, two duplicate breakfasts, dinners and suppers, including all beverages, were collected in one family. One duplicate meal was prepared according to normal food preparing habits, such as mixing, coocking and frying. The edible parts of the meal were separated and cut in smaller pieces with stainless steel cutlery in order to simulate normal eating habits. The other duplicate meal was composed of the same ingredients in

Trace Element – Analytical Chemistry in Medicine and Biology, Vol. 4

the same amounts as in the first duplicate, but collected without any preparation. The edible parts of this second duplicate were separated and cut in smaller pieces with plastic fork and knife. Samples were frozen in plastic containers wich were previously cleaned with 50% nitric acid and double deionized water.

Apparatus

Samples were analysed with a graphite-furnace atomic absorption spectrometer (Perkin Elmer 4000 + HGA 500). Only pyrocoated graphite tubes were used. Plastic labware was used troughout the analysis.

Reagents and standard solutions

Double deionised water (Milli-Q system, Millipore corp.) was used for all solutions and dilutions. Stock solutions were prepared from Titrisol (Merck) ampoules.

Analysis

A stomacher (Seward, London) was used to homogenize the food samples. Using frozen orange juice, it was found that this homogenisation method produced no detectable chromium contamination (2). The homogenized samples were freeze-dried for three days in the same plastic bags that were used for the stomacher. Samples were wet ashed in a decomposition bomb made of aluminium with a teflon liner (Autoclave 3,Perkin-Elmer). In the closed vessel 1g of dried sample was treated with 10 ml of nitric acid (Suprapur, Merck) at 140 °C for 4 hours. After cooling, the volume was made to 25 ml. An aliquot of 20 µl was analysed, using the method of standard additions. All determinations were done in triplo.

Results and Discussion

The chromium contents of the different meals are shown in table 1. The mean chromium intake was 38 µg for the non-treated meals and 83 µg for the treated meals. The largest difference between the two meals was found for the supper (24 vs 90 µg). This can be explained by the fact that the preparation of the supper requires more preparation steps than the other meals. Soup was also included in these suppers ; the use of a mixer with stainless steel mixer blades could be a major source of the added chromium.

Table 1 Influence of food preparation and eating habits on chromium content (µg) of prepared (P) and non-prepared (NP) meals.

	Breakfast		Dinner		Supper		Day (total)	
Day	NP	P	NP	P	NP	P	NP	P
Monday	4.6	5.9	28.3	43.2	19.9	28.5	52.8	77.6
Tuesday	2.0	2.1	21.3	75.0	7.2	17.8	30.5	94.9
Wednesday	2.3	3.5	21.9	61.2	4.5	10.6	28.7	75.3
Mean	3.0	3.8	23.8	89.6	10.5	19.0	37.3	82.6

On the average, more than half (55%) of the chromium found in the prepared meals was not originated from the ingredients but from the preparation and eating habits. These results reveal the errors that can be made by assessing the daily chromium intake based on market basket studies using the chromium content of the different foods.

In an other study (2) we found that the apparent chromium concentration of frozen fruit juice was doubled after 30 seconds of mixing. Kumpulainen (3) found the chromium concentration of coocked meat together with orange juice was 20% higher after 1 minute of mixing in a normal mixer. Anderson (4) suggested release of chromium from stainless steel vessels during processing of beer, since the chromium content of the individual ingredients did not appear to account for the total chromium found in some beer samples. We also found (5) that chromium could be released from stainless steel vessels into wine. Release of significant amounts of chromium from stainless steel during processing of food was reported by Offenbacher (6). These results indicate that a major part of the dietary chromium enters the food during the different stages of food processing. This also means that most of this dietary chromium is probably inorganic and will be poorly absorbed.

References

1. Kumpulainen, J.T., W.R. Wolf, C. Veillon, W. Mertz. 1979. J. Agric. Food Chem. 27, 490.

2. Van Schoor, O., H. Deelstra. 1986. Bull. Soc. Chim. Belge (in press).

3. Kumpulainen, J.T., E. Vuori. 1980. Br. J. Nutr. 44, 257.

4. Anderson, R.A., N.A. Bryden. 1983. J. Agric. Food Chem. 31 308.

5. Van Schoor, O., H. Deelstra. 1986. Belg. J. Food Chem. Biotechnol. (in press).

6. Offenbacher, E.G., F.X. Pi-Sunyer. 1983. J. Agric. Food Chem. 31, 89.

THE CONCENTRATION OF 19 TRACE ELEMENTS IN THE SWISS DIET

A. Wyttenbach, S. Bajo, L. Tobler
Swiss Federal Institute for Reactor Research
CH-5303 Würenlingen, Switzerland

B. Zimmerli
Federal Office of Public Health
CH-3001 Berne, Switzerland

Introduction

As part of a general survey of the Swiss diet (1), we have determined the concentration of 19 elements, concentrating on minor and trace elements (Al, As, Br, Cl, Co, Cr, Cs, Cu, Fe, I, K, Mg, Mn, Mo, Na, Rb, Sc, Se, Zn). In addition to the results presented here, Cd, Ni and Pb in the same samples have been measured else where (2). Elemental distribution functions, correlations and means are discussed in terms of their concentrations in freezedried food and converted to daily intake using measured daily food intake masses.

Sampling

Ten complete daily menues were sampled at 4 different food service kitchens giving a total of 40 samples. The specific locations were: a company cafeteria (A), an hospital (B), a vegetarian restaurant (C) and a military school (D). Each sample was homogenized and freezedried. Details on sampling and composition of the menues are given in (1).

Trace Element – Analytical Chemistry in Medicine and Biology, Vol. 4

Analytical

All determinations were done by neutron activation analysis (NAA) on 100 mg aliquots of the homogenized sample. The specific methods used were:

As, Cu, Mo: extraction of the induced activities after dissolution of the irradiated sample (3).

Br, I: purely instrumental after activation with epithermal neutrons (4).

all others: purely instrumental after activation with reactor neutrons.

All activities were determined by Ge(Li)-γ-spectroscopy. The spectra were evaluated by suitable computer programms and masses determined by comparison with appropriate standards. Details on irradiation and measurement conditions are given in Table 1.

Limits of Detection, Blanks, Precision and Accuracy

Elemental detection limits were much lower than the amounts found except for Cr, Cs and Sc. In the case of Cr, 8 out of 40 samples were below the detection limit of 50 ppb.

Blank corrections due to sample containers were practically nil except for Co and Sc (15%). There are no other blanks involved in NAA. This contrasts favorably to the determination of Cd and Pb in food by nonnuclear methods, which seem to be plagued by blanks (2).

Blanks originating from the homogenisation and freezedrying processes were determined by dissolution of saccarose in distilled water with subsequent treatment identical to the samples. For no element was this blank > 3% of the dietary sample concentrations.

Precision was determined by analyzing 10 samples in duplicate. The resulting coefficients of variation (CV) are given in Table 1. For 9 elements (marked in Table 1 by an asterisk) CV is explained by counting statistics. The remaining 10 elements have CV $\lesssim$ 5%. From this it follows that sample inhomogeneity was < 5% for the 100 mg weight used.

Accuracy was tested by analyzing aliquots (n = 3 to 7) of NBS SRM 1571 (orchard leaves). The 11 certified elements (As, Cr, Cu, Fe, K, Mg, Mn, Ba, Rb, Se, Zn) showed a mean deviation of (4.5 ± 2.1)%, and elements not certified were all close to literature values (5).

Table 1 : Details on Analytical Procedures (Neutron Activation Analysis)

#	Irradiation time	Irradiation neutrons	Measurements decay time	Measurements measuring time	Determinations nuclide	Determinations energy keV	Determinations experimental reproducibility % (4)
1	1 min	epithermal(1)	7 min	20 min	Br-80	617	4
					I-128	443	10 *
2	30 s	thermal (2)	1 min	8 min	Mg-27	844	12 *
					Al-28	1779	10 *
					Cl-38	1642	2
			3 h	20 min	Na-24	1369	3
					K -42	1525	6
					Mn-56	847	5
3	24 h	thermal (3)	60 d	3 h	Sc-46	889	30 *
					Cr-51	320	30 *
					Fe-59	1099	5
					Co-60	1332	5
					Zn-65	1115	5
					Se-75	401	15 *
					Rb-86	1077	10 *
					Cs-134	796	25 *
4	1 h	thermal (2)	1 d	1 h	Cu-64	511	5
				5 h	As-76	559	5
			3 d	10 h	Mo-99	141	10 *

Notes :

(1) Pneumatic transfersystem, sample shielded by BN. Epithermal flux $1*10^{11}$ n/cm^2 s

(2) Pneumatic transfersystem, thermal flux $1.3*10^{13}$ n/cm^2 s

(3) Rotating sample rack, thermal flux $1.2*10^{13}$ n/cm^2 s

(4) Experimental reproducibilities (coefficient of variation) with DF = 10. Values marked by * are completely explained by counting statistics.

Distribution Functions

The whole data set was tested for normal and lognormal distributions by calculation of the mean, the skewness and the kurtosis. Lognormal distribution was found to apply to As, Co and Sc only. All other elements can be considered to be approximately normally distributed; the averaged differences between the means and the geometric means is only 4% in these cases, whereas it was 20% for As and Sc, and 40% for Co.

The distributions were further checked for the presence of extreme outliers at the 99% significance level by the method of the standardized extreme deviation (Pearson-test) (6). Out of the entire data set of 760 values, 24 were found to be extreme outliers; 13 of these were associated with only 3 elements (As, Cr, I). These outliers were excluded from all subsequent calculations and are discussed separately.

Dietary Elemental Compositions at the 4 Different Locations

Table 2 gives the mean elemental concentrations in food at every location, the width of the distributions (as relative standard deviation) and highly significant differences between locations.

Comparison of the relative standard deviations with the experimental reproducibilities (Table 1) shows the latter to be much smaller (with the exception of Cr and Cs); therefore, it can be inferred that the width of the distributions is only slightly influenced by the reproducibily of the analytical determination.

The width of the distribution functions for all normally distributed elements is rather uniform at around 23% (Se: 35%). So the composition of the diet at one location can be expected to show typically a CV of 23% for all elements considered with the exception of As, Co, Sc and Se.

Inspection of the mean abundances shows many highly significant differences between locations. The reasons for these differences are different spicing practices (in the case of Na, Cl), different composition of the food processed and possibly different cooking equipment. Elements with the highest numbers of significant differences are Co, Cu, K, Mn, Mo, Rb.

Table 2 : Mean Values, Relative Standard Deviations and Significant Differences between 4 Different Food Service Kitchens.

Element (1)	Unit	CV % (2)	Mean value (3) at location A	B	C	D	Significant differences at the 99% level (4)
Al	ppm	27	9.12	6.32	9.79	7.24	AB
As	ppb	175	13.3	20.1	13.5	24.1	
Br	ppm	29	4.12	5.49	5.88	4.72	AB
Cl	%	22	1.80	1.52	1.34	1.87	AC, CD
Co	ppb	146	105	54	28	21	AC, AD, BC, BD
Cr	ppb	28	109	120	115	97	
Cs	ppb	27	11.2	13.9	9.0	9.2	BC
Cu	ppm	23	2.58	2.15	3.16	3.45	BC, BD
Fe	ppm	25	22.3	17.0	23.9	20.9	
I	ppb	22	494	594	416	543	BC
K	%	16	0.58	0.69	0.81	0.44	AC, AD, BD, CD
Mg	ppm	21	566	634	758	606	AC
Mn	ppm	20	6.19	6.00	9.93	8.13	AC, AD, BC, BD
Mo	ppb	23	152	202	247	218	AB, AC, AD
Na	%	23	1.19	0.95	0.84	1.20	AC, CD
Rb	ppm	20	3.80	4.78	5.41	3.43	AB, AC, BC
Sc	ppb	152	0.63	0.45	0.95	0.63	BC
Se	ppb	35	179	159	111	152	
Zn	ppm	23	23.6	23.0	17.4	24.5	BD, CD

Notes :

(1) Linear data were used in all calculations except with As, Co and Sc, where log-transformed data were used.

(2) The coefficient of variation (CV) = 100 * (s / $\bar{x}$) was calculated separately for each location.
Since the 4 values for anyone element do not differ significantly, their mean is given in table 2.
In the case of log-transformed data, we report 100 * (antilog s).

(3) Usually n = 10, but excluding limits and outliers.

(4) Significance was tested by the t-test. A pair of letters denotes a significant difference between these 2 locations.

The following generalisations can be made about the elemental means at different locations:

1) The ratio between the highest and the lowest mean for any element is typically 1.5. Higher ratios are found for As, K (1.8), Sc (2.1) and Co (5).
2) The greatest number of extreme values is found in the vegetarian restaurant, where Al, Br, Fe, K, Mg, Mo, Mn, Rb and Sc are highest and Cl, Cs, I, Na, Se and Zn are lowest.
3) The 3 other locations show smaller and roughly equal number of extreme values.

Overall Distribution Widths and Means

All values (excepting limits and extreme outliers) were combined into means (Table 3); this was done despite the differences found between the 4 locations, since such differences will be present in any population.

It can be seen that the width of the distributions is again roughly equal for all elements; CV is about 28%, only As, Co, Sc and Se being much higher. This value is, of course, somewhat larger than the mean value found at the different locations (23%), reflecting the differences between locations. It can, therefore, be stated that the composition of the diet is rather uniform for all elements considered, showing a CV of about 30%. It should be noted that a duplicate portion study (7) in the FRG with n = 104 gave practically identical results for the distribution width, with the exception of Cu, Cs and Se, which had much wider distributions in the German study. The reason for these discrepancies is not known.

Elemental means obtained in this work are best compared with the German results (7), because of the great number of common elements investigated in the two studies (Al, Br, Cr, I, Mo have been determined in the Swiss study only). The ratios of the elemental means found in these studies are given in Table 3 and reveal a striking similarity in the composition of the diets in the two countries. 10 of the comparable elements (excepting Na, Cl, As, Cs) have a

Table 3 : Overall Concentrations and Daily Intake Values.
Comparison with German Food and with Plants.

		concentration values						daily intake, mg/d		
Element	Unit	n	$\bar{x}$	CV %	this work / FRG	diet / plant	correlations P > 99.9 %	x	CV %	Element
Al	ppm	38	8.04	32		0.42		3.7	32	Al
As	ppb	36	17.1	184	0.36	1.00		0.008	194	As
Br	ppm	40	5.05	34		6.7		2.3	35	Br
Cl	%	40	1.63	25	1.47	32	Na, Zn, I	7575	32	Cl
Co	ppb	39	42	207	1.24	1.24	Mn*	0.019	202	Co
Cr	ppb	26	109	28				0.050	27	Cr
Cs	ppb	37	11	32	0.23	0.92		0.005	31	Cs
Cu	ppm	39	2.83	29	1.13	1.23	Mn, Mo	1.3	37	Cu
Fe	ppm	38	21.0	31	0.78	0.81	Sc	9.6	33	Fe
I	ppb	35	510	25			Na, Cl	0.25	32	I
K	%	40	0.63	27	0.98	1.26	Rb, Mg	2825	21	K
Mg	ppm	40	641	24	0.91	0.54	Mn, Rb, K	295	24	Mg
Mn	ppm	40	7.56	29	0.92	0.09	Cu, Mg, Co*	3.5	33	Mn
Mo	ppb	38	206	31			Cu	0.09	33	Mo
Na	%	40	1.04	27	1.35	750	Cl, I, Zn	4851	33	Na
Rb	ppm	40	4.36	26	0.79	1.55	K, Mg	2.0	23	Rb
Sc	ppb	35	0.63	166	0.70	0.37	Fe	0.0003	165	Sc
Se	ppb	38	155	39	1.18			0.070	42	Se
Zn	ppm	40	22.1	27	0.88	0.61	Cl, Na	10.2	33	Zn
mean					0.95	0.90				

mean ratio of 0.95 ± 0.06. Na and Cl are higher in Switzerland by a factor of 1.4 and suggest a more liberal use of NaCl in this country. As is lower in Switzerland by a factor of 0.36, despite the fact that for both countries only meals without fish were taken for this comparison. The reasons for the low factors in As and Cs are not known.

An interesting comparison can be made between the composition of the Swiss diet and that of plant material. Table 3 gives the ratios between the diet and spruce needles (8); the needle values are means from 36 trees in Switzerland. The compositions are strikingly similar in both materials: 11 elements (Al, As, Co, Cs, Cu, Fe, K, Mg, Rb, Sc, Zn) have ratios between 0.4 and 1.5 with a mean of 0.90 ± 0.12. Exceptions to this similarity are Na, Cl, I, Br, Mn and no values are available in spruce needles for Cr, Mo, Se. Na, Cl and I have very large ratios due to the addition of iodized NaCl to the diet. The high Br values might be due to a general enrichment of agricultural soils with Br. The very low factor for Mn is due to the fact that uptake of Mn by plants is highly correlated to the acidity of the soil (8), and since forest soils usually are more acid than agricultural soils, the observed difference seems reasonable. Of course, spruce needles are not a obvious comparator for food, but the results show them to be a good one, and reliable data on trace elements of other plants are scarse.

Correlations

Correlations between elemental daily intake values will be influenced by the amount of food eaten. This is avoided by correlating food elemental concentrations. The results are indicated in Table 3. Only correlations with very high significances ($P > 99.9\%$) are retained and these can be summarized as follows:

1) The most highranking correlations are Na - Cl ($r = 0.98$, DF = 38) and K - Rb ($R = 0.83$, DF = 38).
2) All correlations are positive, except those with Co.
3) Group of elements that are strongly correlated among themselves

are (Na, Cl, I), (Cu, Mo, Mn, Fe) and (K, Rb, Mg).

The first group is obviously due to the use of iodized salt, and the last group to the fact that plants are the most important source for K, Rb and Mg in the diet.

Outliers

The extreme outlier values are explicitly given in Table 4.

Table 4: Outliers

	mean without outliers		outliers	
			n	x
Al	ppm	8.0	2	20, 22
As	ppb	17	4	111, 123, 708, 908
Co	ppb	42	1	623
Cr	ppb	109	4	259, 331, 690, 723
Cs	ppb	11	1	23
Cu	ppm	2.83	1	5.7
Fe	ppm	21	2	60, 74
I	ppb	510	5	1051, 1344, 1546, 1651, 1681
Mo	ppb	206	2	598, 635
Sc	ppb	0.6	2	3.3, 6.8

These outliers are all at least 2 to 3 times higher than their respective mean values and their are no outliers on the low side of the distributions. Many outliers can be traced to menues with special food components:

As (all), I (2) : fish
Sc, Al (all) : spinach
Cu, Fe, Mo (all) : liver.

Daily intake values

Daily intake is obtained by multiplying the concentration values by the daily intake of dry matter. This is done for every sample with subsequent averaging, omitting limits and the outliers mentioned before. The results are given in Table 3.

The mean values for intake (g dry matter/day) were 457 at location A, 428 (B), 420 (C), 536 (D) and 460 ± 67 for the entire data set (1). Since CV in the daily intake of dry matter is only 15%, whereas it was 28% for the concentration data, it can be seen that the latter almost exclusively determines the variation in the daily intake of an element. The expected value is 32%, which corresponds closely to the variation actually found except for As, Co, Sc and Se (see Table 3). A more detailed discussion of daily intake values will be given elsewhere (9).

References

1. Zimmerli, B. and R. Knutti. 1985.
Untersuchung von Tagesrationen aus Schweizerischen Verpflegungsbetrieben, I. Mitt. Gebiete Lebensm. Hyg. 76, 168 - 196.

2. Knutti, R. and B. Zimmerli. 1985.
Untersuchung von Tagesrationen aus schweizerischen Verpflegungsbetrieben, III. Mitt. Gebiete Lebensm. Hyg. 76, 206 - 232.

3. Bajo, S. 1985.
Liquid-liquid extractions with metal diethyldithiocarbamates as reagents. IRNC 90, 51 - 68.

4. Wyttenbach, A., L. Tobler and V. Furrer. 1985. In:
Instrumentelle Multielementanalyse (B. Sansoni, ed.). VCH Weinheim, pp. 137 - 140.

5. Gladney E.S. 1980.
Elemental concentrations in NBS biological and environmental standard reference materials. Anal. Chim. Acta 118, 385 - 396.

6. Wissenschaftliche Tabellen Geigy. 1980. Ciba-Geigy, Basel.

7. Schelenz, R. 1983.
Essentielle und toxische Inhaltsstoffe in der täglichen Gesamtnahrung. BFE-R 82-02, Karlsruhe.

8. Wyttenbach, A., S. Bajo and L. Tobler. 1985.
Major and trace element concentrations in needles of Picea abies. Plant and Soil 85, 313 - 325.

9. Zimmerli, B., L. Tobler, S. Bajo, A. Wyttenbach.
In preparation.

Session III

TRACE ELEMENT INTERACTIONS

Plenary Lectures given by:

R.J. Shamberger

M. Anke

Chairman:

C.F. Mills

P. Brätter

Poster Reports:

P. Brätter

U. Weser

INTERACTIONS OF TRACE ELEMENTS AND ENZYMES IN HUMANS

Raymond J. Shamberger
Department of Biochemistry, The Cleveland Clinic Foundation,
Cleveland, Ohio (USA) 44106

Introduction

Many of the earlier nutritional studies were done with single elements. However, interactions among the trace elements are many times so extensive that nutritional and toxicological studies done with single elements may lead to erroneous conclusions unless the dietary and body tissue levels of interacting elements are known and there is a clear definition of the experimental conditions. The extent to which other elements or compounds are present or absent from the diet or from the body pool may directly influence the absorption, utilization and excretion of many trace elements and also affect their physiological, pharmacological or toxicological interactions within the cell and tissues of the human or animal body. Therefore, a single minimum need of an essential trace element and a single safe dietary level of a toxic element is difficult to define. Minimal requirements or maximum tolerance levels depend on the chemical form of the element, the amount and length of intake, and the rest of the diet, especially the amounts, forms and proportions of other interacting elements and compounds.

There are several important examples of trace element interactions which decrease available trace element levels. Some of these interactions occur at the intestinal absorption level and others occur at the tissue level and impair utilization. Some of the more important trace element interactions are copper and zinc in regard to heart disease, cancer and sickle cell disease and Wilson's disease. Many other interactions can also occur including iron and manganese, iron and zinc, cadmium and copper, cadmium and zinc, copper, molybdenum and sulfate, selenium with either arsenic, cadmium and mercury and fluoride with calcium and phosphate.

Trace Element – Analytical Chemistry in Medicine and Biology, Vol. 4

Copper and Zinc

Copper and zinc have strong interactions at the intestinal absorption levels. Even though the nutritional effect of excess zinc on copper metabolism is greater than the effect of excess copper, there is evidence that excess copper intake decreases zinc absorption. Copper reduces the absorption of Zn^{65} from isolated rat intestinal segments maintained in situ (1). In the intestine of zinc-supplemented but not deficient rats, amounts as low as 1 um Cu^{2+} decrease ^{65}Zn absorption (2). In another study designed to investigate the effect of copper concentration on copper-zinc interrelationships, 2 ppm Cu^{2+} in drinking water decreased serum and kidney zinc levels more than 1 ppm Cu^{2+} when the water contains a constant 25 ppm zinc in both experiments. At even higher zinc levels there was no effect. Even though there is some suggestion that excess copper deters the absorption of zinc, the effect is small and may not be of practical significance.

The interaction between copper and zinc is most obvious when dietary copper is limiting and zinc is in excess. Zinc interfers with copper absorption from the intestine (4) and also will induce signs of copper deficiency in the rat (5-7). The physiological significance is enhanced in subjects with low tissue copper reserves, especially when they are maintained on diets which are marginally adequate in copper content. In rats fed high zinc diets, there is evidence that the decrease in copper absorption is associated with large increases in the content of copper in the metallothionein protein components from the intestinal mucosal cytosol which contains the greatest proportion of total soluble proteins. Copper displaces the zinc and is held in the mucosa until it sloughs off through desquamation of the enterocytes into the intestinal lumen. The loss of the enterocytes in which copper is thus sequestered as copper metallothionein could account for the reported decrease in the efficiency of copper absorption, even though there are apparent increases in mucosal copper content. Zinc has also been observed to reduce copper absorption in dairy calves, in sheep, in Japanese quail and in horses, but there seems to be considerable species and individual variation.

Clinical Significance

Sickle cell anemia

The number of membrane damaged sickle cells circulating in the sickle cell patient has been reduced by adopting a regimen of 25 mg of zinc given every four hours, avoiding food an hour before and an hour after each dose, for a total dose of 150 mg of elemental zinc per day. The gastrointestinal tolerance of the acetate salt of zinc was much better than that of sulfate (8). However, if zinc was administered with food, almost none of the zinc was absorbed. Using the above 150 mg of zinc diet, Brewer et al (9) were able to reduce the number of membrane-damaged sickle cells circulating in the sickle cell patient. However, an unexpected side effect of zinc therapy became apparent - namely copper deficiency. The longest treated patient developed hypochromic, microcytic red cells and leukopenia which are results compared to those found in animals, but the hematologic imbalances were corrected with small amounts of copper supplements (10).

Seven of 13 patients on zinc therapy became chemically copper deficient and had low ceruloplasmin levels in blood (8). In contrast, ceruloplasmin levels in untreated patients are very high in sickle cell anemia. The copper deficiency observed in man on zinc diets has not been observed in previous studies of zinc administration. It is likely that zinc administered without food, rather than with food is better absorbed.

In the sickle cell patient who is untreated, the zinc level is low and the copper and ceruloplasmin level is high. The mechanism of these abnormalities may be related to the failure of the kidney to adequately retain zinc with a resultant zinc deficiency. Because of zinc deficiency, the intestinal metallothionein mechanism is altered and copper absorption is increased. The enhanced copper absorption results with an increase of body copper, especially liver copper, to which the liver responds with increased synthesis of ceruloplasmin. However, when zinc was administered, the process was reversed. Zinc induces intestinal metallothionein which ultimately results in copper deficiency.

Wilson's disease

Wilson' disease is an autosomally recessively inherited abnormally in which there is a positive copper balance leading to copper accumulation in liver and brain. If untreated, Wilson's disease results in an early death. Brewer et al (11) have studied five Wilson's disease patients treated with the zinc diets used in the sickle cell anemia patients. A negative or a neutral copper balance was produced in all five patients. However, other zinc regimens, requiring less frequent zinc administration also work well in controlling the positive copper balance observed in Wilson's disease (8). Penicillamine is also an effective treatment for Wilson's disease, but there is a need for an alternative treatment because of the toxicity of penicillamine. About 10% of Wilson's disease patients eventually develop sufficient side effects so that they are unable to take penicillamine. Zinc therapy seems to be a useful alternative to penicillamine for the treatment of Wilson's disease.

Cardiovascular disease

A dietary deficiency of copper can produce a marked hypercholesterolemia in rats due to an increased synthesis and also clearance of cholesterol into the blood plasma (12). Hypercholesterolemia is thought to be one of the more important contributors to the risk of coronary heart disease which is the leading cause of death in the United States and other countries. A recently completed Coronary Primary Prevention Trial (CPPT) has provided evidence that reducing the total serum cholesterol lowers the risk of heart disease. A 1% reduction in serum cholesterol level in hypercholesterolemic men decreases the coronary heart disease risk by as much as 2% (13). The risk of coronary heart disease is also associated with the distribution of cholesterol among the plasma proteins (14). An increased risk of coronary heart disease is associated with elevations of low density lipoprotein (LDL) bound cholesterol (14). High density lipoprotein (HDL) bound cholesterol (15) has been associated with a reduced risk of coronary heart disease. In addition to a marked elevation in plasma cholesterol, dietary copper has also produced a large increase in plasma triglyceride concentration (16). Even though cholesterol concentration was increased in both the high and low density lipoprotein fractions, copper deficiency induced a significant decrease in the percent of plasma cholesterol associated with high density lipoproteins and also a significant increase in cholesterol associated with low density lipoproteins.

A zinc to copper ratio of 40:1 in drinking water produced a higher concentration of plasma cholesterol in rats compared to rats fed lower ratios (17). In another experiment rats were fed a normal diet and drinking water containing 10 mg Zn/ml with 2 mg Cu/ml had lower plasma cholesterol compared to those not given copper (18). Rats fed high zinc to copper ratios (up to 33:1) had no significant differences between serum cholesterol, HDL-cholesterol, and serum triglycerides (19). After zinc was fed to rats at a constant 120 mg/Kg body weight, an elevation of serum cholesterol was observed in rats fed less than 2 ppm of copper compared to rats fed 18 ppm copper (20).

Copper deficient rats are likely to have low body weight; their arteries abnormal connective tissue and their hearts are likely to be hypertrophied and have abnormal EKG's (21). Abnormal ST segments in EKG patterns; bundle branch blocks, supraventricular beats, ventricular beats and wandering pacemaker have been observed in copper deficient rats. Impaired glucose tolerance and higher glycosylated hemoglobins have also been observed in copper deficient rats.

Klevay has proposed a hypothesis that the incidence of cardiovascular disease may be influenced by the copper and zinc status. Cardiovascular disease may result from copper deficiency due to imbalances resulting from consuming foods low in copper and high in zinc. Klevay has claimed that the majority of diets in the United States contain less than 1.3 mg Cu/day which he suggested is the minimum copper requirement. A mean copper intake of 1.0 mg/day and a mean zinc intake of 8.6 mg/day has ben reported by Holden et al (23) in self-selected diets of 22 subjects (ages 14-64). The estimated safe and adequate daily dietary intake of copper has been listed as 2-3 mg Cu/day (24). Americans consume about the RDA or below for zinc. However, copper deficiency could arise if zinc supplements are used along with a marginal copper intake. In rats a diet Zn:Cu intake of at least 40:1 is required to produce copper deficiency in rats. If the average intake of copper by Americans is in the area of 1 mg/day, at least 40 mg/day of zinc intake would be required to produce cardiovascular abnormalities. Metabolic balance studies would be beneficial in evaluating the role of altered zinc:copper intake on blood levels of ceruloplasmin, blood lipids as well as other indices of copper and zinc status. Marginal dietary copper intakes do not seem to result in obvious copper deficiency symptoms, but the effects may be subtle and appear only on a long-term basis. To date, studies completed on humans in evaluating the copper:zinc hypothesis have been lacking.

Cancer

Abnormally high concentrations of copper have been observed in serum from patients with various malignancies such as Hodgkins disease (25), non-Hodgkins lymphomas (26) and acute leukemia (27). Concentrations of serum zinc have also been observed to decrease in the malignant process. Decreased serum zinc has been useful in the differential diagnosis of breast cancer, cervical cancer, osteosarcoma, lung cancer and intestinal malignant tumors (28-31). Inutsuka and Araki (32) have demonstrated a significantly greater copper/zinc ratio in serum of patients with gastrointestinal cancer than in normals. Lightman et al (33) have measured copper and zinc in serum and tumor tissue from women with a palpable pelvic mass and women who were admitted for suspected ovarian carcinoma. In serum, copper was increased and zinc decreased in the patients who were subsequently proven to have a malignant ovarian tumor. The mean copper/zinc ratio was higher in this group than the benign group. In tumor tissue, the copper/zinc ratio was also significantly higher in the malignant tumors than in the benign tissue. The Cu/Zn ratio in serum may specifically indicate the presence of an advanced ovarian malignant tumor. The zinc concentrations which are decreased and the increased copper concentrations do not seem to result from a shift of zinc or a release of copper out of the malignant tumor tissue. However, the serum copper/zinc ratio should be interpreted cautiously in cases of a suspected ovarian malignancy which is accompanied by acute and chronic infections, chronic liver disease, pregnancy or other conditions which are known to affect copper and zinc concentrations in the serum.

Selenium-Cadmium

Selenium compounds which have been administered to animals have been shown to protect effectively against cadmium induced toxicity (34). In addition, selenium is also effective in protecting against the lethal doses of cadmium (35). In general, selenium appears to have a general detoxifying effect on cadmium. Selenium also protects against cadmium-induced testicular injury, even though it actually increases the cadmium content in the testis (36). Selenium causes a redistribution of cadmium in the testis. Marked diversions of cadmium from the 10,000 and 30,000 molecular-weight proteins to large-molecular-weight proteins have been observed (36) because the target of the cadmium-induced testicular injury appears to be the 30,000-molecular-weight protein, this diversion of cadmium by selenium seems to be one of the more important mechanisms involved in the protection against testicular injury induced by cadmium.

Chen et al (36) found that selenium gave a 22-fold increase in the cadmium content in the blood and selenium doubled the testicular uptake, but decreased the liver uptake by 48% and the kidney uptake by 12% (36). The diversion in the binding of cadmium in the soluble fraction to higher molecular-weight proteins of about 115,000 was also observed for the kidney and the liver. The diversion of cadmium from lower-molecular-weight proteins to higher ones was also observed in the plasma, but these cadmium-binding, selenium-containing proteins in the plasma have an even larger molecular weight than the kidney and liver proteins and appear to be different than the other organs.

Cadmium selenide (CdSe), like cadmium sulfide is almost insoluble in water. This high affinity of Cd^{2+} for Se^{2+} and the stability of CdSe suggest that this complex forms even in the presence of plasma proteins. Colloidal metal preparations have been used to stabilize several substances, including proteins, carbohydrates, and lipids. The formation of a colloidal Cd-Se complex whose stability is dependent upon proteins seems to be a reasonable possibility. The fate of the Cd-Se complex has not been completely studied. After 24 hours the amount of the complex is about half of its peak value. Cadmium inhibits the urinary excretion of selenium, even though six times as much selenium as cadmium is excreted by 48 hours. Cadmium may deposit slowly in other tissues as the Cd-Se complex breaks down. When selenium is given simultaneously the liver burden of cadmium is less and therefore other tissues must have greater levels. Because cadmium is released at a slower rate, the synthesis of metallothionein may afford secondary protection. Selenium itself has caused a marked alteration in the metallothionein levels induced in rat liver and kidney by cadmium. Rats fed a basal, low-selenium diet, or this diet supplemented with 0.1 ppm and 1.0 ppm selenium and treated with cadmium have significant reductions in the activity of the selenoenzyme glutathione peroxidase in kidney and liver (37). Cadmium treatment resulted in a significant increase in the activity of selenium-independent glutathione peroxidase activity in the liver of selenium supplemented rats. No increased lipid peroxidation was observed. However, cadmium cardiotoxicity resulted from cadmium-induced activity of the selenoenzyme, glutathione peroxidase, and superoxide dismutase (SOD) in the rat heart. The reduction in the activities of these enzymes was accompanied by significant increases in lipid peroxidation.

Gerhardsson et al (38) has found a very high Cd/Se ratio in the lung tissue of smelter workers who died of lung cancer. In contrast, workers who died from gastrointestinal cancer and other malignancies had a considerably smaller Cd/Se ratio. Shamberger and Bratush (39) found a significantly greater Cd/Se ratio in the kidney cortex from people who died from various diseases. Shamberger et al (40) have observed a negative significant correlation between selenium intake and ischemic heart disease and the greatest positive correlation was observed between the Cd/Se ratio and ischemic heart disease.

Selenium-Arsenic

Arsenic was first found to protect against the chronic toxicity of seleniferous grains or selenite in rats (41). Arsenic also decreased the chronic toxicity of seleniferous grains or of selenite in dogs (42), cattle (43) and swine (44). Arsenic has also decreased the growth inhibitor caused by excess selenium and improved the poor hatchability of eggs laid by selenized birds (45) by apparently reducing the amount of selenium that is incorporated into the egg. At levels of chronic selenium toxicity the absorption and retention of selenium did not seem to be affected by protective levels of arsenite (46). After rats were injected with single subtoxic doses of selenite, arsenic increased the amount of selenium in blood and decreased it in the liver (47). In addition, the amount of selenium exhaled as volatile compounds following selenite injection was greatly decreased when selenite was present (48). This latter finding was a surprise because the formation of volatile selenium compounds has long been regarded as a detoxification process. It is possible that arsenite blocks the methyltransferase reaction which produces hydrogen selenide. This decrease in exhaled selenium would be expected to increase rather than decrease the symptoms of selenium toxicity. However, when subacute injections of arsenic are given with the selenium, excretion of selenium into the gastrointestinal tract via the bile fluid is decreased, resulting in less retention by the tissues. If the methyltransferase reaction were blocked, the excess hydrogen selenide may react with arsenite to form a detoxification conjugate, selenoarsenite, which might be excreted into the bile. Arsenic might react directly with selenium in vivo, possibly as Se^{2-} to form $As^{+}Se^{-}$ complexes of a particle size and solubility that favors biliary excretion. Selenium and arsenic differ by only one atomic number and one electron in the outer ring which may relate to their metabolic interaction. Hill and Matrone (49) have suggested that trace elements with similar electron structures such as Se^{4+}, As^{3+} and Te^{4+} are likely to be biological antagonists. In this connection, Hill (50) found that selenate counteracted the effect of arsenate in the uncoupling of oxidative phosphorylation.

There is a possibility that arsenic could be used as an antidote against selenium poisoning or conversely selenium could be used as an antidote against arsenic poisoning. In the cases of those exposed to excessive arsenic in well water, perhaps selenium could help alleviate the symptoms of chronic arsenic poisoning.

Selenium-Mercury

When selenium interacts with inorganic mercury, a decreased uptake of mercury occurs at the site of its administration and there is a decreased mercury excretion in the urine and feces. The amounts of mercury in the liver and blood are significantly increased, but at the same time the mercury content of the kidney is significantly decreased. The interaction between selenium and inorganic mercury is characterized by a clear change of mercury from low molecular to high molecular weight kidney proteins. This change is accompanied by about a five-fold decrease of mercury concentration in the kidneys and inhibition of metallothionein biosynthesis. At the same time there is about a four-fold increase of the metal in the mitochrondrial and nuclear fractions and also the mercury in blood increases markedly.

When mercury and selenium are administered in at least equimolar doses, these changes are most evident. The sequence of administration of both of these elements is also important. Selenium's effect on mercury distribution is smaller when the selenium is administered after mercury than in the case where these metals are administered simultaneously.

Because the mercury content of the kidney is significantly reduced, sodium selenite protects against the nephrotoxic effect of inorganic mercury and also has a beneficial effect in protecting against methyl mercury poisoning. Administration of sodium selenite results in delayed occurrence of symptoms of neurological and histological disturbances caused by methylmercury and results in increased life expectancy of exposed animals. Selenium has been shown to increase whole-body retention of methylmercury, increase levels of methylmercury in brain, liver, blood, and spleen while reducing the mercury content of kidneys. The brain effect is of interest because no symptoms of intoxication were seen in animals who received sodium selenite at the same time as methylmercury, even though the concentration of methylmercury in their brain exceeded critical values. However, there are reports showing that selenium reduces the level of methylmercury in the blood. This decrease may result from a selenium-induced decrease in the affinity of red blood cells for methylmercury. The difference between these reports and those earlier may relate to concentration of each element. Naganuma and Imura (51) have reported that bis(methylmercuric) selenite was formed from methylmercury and selenite in rabbit blood in the presence of reduced glutathione. The formation of bis(methylmercuric) selenite may play an important role in the modifying effect of selenium on methylmercury toxicity. There is a possibility that the increase of brain-mercury by brain-mercury concentration by the administation of selenite reported by several investigators is due to the formation of the complex bis(methylmercuric) selenite from methylmercury and selenite. Some methylmercury is bound to the proteins of the soluble fraction of the kidneys which indicates a possible participation of metallothionein-like proteins. This probably results in a higher biotransformation of methylmercury to inorganic mercury in rat kidneys compared to the liver. This process seems to be dependent on the dose of methylmercury.

Iron-Manganese

Dietary manganese at high levels have been demonstrated to interfere with iron metabolism and to decrease hemoglobin formation in sheep (52), cattle (53), rabbits and pigs (54). The rate of hemoglobin synthesis was greatly decreased and serum iron was depressed in anemic lambs fed diets containing 1000 ppm or 2000 ppm Mn. Decreased concentrations of iron were observed in the liver, kidney and spleen (52) of normal lambs fed levels up to 5000 ppm Mn. Manganese as well as iron is more efficiently absorbed by man and rats that are iron-deficient (55,56).

The exact mechanism by which manganese affects iron metabolism and hemoglobin formation is likely an interference with iron absorption rather than with hematoporesis seems to better fit the experimental data. Manganese supplements reduce iron absorption and on the other hand iron supplements reduce manganese absorption. Manganese levels used in the experiments with lambs, pigs and rabbits are so great that they are never likely achieved under normal conditions. Matrone et al (54) have estimated that the minimal level of dietary manganese that is capable of adversely affecting hemoglobin formation is about 45 ppm for anemic lambs and somewhere between 50 and 125 ppm for mature rabbits and baby pigs. These manganese levels are quite common in livestock diets. Therefore, the Mn:Fe interaction may have a greater importance than commonly considered.

Iron-Zinc

Zinc at high level decrease iron bioavailability, but it is not entirely clear whether this is a direct or an indirect effect. Very high zinc supplements at first induce a copper deficiency with a markedly reduced ceruloplasmin activity. Because ceruloplasmin acts as a ferroxidase, the ultimate consequence is iron anemia. If even greater amounts of zinc supplements are given, the zinc:iron ratio in the diet, iron storage and iron parameters of blood may be directly affected. The life-span of red cells is shortened by two-thirds with toxic zinc diets in comparison to a normal zinc supply and anemia and faster iron turnover by the liver is observed. Zinc interfers with the iron uptake by the liver and the iron storage form, ferritin, is decreased. Decreased iron contents in the liver and kidneys in response to excessive zinc supplementation is often observed. The interaction of zinc and iron may be due to competition for binding sites on transferrin which is important for both the transport of iron and of zinc.

When the zinc intake of adolescent girls was increased from 11.5 to 14.7 mg per day, the iron balance decreased from 1.4 to 1.0 mg day (57). Data that seems to support a direct interaction is the observation that zinc absorption is reduced by nonheme iron (58).

Iron and zinc interact at the level of the intestinal mucosa, and zinc absorption is decreased by iron (58). Ferrous iron inhibits zinc absorption more so than does ferric iron (59).

Cadmium-Copper

Cadmium impairs copper absorption and utilization in growing chicks (60) and cadmium also impairs the absorption of copper from the rat intestine (61). In these animals as well as goats the copper concentrations in the liver and other tissues decreased. A significant decrease in ceruloplasmin is also observed. Cadmium increases the proportion of copper which is tied up in the intestinal mucosa where copper is associated with metallothionein. The interaction between cadmium and copper depends not only on the level of the cadmium supply, but depends also upon the level of the copper supply. Campbell et al (62) have demonstrated that copper-cadmium interactions are especially obvious during suboptimum copper intake.

Cadmium-Zinc

From the chemical similarities between cadmium and zinc a metabolic antagonism between them might be expected. Both zinc and cadmium are members of group IIB of the periodic table and a similar tendency to form tetrahedral complexes. Cadmium has been shown in vitro to have an effect on the activity of zinc-containing enzymes such as carboxypeptidase.

The toxic effects of cadmium can be decreased by appropriate dietary intakes of zinc, and the toxic effects of zinc can be also decreased by appropriate dietary intakes of cadmium. In addition, at marginally deficient intakes of zinc, cadmium can produce overt signs of zinc deficiency in the animal.

The interaction between cadmium and zinc might be explained by competition for protein-binding sites at the intestinal mucosal cell level as well as the tissues. Reduced absorption of ^{65}Zn has been observed in the chick when cadmium is fed (63) and the intracelluar distribution of ^{65}Zn in mouse liver cells is altered by injected cadmium or cadmium feeding (64).

The competition between zinc and cadmium for protein-binding sites may include the zinc metalloenzymes such as metallothionein. There is a rapid rise in metallothionein synthesis following the oral or intraperitoneal administration of cadmium to rabbits, rats and chicks. One important action of metallothionein seems to be the protection of the organism from harmful cadmium ions.

The metabolic competition between cadmium and zinc is apparent from the protection from injected zinc against the toxic effects of injected cadmium on the testes (65) and on pregnant rats (66). Zinc and cadmium may compete for essential binding sites on essential enzymes needed for gametogenesis. Rats protected by zinc salts show no reduction in blood cadmium and the concentration of liver cadmium is actually increased. Zinc may be reducing cadmium toxicity through an increased excretion of cadmium from the organism by the kidneys, intestine or biliary tract.

Copper, Molybdenum and Sulfate

Another copper interaction of marked importance in ruminant animals, but of little significance in man and simple-stomached animals, is the three-way relationship of copper, molybdenum and sulfur. Molybdenum decreases copper bioavailability in both ruminants and monogastric animals. Marciless et al (67) has studied ruminants with labeled copper. Both molybdenum and sulfate supplementation decreased the incorporation of copper into the liver and ceruloplasmin synthesis. Copper supplements at high concentrations were able to counteract the effects of molybdenum and sulfate. Ruminants, in contrast to monogastrics are able to reduce sulfate to sulfide to a major extent in the rumen because of the action of Desulfovibrio bacteria. Sulfide thus produced may be either absorbed and detoxified in the liver, or incorporated into sulfur-amino acids, or precipitated as heavy metal sulfide. In the presence of copper, insoluble

CuS may be formed in the rumen, so the copper becomes less absorbable. Sulfide oxidase is dependent upon content of available copper in the liver and is responsible for the oxidation of sulfide to sulfate. Reduced sulfide oxidase activity leads to the accumulation of sulfide in the liver and may lead to precipitation of copper as insoluble nonavailable, CuS. In ruminants dietary sulfur also accentuates the formation of insoluble cupric thiomolybdate ($CuMoS_4$) and related oxythiomolybdates. Tetrathiomolybdate impairs copper absorption by reacting with copper ions and prevents copper utilization within tissues and induced a conditioned copper deficiency. Ruminants differ mainly from monogastrics in that more sulfide is produced in the forestomach and copper sulfide is formed there. More sulfate is passed to the intestine in monogastrics and less molybdate is absorbed because sulfate inhibits molybdate absorption and hence less molybdenum may reach the liver of monogastrics. Finally molybdenum inhibits sulfide oxidase so that more sulfide may accumulate in the livers of ruminants than in those of monogastrics. Since sulfide reacts with copper directly or is needed for the copper-thiomolybdate complex, less copper would be available in ruminants. Since the sulfide supply is higher in ruminants than in monogastrics, a greater portion of the copper may be available as the poorly utilizable copper-thiomolybdate complex. This would lead to the aggravation of the copper deficiency in ruminants.

Fluoride, Calcium and Phosphate

Animal studies have shown that calcium, phosphorous, magnesium and aluminum can reduce the intestinal absorption of fluoride. A high calcium intake has decreased the absorption of fluoride in dogs and in rats and both calcium and phosphorous reduce the skeletal uptake of fluoride in rats more than did a high calcium intake alone. Fluoride stabilizes the bone crystal and fluoride has been shown to decrease bone resorption. Fluoride has been shown to be a desirable form of treatment for osteoporosis. Aluminum antacids complex both phosphorous and fluoride in the intestine. This complexation can also lead to a decrease of the intestinal absorption of fluoride. The inhibitory effect of the intestinal absorption of fluoride in conjunction with the antacid-induced calcium loss may have an increased effect in promoting the development and progression of bone loss. This may become important to middle-aged and elderly people who may use unknown amounts of aluminum-containing antacids and may also have an inadequate intake of calcium.

References

1. Van Campen, D.R.: J. Nutr. 97, 104-8 (1970).
2. Evans, G.W., Grace, C.I., Hahn, C.: Bioinorg. Chem. 3, 115-120 (1974).
3. Murthy, L., Klevay, L.M., Petering, H.G.: J. Nutr. 104, 1458-1465 (1974).
4. Van Campen, D.R., Scaife, P.U.: J. Nutr. 91, 473-6 (1976).
5. Ketcheson, M.R., Barron, G.P., Cox, D.H.: J. Nutr. 98, 303-311 (1969).
6. O'Dell, B.L., Reeves, P.G., Morgan, R.J.: In: Hemphill, D.D., ed. Trace Sub. in Environ. Health. Columbia: University of Missouri X, 411-442 (1976).
7. Hall, A.C., Young, B.W., Bremmer, I.: J. Inorg. Biochem. 11, 57-66 (1979).
8. Brewer, G.J., Hill, G.M., Dick, R.D., Prasad, A.S., Cossack, Z.T.: J. Amer. Coll. Nutr. 4, 33-38 (1985).
9. Brewer, G.J., Brewer, L.F., Prasad, A.S.: J. Lab. Clin. Med. 90, 549-554 (1977).
10. Prasad, A.S., Brewer, G.J., Schoomaker, E.B.: JAMA 240, 2166-2168 (1978).
11. Brewer, G.J., Hill, G.M., Prasad, A.S., Cossack, Z.T., Rabbani, P.: Ann. Intern. Med. 99, 314-320 (1983).
12. Allen, K.G.D., Klevay, L.M.: Atherosclerosis 29, 81-93 (1978).
13. Lipid Research Clinics Program: JAMA 251, 351-364 (1984).
14. Kannel, W.B., Castellani, W.P., Gordon, T., McNammara, P.M.: Ann. Int. Med. 74, 1-12 (1971).
15. Miller, G.J., Miller, N.E.: Lancet 1, 16-19 (1975).
16. Allen, K.G.D., Klevay, L.M.: Nutr. Rep. Int. 22, 295-299 (1980).

17. Klevay, L.M.: Am. J. Clin. Nutr. 26, 1060-1068 (1973).

18. Klevay, L.M.: Nutr. Rep. Internal 22, 617-621 (1980).

19. Woo, W., Gibbs,D., Hooper, P.L., Garry, P.J.: Am. J. Clin. Nutr. 34, 120-121 (1981).

20. Lei, K.: Nutr. Rep. Internal 15, 597-605 (1977).

21. Medeiros, D.M.: Biochem. Archiv. 1, 67-73 (1985).

22. Klevay, L.M.: Am. J. Clin. Nutr. 28, 764-774 (1973).

23. Holden, J.M., Wolf, W.R., Mertz, W.: J. Am. Dietet. Assoc. 75, 23-28 (1979).

24. National Research Council, Recommended Dietary Allowances, Washington, D.C., 1980.

25. Thorling, E.B., Thorling, K.: Cancer 38, 225-231 (1976).

26. Hrgovcic, M., Tessmer, CF., Thomas, F.B. Ong, P.S., Gamble, J.F., Schullenberger,C.C.: Cancer 32, 1512-1524 (1973).

27. Alexander, F.W., Delves, H.T., Lay, T.: Arch. Dis. Child. 47, 671-675 (1972).

28. Davis, L.J.T., Musa, M., Dormany, T.L.: J. Clin. Pathol. 21, 359-65 (1968).

29. Andrews, G.S.: J. Clin. Pathol. 32, 325-333 (1979).

30. Strain, W.H., Mansour, EG., Flynn, A., Pories, W.J., Tomaro, A.J., Hill, O.A.: Lancet 1, 1021-1022 (1972).

31. Fisher, G.L., Byers, V.S., Shifrine, M., Levin, A.S.: Cancer 37, 356-362 (1976).

32. Inutsuka, S., Aaraki, S.: Cancer 42, 626-631 (1978).

33. Lightman, A., Brandes, J.M., Binur, N., Drugan, A., Zinder, O.: Clin. Chem. 32, 101-103 (1986).

34. Parizek, J., Ostadalova, I., Benes, I., Pitha, J.J.: Reprod. Fert. 17, 559-592 (1968).

35. Gunn, S.A., Gould, T.C., Anderson, W.A.D.: Proc. Soc. Exp. Biol. Med. 128, 591-595 (1968).

36. Chen, R.W., Whanger, P.D., Weswig, P.H.: Bioinorg. Chem. 4, 125-133 (1975).

37. Jamall, I.S., Smith, J.C.: Arch. Toxicol. 58, 102-105 (1985).

38. Gerhardsson, L., Brune, D., Nordberg, I.G.F., Wester, P.O.: Brit, J. Indust. Med. 42, 617-626 (1985).

39. Shamberger, R.J., Bratush, C.M.: Trace Substances in Environmental Health Hemphill, DD., Ed., Columbia, Missouri XIV, 203-210 (1980).

40. Shamberger, R.J., Gunsch, M.S., Willis, C.E., McCormack, L.J.: Trace Substances in Environmental Health. Hemphill, DD., Ed., Columbia Missouri XII, 48-52 (1978).

41. Dubois, K.P., Moxon, A.L., Olson, O.E.: J. Nutr. 19, 477-482 (1940).

42. Rhian, M., Moxon, A.L.: J. Pharmacol. Exp. Therap. 78, 249-264 (1943).

43. Moxon, A.L., Rhian, M.A., Anderson, H.D., Olson, O.E.: J. Anim. Sci. 3, 299-309 (1944).

44. Wahlstrom, R.C., Kamstra, L.D., Olson, O.E.: J. Anim. Sci. 14, 105-110 (1955).

45. Moxon, A.L., Wilson, W.O.: Poult. Sci. 23, 149-151 (1944).

46. Klug, H.L., Lampson, G.P., Moxon, A.L.: Proc. S. Dak. Acad. Sci. 29, 57-65 (1950).

47. Palmer, I.S., Bonhorst, C.W.: J. Agr. Food Chem. 5, 928-930 (1957).

48. Kamstra, L.D., Bonhorst, C.W.: Proc. S. Dak. Acad. Sci. 32, 72-74 (1953).

49. Hill, C.H., Matrone, G.: Fed. Proc., 29, 1474-1481 (1970).

50. Hill, C.H.: In: Trace Elements in Human Health and Disease, II (A.S. Prasad, ed.) Academic Press, New York p. 281-300, 1976.

51. Nogamura, A., Imura, N,: Res. Commun. Chem. Pathol. Pharmacol. 27, 163-173 (1980).

52. Hartman, R.H., Matrone, G., Wise, G.H.: J. Nutr. 57, 429-439 (1955).

53. Robinson, N.W., Hansard, S.L., Johns, D.M., Robertson, G.L.: J. Anim. Sci. 19, 1290 (1960).

54. Matrone, G., Hartman, R.H., Clawson, A.J.: J. Nutr. 67, 309-317 (1959).

55. Thomson, A.B.R., Valberg, L.S., Sinclair, D.G.: J. Clin. Invest. 50, 2384-2394 (1971).

56. Pollack, S., George, J.N., Reba, R.C., Kaufman, R.M.: J. Clin. Invest. 44, 1470-1473 (1965).

57. Greger, J.L., Zaikes, S.C., Abernathy, R.P., Bennett, O.A., Huffman, J.: J. Nutr. 108, 1449-1456 (1978).

58. Solmons, N.W., Jacob, R.A.: Am. J. Clin. Nutr. 34, 475-482 (1981).

59. Solmons, N.W., Pineda, O., Viteri, F., Sandstead, H.H.: J. Nutr. 113, 337-349 (1983).

60. Hill, C.H.: In: Trace Elements in Human Health and Disease, Prasad, A.S., ed. Vol. II, New York: Academic Press p. 281-309 (1976).

61. Davies, N.T., Campbell, J.K.: Life Sci. 20, 955-960 (1977).

62. Campbell, J.K., Davies, N.T., Mills, C.F.: In: Trace Element Metabolism in Man and Animals - 3. M. Kirchgessner, ed., ATW, Weihenstephan, Germany. p. 553-556 (1978).

63. Lease, J.G.: J. Nutr. 96, 294-302 (1969).

64. Cotzias, G.C., Papavasilou, P.S.: Amer. J. Physiol. 206, 787-792 (1964).

65. Parizek, J.: J. Endocrinol. 15, 56-63 (1957).

66. Parizek, J., Benes, I., Kalowskova, J., Babicky, A., Lener, J.: Physiol. Bohemoslor 18, 89-93 (1969).

67. Marilese, N.A., Ammerman, C.B., Valsecchi, R.M., Dunavant, B.G., Davis, G.K.: J. Nutr. 99, 177-183 (1969).

TOXIC ACTIONS OF ESSENTIAL TRACE ELEMENTS (Mo, Cu, Zn, Fe, Mn)

Manfred Anke, Bernd Groppel

Wissenschaftsbereich Tierernährungschemie der Karl-Marx-Universität Leipzig, G.D.R.-6900 Jena

Introduction

In the course of the absorption of inorganic soil, water, and food components by flora and fauna which lasted millions of years, the majority of these elements have probably taken over biological functions as components or activators of enzymes, proteins or other organic compounds. This hypothesis is supported by the fact that biological functions have been demonstrated experimentally with semisynthetic rations in the elements As (18), Pb (87) and Cd (131, 4, 23), which had only been known as highly toxic.

With the exception of Co, which is needed as a component of vitamin B_{12}, deficiency symptoms in humans were shown in classical trace elements (table 1) (133).

Table 1: Essential trace elements

classical trace elements:	Fe, J, F, Cu, Mn, Zn, Co, Mo, Se, Cr
new trace elements:	Si, Sn, V, Ni, As, Cd, Li, Pb

These deficiency symptoms only occurred in some of them in the case of parenteral nutrition or gene defects (Mo, Mn). It is astonishing that the knowledge of the essentiality of F is also extremely fragmentary. According to the present level of knowledge, the new trace elements are available so plentifully in feed and food stuffs that no primary deficiency symptoms are to be expected in case of normal nutrition (Ni, Li, Cd, Pb, V) (6).

In the case of elements essential for the fauna it must be distinguished between a deficient offer, an optimum supply and a toxic exposure with the corresponding transition ranges (100).

Trace Element – Analytical Chemistry in Medicine and Biology, Vol. 4

The pharmaco-dynamic or therapeutic range between the optimum and toxic trace element offer in some elements is dependent on the species (Fig. 1). Well-known examples for this effect of inorganic components of the diet are As in poultry and Cu in pigs. This therapeutic efficiency stimulus exceeding the nutritive effect has not yet been described for other elements and species of animals.

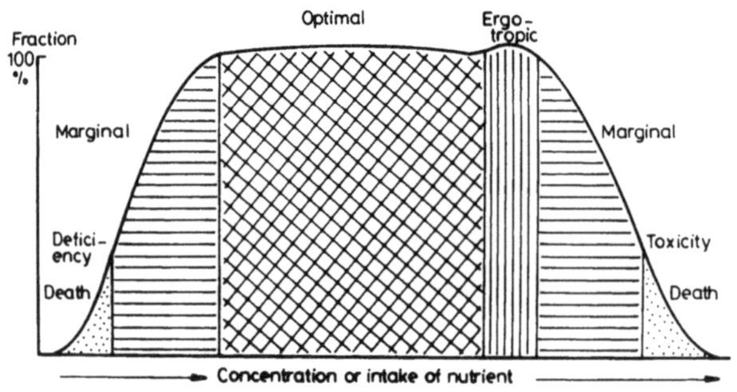

Fig. 1: Dependence of performance on the supply with trace elements

The toxic effect of elements is species-specific. The results obtained in one species cannot be transferred to another without experimental testing. Cattle e.g. are extremely sensitive to an exposure of 10 ppm Mo in the ration and react with molybdenosis, sheep tolerate the threefold, goats the thirtyfold Mo amount without difficulties (14). Furthermore, tolerance of and sensitivity to trace element exposures are influenced by genetic polymorphism of species and various gene defects.

Antagonistic interactions between the elements but also influences of organic compounds vary the trace element utilization highly significantly (Fig. 2) and are practically efficient. The S, Mo, Ag, Cd, Fe, and Zn offer e.g. is influenced by the effects of a Cu exposure of sheep. Therefore, the knowledge of the offer of the trace elements only renders restricted information about the supply or needs to be expected or about exposure. Details of antagonistic influences are dealt with in individual essential elements. The possibilities shown in Fig. 2 are not exhausted with that (K, Na, Li, Rb etc.) (7).
In various cases, the toxicity of essential trace elements is also drastically changed by their biotransformation. Trimethyl-

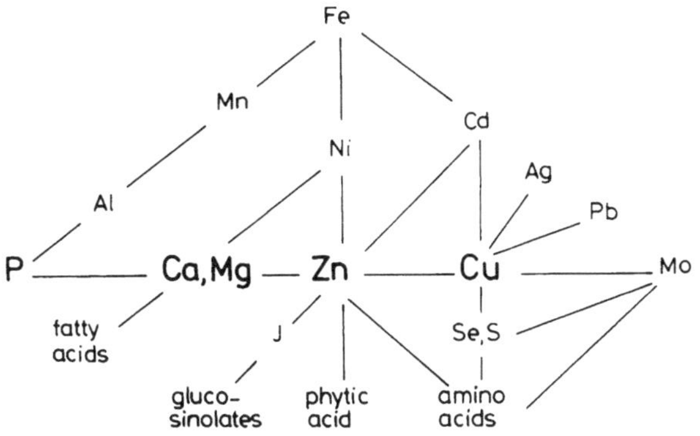

arsine e.g. is highly toxic, whereas the As of the food chain of oceans is transformed into arsenobetaine by their flora which is harmless for the fauna and excreted renally (6).

Fig. 2: Schematic representation of possible interaction between elements and other components of nutrition

The danger of trace element exposure of animals and humans has increased in the 19th and 20th centuries due to the industrial utilization and emission of these elements. Most various trace elements became components of commodities and packing materials without the effects of this utilization having been tested over a long period under the aspect of biotransformation.

Molybdenum

The heavy metal Mo became important for the fauna when Ferguson et al. (1938) found out that the teat disease of cattle of certain sites in England was caused by an ample Mo absorption.

The endemic disease is characterised by a severe scouring and loss of condition in cattle. Mo exposure leads to Cu depletion and secondary Cu deficiency. In this connection, the Cu and S supply of animals is of great importance. Inorganic Mo compounds, in the presence of sulphide, are converted by micro-organisms to thiomolybdates (e.g. tetrathiomolybdate MoS_4^{-2}) in the rumen. Thiomolybdates rapidly react with particulate matter and proteins to form complexes that bind Cu strongly (141, 50, 101, 62). Rat bioassays indicated that Cu became completely unavailable (117).

The influence of high Mo concentrations in the ration can lead to bone deformities in cattle, sheep, rabbits, and rats (44, 79, 76, 153, 12). This skeleton damage occurs in ruminants in molybdenosis areas. It can also be produced experimentally. The reason for its occurence is secondary Cu deficiency as could be shown in cattle (table 2). Cu deficiency leads to such skeleton damage. Mo exposure also reduces the fertility of heifers and cows (112). But this does not seem to be due to an effect of secondary Cu deficiency, but to a particular effect of Mo exposure. The reduction of the ovulatory luteinizing hormone peak suggests that events controlling oestrus have been altered by Mo either by affecting LH secretion directly or as an indirect consequence of the inhibitory effect of molybdate upon oestrogen receptor activity demonstrated by in vitro studies (48, 134).

Table 2: Copper content of liver, cerebrum and serum of unexposed and Mo-exposed milk cows (mg/kg dry matter) (12)

	unexposed		Mo-exposed			
organ, n	s	$\bar{x}$	$\bar{x}$	s	p	%[1)]
liver (100; 15)	28	160	7,1	4,0	<0,001	4
cerebrum (100; 15)	4,8	11	6,4	2,3	<0,001	58
blood serum (100; 15)	0,32	0,91	0,46	0,22	<0,001	51

1) unexposed ≙ 100%, Mo-exposed ≙ x %

Apart from the described Mo exposure and secondary Cu deficiency symptoms, Mo offers to bulls led to the loss of the Libido sexualis, to damage of interstitial cells and germinal epithelium (142). These investigations must be confirmed experimentally . Mo seems to become effective in vitro via the inhibition of androgen receptor complexes as well (147).

According to the same scheme, high Mo offers in vitro also inactivate non-activated and activated glucocorticoid receptor complexes of rat livers (106). These Mo influences need further investigations and are contradictory. The Mo effect on the glucocorticoid receptor was only registered in cell-free systems and not in intact lymphocytes (107).

There are only limited and not very reliable data on the effects of Mo exposure on humans. The population of Ankavan (Armenian Socialist Republic) consume twice to 10 times more Mo via local food-stuffs (10 to 15 mg Mo/day) than people in areas with normal Mo supply (92). Apart from increased Mo serum values, these people also had an increased xanthine dehydrogenase activity and uric acid concentration in the blood. Gout symptoms were connected with the higher uric acid amounts in the blood of the majority of examined persons (92). Workers of a molybdenite roasting plant who absorbed Mo with dust had obviously increased ceruloplasmin and serum uric levels. Mo absorption from the dust was demonstrated by increases in plasma and urinary Mo levels. Further investigations into the effects of along-term Mo exposure are necessary (151).

The high incidence of esophageal cancer among Bantu in Transkei in Southern Africa was attributed to the consumption of the food locally gronw in soil deficient in Mo.
Further epidemiological studies in the People's Republic of China showed the Mo content of the drinking-water and food of the areas

Table 3: The biological importance of molybdenum poisoning

Mo exposure toxic effects on animals	Ferguson et al, 1938 - Pasture diarrhoea, teat, molybdenosis of cattle - >3 ppm Mo -> secondary Cu deficiency, dependent on Cu and S content of the ration - Bone deformities - Infertility in heifers and cows - Inhibition of oestrogen receptor activity in vitro - Lack of sexual interest Damage to interstitial cells and germinal epithelium - Inhibition of androgen receptor complexes in vitro

(Table 3 continued)

toxic effects on humans	- Endemic gout in the Ankavan Province (Armenian Socialist Republic)
	- High serum uric acid level and tissue xanthine oxidase activity, gout? Inhibition of carcinogenesis?
Species specificity	- Cattle 3 ppm, > sheep, > goat (300 ppm), > pig (1000 ppm)
antagonists	W, Cu
Mo requirement	- < 100 µg/kg ration dry matter
Mo absorption humans	- < 350 µg/day
indicator organs	- Serum, spleen, liver, kidney, hair, feathers

high in esophageal cancer incidence to be lower than that of low-incidence areas (27). Yang et al. (1985) were able to produce inhibited NSEE (N-nitrososarcosine ethyl ester) induced esophageal and forestomach carcinogenesis by giving 2 and 20 ppm Mo to mice. 10 ppm Mo showed the same effect in virgin inbred female Sprague Dawley rats with 50 mg N-nitroso-N-methylurea (NMU)kg body weight. The Mo offer demonstrates the inhibitory effect on mammary carcino-genesis. This Mo influence needs further investigations as well.

The effect of Mo exposure is species-specific. Cattle reacted to Mo exposure most sensitively. They were followed by sheep and other kinds of wild ruminants. Horse and pig tolerated more Mo. In our experiments goats proved to be particularly Mo-tolerating. Lucerne with 300 ppm Mo, which immediately causes molybdenosis in moufflons (Ovis musimon), was consumed without any negative effects as were rations with 1000 ppm Mo for weeks. The high tolerance of goats towards the excessive dietary Mo does not result from insufficent Mo absorption (14) (table 4).

Table 4: The reflection of molybdenum exposure (1000 mg Mo/kg ration dry matter) by several organs of adult he-goats (mg/kg dry matter) (15)

organ, n	control animals s	$\bar{x}$	Mo-exposed $\bar{x}$	s	p	multiplication
blood serum (4;5)	0,003	0,018	21	8,7	<0,001	1167
spleen (33;5)	0,29	0,55	109	23	<0,001	198
rib (23;6)	0,49	0,33	55	28	<0,001	167
skeleton muscle (31;5)	0,051	0,067	9,0	3,1	<0,001	134
kidneys (33;5)	0,44	0,90	111	35	<0,001	123
testicles (5;5)	0,24	0,49	52	14	<0,001	106
cerebrum (12;5)	0,11	0,22	23	18	<0,001	105
cardiac muscle (11;5)	0,14	0,24	24	10	<0,001	100
pancreas (8;5)	0,15	0,38	35	18	<0,001	92
lungs (26;5)	0,35	0,63	54	21	<0,001	86
liver (35;4)	0,78	1,2	70	74	<0,001	58

All organs of Mo-exposed goats and the blood serum accumulated Mo. The mechanism of the Mo tolerance of goats is still to be explained. It is not connected with Cu metabolism.

W and Cu have antagonistic effects on Mo metabolism. It is true that long-term W offers to Mo-exposed ruminants inhibit molybdenosis, but they lead to anaemia at the same time. Therefore, their long-term application cannot be recommended (64).

The Mo requirement of humans and animals amounts to <100 µg/kg ration dry matter. In Central Europe this requirement is met by food consumption. Therefore, primary Mo deficiency of the fauna is not to be expected in this area (16). 20 years ago the Mo consumption of adults amounted to <350 µg/day in the U.S.A. (129). Babies from India who were suckled or fed on commercial milk formulae, took in 6.4 or 12 to 14 µg Mo/day at the age of 11 or 14 weeks. The Mo content of water and the Mo absorption via water varies considerably (0.2 - 19 µg/day) (49).

In Central Europe the weathering soils of granite, gneiss, the lower strata of new red sandstone, slate and moor or peat can produce a Mo-rich vegetation and can lead to Mo exposures of domestic and wild ruminants (12, 26). Mo emission by Mo-processing or -using industrial enterprises is possible as well.

The antagonistic effect of Mo on Cu metabolism and its species-specifity are particularly characteristic of Mo exposure.

Copper

Cu intoxications of the fauna were investigated best in sheep since this species reacts particularly sensitively. Acute Cu poisoning occurs in them after intake of 20 mg Cu/kg live weight. Death can occur within a short time.

In the case of chronic Cu poisoning, the pathways are distinguished by which excessive Cu can be stored in the liver.

In "simple chronic copper poisoning", the source of the excess Cu can be determined. Some of the more common sources are: forages high in Cu, mineral mixes formulated for pigs, footpath containing copper sulfate, or drinking water that has been treated to control algae.

In "hepatogenous chronic Cu poisoning", the liver has an increased affinity for, and increased storage of Cu because of injury by toxic alkaloid of plant origin. This type is seen only in sheep. The plant most often involved is Heliotropium europaeum, but Senecio jacobaea is also implicated. A specific metabolic type of liver damage occurs by which the affinity for Cu increases. Due to mycotoxines of fungus Phomopsis leptostromiformis there can be also liver damage, Cu accumulation and Zn depletion if the narrow-leaved sweet lupine is fed (46, 47).

"Phytogenous chronic Cu poisoning" occurs in animals grazing on pastures containing normal concentrations of Cu (10 ppm Cu), but a deficiency of Mo. the absence of Mo increases the absorption of Cu. 1 to 6 weeks before the crisis, serum aspertate transaminase concentration rises and spongy degeneration of the midbrain, pons, and cerebellum occurs. The Cu concentration in most of the hemolytic crises reaches 1 to 3.000 ppm in the liver. Methemoglobin concentration rises to 5 g/100 ml. An abrupt increase in blood Cu 24 to 48 hours before clinical signs appear and an elevated serum glutamicoxacetic transaminase concentration is seen in chronic cases 4 to 6 weeks in advance of the crises. The clinical course is 1 to 4 days (35, 108).

Hardy et al. (1975) point to the increased occurrence of hepatic abnormalities ranging from focal hepatitis to micro- and macronodular cirrhosis and abnormally increased concentrations of Cu in the liver of Bedlington terriers (148). Johnson et al. (1980) investigated this disorder's genetic basis. The abnormally high concentration of hepatic copper (>850 ppm) underlying the disorder was found to follow an autosomal recessive pattern of inheritance. The bulk of this excess Cu was sequestered in lysosomes. When the Cu concentration exceeded 2.000 ppm in the liver, progressive signs of functional and morphologic disturbance appeared as focal hepatitis, chronic active hepatitis and ultimately as cirrhosis (148). The disease was latent in some dogs for many years, but led to acute or chronic hepatic disease and death in others early in life. In the meantime, hereditary Cu toxicosis has also been detected in West Highland white terriers (144) and doberman pinschers (145). The diagnosis of Cu toxicosis is carried out by liver biopsy. Affected dogs show obvious Cu-containing granules in the centrilobular and midzonal hepatocytes. Treatment is possible with penicillamine.

Toxic milk is a new autosomal recessive mutation in mice (120). The syndrome induces poor growth and ultimately death at 2 weeks of age. These features are consistent with Cu deficiency. Such infants are rescued by the administration of Cu. The mutants themselves amass high Cu concentrations of hepatic Cu.

The same gene defect occurs as Wison disease in humans (McKusick 27790). Before puberty the initial signs and symptoms are almost always realted to liver damage with or without hemolysis. There are reduced Cu and coeruloplasmin concentrations in the plasma.

After puberty neurological signs become increasingly prominent. The most characteristic physical sign is the Kayser Fleischer corneal pigment ring from deposition of Cu in Descemete membrane. The disease seems to result from a failure by the hepotocytes to excrete Cu via the bile. As a result this metal accumulates first in the liver, then, when this organ is saturated, in the brain, the cornea and, to a lesser extent, in the kidneys (154).

Another disease connected with an extreme Cu accumulation in the liver is the Indian childhood disease described by Portmann et al. (1978). But it can also occur in white U.S. Americans (95). It leads to a particularly high Cu content in the liver and is probably hereditary as well. The clinical and pathological features, especially the age at onset and the absence of Kayser Fleischer corneal rings distinguish the Indian childhood disease from the Wilson disease.

The Cu toxicity is extremely species-specific. In the case of normal amounts of Cu antagonists, sheep tolerate 10 to 20 ppm Cu. Cattle are by far less sensitive. Szücs et al. (1982) fed 100 ppm

Table 5: Liver copper concentrations in various conditions (mg/kg dry matter) (25,95)

normal, adult	22
newborn (1st year of life)	81
Wilson's disease	728
Indian childhood cirrhosis	1832

Cu to fattening bulls without Cu intoxications occurring. If 250 mb Cu are supplemented to the ration, pigs show a mean increased weight gain of 8%. In this species, there is an ergotropic effct of Cu within this range (31). Hens tolerate 300 ppm Cu without the laying performance being reduced. $CuSO_4$ decreased feed-intake and all performance criteria more than CuO (81,138). Geese and swans seem to be more sensitive to Cu exposure. 600 ppm Cu in the ration led to acute intoxications in Canada geese (Branta canadensis) (75). Wild mute swans (Cyngus olor) also died of acute Cu intoxication with liver copper values of 208 to 9880 ppm

(105). Ducks store much more Cu in the liver than hens in case of the same Cu-offer (156). Rats have been shown to maintain normal growth and health on diets containing 500 ppm Cu (149). The Cu toxicity in the marine fauna and amphibians varies in a similar way.

Cu utilization is influenced by a number of Cu antagonists. They can produce secondary Cu deficiency or limit the damaging effect of Cu exposure.

S is the practically most effective Cu antagonist. The S of the feed is microbiologically metabolised into sulphide in the rumen of ruminants (Fig. 3). Due to the over-supply with sulphide S almost all of the Cu is bound as CuS. CuS cannot, or to a limited extent only, be absorbed. Cu resorption takes place in the stomach and the first parts of the small intestine. Therefore, only species with microbial rumen digestion suffer from S-induced deficiency. Monogastric herbivores (horse, donkey, rabbit, hare) do not fall ill with this form of secondary Cu deficiency since the metabolism of dietary S into sulphide S only takes place in the appendix, i.e. after the absorption place of Cu. The same S-effect can avoid Cu intoxication in sheep (140). Feed- and foodstuffs from the vicinity of copper mies or Cu-processing enterprises can considerably accumulate Cu and contain the three- to fourfold Cu amount (table 6).

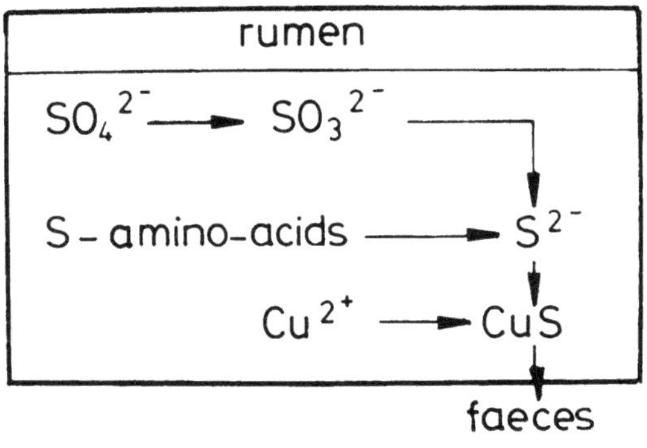

Fig. 3: Model of the interrelations between S and Cu in ruminants

Table 6: The copper content of several species of plants from copper-exposed and non-exposed areas (mg/kg dry matter) (12)

species, n	non-exposed s	$\bar{x}$	Cu-exposed $\bar{x}$	s	p	%
meadow grass (53;21)	4,1	11	44	41	<0,001	436
green rye (280;8)	1,3	3,9	14	9,5	<0,001	359
green wheat (314;14)	1,6	3,9	12	7,8	<0,01	308
acre red clover (3185;15)	3,0	9,4	27	14	<0,001	287
lettuce (19;9)	5,0	14	36	13	<0,001	257

In spite of this Cu exposure, sheep from these areas do not suffer from Cu intoxications since, due to the emission of S, Mo, Cd, Pb and Zn, feed-stuffs accumulate these Cu antagonists at the same time (12). Liver, cerebrum and wool of these sheep showed normal in one case also subnormal Cu concentrations (table 7). Only the Cu content of one flock was high. This finding points to the exogenous contamination of the fleece.

The high Cu, S, Fe, and Zn content of fertilized pig slurry containing high levels of Cu has similar effects on sheep pastures. In spite of the high Cu content in grass (30 - 100 ppm), sheep fed on it did not fall ill with chronic Cu poisoning (91, 114). The same effect of high S, Mo, and Zn amounts in the ration on trhe prophylaxis of Cu poisoning in sheep was found when seaweed was fed (97). Methionine offers reduce the accumulation of Cu in

Table 7: The copper content of several indicator organs of unexposed sheep (n 88) and of those from copper exposed areas (n=16) (mg/kg dry matter) (12)

organ		non-exposed s	$\bar{x}$	Cu-exposed $\bar{x}$	s	p	%
	A			40	16	>0,05	18
liver		251	220				
	B			290	86	>0,05	132

(Table 7 continued)

	A			19	5,4	>0,05	106
cerebrum		7,1	18				
	B			22	6,2	>0,05	122
	A			22	7,4	>0,01	220
wool		8,2	10				
	B			12	3,0	>0,05	120

the liver of monogastric animals as well (51, 124, 84).

Cd influences the Cu accumulation in liver and cerebrum of most various species very intensively (19, 20). Cd is a distinct Cu antagonist which can induce Cu deficiency symptoms. It cannot be used as prophylaxis for Cu poisoning because of its damaging effect. It accumulates in kidneys and liver independent of the Cu offer (19, 20, 99).

The influence of high Fe supply on Cu storage was summarized by Kirchgessner et al. (1979). High Fe amounts in the ration generally reduced the Cu level of the liver (67, 81, 36).

The Zn influence on Cu storage is much more obvious than that of Fe. Bremner et al. (1976) were able to avoid Cu poisoning by 420 mg Zn/kg lamb ration. These findings were repeatedly confirmed and are of importance for humans as well. Hypocupremia occurred in an adult with sickle cell anemia when high Zn amounts were given (116). The same happened in patients with Acrodermatitis Enteropathica after Zn therapy (80). Zn exerts its antagonistic effect by inducing the synthesis of copper-binding ligand, probably a thionein, in the mucosal cells which sequesters Cu from the nutrient medium, making it unavailable for seresal transfer (55). Zn also proved a success in the treatment of the Wilson disease (77, 40, 39). Zn also protects from pyrrolizidine alkaloid-induced hepatotoxicity from transy ragwort (Senecio jacobaea). The protective effect was associated with a marked increase in the liver metallothionein (104).

Ag supplementation can also reduce Cu incorporation. Ag belongs to the distinct Cu antagonists. Its practical application remains controversial (90, 93, 69).

The Cu requirement of ruminants is the highest with 8 mg/kg ration dry matter. Monogastric animals can do with 4 to 5 mg/kg. Infants need 50 to 80 µg Cu/kg body weight, adults 30 µg/kg body weight. Cu antagonists influence Cu utilization to such a degree that it is recommendable to reckon with the so-called Cu minimum requirement. It only indicates that the possibility of Cu deficiency must be taken into account if the Cu concentrations are lower than the minimum requirement (6, 5).

Adolescents and adults can do with 2 mg Cu/day if the offer of Cu antagonists is low. Most daily rations from England, New Zealand and the U.S.A. contain 2 to 4 mg Cu. In the GDR, the FRG, Netherlands and Scotland 1 to 2 mg Cu were taken in by adults without Cu deficiency becoming manifest. Patients with parenteral nutrition should be given 0.3 mg Cu/day (6, 127).

Table 8: The biological importance of copper poisoning

Cu poisoning, acute:	Gastroenteritis, passage of bluish green fluid feces, shock and death of sheep (20 mg Cu/kg live weight)
chronic:	<500 ppm Cu in liver is normal, 800 ppm Cu. → possibility of a hemolytic crisis, accelerated breathing, weakness, excessie thirst, dark brown urine, skin and membranes colored in a bright yellowish orange, 1 to 3.000 ppm Cu in liver
inherited Cu toxicosis/animal:	Bedlington terriers, West Highland white terriers, Doberman pinschers with 850 ppm Cu in liver, normal concentration 200 ppm, in Bedlington terriers 66% of the population, toxic milk in mice

(Table 8 continued)

humans:	Morbus Wilson in humans, first symptoms at the age of 10 to 15 (5 to 40 years), incidence no less than 29 per million; treatment: penicillamine, triethylene tetramine, Zn; Indian childhood cirrhosis, absence of Kayser Fleischer corneal rings
species specifity:	sheep 10 - 20 ppm, cattle (100 ppm), pig (250 ppm), hen (300 ppm), rat (500 ppm)
antagonists:	S, Mo, Cd, Fe, Zn, Ag
Cu requirements:	ruminants 8 mg/kg ration dry matter, monogastric animals 5 mg/kg
Cu intake by humans:	2 - 3 mg/day for adolescents, adults 1 ... 4 mg/day
indicator organs:	liver, cerebrum, blood serum, hair

Zinc

Zn belongs to the elements which are almost harmless to mammals and birds. Zn poisoning has been reported in man as a result of consuming acid foods which were cooked in galvanized vessels and in piglets after being fed on skim milk passed through galvanized pipes (149). The intakes necessary to produce deleterious effects are so large that a wide margin of safety exists with this element.

Thus Zn intakes of 2500 ppm are without discernible effects on rats over several generations; at 5000 ppm Zn, growth is severely depressed with heavy mortality in young animals when ingested as chloride and it is slightly depressed with little mortality when ingested as oxide (73). Rats with 7500 ppm Zn showed growth depression, reduced liver values in Cu and Fe and decreasing hemoglobin concentrations. The Ca, Mg, and P content of their skeletons were significantly reduced. The decrease of the Ca and Mg

content in bones remained even if the Zn supply was normal. The other parameters, however, normalized (132). Excess dietary Zn has been reported to cause histological changes in the adrenal glands of rats (125, 118).
These findings were confirmed in C3H mice. the Zn content of liver, spleen and skin was not altered. Plasma insulin and glucose concentration were not significantly different from control values (29). Swine are not influenced by 500 ppm Zn and tolerate 5000 ppm Zn as ZnO; their copper metabolism was negatively influenced (78).

Broilers and laying hens exhibited a similar tolerance to Zn of 1200 to 1400 ppm of the diet. At 3000 ppm Zn, growth and appetite depression occur (110). In adult minks 1000 ppm Zn in the ration did not lead to Zn poisoning and secondary Cu deficiency. The same Zn amount, however, caused achromatrichia, lymphogenia, alopecia and a reduced growth rate in the offsprings of Zn-treated mothers. Ferrets died with manifest nephrotoxicity after offers of 3000 ppm Zn (139).

After deficient dialysis, subacute Zn toxicity and anaemia may occur in humans which disappear after the removal of Zn contamination (111). An increase of the serum zinc concentration to 7000 µg/l was observed in other dialysis patients. Nauses, vomiting, abdominal pains, fever and heavy anaemia occurred after dialysis (60).

Symptoms of an oral Zn poisoning in humans are fever, vomiting, gastric spasms, and diarrhoea. The inhalation of Zn-rich smoke leads to "metal smoke fever". Clinical symptoms are fever and shivers. Repeated exposure, however, does often not lead to further reactions (103).

There is apparently a heritable hyperzincemia in humans in Zn as well. Members of the family had Zn concentrations of 2.5 to 4.4 mg/l without the organ zinc values considerably deviating from the standard. There were no clinical symptoms (135). A patient with the Chediak-Higashi-syndrome also had increased plasma, erythrocyte, lymphocyte and granulocyte Zn concentrations (109).

In fish (salmon), too, the Zn content of water (242 µg/l) did not influence survival, fertility, fecundity and growth (52, 43).

It is possible that Zn exposure inhibits the calmodulin of cells. Data could be found in vitro. Ca stimulates calmodulin (38).

Compared to Mo and Cu, species specificity on Zn exposure is not at all prominent. The registered differences are hardly reproducible. The homeostatic control of the Zn level also avoids acute Zn poisoning without human influence.

The effects of Zn antagonists was summarized by Mills et al. (1985). Ca multiplies the antagonistic influence of phytine in monogastric animals and humans. As a rule, Cd, Cu, and Fe influence Zn utilization antagonistically (1).

Under the influence of Zn antagonists, the Zn requirement of humans and animals varies extraordinarily. It may be overstated (table 9) (5, 6).

All over the world, the Zn consumption of women varies between 10 and 15 mg/day and thus is below the assumed requirement. The same Zn amounts of 11 and 11 - 12 mg Zn/day are taken in in the FRG and the GDR (21, 127).

Due to the intensive homeostatic control it is complicated to determine the Zn status. The Zn content of skeleton, testicles, and hair of animals gives certain information. Blood reflects the Zn status only in case of stress-free blood-letting (24, 89, 124).

The identification of Zn exposure by organ analysis might be as uncertain as that of a marginal Zn deficiency. There was no statistically relevant increased Zn accumulation in ribs and hair of cattle, sheep and horse in spite of a threefold higher Zn offer via the feed-stuffs of the living area (table 10).

Table 9: The biological importance of zinc poisoning

Zn poisoning:	Heller and Burke (1927)
animals:	2500 mg Zn/kg ration dry matter without influence in rats over several generations; growing rats show disturbances of the Cu, Fe, Ca, Mg, and P metabolism; histological changes in the adrenal gland; poultry reacts with appetite depression at 3000 mg Zn/kg ration; the offsprings of minks with 1000 mg Zn/kg ration dry matter showed worse growths; ferrets (Mustella putoris furo) died with manifest nephrotoxixicity after offers of 3000 mg Zn/kg ration;
humans:	Due to deficient dialysis, anaemia caused by subacute Zn toxicity can occur in patients; fever, vomiting, gastric spasms, diarrhoea;
heritable hyperzincemia:	Zn values of the plasma of this family from 2.5 to 4.4 mg/l; no deviation from organ zinc concentrations;
species specifity:	not prominent; poultry, >minks, >rats, >pigs, academic differences;
antagonists:	phytine, Ca, Cd, Cu, Fe;
Zn requirements:	about 30 - 40 mg/kg ration dry matter, pig, hen 40 mg/kg, babies 3 mg/day, 1.0 - 10 years of age 10 mg/day, adults 15, pregnant women 20 mg, nursing women 25;
Zn intake of humans:	10 to 15 mg/day; 10 - 24 mg/day;
indicator organs:	rib, hair, serum.

Table 10: Zinc content of ribs and hair of cows, sheep, and horses from unexposed and zinc exposed living areas (mg/kg dry matter) (12)

organ	species, n		unexposed s	unexposed $\bar{x}$	Zn-exposed $\bar{x}$	Zn-exposed s	p	%
	cows	(273; 59)	24	59	53	14	>0,05	90
rib	sheep	(73; 15)	15	78	97	15	<0,001	124
	horse	(66; 4)	14	78	70	16	>0,05	90
top hair	cows	(241; 47)	32	129	142	32	<0,05	110
wool	sheep	(45; 15)	33	143	143	24	>0,05	100
mane hair	horse	(72; 4)	22	143	130	13	>0,05	91

The Zn analysis of feed- and food-stuffs (table 11) showed that leafy feed-stuffs quadruple to decuple their Zn content, but most only double it. There is hardly any danger from this Zn accumulation.

The faeces of cattle show the probably existing quadruplication of Zn intake compared to Zn-unexposed areas.

The danger of hyperzincemia is principally reduced to exceptional cases. It might not occur if the nutrition is hygienically unobjectionable. Prophylaxis and therapy are to be concentrated on the elimination of the Zn source.

Table 11: Zinc content of several feed- and food-stuffs from unexposed and zinc rich living areas (mg/kg dry matter) (12)

species	unexposed s	unexposed x	Zn rich x	Zn rich s	p	% [1]
apples	0,3	3	4	1	>0,05	133
potatoes	4	17	25	6	<0,05	147
carrots	8	24	41	10	<0,01	171
turnips	4	24	115	61	>0,05	479
tomatoes	6	26	34	8	>0,05	131

[1] unexposed $\hat{=}$ 100%, Zn rich $\hat{=}$ x%

(Table 11 continued)

	unexposed		Zn rich			
species	s	$\bar{x}$	$\bar{x}$	s	p	%[1)]
onions	10	26	43	13	<0,05	165
oats, grain	3	26	66	27	<0,001	254
maize, in blossom	6	26	107	67	<0,05	412
kohlrabi	4	29	68	24	<0,01	234
barley, grain	9	31	56	31	>0,05	181
acre red clover	15	32	64	16	<0,01	200
green beans	6	33	47	8	<0,01	142
wheat, grain	3	35	52	15	<0,05	149
turnip leaf	18	36	404	387	>0,05	1122
meadow grass	24	39	184	69	<0,001	472
faeces of cow	30	69	306	74	<0,001	443
cow's milk	3	23	36	8	<0,001	157
Fp site	100%		294%		<0,001	
Fp species of plants		<0,001			-	

Iron

There are 4 types of Fe exposure in animals and humans according to the present level of knowledge.

The first type is an Fe intake by domestic animals and game via feed-stuffs and fodder, which considerably exceeds the requirement.This reduces feed consumption and influences various metabolic pro-cesses without acute "poisoning symptoms". The second type is acute Fe poisoning. The third type is a specific form of chronic Fe over-load, hereditary hemochromatosis. The fourth type is the chronic Fe overload resulting from multiple blood transfusions for chronic anemia. When ground or contaminated feed-stuffs or silage are fed, the rations of domestic animals contain the ten- to hundredfold Fe amount of the requirement. The same is true when animal excrements are applied. Even without these contaminations, the feed-stuffs of domestic animals, apart from milk, contain the manifold Fe amount of the Fe requirement (2, 123). Game take in less Fe via fodder than domestic animals, since they can select it better than domestic animals. Astonishingly, low Fe amounts were found in the stomach content of wild boars (8).

Fe supplementation of 500 to 2000 mg/kg ration dry matter reduced feed consumption and live weight gain in cattle, sheep, horse and broiler (table 12) (45, 94, 72, 136, 53, 56, 68).

Table 12: Dry matter consumption and live weight gain of male lambs dependent on the iron content of the ration (g animal/day) (56)

		Fe content of the ration (mg/kg dry matter)			minimum limit difference
		250	1000	1750	
dry matter consumption	g animal/day	968	886	726	185
	%	100	92	75	
live weight gain	g animal/day	320	222	183	70
	%	100	69	57	

$FeSO_4$ influences feed intake particularly intensively. Live weight decreases more strongly than it was to be expected after the reduced feed intake. This finding is in connection with Fe-induced metabolism disturbances. Under the influence of Fe supplementation Cu accumulation in the liver was significantly reduced (67, 112) (table 13). The Mn content of the liver of goats or heifers, which were given rations with 75 or 750 mg/Fe and with about the same Mn offer (68 or 74 mg/kg ration dry matter), was also significantly reduced. Their reproduction performance was worse (table 14).

The inhibition of Mn absorption by Fe was repeatedly found (143, 42, 65, 66). The registered reproduction disturbances are typical of Mn deficiency in connection with reduced Mn liver values (3, 17, 24).

Table 13: The copper content of the liver of sheep in case of different iron supply (mg/kg dry matter) (67)

organ		Fe content of the ration (mg/kg dry matter)			minimum limit
		250	1000	1750	
liver of sheep	$\bar{x}$	160	124	73	63
	s	53	48	48	

The first real assessment of the dangers of acute Fe intoxication was the report of Forbes (1947), in which he described the death of two children after the ingestion of a micture containing ferrous sulfates. Death may occur from ingestion of as little as 2 to 4 g ferrous sulfate by small children. The symptoms of Fe poisoning are summarized in table 15.

Table 14: Influence of iron exposure on reproduction performance and manganese of liver (9, 10)

species, parameter		control animals	Fe-exposed animals	p
goats	success of first insemination	100	82	>0,05
	abortion rate	0	44	<0,05
	Mn content of liver	8,6	5,0	<0,05
heifers	success of first insemination	63	48	<0,05
	conception rate	78	59	<0,05
	Mn content of liver	10	7,9	<0,01

Hemochromatosis was first described and named approximately a hundred years ago (146, 121). Hereditary hemochromatosis presents the pattern of a subtle build-up of Fe overload with associated tissue damage that may not be recognized for 50 years or more. The second major category of chronic Fe excess is transfusional Fe overload. The prototype of this disorder is ß-thalassemia, a hereditary anaemia usually discovered during infancy that requires lifelong transfusion therapy. Further diseases making blood transfusions necessary are represented in table 15 (41).

Table 15: The biological importance of iron exposure

Fe poisoning:	Forbes (1947)
Fe exposure animals:	500 mg Fe/kg ration dry matter reduce feed consumption and growth in various species of animals, decrease the Mn and Cu content of the liver, reproduction disorder;
humans, acute Fe poisoning:	vomiting, diarrhoea, coma; bleedings in the gastro-intestinal tract 6 to 24 h, fever, leucocytosis, metabolic acidosis, blood clotting disorders, liver and kidney damage, cicatrices in the gastro-intestinal tract, application of desferrioscamin;
hereditary hemochromatosis:	Fe overload with associated tissue damage;
transfusional Fe overload:	ß-thalassemia, some sickle cell disease, aplastic anaemia, myelofibrosis, desferoxamine therapy;
species specifity:	no experimental data;
antagonists:	Mn, Cu, Zn
Fe requirement:	40 - 75 mg/kg ration dry matter in domestic animals;
Fe intake of humans:	men, adolescents 12 mg/day, women 15 mg/day, babies 0,8 mg/kg bodyweight, 8 - 22 mg/day, as a rule 10 to 20 mg/day;
indicator organs:	hemoglobin, hematocrit, spleen, liver.

No species-specific differences could be found in Fe intoxications which were limited to accidents. Fe intake via feed- and food-stuffs excludes acute Fe intoxications according to the present level of knowledge.

High Fe offers reduce the Cu and Mn content of the liver significantly (10, 9, 67, 65, 59). The influence of abundant Fe amounts on Zn utilization has not been clearly secured yet (126).

Dependent on species and development specifity, the Fe requirement of domestic animals amounts to 40 to 75 mg/kg ration dry matter. It is not met in suckling piglets during the suckling period. Milk is Fe poor. The Fe requirement of humans is represented in table 15.

Young women consumed 12 to 17 mg Fe/day in the GDR, 11 mg/day in the FRG and 15 mg/day in Sweden. The different Fe supply results from the different intake of milk products, meat and sausage. Female students with vegetarian diets (milk, eggs, vegetables) took in only 8.5 or 10 mg Fe/day. In spite of that they had normal hemoglobin and hematocrit values. The literature demonstrates that 10 to 20 mg Fe/day are consumed as a rule (149, 21, 127, 32, 96).

The Fe exposure of domestic animals results from the Fe contamination of feed-stuffs. In humans Fe poisoning or Fe exposure can result from carelessness when dealing with Fe preparations, from hereditary hemochromatosis and multiple blood transfusions (58). The effects of high Fe intake on the Mn and Cu metabolism are to be taken into consideration.

Manganese

The effects of high Mn offers (>500 mg/kg ration dry matter) on the fauna were registered early. Rats tolerated 1000 to 2000 mg Mn/kg ration dry matter without growth depression. Higher doses influence P retention. Hens tolerated 1000 mg/kg dry matter without symptoms (61). 4800 mg/kg dry matter were toxic for broilers (74). Growing pigs are more sensitive. 500 mg Mn/kg dry matter led to reduced feed intake and growth (70). These results remained unconfirmed.

Sheep tolerated 500 or 1500 mg Mn/kg ration dry matter over two years without growth and reproduction disorders (13). Their lambs also developed normally. Wild sheep from acid wood sites may be sensitive to a massive Mn intake >1000 mg/kg fodder dry matter (22). They accumulated excessive Mn amounts in liver and wool.

Similar to roe deer, fallow deer and red deer, cattle and goats tolerate >1000 mg Mn/kg dry matter. Under natural conditions an Mn-content of the ration of >1000 mg/kg is not to be reckoned with on the average of longer feeding periods. The pasture grass from extremely acid, Mn-rich mountainous sites contained <500 mg Mn/kg dry matter, the normal pasture grass 100 mg Mn/kg dry matter (130). The oral application of Mn with the ashes of coal gasification did also not cause damage (119). Mn toxicity (Mn injection) led to an increase of blood glucose values (86). Schroeder et al. (1966) mention the following toxic Mn concentrations of rations: fresh water fish 2420 to 3450, birds 4800, chicks 4800, rats >2000, rabbits 1250 to 6000, pigs 500 - 2000, lambs 5000.

The first description of toxic effects of Mn on the central nervous system of humans was already carried out by Couper in 1837 (WHO 1980). the prerequisite for the damaging of the central nervous system by Mn is a certain Mn content of the air and sufficient exposure time (months to years). The symptoms are different in different regions. Mn intoxications in Chilean mines led to crazy notions, hallucinations, absentmindedness as first symptoms. This phase was already described as Mn insanity. After the unspecific cerebral general symptoms, there are disorders of the psychomotoric coordination and of the extrapyramidal nervous system (158). The prognoses for an improvement of the cerebral symptoms after the end of Mn exposure via lungs are not encouraging.

Rats chronically treated with Mn showed decreased concentrations of dopamine and increased concentrations of gamma-amino butyric acid (GABA) in brain (34).

Mn intoxications are certainly not species-specific. The hitherto registered differences between pigs and other species could not be confirmed in every experiment. Furthermore, no Mn content of the ration of 500 or 1000 mg/kg dry matter is to be expected in pigs without human influence.

Mn utilization is influenced by Fe antagonistically (143, 42, 65, 66). Older data also showed the influence of extremely high Mn offers on P metabolism (149).

Patients treated with Al had a reduced Mn content in ribs and cerebrum (137). The Mn content of different brain parts was also reduced in patients with the Alzheimer syndrome (98). There may also be antagonistic relations between Mn and Ni (88).

The Mn requirement of different species of animals is different. It varies between 25 and 60 mg/kg ration dry matter (table 16). The estimated Mn requirement of humans amounts to 0.035 to 0,070 mg/kg bodyweight. This is equivalent to a daily gross requirement of adults of 2 to 5 mg and of children of 1 to 2 mg.

In the different countries of the world, adults take in 2.7 to 4 mg Mn via their diets. In the GDR, female students with different types of rations took in 3.0 or 3.8 mg Mn/day; in Scotland 3.6 or 3.7 mg/day (21, 96). Sufficient Mn is taken in via wholemeal bread, leafy vegetables and tea.

In the fauna there is a considerable gene defect with reduced Mn utilization. The so-called "pallid" mouse cannot swim and has anomalously developed statoliths. Mn offers of 1500 to 2000 mg Mn/kg ration dry matter normalised the statoliths. The same gene defect was described in a mink mutant (82, 11).

Table 16: The biological importance of manganese exposure

Mn exposure, animals:	pigs tolerate 500, all other tested species >1000 mg Mn/kg ration dry matter;
humans:	miners working with Mn ores; severe psychiatric disorders (locura manganica) resembling shizophrenia, followed by a permanently crippling neurological disorder, clinically similar to

(Table 16 continued)

	"Parkinson's disease", $Na_2CaEDTA$, L-DOPA;
species specifity:	not known up to now;
antagonists:	Fe, P, Al;
Mn requirement:	cattle 60, hen 60, pig 25 mg/kg ration dry matter, humans 0.035 to 0.070 mg/kg bodyweight, gross requirement of adults 2 to 5 mg;
Mn intake of humans:	2.7 to 4 mg/day all over the world;
indicator organs:	hair, cerebrum, liver, kidneys.

According to long-term investigations into sheep and their lambs, Mn exposure is best reflected by wool (table 17). The same result (hair) was obtained in humans (25).

Table 17: The manganese content of several organs of adult sheep with normal and extremely rich manganese offer (mg/kg dry matter)

organ, n	55 ppm	Mn	1500 ppm	Mn	p	%
	s	x	x	s		
wool (7; 6)	2,6	4,2	74	49	<0,01	1762
cerebrum (7; 6)	0,42	1,3	3,6	1,1	<0,001	277
rib (6; 6)	1,1	2,5	5,7	1,6	<0,01	228
liver (6;6)	1,9	7,6	12	1,6	<0,01	158
kidneys (6; 6)	0,8	4,6	7,0	1,6	<0,01	152

The offer of 1500 mg Mn/kg ration dry matter did not lead to remarkable Mn incorporation into the organs. The differences to control animals are significant.

Their lambs, the Mn status of which was determined at the end of the suckling period, showed the same result in the tendency (table 18). In accordance with the findings of Weigand and Kirchgessner (1985), the body tries to keep the Mn status of the organs stable even in case of Mn exposure.

Table 18: The manganese content of several organs of lambs, the mothers of which were given normal and extremely manganese rich rations (mg/kg dry matter) (13)

organ, n	55 ppm	Mn	1500 ppm	Mn	p	%
	s	x	x	s		
wool (5; 6)	1,0	3,0	37	27	<0,01	1233
cerebrum (5; 4)	0,42	1,4	3,3	0,86	<0,05	236
liver (5; 6)	1,1	7,8	14	2,7	<0,001	179
kidneys (5; 7)	0,41	3,8	5,8	1,4	<0,01	153
rib (5; 6)	3,1	6,0	6,1	1,8	>0,05	102

The intake of Mn amounts typical of diets has not led to Mn intoxications in humans and animals. They are also not to be expected. Erroneous oral consumption of extremely high Mn amounts led to disorders of the central nervous system after 2 days.

Conclusion

The toxicity of the five essential trace elements Mo, Cu, Zn, Fe, and Mn is extremely different and partly species-specific. In practice, Mo surplus can lead to Mo intoxication in cattle and induce secondary Cu deficiency. The endangering of humans by Mo seems to be insignificant.

Cu toxicoses are frequent in sheep. Hereditary Cu intoxications occur in dogs and humans.

Zn intoxications of animals and humans are concentrated on accidents.

There are 4 types of Fe exposure (chronic Fe exposure in animals with performance depressions, acute Fe poisoning by erroneous Fe intake, hereditary hemochromatosis, Fe overload from blood transfusions). Mn intoxications only occur in humans after Mn intake via lungs. No Mn poisoning is to be expected via feed- and foodstuffs.

Literature

1. Aggett, P. J.: In: Gladke, E., Heimann, G., Lombeck, I., Eckert, I. (eds) Spurenelemente, Georg Thieme Verlag Stuttgart-New York p. 213-217 (1985)

2. Anke, M.: Jb. Arbeitsgemeinschaft für Fütterungsberatung 4, 257-279 (1961/62)

3. Anke, M.: Arch. Tierernähr. 16, 199-213 (1966)

4. Anke, M.: In: Anke, M., Schneider, H.-J. (eds) Cadmium, Spurenelementsymposium 2 - Wissenschaftl. Publikationen F.-Schiller-Univ. Jena p. 193-194 (1977)

5. Anke, M.: In: Püschner, A., Simon, O. (eds) Grundlagen der Tierernährung, VEB Gustav Fischer Verlag Jena p. 49-67 (1982)

6. Anke, M.: In: Macholz, R., Lewerenz, H. J. (eds) Lebensmitteltoxikologie, Akademieverlag Berlin, im Druck (1986a)

7. Anke, M.: Z. gesamte Inn. Med. 41, 106-111 (1986b)

8. Anke, M.: Beiträge zur Jagd- und Wildforschung 14, im Druck (1986c)

9. Anke, M., Flachowsky, G., Grün, M., Kronemann, H., Stubendorff, G.: Tierzucht 32, 91-93 (1978)

10. Anke, M., Flachowsky, G., Hennig, A., Partschefeld, M.: In: Wiss. Tagung der Sektion Tierproduktion und Veterinärmedizin, Karl-Marx-Univ. Leipzig p. 309-319 (1975)

11. Anke, M., Fuchs, M., Lösch, E.: In: Hattenhauer, H. (ed) 1. Internationales Pelztiersymposium, Sektion Tierproduktion und Veterinärmedizin, Karl-Marx-Univ. Leipzig p. 167-179 (1983)

12. Anke, M., Groppel, B.: In: Legutko, S. (ed) Protection of the Natural Environment and Development of Animal Production, Krakow Instytut Zoótechniki p. 41-63 (1985)

13. Anke, M., Groppel, B.: Mengen- und Spurenelemente 6, im Druck (1986)

14. Anke, M., Groppel, B., Grün, M.: In: Mills, C. F., Bremner, I., Chesters, J. K. (eds) Trace Elements in Man and Animals, Commonwealth Agricultural Bureaux U.K. 5, 154-157 (1985)

15. Anke, M., Groppel, B., Kronemann, H., Grün, M.: Wiss. Z. Karl-Marx-Univ. Leipzig, Math.-Naturwiss. R. 33, 148-156 (1984)

16. Anke, M., Groppel, B., Kronemann, H., Grün, M.: Nutr. Res. Suppl.I p. 180-186 (1985)

17. Anke, M., Groppel, B., Reissig, W., Lüdke, H., Grün, M., Dittrich, G.: Arch. Tierernähr. 23, 197-211 (1973)

18. Anke, M., Grün, M., Partschefeld, M.: In: Hemphill, D. D. (ed) Trace Substances in Environmental Health, Univ. Missouri, Columbia, USA 10, 403-409 (1976)

19. Anke, M., Hennig, A., Grün, M., Groppel, B., Lüdke, H.: In: Hemphill, D. D. (ed) Trace Substances in Environmental Health, Univ. Missouri, Columbia, USA 10, 105-111 (1976)

20. Anke, M., Hennig, A., Schneider, H.-J., Groppel, B., Grün, M., Partschefeld, M., Lüdke, H.: Wiss. Z. Karl-Marx-Univ. Leipzig, Math.-Naturwiss. R. 25, 241-269 (1976)

21. Anke, M., Klinger, G., Grün, M., Krause, G.: Dt. Gesundh.-Wesen 35, 1967-1972 (1980)

22. Anke, M., Kronemann, H., Dittrich, G., Neumann, A.: Arch. Tierernähr. 29, 845-858 (1979)

23. Anke, M., Kronemann, H., Groppel, B., Riedel, E.: Wiss. Z. Karl-Marx-Univ. Leipzig, Math.-Naturwiss. R. 33, 157-165 (1984)

24. Anke, M., Risch, M.: Haaranalyse und Spurenelementstatus, VEB Gustav Fischer Verlag Jena (1979)

25. Anke, M., Schneider, H.-J., Grün, M., Groppel, B., Hennig, A.: Zbl. Pharmazie 17, 688-712 (1978)

26. Anke, M., Schwark, H. J., Groppel, B., Dittrich, G.: Beiträge zur Jagd- und Wildforschung 14, im Druck (1986)

27. Anonym: Atlas of Cancer Mortality in The People's Republic of China p. 75-78 (1979)

28. Anonym: Recommended Health-Based Limits in Occupational Exposure to Heavy Metals, WHO Technical Report Series 647, Genf (1980)

29. Aughey, E., Grant, L., Furman, B. L., Dryden, W. F.: J. Comp. Path. 87, 1-14 (1977)

30. Badawy, A. B. N., Shakour, A. A.: In: Mills, C. F., Bremner, I., Chesters, J. K. (eds) Trace Elements in Man and Animals, Commonwealth Agricultural Bureaux U.K. 5, 261-263 (1985)

31. Barber, R. S., Braude, R., Mitchell, K. G.: Brit. J. Nutr. 9, 378-381 (1955)

32. Bindra, G. S., Gibson, R. S.: In: Mills, C. F., Bremner, I., Chesters, J. K. (eds) Trace Elements in Man and Animals, Commonwealth Agricultural Bureaux U.K. 5, 779-781 (1985)

33. Bleavins, M. R., Aulerich, R. J., Hochstein, J. R., Hornshow, T. C., Napolitano, A.: J. Nutr. 113, 2360-2367 (1983)

34. Bonilla, E., Dietz-Ewald, M.: J. Neurochemistry 22, 297-299 (1974)

35. Bostwick, J. L.: JAMA 180, 386-387 (1982)

36. Bremner, I., Price, J.: In: Mills, C. F., Bremner, I., Chesters, J. K. (eds) Trace Elements in Man and Animals, Commonwealth Agricultural Bureaux U.K. 5, 374-376 (1985)

37. Bremner, I., Young, B. W., Mills, C. F.: Brit. J. Nutr. 36, 551-561 (1976)

38. Brewer, G. J., Aster, J. C., Knutsen, C. A., Kruckeberg, W. C.: Am. J. Hematology 7, 53-60 (1979)

39. Brewer, G. J., Hill, G. M., Dick, R. D., Piko, C. R., Prasad, A. S., Cossack, Z. T.: Nutr. Res. Suppl. I p. 478-481 (1985)

40. Brewer, G. J., Hill. G. M., Prasad, A. S., Cossack, Z. T., Rabbani, P.: Ann. Internal Medicine 99, 314-320 (1983)

41. Brown, E. B.: In: Sarkar, B. (ed) Biological Aspects of Metals and Metal-Related Diseases, Raven Press, New York p. 263-278 (1983)

42. Chandra, S. V., Tandon, S. K.: Environm. Physiol. Biochem. 3, 230-235 (1973)

43. Chapman, G. A.: Trans. Am. Fish. Soc. 107, 828-836 (1978)

44. Cook, G. A., Lesperance, A. L., Bohman, V. R., Jensen, E. H.: J. Animal Sci. 25, 96-101 (1966)

45. Coup, M. R., Campbell, A. G.: N.Z.J. agric. Res. 7, 624-630 (1964)

46. Croker, K. P., Allen, J. G., Masters, H. G., Petterson, D. S.: Aust. J. Agric. Res. 30, 929-938 (1979)

47. Croker, K. P., Allen, J. G., Petterson, D. S., Masters, H. G., Frayne, R. F.: Aust. J. Agric. Res. 30, 551-564 (1979)

48. Dahmer, M. K., Hausley, P. R., Pratt, W. B.: Ann. Rev. Physiol. 46, 67-81 (1984)

49. Dang, H. S., Jaiswal, D. D., Wadhwani, C. N., DaCosta, H., Sowasundaram, S.: In: Brätter, P., Schramel, P. (eds) Trace Element Analytical Chemistry in Medicine and Biology 2, 37-43, Walter de Gruyter, Berlin, New York (1983)

50. Dick, A. T., Dewey, D. W., Gawthorne, J. M.: J. Agric. Sci., Camb. 85, 567-576 (1975)

51. Ekperigin, H. E., Vohra, P.: J. Nutr. 111, 1630-1640 (1981)

52. Farmer, G. J., Ashfield, D., Samant, H. S.: Environm. Pollution 19, 103-117 (1979)

53. Fekete, J. M. T., Surány, E.: In: Mills, C. F. (ed) Trace Element Metabolism in Animals, E. and S. Livingstone, Edinburgh and London p. 123-125 (1970)

54. Ferguson, W. S., Lewis, A. H., Watson, S. J.: Nature 141, 553-560 (1938)

55. Fischer, P. W. F., Giroux, A., L'Abbé, M. R.: Amer. J. Clin. Nutr. 34, 1670-1675 (1981)

56. Flachowsky, G., Hennig, A., Löhnert, H.-J., Grün, M.: Arch. Tierernähr. 26, 765-771 (1976)

57. Forbes, G.: Brit. Med. J. 1, 367-370 (1947)

58. Forth, W., Rummel, W.: In: Forth, W., Henschler, D., Rummel, W. (eds) Allgemeine und spezielle Pharmakologie und Toxikologie, Bibliographisches Institut Mannheim/Wien/Zürich B. I. - Wissenschaftsverlag (1984)

59. Fransson, G.-B., Hoffman, B., Lönnerdal, B., Keen, C. L.: In: Mills, C. F., Bremner, I., Chesters, J. K. (eds) Trace Elements in Man and Animals, Commonwealth Agricultural Bureaux U.K. 5, 512-515 (1985)

60. Gallery, E. D. M., Blomfield, J., Dixon, S. R.: Brit. med. J. 4, 331-333 (1972)

61. Gallup, W. D., Norris, L. C.: Poultry Sci. 18, 76-83 (1939)

62. Gawthorne, J. M., Allen, J. D., Nader, C. J.: In: Mills, C. F., Bremner, I., Chesters, J. K. (eds) Trace Elements in Man and Animals, Commonwealth Agricultural Bureaux U.K. 5, 346-351 (1985)

63. Gough, L. P., Shacklette, H. T., Case, A. A.: Element concentrations toxic to plants, animals, and man. Geological Survey Bulletin 1466, Washington 1979

64. Graupe, B.: In: Bergmann, W., Hennig, A. (eds) Mineralstoffversorgung von Pflanze und Tier, Tagungsbericht der Dt. Akad. d. Landwirtschaftswissenschaften zu Berlin 85, 309-317 (1966)

65. Gruden, N.: Nutr. Metab. 21, 305-309 (1977)

66. Gruden, N.: In: Mills, C. F., Bremner, I., Chesters, J. K. (eds) Trace Elements in Man and Animals, Commonwealth Agricultural Bureaux U.K. 5, 515-516 (1985)

67. Grün, M., Anke, M., Hennig, A., Seffner, W., Partschefeld, M., Flachowsky, G., Groppel, B.: Arch. Tierernähr. 28, 341-347 (1978

68. Grün, M., Flachowsky, G., Hennig, A., Anke, M.: Mengen- und Spurenelemente 4, 297-304 (1984)

69. Grün, M., Flachowsky, G., Partschefeld, M., Anke, M., Hennig, A. In: Pais, I. (ed) New Results in the Research of Hardly Known Trace Elements, Univ. of Horticulture Budapest p. 104-114 (1985)

70. Grummer, R. H., Bentley, O. G., Phillips, P. H., Bohstedt, G.: J. Animal Sci. 9, 170-179 (1950)

71. Hardy, R. M., Stevens, J. B., Stowe, C. M.: Minn. Vet. 15, 13-24 (1975)

72. Harmon, B. G., Becker, D. E., Jensen, A. H., Norton, H. W.: J. Animal Sci. 27, 418-427 (1968)

73. Heller, V. G., Burke, A. D.: J. Biol. Chem. 74, 85-92 (1927)

74. Heller, V. H., Penquite, R.: Poultry Sci. 16, 243-251 (1937)

75. Henderson, B. M., Winterfield, R. W.: Avian Diseases 19, 385-387 (1975)

76. Hidiroglou, M., Morris, G., Ivan, M.: J. Dairy Sci. 65, 619-624 (1982)

77. Hill, C. H., Matrone, G.: Fed. Proc. 29, 1474-1481 (1970)

78. Hill, G. M., Miller, E. R.: J. Animal Sci. 57, 106-113 (1983)

79. Hogan, K. G., Money, D. F. L., White, D. A., Walker, R.: New Zealand J. of Agricult. Res. 14, 687-701 (1971)

80. Hoogenraad, T. M., Dekker, A. W., Van den Hamer, C. J. A.: Sci. Total Environm. 42, 37-43 (1985)

81. Humphries, W. R., Bremner, I., Phillippo, M.: In: Mills, C. F., Bremner, I., Chesters, J. K. (eds) Trace Elements in Man and Animals, Commonwealth Agricultural Bureaux 5, 371-374 (1985)

82. Hurley, L. S.: In: Hemphill, D. D. (ed) Trace Substances in Environmental Health, Univ. Missouri, Columbia, USA 2, 41-51 (1968)

83. Jackson, N., Stevenson, M. H.: Brit. J. Nutr. 45, 99-110 (1981)

84. Jensen, L. S., Maurice, D. V.: J. Nutr. 109, 91-97 (1979)

85. Johnson, G. F., Sternlieb, I., Twedt, D. C., Grushoff, P. S., Scheinberg, I. H.: Amer. J. Vet. Res. 41, 1865-1866 (1980)

86. Keen, C. L., Baly, D. L., Tamai, K. T., Lönnerdal, B.: In: Mills, C. F., Bremner, I., Chesters, J. K. (eds) Trace Elements in Man and Animals, Commonwealth Agricultural Bureaux U.K. 5, 258-261 (1985)

87. Kirchgessner, M., Reichlmayr-Lais, A. M.: In: McC Howell, J., Gawthorne, J. M., White, C. L. (eds) Trace Element Metabolism in Man and Animals, Austral. Acad. Sci. Canberra 4, 390-393 (1981)

88. Kirchgessner, M., Reichlmayr-Lais, A. M., Maier, R.: In: Mills, C. F., Bremner, I., Chesters, J. K. (eds) Trace Elements in Man and Animals, Commonwealth Agricultural Bureaux U.K. 5, 147-151 (1985)

89. Kirchgessner, M., Schwarz, F. J.: In: The Use of Isotopes to Dedect Moderate Mineral Imbalances in Farm Animals, IAEA-TEC DOC 267, 69-90 (1982)

90. Kirchgessner, M., Schwarz, F. J., Grassmann, E., Steinhart, H.: In: Nriagu, J. O. (ed) Copper in the Environment, Part II, John Wiley and Sons, Inc. New York pp. 433-471 (1979)

91. Kneale, W. A., Smith, P.: Experimental Husbandry 32, 1-7 (1977)

92. Kovalskij, V. V.: Geochemische Ökologie, Biogeochemie, VEB Dt. Landwirtschaftsverlag Berlin, pp. 233-238 (1977)

93. Kronemann, H., Hennig, A., Grün, M., Anke, M.: In: Mehlhorn, G. (ed) Emission und Immission von Schadstoffen in der Tierproduktion, Karl-Marx-Univ. Leipzig pp. 475-479 (1979)

94. Lawlor, N. J., Smith, W. H., Beeson, W. M.: J. Animal Sci. 24, 742-749 (1965)

95. Lefkowitch, J. H., Homig, C. L., King, M. E., Hagstrom, J. W. C. New Engl. J. Medicine 307, 271-277 (1982)

96. Lyon, T. D. B., Smith, L. B., Leniham, J. M. A.: In: Mills, C. F Bremner, I., Chesters, J. K. (eds) Trace Elements in Man and Animals, Commonwealth Agricultural Bureaux U.K. 5, 625-627 (1985

97. Maclachlan, G. K., Johnston, W. S.: The Veterinary Record, Sept. 25, 299-301 (1982)

98. Markesbery, W. R., Ehmann, W. D., Hossain, T. I. M., Alauddin, M.: Neurotoxicology 5, 49-56 (1984)

99. Massie, H. R., Aiello, V. R.: Mechanisms of Ageing and Development 26, 195-203 (1984)

100. Mertz, W.: Science 213, 1332-1336 (1981)

101. Mills, C. F., Bremner, I., El-Gallad, T. T. T., Dalgarno, A. C., Young, B. W.: In: Kirchgessner, M. (ed) Trace Element Metabolism in Man and Animals 3, 150-158 (1978)

102. Mills, C. F., Davies, N. T., Quarterman, J., Aggett, P. J.: Nutrition Research, Suppl. I, 471-477 (1985)

103. Mills, B. J., Lindeman, R. D.: In: Zumkley, H. (ed) Spurenelemente, Georg Thieme Verlag Stuttgart-New York, pp. 197-214 (1983)

104. Miranda, C. L., Henderson, M. C., Reed, R. L., Schmitz, J. A., Buhler, D. R.: J. Toxicol. Environm. Health 9, 359-366 (1982)

105. Molnar, J. J.: Arch. Pathol. Lab. Med. 107, 629-632 (1983)

106. Murakami, N., Moudgil, V. K.: Biochem. J. 198, 447-455 (1981)

107. Naray, A.: Biochemica et Biophysica Acta 756, 328-334 (1983)

108. Nederbragt, H., Van den Ingh, T. S. G. A. M., Wensvoort, P.: The Vet.Quarterly 6, 179-235 (1984)

109. Nishi, Y., Sawano, K., Sakano, T., Kagosaki, Y., Tanaka, Y., Kobayashi, Y., Usui, T.: Eur. J. Pediatr. 134, 9-11 (1980)

110. Pensack, J. M., Klussendorf, R. C.: Poultry Nutr. Conf. Atlantic City, New Yersey (1956)

111. Petrie, J. J. B., Row, P. G.: The Lancet June 4, pp. 1178-1180 (1977)

112. Phillippo, M., Humphries, W. R., Bremner, I., Atkinson, T., Henderson, G.: In: Mills, C. F., Bremner, I., Chesters, J. K. (eds) Trace Elements in Man and Animals, Commonwealth Agricultural Bureaux U.K. 5, 176-180 (1985)

113. Pitt, M., Fraser, J., Thurley, D. C.: J. Comp. Pathol. 90, 567-576 (1980)

114. Poole, D. B. R., McGrath, D., Flennig, G. A., Sinnott, J.: Ir. J. Agric. Res. 22, 1-10 (1983)

115. Portmann, B., Tanner, M. S., Mowat, A. P., Williams, R.: Lancet 1: 1338-1340 (1978)

116. Prasad, A. S., Brewer, G. J., Schoomaker, E. B., Rabbani, P.: JAMA 240, 2166-2168 (1978)

117. Price, J., Chesters, J. K.: In: Mills, C. F., Bremner, I., Chesters, J. K. (eds) Trace Elements in Man and Animals, Commonwealth Agricultural Bureaux U.K. 5, 682-685 (1985)

118. Quarterman, J.: In: Hoekstra, W. G., Suttie, J. W., Ganther, H. E., Mertz, W. (eds) Trace Element Metabolism in Animals, University Park Press, Baltimore, London, Tokyo 2, 742-744 (1974)

119. Rabar, I., Kostial, K.: In: Mills, C. F., Bremner, I., Chesters, J. K. (eds) Trace Elements in Man and Animals, Commonwealth Agricultural Bureaux U.K. 5, 510-512 (1985)

120. Rauch, H.: J. Heredity 74, 141-144 (1983)

121. Recklinghausen, v. H.: Berlin. Klin. Wochenschr. 26, 925 (1889)

122. Reinhardt, F., Gabsch, H.-Ch., Nelz, P.: Mengen- und Spurenelemente 5, 178-185 (1985)

123. Riedel, E., Anke, M., Kronemann, H.: Mengen- und Spurenelemente 2, 57-66 (1982)

124. Robbins, K. R., Baker, D. H.: Canad. J. Animal Sci. 60, 201-202 (1980)

125. Roventa, E., Pora, E. A., Sahleanu, V., Vaduva, E.: Studia Universitos Bubes-Bolyai, Seria Biologica 13, 107-114 (1968)

126. Sandström, B., Davidsson, L., Cederblad, Å., Lönnerdal, B.: In: Mills, C. F., Bremner, I., Chesters, J. K. (eds) Trace Elements in Man and Animals, Commonwealth Agriculture Bureaux U.K. 5, 414-416 (1985)

127. Schelenz, R. F. W., Hartmuth-Hoene, A.-E.: In: Mills, C. F., Bremner, I., Chesters, J. K. (eds) Trace Elements in Man and Animals, Commonwealth Agriculture Bureaux U.K. 5, 620-622 (1985)

128. Schroeder, H. A., Balassa, J. J., Tipton, I. H.: J. Chronic Diseases 19, 545-571 (1966)

129. Schroeder, H. A., Balassa, J. J., Tipton, I. H.: J. Chronic Diseases 23, 481-499 (1970)

130. Schubert, H.: Untersuchungen zur Mangan-, Zink-, Kupfer- und Cadmiumversorgung weidender Jungrinder... Diss. Jena, Mathem.-Naturwiss.-Techn. Fakultät der Fr.-Schiller-Univ. (1972)

131. Schwarz, K., Spallholz, J. E.: In: Anke, M., Schneider, H.-J. (eds) Cadmium, Spurenelementsymposium 2- Wissenschaftl. Publikationen F.-Schiller-Univ. Jena pp. 188-193 (1977)

132. Scott, A. S., Magee, A. C.: Nutr. Report International 20, 549-559 (1979)

133. Shenkin, A., Fell, G. S.: Nutr. Res. Suppl. I, pp. 193-202 (1985)

134. Shyamala, G., Leonhard, L.: J. Biol. Chem. 255, 6028-6031 (1980)

135. Smith, jr. J. C., Zeller, J. A., Brown, E. D., Ong, S. G.: Science 193, 496-498 (1976)

136. Standish, J. F., Ammerman, C. B., Simpson, F. C., Palmer, A. Z.: J. Animal Sci. 29, 496-503 (1969)

137. Stein, G., Anke, M., Fünfstück, R., Schneider, H.-J.: Z. gesamte Inn. Med. 34, 648-654 (1979)

138. Stevenson, M. H., Jackson, N.: Brit. J. Nutr. 46, 71-76 (1981)

139. Straube, E. F., Walden, N. B.: Laboratory Animals 15, 45-47 (1981)

140. Suttle, N. F.: Proc. Nutr. Soc. 33, 299-305 (1974)

141. Suttle, N. F.: Brit. J. Nutr. 32, 559-568 (1974)

142. Thomas, J. W., Moss, S.: J. Dairy Sci. 34, 929-934 (1951)

143. Thomson, A. B. R., Valberg, L. S.: Am. J. Physiol. 223, 1327-1329 (1972)

144. Thornburg, L. P., Polley, D., Dimmit, R.: Canine Practice 11, 36-39 (1984)

145. Thornburg, L. P., Rottinghaus, G., Koch, J., Hause, W. R.: J. Amer. Animal Hospital Ass. 20, 1003-1005 (1984a)

146. Trosier, M.: Bull. Soc. Anat. (Paris) 46, 231-235 (1971)

147. Tsai, Y.-H., Steinberger, A.: J. steroid Biochem. 17, 131-136 (1982)

148. Twedt, D. C., Sternlieb, I., Gilbertson, S. R.: J. Am. Vet. Med. Ass. 175, 269-275 (1979)

149. Underwood, E. J.: Trace Elements in Human and Animal Nutrition, Academic Press Inc. New York and London (1962)

150. Van der Schee, W., Schotman, A. J. H., van der Berg, R.: Zbl. Vet. Med. A 30, 664-673 (1983)

151. Walravens, P. A., Moure-Eraso, R., Solomons, C. C., Chappel, W. R., Bentley, G.: Arch. Environm. Health 34, 302-308 (1979)

152. Walsh, F. M., Crosson, F. J., Bayley, U., McReynolds, J., Pearson, B. J.: Amer. J. Disease Childrens 31, 149-151 (1977)

153. Walsh, T. O., O'Moore, L. B.: Nature 171, 1166 (1953)

154. Walshe, J. M.: J. Inher. Metab. Dis. Suppl. 1, 51-58 (1983)

155. Weigand, E., Kirchgessner, M.: In: Mills, C. F., Bremner, I., Chesters, J. K. (eds) Trace Elements in Man and Animals, Commonwealth Agricultural Bureaux U.K. 5, 506-509 (1985)

156. Wood, E. C., Worden, A. N.: J. Sci. Fol. Agric. 24, 167-174 (1973)

157. Yang, S. I., Luo, X.-M., Wei, H.-J.: In: Mills, C. F., Bremner, I., Chesters, J. K. (eds) Trace Elements in Man and Animals, Commonwealth Agricultural Bureaux U.K. 5, 160-161 (1985)

158. Zidek, W.: In: Zumkley, H.: Spurenelemente, Georg Thieme Verlag Stuttgart-New York, pp. 140-151 (1983)

ADVANCES IN THE UNDERSTANDING OF ZINC-COPPER ANTAGONISM AND IT'S APPLICATIONS TO THE TREATMENT OF WILSON'S DISEASE

ZAFRALLAH T. COSSACK

Odense University, Human & Clinical Nutrition Programme,
Institute of Biochemistry, Campusvej 55, DK-5230 Odense M, Denmark

Introduction

Early studies by Schouwink (1961) have shown that Zinc in therapeutic dosages when given to patients with Wilson's disease, resulted in a significant increase in fecal excretion of copper. The mechanism involved was not yet established. Studies in experimental animals and in human subjects suggest that Zinc exerts its antagonistic effect by inducing the synthesis of copper binding ligand, a thionine, in the mucosal cells which sequesters copper from the nutrient medium, making it unavailable for serosal transfer and, thus, decreases copper absorption leading to a decrease in copper status. We have examined the issue of Zinc-copper interaction in Wilson's patients treated with oral zinc. In addition, we conducted several investigations using experimental animals to substantiate our findings. A summary of the clinical and experimental studies on oral zinc therapy for Wilson's disease, is presented here, followed by general conclusions.

Penicillamine Therapy for Wilson's Disease

Wilson's disease (or hepatolenticular degeneration) is an autosomal, recessive, inborn error of copper metabolism in which excessive amounts of copper accumulate in tissues such as the liver, brain, kidney and the cornea. Biochemically, the disease is characterized by a low serum cerulloplasmin level. Kayser-Fleischer rings, cirrhosis and neurological disorders are the major clinical manifestations of this disease which subsequently leads to death. The primary defect seems to

Trace Element – Analytical Chemistry in Medicine and Biology, Vol. 4

involve the liver handling of copper, resulting in a failure to excrete an adequate amount of copper in the bile (1).

The standard treatment of Wilson's disease involves the use of the chelating agent -D-penicillamine (2,3,4) which mobilizes copper for its renal excretion. However, it has been shown that this treatment results in many adverse effects including nephrotic syndrome (5), Goodpasture's syndrome (4), optic neuritis (6), retinal hemorrhages (7), conjunctival inflammation (8), follicular conjunctivitis (9), a lopus-like syndrome (10), a myasthenia-like syndrome (11), aplastic anemia (10,12,13), and episodes of acute polyarthritis (10). In addition, it has been reported (14) that intolerance to penicillamin has resulted in the discontinuation of this therapy in many cases.

Oral Zinc Therapy for Wilson's Disease

It has been shown that high levels of dietary zinc can interfere with copper absorption (15,16). Data from several investigations (17,18,19) suggested that zinc exerts its antagonistic effect by inducing the synthesis of copper-binding ligand, probably a thionein, in the mucosal cells which sequesters copper from the nutrient medium, making it unavailable for serosal transfer and, thus, decreases copper absorption leading to a decrease in copper status. This phenomenon has been utilized to reduce the accumulation of copper in Wilson's patients.

Schouwink reported (20) the results of the first study on oral zinc therapy for Wilson's disease in 1961. He observed higher amounts of fecal copper in 2 patients when given high dosages of zinc sulfate orally (200 mg, three times daily for 1.5 year). His work was presented in a doctoral degree dissertation submitted to the University of Amsterdam (fig. 1).

DE
HEPATO-CEREBRALE
DEGENERATIE

(MET EEN ONDERZOEK VAN DE ZINKSTOFWISSELING)

ACADEMISCH PROEFSCHRIFT

TER VERKRIJGING VAN DE GRAAD VAN DOCTOR IN DE GENEESKUNDE
AAN DE UNIVERSITEIT VAN AMSTERDAM,
OP GEZAG VAN DE RECTOR MAGNIFICUS DR J. KOK,
HOOGLERAAR IN DE FACULTEIT DER WISKUNDE EN NATUURWETENSCHAPPEN,
IN HET OPENBAAR TE VERDEDIGEN IN DE AULA DER UNIVERSITEIT OP
DONDERDAG 19 OCTOBER 1961, DES NAMIDDAGS
TE 4 UUR PRECIES

DOOR

GERRIT SCHOUWINK

GEBOREN TE ENSCHEDE

DRUK: G. W. VAN DER WIEL & CO. - ARNHEM

Fig. 1 : From Schouwink, G. 1961. Hepatolenticular degeneration: Treatment by oral zinc. a Ph.D. dissertation, University of Amsterdam, The Netherlands.

Hoogenraad et al., reported (21) the clinical and biochemical changes in one of the patients described in Schouwink's study (20) who continued on oral zinc treatment for a period of 14 years. They observed the disappearance of the neurological symptoms of the disease and the disappearance of Kayser-Fleischer rings. This study was followed by other reports where ^{64}copper-loading test (22,23) or copper balance (24,25) were used to assess the efficacy of zinc therapy for Wilson's disease. Recently, Cossack and Bouquet reported (25) the first study on the use of zinc in the treatment of children with this disease. They observed that in children treated with 85 to 137 mg of elemental zinc per day, copper balance was negative as compared to controls (table 1). However, as shown in this table, patients treated with penicillamine showed a significantly higher negative copper balance as compared to zinc treated patients.

Table 1:

Copper Balance in Children with Wilson's disease and in controls

Mg/day	Zinc Therapy	Penicillamine Therapy	Controls
Copper Intake	0.57 ± .07	0.57 ± .03	0.61 ± .01
Urinary Copper	0.09 ± .01	0.39 ± .07	0.0
Fecal Copper	0.56 ± .07	0.37 ± .01	0.42 ± .01
Copper Balance (±)	-0.08 ± .01	-0.19 ± .09	+.19 ± .02

From: Cossack, Z.T. and J. Bouquet. Acta. Pharmacol. Toxicol. (reference 25; in press).

Endogenous Zinc Status and Copper Absorption

a) Observations in Wilson's disease patients:

Recently, we observed (table 2) that inadequate pretreatment with zinc (for a period of 3 weeks) was not sufficient to produce zero or negative copper balance in adult patients. However, when pretreatment continued for 7 weeks, we then observed a negative copper balance. We concluded, from this observation, that the inhibition of copper absorption by oral zinc is related not to the intestinal cell turnover time which is 4-6 days (in human), but to the restoration of an acceptable zinc status in the body. If this holds true, then we would expect that in a subject previously loaded with zinc, the inhibition of copper absorption should continue for a measurable period of time following the withdrawal of zinc therapy while his/her dietary intake of zinc is negligible. Van den Hamer and co-workers (26) have demonstrated this in patients with Wilson's disease and in controls (fig. 2).

Table 2:

COPPER BALANCE STUDIES IN ADULTS WITH WILSON'S DISEASE

	Inadequate Zinc Pretreatment	Adequate Zinc Pretreatment
Weeks of Zinc Pretreatment	3	7
Days of Balance on Zinc Therapy	17	12
Balance Results on Zinc Therapy, Mg/day	+0.54	-0.02

From: Brewer, G.J., G.M. Hill, A.S. Prasad, Z.T. Cossack and P.I. Rabbani: Ann. Intern. Med. 99, 311-320, 1983 (reference 24).

In their study, copper absorption was measured by the appearance, in plasma, of orally administered ^{64}copper. They observed that in patients pretreated with zinc (3 x 100 mg $ZnSo_4$ daily for 3 years in one patient and 3 x 200 mg $ZnSo_4$ daily for 6 years in a second patient) the adverse effect on copper absorption continued for a period of 12-24 days following the termination of zinc therapy. They observed the same trend in a normal volunteer who took zinc (600 mg $ZnSo_4$/day for 2 weeks).

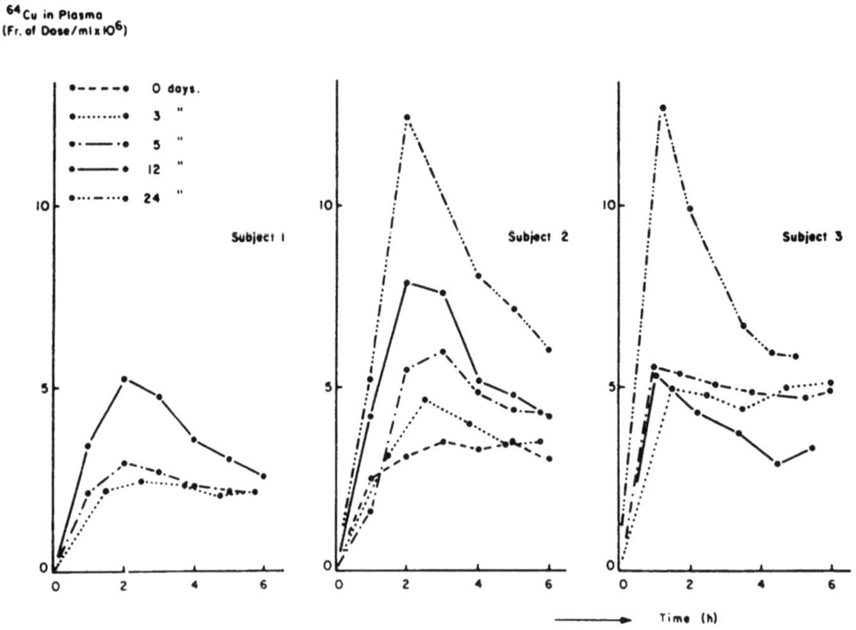

Fig. 2: ^{64}Cu-loading test as a function of timer. All subjects were tested at 0t and on the 3rd, 5th, and 12th day following the withdrawal of zinc therapy. Subjects 2 and 3 were also tested on the 24th day. Subject 1 and 2 are Wilson's disease patients and subject 3 is a control. [From: Van den Hamer, C.J.A. et al. Trace Elem. Med. 1, 88-90, 1984. reference 26].

b) Animal Studies

Our observations in Wilson's patients (table 2) and the observations of van den Hamer et al. (fig. 2) may indicate that the inhibition of copper absorption is mediated through high endogenous zinc status. We tested this hypothesis in a reecent study using mice (27). In this study, 2 groups of mice were used: the experimental group was fed a diet containing 1000 ppm of zinc for a period of 35 days (repletion) afterwhich mice were switched to the basal diet, which contained 6 ppm of zinc, for the following 35 days (depletion). The control group was fed the same basal diet with 30 ppm of zinc throughout the experimental period. Endogenous fecal zinc was determined and ^{64}copper loading test was also conducted at intervals throughout the experimental period. As shown in fig. 3, copper absorption (measured by the ^{64}Cu-loading technique) was still impaired in the experimental group (significantly lower as compared to control) for up to 21-35 days following the withdrawal of zinc supplement (depletion phase).

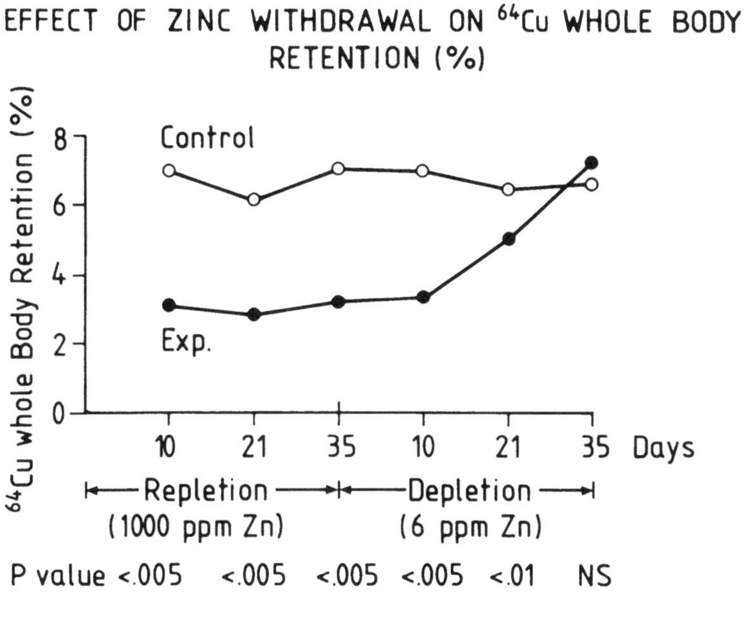

Fig. 3: From Cossack, Z.T., and C.J.A. Van den Hamer (reference 27).

We then correlated the data from ^{64}Cu-loading test with those of endogenous fecal zinc (fig. 4). As shown, endogenous fecal zinc (mg Zn/d) correlated ($r = -0.92$, $p < 0.005$) significantly with ^{64}Cu-retention. This finding indicates that the inhibition of copper absorption occurs when animals are in high status of endogenous zinc regardless of the level of dietary zinc.

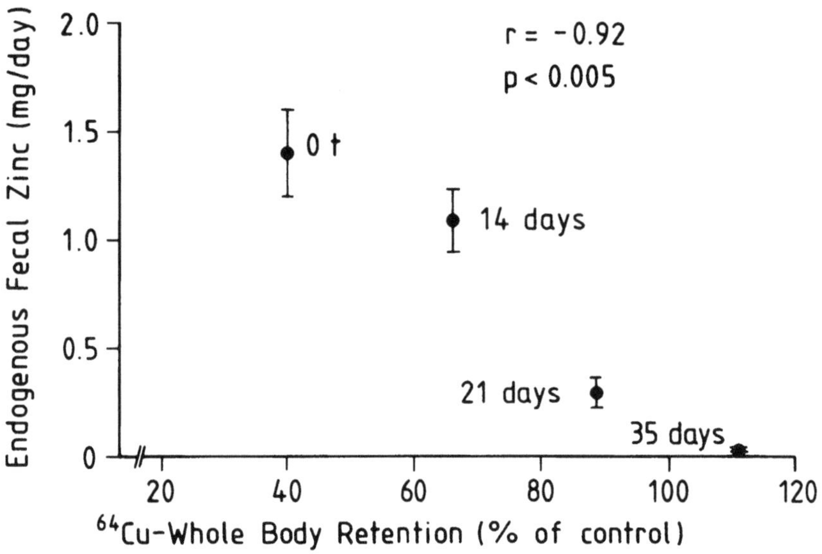

Fig. 4: From Cossack, Z.T., and C.J.A. Van den Hamer (reference 27).

Conclusions

From human and animal data we conclude the followings: 1) in adults and children with Wilson's disease, zinc therapy does not completely inhibit the gastrointestinal absorption of copper if we consider the total amount of copper present in the GIT (dietary + gastric and salivary secretions), 2) the inhibition of copper absorption, in human and experimental animals, is dependent on the maintenance of high endogenous status of zinc in the body: metabolic fecal zinc correlated significantly ($r = -0.92$, $p < 0.005$) with whole body retention of orally administered ^{64}copper. This suggests that endogenous zinc (and not dietary zinc per se) stimulates the synthesis of metallothionein and subsequently inhibits copper absorption and, 4) in children, early diagnosis and the institution of zinc therapy were effective in controlling Wilson's disease.

References

1. Gibbs, K., and J.M. Walshe. 1980. Lancet 2, 538-539.

2. Bern, A.G. 1972. In: Stanbury, J.B., Wynngaarden, J.B., Frederickson, D.S., editors. The Metabolic Basis of Inherited Disease. 3rd edition. New York: Mc Graw-Hill Book Co., 1033-50.

3. Walshe, J.M. 1963. Practitioner 191, 789-795.

4. Sternlieb, I. 1975. In: Poppe, H., Bceker, K., editors. Collagen Metabolism in the liver. New York: Stratton International Corp., 183-190.

5. Hirschman, S.Z., and K.J. Issel Bacher. 1965. Ann. Intern. Med. 162, 1297-1300.

6. Tu, J., R.Q. Blackweli and P. Lee. 1963. JAMA 185, 83-86.

7. Bigger, J.F. 1968. Am. J. Opthalmol. 66, 954-955.

8. McDonald, J.E. and P.H. Henneman. 1965. N. Engl. J. Med. 273, 578-583.

9. Dingle, J. and W.H. Havener. 1978. Ann. Opthalmol. 10, 1227-1230.

10. Golding, D.N.and J.M. Walshe. 1977. Ann. Rheum. Dis. 36, 99-111.

11. Czlonkowska, A. 1975. Br. Med. J. 2, 726-727.

12. Bird, E.D. 1974. Postgrad. Med. J. 50, (suppl. 2), 73-74.

13. Cramer, K., J.R. Golding and I.A. Jaffe et al. 1974. Postgrad. Med. J. 50, 80-83.

14. Borthwick, T.R., G.D. Benson and J.H. Schugar. 1980. J. Lab. Clin. Med. 95, 575-580.

15. Magee, A.C. and G. Matione. 1960. J. Nutr. 72, 233-242.

16. Van Campen, D.R. and P.U. Scaife. 1967. J. Nutr. 91, 473-476.

17. Fischer, P.W.F., A. Giroux and M.R. L'Abbe. 1981. Am. J. Clin. Nutr. 34, 1670-1675.

18. Ogiso, T., N. Ogawa and T. Muira. 1979. Chem. Pharm. Bull. 27, 515-521.

19. Richards, M.P.and R.J. Cousins. 1976. Proc. Soc. Exp. Biol. Med. 153, 52-56.

20. Schouwink, G. 1961. Thesis, University of Amsterdam.

21. Hoogenraad, T.U., R. Loevoet and E.G.M. De Ruyter Korver. 1979. Eur. J. Neurol. 18, 205-211.

22. Van den Hamer, C.J.A., and T.U. Hoogenraad. 1984. Trace Elem. Med. 1, 84-87.

23. Hoogenraad, T.U., and C.J.A. Van den Hamer. 1983. Acta. Neurol. Scand. 67, 356-364.

24. Brewer, G.J., G.M. Hill, A.S. Prasad, Z.T. Cossack and P.R. Rabbani. 1983. Ann. Intern. Med. 99, 311-320.

25. Cossack, Z.T. and J. Bouquet. 1986. Acta Pharm. Toxicol. (in press).

26. Van den Hamer, C.J.A., T.U. Hoogenraad and E.R.K. Klompjan. 1984. Trace Elem. Med. 1, 88-90.

27. Cossack, Z.T. and C.J.A. Van den Hamer. (submitted).

AN ORAL ZINC LOADING TEST USING AN ENRICHED STABLE ZINC ISOTOPE

C.J.A. Van den Hamer, J.J. Kroon, P.S. Tjioe
Interuniversity Reactor Institute, 2629 JB Delft, The Netherlands

C. Cornelisse
St. Ignatius Hospital, 4800 RK Breda, The Netherlands

Introduction

An oral zinc loading test is a tool used to gain insight in the zinc-metabolism under special conditions, such as disease and particular dietary habits (e.g., 1, 2). In this test an amount of zinc is given orally and the appearance in and the clearance from the plasma as function of time post dose is measured. Use of radio-zinc (e.g., ^{69m}Zn, phys.$t_{1/2}$=13.9 h) has the advantage that its appearance in the plasma is not superimposed on the "natural" zinc already present. Moreover, its use offers the opportunity to determine the specific activity of the zinc in the plasma and to estimate from this with how much "natural" zinc the radio-zinc had been diluted before appearing in the plasma. This in turn is a measure of the zinc status of the patient (3).In the following the use of enriched, stable ^{68}Zn (natural abundance: 18.6%) instead of radio-zinc will be described.

Materials

^{68}Zn (98.8 % enriched) was purchased from Rohstoff-Einfuhr GmbH (Düsseldorf, FRG) as ZnO. It was dissolved in a slight excess of 1 N HCl and diluted with 0.1 M Na-acetate buffer pH 5.6.

HNO_3, H_2SO_4, NH_4OH and H_2O_2 (Suprapur; Merck, Darmstadt, FRG) were used undiluted. APDC (ammonium pyrolidine dithiocarbamate; Spectrosol) was obtained from BDH Chemicals Ltd, Poole, U.K. The

Trace Element – Analytical Chemistry in Medicine and Biology, Vol. 4

water used was demineralized water, twice distilled in a quartz apparatus.

The quartz used for the irradiation vials (ϕ 8 mm; length 70 mm) was of synthetic quality (Suprasil, Heraeus Quarz-Schmelze, Hanau, FRG).

Zinc Loading Test

The patient, fasting since 23:00 the previous day till the end of the test, received at 9:00 a dose of 50 mg ^{68}Zn, dissolved in about 25 ml 0.1 M Na-acetate buffer pH 5.6. Samples of heparine blood were drawn just before administering the dose and at 1, 2, 3, 4, 5 and 6 h post dose and centrifuged. Urine was collected in separate portions, usually 0-3, 3-6, 6-9 and 9-24 h post dose. ^{68}Zn and ^{64}Zn were determined by destructive neutron activation analysis in all samples.

Determination of ^{68}Zn and ^{64}Zn

The samples were subjected to chemical separation before (to remove, e.g., the Na) and after irradiation (to separate the radiozinc from other isotopes produced) (Fig.1).

The pre-separation for serum and urine, described earlier by Janghorbani et al. (4),was as follows:
1 ml sample was destructed overnight with 1 ml HNO_3 at 140°C in a closed Teflon destruction bomb. After cooling the digest was transferred with water to a centrifuge tube and adjusted to pH 2.0 with NH_4OH. After addition of 0.4 mg Pb (as acetate) as carrier, the Zn was co-precipitated using APDC.
The precipitate was dissolved in 1 ml HNO_3 and transferred to a quartz irradiation vial. After evaporation of the HNO_3 under an infrared lamp the vial was sealed and irradiated for 12 h in a thermal neutron flux of $10^{13}.s^{-1}.cm^{-2}$.

After the irradiation the Zn-nuclides (^{69m}Zn and ^{65}Zn) were further purified as described before (5):
After opening the vial its content was taken up in 5 ml H_2SO_4 and destructed at 250°C with H_2O_2 till clear. After evaporation of the H_2SO_4 the residue was taken up in 8 M HCl and passed over a Dowex 2x8 column (4 ml). After washing the column with 20 ml 0.5 M HCl and eluting the zinc with 20 ml 0.02 M HBr, the zinc was precipitated by adding a solution containing 20 mg $ZnSO_4 \cdot 7H_2O$, 4.5 g NH_4Cl and 0.5 g Na-acetate, heating to boiling point and adding 5 ml 10 % $(NH_4)_2HPO_4$. After 30 min the precipitated Zn-phosphate

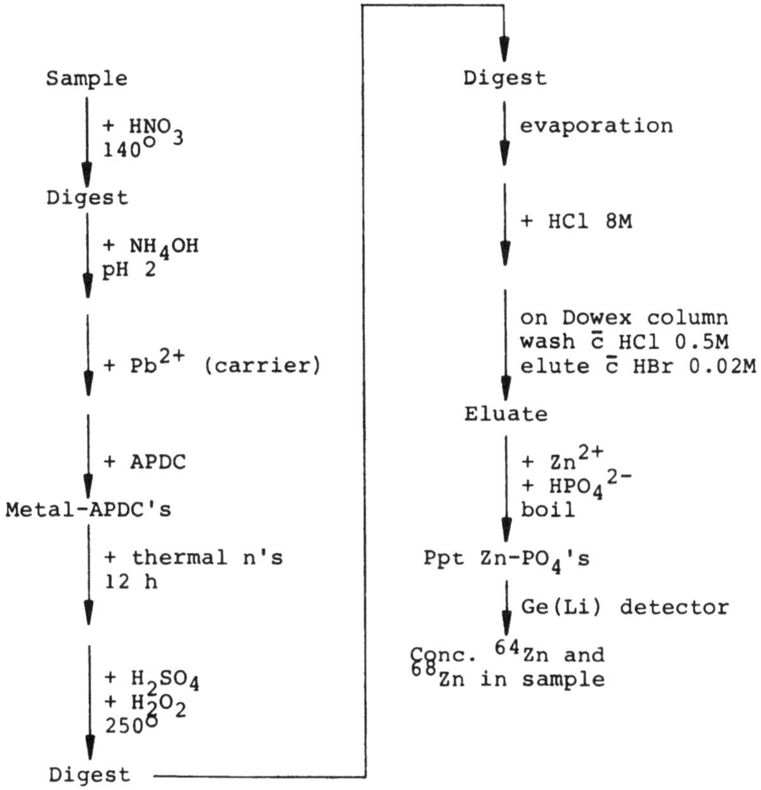

Fig.1. Pre- and post-irradiation separation of plasma and urine samples.

was filtered over a Millipore filter and the ^{69m}Zn and ^{65}Zn, originating from the ^{68}Zn and ^{64}Zn, respectively, were counted on a Ge(Li) detector.

Calculations

Since almost no ^{64}Zn (viz., 0.56 %) was present in the enriched ^{68}Zn, all ^{64}Zn must have come from other sources than the dose. Given the natural abundance of the various isotopes, one can calculate from the ^{64}Zn how much ^{68}Zn came from endogenous origin and, knowing the total amount of ^{68}Zn in plasma or urine, how much came from the dose.

The γ-spectrum showed 2 main Zn-peaks, viz., at 0.44 MeV from the ^{69m}Zn and at 1.12 MeV from the ^{65}Zn.

Results

Figs. 2 and 3 show examples of the changes in the plasma of the zinc from the dose and from other (endogenous) sources as function of time post dose during a Zn-loading test in case of a control person and of a patient with a disturbed Zn-metabolism, respectively.

Discussion

The curve of the total Zn in plasma during a normal Zn-loading test (Fig.2) has a maximum at 2 h post dose. This is in agreement with most curves published by others. Also the maximal value, about 2 - 2.5 µg/ml, is a common value for Zn-loading tests in which a dose of 50 mg Zn has been given.

The ^{68}Zn originating from the dose can be expressed as fraction of the total plasma-Zn. For the control curve shown in Fig.2, at 2 h post dose, i.e., at the maximal Zn-value in the plasma, this is about 0.27. However, it would be more appropriate to look at this

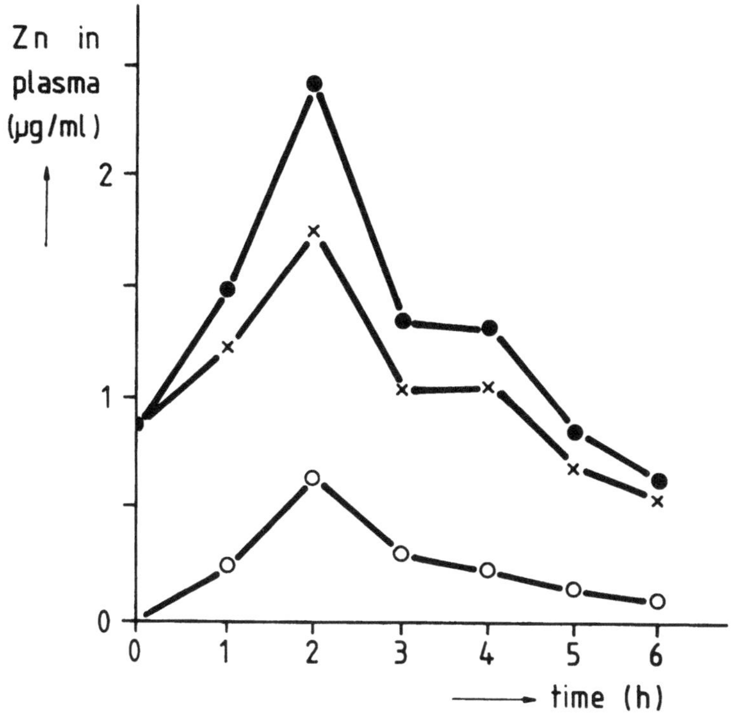

Fig.2. Plasma-Zn as function of time after a dose of 50 mg Zn as acetate, given to a healthy control. X——X: Zn from endogenous sources; O——O:Zn from the dose; ●——●: total Zn.

^{68}Zn as fraction of the <u>increase</u> of total Zn. From Fig.2 a value of 0.42 for this fraction is estimated. This value agrees quite well with the one estimated earlier, when ^{69m}Zn instead of ^{68}Zn was used (3), but when all other conditions were comparable. As has been argued before, this lowering of the "specific activity" of the ^{68}Zn from the dose must be due to mixing of this Zn with endogenous Zn. However, it cannot be just mixing with Zn present in the gut: to lower the "specific activity" to 0.42, the dose of 50 mg would have to be mixed with 70 mg Zn! The main dilution of the dose with endogenous Zn will probably take place in the mucosal layer of the intestinal wall: only a small percentage of the dose will be taken up in this layer. Details of the uptake of Zn are not yet known, but there is enough Zn in this layer (6) to explain the extent of the dilution at least in order of magnitude.

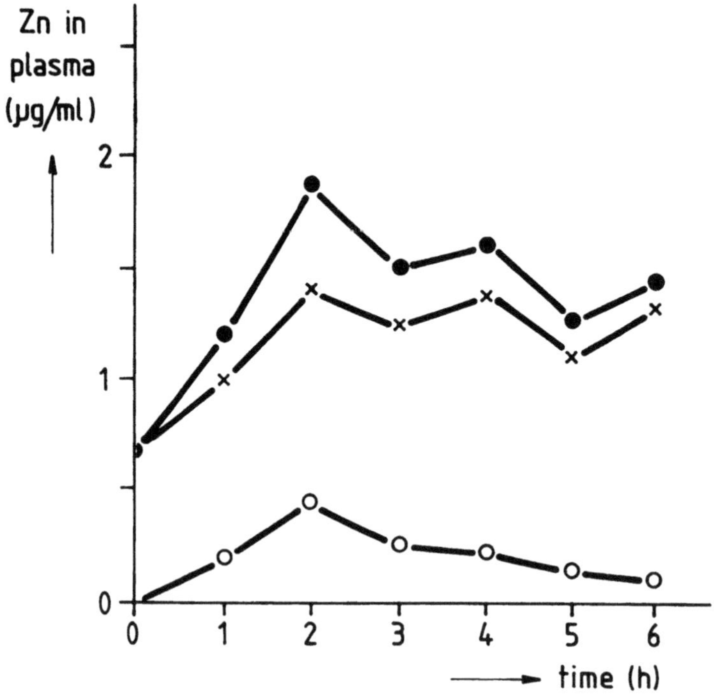

Fig.3. Plasma-Zn as function of time after a dose of 50 mg as acetate, given to a patient with yet unknown disturbance of her Zn-metabolism. Symbols as in Fig.2.

If this reasoning is correct, this dilution will be a measure of the Zn present in the mucosal wall which, in turn, will depend on the Zn-status of the patient.

The Zn-loading test of a patient with a disturbed Zn-metabolism was clearly abnormal: the clearance of the Zn from the plasma was delayed (Fig.3). The precise nature of the defect is not yet known but several other parameters of the patient's Zn-status were abnormal and she had given birth to a child who was Zn-deficient but who quickly responded to Zn-supplementation.

Conclusion

The oral zinc-loading test has frequently been used to study the zinc-metabolism (often in combination with various other parameters). Use of radio-zinc opens the possibility to measure the specific activity of the zinc entering the blood circulation. The present article shows that - instead of natural zinc or radio-zinc - an enriched stable isotope can also be used. This is particularly of interest in those cases in which the use of radio-nuclides is unacceptable, e.g., for small children or during pregnancy.

A procedure has been described for the determination of ^{68}Zn in plasma and urine and for the determination of the extent of dilution of the dose, using neutron activation analysis. Of course, use of a mass-spectrograph could be substituted for the activation analysis.

References

1. Dinsmore, W.W., J.T. Alderdice, D. McMaster, C.E.A. Adams and A.H.G. Love. 1985. Lancet i, 1041-1042.

2. Valberg, L.S., P.R. Flanagan, J. Brannan and M.J. Chamberlain. 1985. Amer.J.Clin.Nutr. 41, 37-42.

3. Van den Hamer, C.J.A., C. Cornelisse, T.U. Hoogenraad and J.P. Van Wouwe. 1985. In: Trace Element Metabolism in Man and Animals (C.F. Mills, I. Bremner and J.K. Chesters, eds.). Commonwealth Agricultural Bureau, U.K., p.689-691.

4. Janghorbani, M., B.T.G. Ting, N.W. Istfan and V.R. Young. 1981. Amer.J.Clin.Nutr. 34, 581-591.

5. Tjioe, P.S., J.J.M. De Goeij and J.P.W. Houtman. 1977. J.Radioanal.Chem. 37, 511-522.

6. Iyengar, G.V., W.EW. Kollmer and H.J.M. Bowen. 1978. The Elemental Composition of Human Tissues and Body Fluids. Verlag Chemie, Weinheim, FRG.

METAL-DEFICIENT DIETS CAUSE TISSUE-SPECIFIC ALTERATIONS OF TRACE ELEMENTS

J. Vandenhaute, W. Maenhaut
Institute for Nuclear Sciences, University of Ghent, B-9000 Ghent, Belgium.

H.A. Van Rinsvelt, R.W. Hurd, J.M. Andres
Departments of Physics, Food Science and Human Nutrition, and Pediatric Gastroenterology, University of Florida, Gainesville, Florida 32611, U.S.A.

Introduction

Numerous studies in the past have shown that certain diseases are closely correlated with alterations of trace element levels. For the past 5 years, our research groups have been involved in studies connected with Reye's syndrome and other neurological disorders (1, 2). Reye's syndrome (RS) is a childhood illness characterized by fatty deposition in viscera (liver, heart, kidney) and encephalopathy (3). The syndrome is frequently preceded by illness, usually influenza or chicken pox (4), and it is associated with decreased serum Se and Cu levels. Also, aspirin intake has been highly correlated with RS (5) and aspirin is a potent metal chelator (6). On the other hand, several studies established biological interactions among elements. Se decreases the toxicity of Cd, Hg, As and Cu (7), while the effects of toxic concentrations of Se can in turn be prevented by some of these elements (8). Interactions of Cu with other elements are also recognized.

In this study particle induced X-ray emission analysis (PIXE) was used to investigate the effect of alterations in diet composition (mainly Cu and Se deficiency) on the trace element levels of several tissues in mice. Subgroups of the mice received influenza A (H_1N_1) virus and/or salicylate in addition to the metal-deficient diet. The study included a microscopic examination of heart and liver sections, and elemental determinations in up to 7 tissues (i.e. blood, kidney, liver, spleen, thymus, heart and pancreas). However, only the results of the PIXE analyses of liver, spleen and thymus will be reported here.

Trace Element – Analytical Chemistry in Medicine and Biology, Vol. 4

Materials and Methods

The investigations were carried out on 108 CD1 male mice (Charles River, MA) of 22 to 24 days old (weight 20 g). The mice were divided into 12 experimental groups (A to M), and 1 control group (CON) of 12 animals. They were housed in stainless steel metabolic cages containing 3 animals each, and were kept in the Infectious Disease Quarantine Section of the Animal Department, Shands Hospital, University of Florida. The mice had at libitum access to diet and distilled water. The 5-week treatments of the different groups are shown schematically in Table 1.

Table 1 : Treatments (5 weeks) of the mice.

Groups :	0		1			2			3			4	
Subgroups :	CON	A	B	C	D	E	F	H	I	J	GK	L	M
Diet 1 (1)	X												
Diet 2 (2)		X	X	X	X	X	X						
Diet 3 (3)								X	X	X	X	X	X
+ Cu in H_2O		X	X	X									
+ Se in H_2O											X	X	X
+ virus			X	X		X	X		X	X		X	X
+ salicylate				X			X			X			X

(1): normal lab chow
(2): mainly Cu-deficient diet
(3): mainly Se-deficient diet

Diet 1 (rodent lab chow #5001) was purchased from Ralston Purina Co., St. Louis, MO, while diets 2 and 3 originated from ICN Nutritional Biochemicals, Cleveland, OH. The Cu content of the Cu-containing drinking water was about 5 mg $CuSO_4$ per liter, while Se was added in a quantity of 0.1 mg Na_2SeO_3 per liter to the Se-containing water. Mice were inoculated intranasally with influenza A (H_1N_1) virus solution as described by Reuman et al. (9). Sodium salicylate (# 53007 from Sigma Co., St. Louis, MO) was administered daily via an intraperitonial injection (50 mg/Kg), starting 5 to 6 days before sacrifice. As virus and salicylate administration did not seem to alter the trace element levels (see Figures 1 and 2), all experimental subgroups of the same diet and drinking water were combined. This results in 4 main experimental groups (numbered 1 through 4 in Table 1). The animals were killed by decapitation and allowed to bleed out. Their weight just before death varied from 22 to 31 g. Blood, kidney, liver, spleen, thymus, heart and pancreas were collected, using stainless steel dissection tools. All organs were rinsed in physiological saline solution, blotted dry, weighed and finally lyophylized.

The 3 diets and all organs were analyzed by PIXE for up to 16 elements. The diets were also analyzed by instrumental neutron activation analysis (INAA), using long-lived isotopes. For each sample analyzed by PIXE, an internal standard was added, 3 targets were prepared and each target was bombarded for 40 μC with 2.4 MeV protons. Two different sample/target preparation procedures were applied. The "brittle fracture technique" was used for doping, homogenizing and target preparation of all liver samples. The spleen and thymus samples, on the other hand, were too small to be handled with this technique (less than 10 mg), and the "wet digestion technique" was applied in these cases. The experimental PIXE set-up and analytical procedures have been described in detail previously (10 - 13). The INAA procedure used for the diet samples was similar to that described by Schutyser et al. (14).

Results and Discussion

The results of the analysis of the 3 diets and the recommended minimal dietary allowances (RDAs) for mice (15) are given in Table 2. The RDAs are expressed as concentrations needed in the diet. Although mice have been used extensively as laboratory animals, research on nutrient requirements for this species has received little attention (15). As rats are very similar to mice, and because these animals were studied more carefully, the RDAs for rats are also included in Table 2. An additional problem is formed by the numerous genetic strains of mice, who all have different dietary requirements (16). Genetic factors may also affect the degree of response to a nutritional deficiency (17). The availability of elements is affected by the chemical form of these elements (i.e. their speciation), as well as by the concentration of other nutrients.

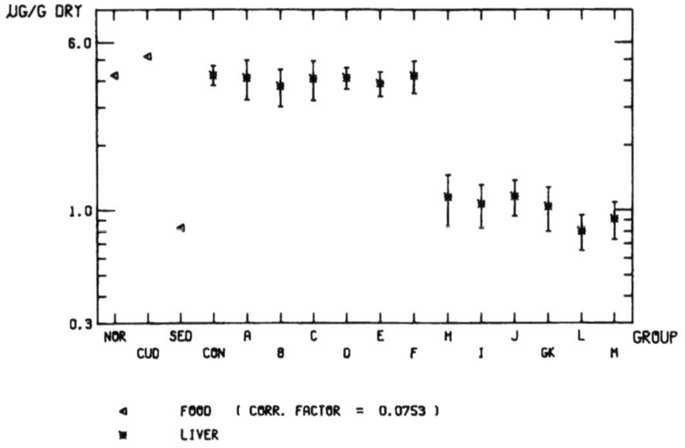

Fig. 1 : Mn concentrations in the three diets and in the livers of the various animal groups. Error bars, associated with the diet data, represent the analytical standard deviation. Data for the animal groups are $\bar{x} \pm s$, with $\bar{x}$ the mean value and s the standard deviation, as obtained by averaging over the various mice of each group. The diet Mn concentrations plotted are the product of the actual levels and a food correction factor (this factor makes the plotted value for diet 1 equal to the control group liver concentration).

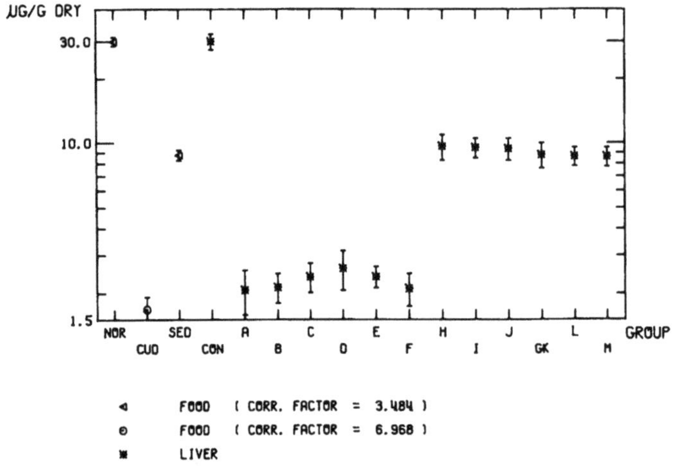

Fig. 2 : Rb concentrations in the three diets and in the livers of the various animal groups. See caption of Fig. 1 for more explanation.

Table 2 : Elemental concentrations[a] in the 3 diets (in µg/g dry weight). Comparison with recommended dietary allowances (RDAs).

	diet 1	diet 2	diet 3	RDAs mice	RDAs rats
Elem.	µg/g	µg/g	µg/g	µg/g	µg/g
Na	3640	2810	1614		500
K	14590	6040	12300	2000	3600
Ca	13000	5960	13460	4000	5000
Ti	24.9	< 1.0	< 1.0		
Cr	0.857	1.351	1.158	2.0	0.30
Mn	56.2	68.7	11.02	45	50
Fe	269	48.0	355	25	35
Co	0.339	0.023	0.099		
Ni	2.14	0.420	0.500		
Cu	13.72	1.680	11.85	4.5	5.0
Zn	80.7	30.5	36.1	30	12
As	0.417	< 0.17	< 0.26		
Se	0.403	< 0.12	< 0.12		0.10
Br	9.62	1.938	1.520		
Rb	8.61	0.240	2.51		
Sr	22.2	3.10	7.59		
Mo	1.67	< 0.80	< 0.80		
Cd	< 1.8	< 1.27	< 2.5		
Cs	0.0498	0.0127	0.0375		
Ba	15.5	< 4.2	< 8.4		
Pb	1 *	1.71	< 0.8		

a : Analytical errors are typically < 5% for Na, Ca, Cr, Mn, Cu, Br and Sr, 5-10% for K, Ti, Fe, Co, Zn and Rb, and > 10% for the other elements.
* : for information only (may be inaccurate).

Table 2 indicates that there are significant differences between the different diets. Also, for certain elements, concentrations in the deficient diets are below the RDA values. Diet 3 is deficient in Mn, and the Cu-deficient diet (no. 2) is also too low in Se. The Zn level in diets 2 and 3 is close to the RDA value. However, Zn is an important trace element because of its role in numerous organic and protein complexes and its interactions with other elements. One can conclude that the mineral mixtures, which are used in commercial synthetic diets, often exhibit deleterious imbalances and inadequacies, as was already indicated by Greenfield and Briggs (18).

The results of the PIXE analysis for liver, spleen and thymus are summarized in Tables 3, 4 and 5. Numeric results and standard deviations are reported for control animals only. For the other groups the percentage deviation relative to the control group is given.

Although the K and Ca levels are significantly different in the 3 diets, the concentrations in the organs remain generally unchanged. This can be explained by the fact that K and Ca are strongly homeostatically controlled and are available well above RDAs in the diets. The high Ca concentration (+114%) in thymus for group 2 (Cu and Se deficient diet) is, however, remarkable. Mn is significantly lowered in liver of groups 3 and 4 (diet 3), and the percentage decrease is very similar as for the element in the diet (see also Figure 1). There is a slight effect of diet on the Mn level of thymus, but no effect on spleen can be recognized.

Table 3 : PIXE results for LIVER, expressed in μg/g dry weight.

	control group	% deviation for each of the 4 groups, relative to the control group.[a]			
Elem.	$\bar{x} \pm s$ (N=12)	1	2	3	4
K	10390 ± 340	-0.5	-2.1	7.8 +	1.3
Ca	110 ± 19	-3.0	-0.3	-3.6	-6.1
Mn	4.23 ± 0.44	-5.8	-3.8	-73.4 *	-78.3 *
Fe	860 ± 200	-13.6 -	165 *	19.7	3.3
Cu	14.23 ± 0.60	-9.6 -	-58.3 *	-3.7	-6.9 +
Zn	89.4 ± 4.6	-3.9 -	-5.1 +	-3.8	-17.1 *
Se	4.58 ± 0.30	-83.4 *	-77.8 *	-91.6 *	-31.3 *
Br	4.70 ± 0.39	-49.3 *	-37.9 *	-60.5 *	-68.9 *
Rb	30.0 ± 2.6	-92.7 *	-92.1 *	-68.3 *	-71.0 *
Sr	0.51 ± 0.21	-26.8	5.9	34.0	-33.0 -
Mo	2.93 ± 0.70	-66.2 *	-56.8 *	-65.6 *	-64.5 *

a : -, +, * after the % deviation indicate that the concentration is significantly different ($p < 0.05$) from the control value in 1, 2 or all 3 subgroups of the experimental group under consideration.

Table 4 : PIXE results for SPLEEN, expressed in μg/g dry weight.

Elem.	control group $\bar{x} \pm s$ (N=12)	% deviation for each of the 4 groups, relative to the control group.[a] 1	2	3	4
K	18730 ± 990	-5.4 -	-3.0 -	-5.1	-4.3
Ca	164 ± 46	-15.4 -	32.9 -	-12.4 -	-2.4
Mn	1.42 ± 0.26	-13.6	-4.7 -	1.2	6.3
Fe	5300 ± 860	-40.4 *	-44.2 *	5.9	11.3 -
Cu	4.22 ± 0.48	-8.8 -	-78.7 *	-2.1	4.7
Zn	84.0 ± 3.9	-0.2	15.0 +	0.2	1.4
Se	2.71 ± 0.14	-48.5 *	-43.5 *	-60.9 *	-11.2 +
Rb	33.4 ± 3.4	-93.1 *	-93.1 *	-66.9 *	-68.7 *
Sr	0.21 ± 0.53	-38.1	-30.2	66.7	100
Mo	0.82 ± 0.42	-75.2 *	-59.8 +	-49.6 -	-57.7 +

a : see footnote a of Table 3.

Table 5 : PIXE results for THYMUS, expressed in μg/g dry weight.

Elem.	control group $\bar{x} \pm s$ (N=12)	% deviation for each of the 4 groups, relative to the control group.[a] 1	2	3	4
K	13200 ±2300	13.1	-13.4 +	9.1	20.1 -
Ca	116 ± 21	7.2	114 +	15.5	-4.6
Mn	0.67 ± 0.23	31.8 +	67.2 +	-25.9	-32.0 +
Fe	235 ± 50	-25.5 +	7.8	-1.3	0.7 -
Cu	4.1 ± 1.2	-5.7	-38.4 *	8.0	-3.4
Zn	58.4 ± 8.4	14.2 -	-4.1	5.5	15.3 +
Se	0.56 ± 0.20	-13.1	-18.5	-53.2 *	16.1
Rb	25.1 ± 4.2	-92.0 *	-94.1 *	-64.7 *	-64.1 *
Sr	0.12 ± 0.13	-36.1	189	55.6	-27.8
Mo	0.15 ± 0.20	-66.7	31.1	-28.9	-48.9

a : see footnote a of Table 3.

The Fe content of diet 2 is rather low (of the same order as the RDA level). This resulted in a decrease of Fe in spleen for groups 1 and 2. The concentrations in other organs were not lowered significantly. The important increase of Fe in the liver of group 2 (+165%) is most likely caused by the very low Cu level in the diet. As Fe and Cu are essential for haemoglobin production, low Cu concentrations possibly inhibit this synthesis, so that an accumulation of Fe in the organ may be expected. Williams et al. (19) found that anaemia develops in Cu-deficient rats, and that there is a selective accumulation of hepatic Fe. Supplementation of Fe, either parenterally or orally, incompletely corrected the anaemia (19).

The results for Cu and Se will be discussed in detail elsewhere. The Zn concentrations did not differ from the control values for all groups and all organs, except for the livers of group 4. As Br is volatilized during the wet digestion, this element is reported for liver only. Br is lowered by 40% to 70% in the various experimental groups. This element is a suspected essential trace element (20). Huff et al. (21) found that 3.75 mg/Kg of bromine in the diet reversed the growth depression caused by feeding iodinated casein. The levels of Br in diets 2 and 3 were lower than this value. The Rb content for each of the three organs in groups 1 through 4 was decreased in the same proportion as for the diets (i.e. by 90% for diet 2, and by 70% for diet 3). The Rb levels for liver in the various mice subgroups are compared with the Rb concentrations in the diets in Figure 2. The observations for this element suggest that, for the concentration range of 0.2-10 mg Rb per Kg diet, there is a close correlation between Rb intake and tissue content for liver, spleen and thymus. Rb is present in tissues in cationic form and biological interactions are not known. The Mo concentrations are significantly lowered in liver and spleen. The concentrations in thymus were too low for precise and accurate PIXE analysis.

This study demonstrates that the levels of certain trace elements in different organs depend strongly on the diet composition. Another important conclusion is that there is a large difference between the composition of commercial synthetic diets and normal lab chow. In experimental work with animals a lot of parameters must be controlled thoroughly. Diet seems to be an often neglected experimental variable, whereas it should receive primary attention (22). When experiments are planned with diets which are deficient in one or more nutrients, the composition of control and deficient diets must be similar for all other nutrients, so that reliable conclusions can be drawn concerning possible interactions among trace elements. In general, data from controlled experiments, designed to establish the minimum nutrient requirements for laboratory animals, are needed to increase the efficiency, and in some cases the validity of biomedical research.

Acknowledgements

We thank Prof. J. Hoste for his continued interest and support, and J. Cafmeyer for technical assistance. We benefited from financial support provided by the Belgian "Nationaal Fonds voor Wetenschappelijk Onderzoek" and the "Interuniversitair Instituut voor Kernwetenschappen".

References

1. Hurd, R.W., H.A. Van Rinsvelt, B.J. Wilder, B. Karas, W. Maenhaut and L. De Reu. 1984. Selenium, zinc and copper changes with valproic acid : Possible relation to drug side effects. Neurology 34, 1393-1395.

2. Van Rinsvelt, H.A., R.W. Hurd, J.M. Andres and B.J. Wilder. 1986. Review of PIXE studies of metal involvement in Reye's syndrome. Nucl. Instr. and Meth. B14, 221-225.

3. Reye, R.D.K., G. Morgan and J. Baral. 1963. Encephalopathy and fatty degeneration of the viscera. A disease entity in childhood. Lancet 2, 749-752.

4. Linneman, C.C. Jr., L. Shea and C.A. Kaufman. 1974. Association of Reye's Syndrome with virus infection. Lancet 2, 179-182.

5. Glezen, W.P.. 1982. Aspirin and Reye's Syndrome. Am. J. Dis. Child. 136, 971-972.

6. Schubert, J.. 1960. The Relationship of Metal-Binding to Salicylate Action. In: Metal-Binding in Medicine (M.J. Sevin and L.A. Johnson, eds.). Lippincott, Philadelphia. pp. 325-328.

7. Magos, L. and M. Webb. 1980. The interactions of selenium with cadmium and mercury. CRC Critical Reviews in Toxicology 8, 1-42.

8. Hill, C.H.. 1975. Interrelationships of selenium with other trace elements. Fed. Proc. 34, 2096-2100.

9. Reuman, P.D., E.M. Ayoub and P.A. Small Jr.. 1983. Influenza infection in infant mouse. Pediatr. Res. 17, 338-343.

10. Maenhaut, W., L. De Reu, H.A. Van Rinsvelt, J. Cafmeyer and P. Van Espen. 1980. Particle-induced X-ray emission (PIXE) analysis of biological materials : Precision, accuracy and application to cancer tissues. Nucl. Instr. and Meth. 168, 557-562.

11. Maenhaut, W. and H. Raemdonck. 1984. Accurate calibration of a Si(Li) detector for PIXE analysis. Nucl. Instr. and Meth. B1, 123-136.

12. Maenhaut, W., L. De Reu and J. Vandenhaute. 1984. Comparison of different sample and target preparation procedures for PIXE analysis of biological materials. Nucl. Instr. and Meth. B3, 135-140.

13. Duflou, H., W. Maenhaut and J. De Reuck. 1985. Application of PIXE analysis to the study of the regional distribution of trace elements in normal human brain. Biol. Trace Elem. Res. (submitted).

14. Schutyser, P., W. Maenhaut and R. Dams. 1978. Instrumental neutron activation analysis of dry atmospheric fall-out and rainwater. Anal. Chim. Acta 100, 75-85.

15. National Research Council. 1978. Nutrient Requirements of Laboratory Animals; no. 10, Third revised edition, 1978. National Academy of Sciences, Washington, D.C.

16. Lee, Y.C.P., J.T. King and M.B. Visscher. 1953. Strain differences in vitamin E and B12 and certain mineral trace-element requirements for reproduction in A and Z mice. Am. J. Physiol. 173, 456-458.

17. Hurley, L.S. and L.T. Bell. 1974. Genetic influence on response to dietary manganese deficiency. J. Nutr. 104, 133-137.

18. Greenfield, H. and G.M. Briggs. 1971. Nutritional methodology in metabolic research with rats. Annu. Rev. Biochem. 40, 549-572.

19. Williams, D.M., F.S. Kennedy and B.G. Green. 1983. Hepatic iron accumulation in copper-deficient rats. Br. J. Nutr. 50, 653-660.

20. Mertz, W.. 1980. Trace elements in nutrition. In: Trace Element Analytical Chemistry in Medicine and Biology (P. Brätter and P. Schramel, eds.). Walter de Gruyter, Berlin. pp. 727-744.

21. Huff, J.W., D.K. Bosshardt, O.P. Miller and R.H. Barnes. 1956. A nutritional requirement for bromine. Proc. Soc. Exp. Biol. Med. 92, 216-219.

22. Nielsen, F.H.. 1985. Critical Review : The importance of diet composition in ultratrace element research. J. Nutr. 115, 1239-1247.

MOBILIZATION OF ZINC AND STRONTIUM FROM BONE IN ZN DEFICIENT RATS.

D. Berg, W.E. Kollmer

Abt. für Nuklearbiologie, Gesellschaft für Strahlen- und Umweltforschung m.b.H. München, Ingolstädter Landstraße 1, D-8042 Neuherberg

Introduction

We have shown previously (1), that in rats fed a Zn deficient diet the percentage Zn resorption from the diet was elevated along with a decreased excretion of Zn. Other investigators have demonstrated that during a period of Zn depletion the amount of Zn in bone decreases rather early (2-6) whereas the level of Zn in the soft tissues remains constant for some time (4,6-8). Since approximately 27% of the total body Zn of the rat is found in the skeleton (9), bone may be assumed to function as a storage organ for Zn which may be mobilized during periods of Zn deficiency (10-14) to maintain a constant level in the soft tissues.

This investigation was designed to check especially the influence of Zn deficiency on the release of Zn and alkaline earth elements from the deep layers of bone mineral.

Material and Methods

Male rats of a Wistar derived strain received an intraperitoneal injection of 0.56 MBq Sr 85 along with Zn 65. The specific activities were 0.54 MBq/μg Sr and 12.3 MBq/μg Zn. The animals were distributed to 4 experimental groups. Groups 1 and 3 were fed a standard commercial diet (Altromin) containing 80 μg Zn/g which is relatively high. - Approximately 8 μg/g to 15 μg/g Zn/g have been found sufficient for the rat (15). - The other 2 groups were fed a Zn deficient diet containing 2 µg Zn/g. Group 2 received this Zn deficient diet from day 49 until day 82 after the injection of the tracers, group 4 from day 112 until day 187. Whole body measurements of radioactivity were made during the entire period of the experiment. At the end of the Zn deficient period the animals were killed along with those of a group fed the standard diet. Liver, kidneys, pancreas, spleen, heart, tibia, femur, parietal bone and the fur were recovered for radioactive measurements and stable calcium and zinc analyses. Total body radioactivity for both radioisotopes was measured in a whole body counter equipped with two 5"x4"NaJ(Tl) cristals. The radioactivity of the various organs was measured in a well type counter. In both cases γ-spectrometry was applied. Stable calcium and zinc were determined with flame atomic absorption spectrometry. Statistical significance was calculated by way of Students t-Test.

Results

The mean body weight of the animals on the standard diet was 271.8 $\pm$ 2.0 g at day 49 and 332.4 $\pm$2.0 g at day 112 after injection of the radioisotopes. At the end of Zn deficiency the mean body weight was significantly lower compared to that of the controls. Group 2 exhibited a mean weigt of 295.5 $\pm$ 2.0 g compared to 330.3 $\pm$ 2.8 g of group 1 (control), group 4 a weight of 366.6 $\pm$ 2.7 g compared to 402.0 $\pm$ 2.7 g of group 3 (control). The mean weights of the soft tissues were lower in the Zn deficient groups compared to those of the controls. No significant differences were found in the weights of the hard tissues.

Figure 1 depicts the whole body retentions of Zn 65. In all groups the biological halftime of the tracer increased towards the end of the experiment which may be adequately approximated by a simple exponential function. During the zinc deficiency the mean retentions exhibited a significant deviation from the controls. In the early period of Zn deficiency the halftime in-

Table 1 : Zn 65 in the soft tissues ($\bar{x} \pm s_{\bar{x}}$, N=5) as % of the dose per gram fresh weight 82 and 187 days after i.p. injection of the tracer

group	kidney 10^{-3}	liver 10^{-3}	heart 10^{-3}	spleen 10^{-3}	pancreas 10^{-3}	fur
control 82 d	33.76 ±.46	203.3 ±4.4	12.21 ±.29	9.83 ±.44	8.76 ±.66	7.00 ±.31
Zn def. 82 d	107.00 ±1.60 **	674.0 ±19.0 **	32.21 ±.49 **	37.30 ±1.20 **	26.80 ±1.60 **	7.81 ±.24 **
control 187 d	4.69 ±.14	26.0 ±.6	1.36 ±.06	1.41 ±.06	1.25 ±.04	2.95 ±.12
Zn def. 187 d	30.30 ±9.00 **	215.0 ±6.0 **	7.82 ±.67 **	9.15 ±.23 **	9.21 ±.44 **	3.58 ±.19 *

* 2p < 0.05 and ** 2p < 0.001 (t-test)

Table 2 : Stable and radioactive Zn ($\bar{x} \pm s_{\bar{x}}$,(N))as µg/g fresh weight and % of dose/g 82 and 187 days after i.p. injection of the tracer

group	tibia		femur		os parietale	
	st.Zn µg/g (10)	Zn65 %/g (5)	st.Zn µg/g (10)	Zn65 %/g (5)	st.Zn µg/g (10)	Zn65 %/g (5)
control 82 d	148.5 ±3.7	1.233 ±.046	192.9 ±4.2	0.982 ±.048	148.5 ±3.4	0.876 ±.024
Zn def. 82 d	117.4 ±3.0 *	1.191 ±.076	136.1 ±5.6 *	0.705 ±.010 **	111.3 ±5.0 *	0.918 ±.061
control 187 d	182.6 ±5.1	0.900 ±.044	197.0 ±12.0	0.703 ±.033	155.8 ±6.9	0.549 ±.009
Zn def. 187 d	154.7 ±3.5 *	0.996 ±.019	180.0 ±13.0 **	0.778 ±.021	118.9 ±5.1 **	0.517 ±.056

* 2p < 0.01 and ** 2p < 0.005 (t-test)

Table 3 : Stable calcium and Sr 85 in bone ($\bar{x} \pm s_{\bar{x}}$,(N)) in mg/g fresh weight and % of dose/g 82 and 187 days after i.p. injection of the tracer

group	tibia		femur		os parietale	
	Ca mg/g (10)	Sr85 %/g (5)	Ca mg/g (10)	Sr85 %/g (5)	Ca mg/g (10)	Sr85 %/g (5)
control 82 d	181.2 ±7.7	3.36 ±.12	135.1 ±3.9	2.36 ±.10	190.0 ±11.0	4.07 ±.20
Zn def. 82 d	163.6 ±4.0	3.36 ±.21	135.2 ±5.0	2.35 ±.07	164.0 ±10.0	4.05 ±.14
control 187 d	184.1 ±5.0	2.64 ±.11	148.2 ±4.6	1.54 ±.12	161.2 ±6.6	2.67 ±.16
Zn def. 187 d	184.6 ±6.7	2.63 ±.12	144.9 ±7.2	1.54 ±.16	150.3 ±7.2	2.54 ±.17

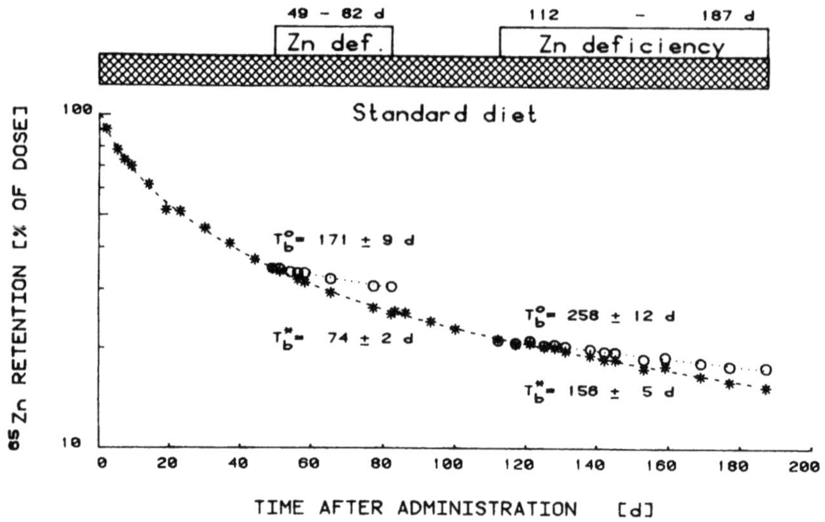

Figure 1: Whole body retention of Zn 65 ($\bar{x}$) in rats fed with standard (*) and those fed Zn deficient diet (o) after i.p. injection of the tracer. T_b= biological halftime

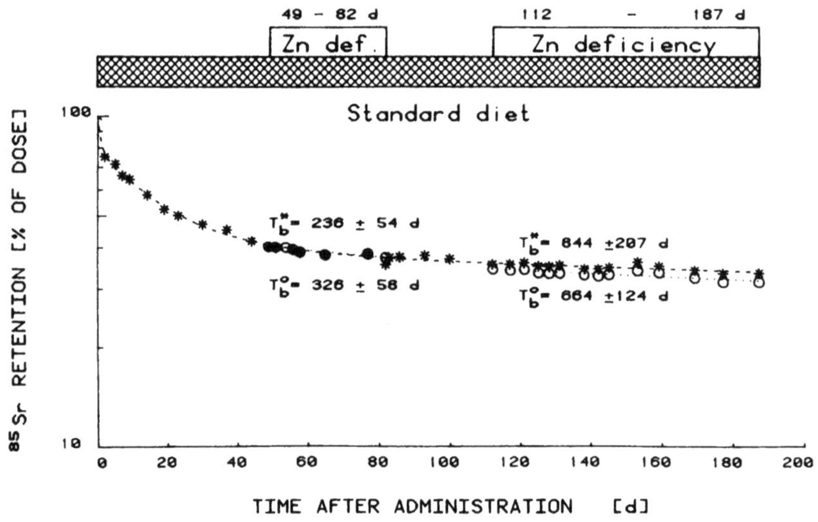

Figure 2: Whole body retention of Sr 85 ($\bar{x}$) in rats fed with standard (*) and those fed Zn deficient diet (o) after i.p. injection of the tracer. T_b= biological halftime

creased from 74 ± 2 d (controls, group 1) to 171 ± 9 d, (deficient animals, group 2), in the later one from 156 ± 5 (controls, group 3) to 258 ± 12 d (deficient group 4). Thus, the level of the Zn tracer relative to the controls became significantly elevated in the total body of the zinc deficient animals. This finding was also true for the kidneys, liver, heart, spleen, pancreas and the fur (Table 1). In bone (Table 2) Zn 65 was not different except of a minor decrease in the femur of the deficient animals. Stable zinc, however, (Table 3) was significantly decreased in the bones of the zinc deficient animals. In the liver no significant difference could be found between the deficient animals (23.1 ± 0.8 µg/g 82 days p.i. and 25.6 ± 1.7 µg/d 187 days p.i.) and the controls (25.0 ± 0.8 µg/g and 24.9 ± 1.7 µg/g respectively).

The whole body radioactivity of Sr 85, a tracer of bone mineral, is shown in Fig. 2. During the periods of zinc deficiency the total body retention of Sr 85 could also be described by a simple exponential function in the deficient animals as well as in the controls. No difference was found between these. The concentration of Sr 85 as well as the content of stable calcium in the hard tissues (Table 3) were also not significantly different between the deficient animals and those fed the normal diet. In the kidney and the liver the level of Sr 85 was only about 2×10^{-4} % of the dose and 4×10^{-3} % of the dose respectively. Since this was only slightly above the detection limit, these data cannot be used for detecting differences between the groups. In all other tissues examined the radioactivity of Sr 85 was below the detection limit.

Discussion

The curve describing the whole body retention of Zn 65 of all untreated animals between its injection and the end of the experiment is the sum of at least three exponential functions (Fig. 1). This points to the existence of three major compartments with a different turnover of zinc. As we have

shown earlier (1) the slowest of these compartments represents the skeleton or part of it. The interruption of the decrease of radioactive Zn in the whole body at the start of the periods of Zn deficiency demonstrates a decreased excretion of Zn due to homeostatic control. Zinc in the soft tissues is readily exchangeable. We have found previously (1) a halftime of 10-20 d in the soft tissues in rats on a standard diet. Due to the long time elapsed between the administration of the tracer and the start of Zn deficiency the tracer in the soft tissues had become very low before the start of Zn deficiency. In the bone samples taken at the end of the experiment practically the same amount of Zn 65 was found in the Zn depleted animals as in the controls. The differential loss of tracer from the whole body must therefore be attributed to loss from the soft tissues.
The level of stable zinc in the tibia, the femur and the parietal bone was significantly decreased during the periods of the zinc deficiency (Table 2) whereas the level in the liver was practically not influenced. This is consistent with the observations of other investigators (16) and indicates a considerable shift of Zn from the skeleton to the soft tissues due to homeostatic control processes.

The apparent contradiction between the behaviour of the tracer and that of stable Zn may be explained by the existence of at least two different fractions (pools) of zinc in bone, one being mobilized (depleted) during zinc deficiency (fraction I) and one remaining untouched (fraction II). During the periods of Zn depletion in the present experiment the tracer may have been localized exclusively in the fraction of bone which was not subjected to mobilization. In bone 40-50% of the sodium and up to 30% of the magnesium may be mobilized (17) in deficiency. The part which may be mobilized belongs to the exchangeable fraction while the part which is not mobilized belongs to the unexchangeable fraction which is localized in the deep layers of bone mineral. It may be expected that bone exhibits similar properties with respect to zinc.
Radioactive Sr is a good tracer for calcium in bone because its uptake and distribution in bone mineral is very similar to that

of calcium while it is much faster eliminated from the soft tissues due to its fast excretion in the urine. At the time of the start of Zn depletion Sr 85 must have been localized almost exclusively in the "deep" compartment of bone which normally has a very low turnover. Its retention followed a curve (Fig.2) which was practically the same in the Zn deficient animals and in the controls. In accordance with these findings of whole body measurements there was no difference between Sr 85 in the samples of bone taken at the end of the experiment in the Zn depleted animals and in the controls. The concentrations of stable calcium in the respective bone samples were also practically equal. Just as the findings of radioactive Zn in bone these findings indicate that the mineral in the deep part of bone was not influenced by Zn deficiency. Thus the mobilisation of stable Zn from bone was not accompanied by an enhanced resorption of mineral in the deep layers of bone. This is in agreement with Murray et al. (16) who found no influence of zinc status on the rate of bone resorption measured by the loss of H 3 labelled tetracycline. Our procedure was not suitable to detect a minor decrease in bone resorption or an influence on the rapidly exchangeable fraction of bone mineral. Thus it is not contradictory to Hurley et al. (8) who described a decreased resorption of Ca from bone in Zn deficient rats in the frame of a generally decreased turnover. Obviously Zn deficiency itself does not enhance bone resorption. If Zn deficiency is accompanied by Ca deficiency an enhanced release of Zn from bone may occur, however, due to enhanced bone resorption (18).

Summary

Young rats received an intraperitoneal injection of the radioactive tracers Zn 65 and Sr 85. 49 or 112 days later some of them were fed a Zn deficient diet for 33 or 75 days respectively. At the end of the experiment the stable Zn concentration in bone was lower in the Zn deficient animals compared to the controls. With contrast to stable Zn the radioactive tracer Zn

65 in bone was not influenced by Zn deficiency. Whole body measurements of Zn 65 indicate a decreased excretion of Zn during the periods of Zn depletion.At this stage apparently only radioactive Zn 65 contained in the soft tissues was available for excretion. Neither the whole body measurements of radioactive Sr 85 nor the concentrations of stable calcium and radioactive Sr 85 in bone were influenced by Zn depletion.

The findings indicate that bone mav contain at least two fractions of Zn, one functioning as a depot which may be depleted for regulating the Zn content in the soft tissues in the ease of dietary deficiency and another one which is not available for this. The mobilizable fraction may be identical with the readily exchangeable pool of Zn in bone whereas the unavaible Zn seems to be located in the deep fraction of bone mineral whose molecules are not readily exchangeable. Thus large amounts of Zn may be removed from bone while Zn and alkaline earth elements in the deep compartment of bone are not influenced. Our findings also indicate that Zn deficiency despite of its influence on Zn in bone does not lead to an increased resorption of bone.

Acknowledgement

For technical assistance we are grateful to Mrs. C. Fries, Mrs. M. Hoffmann, Mr. H. Ruhstorfer and Mrs. R. Schmidt.

References

1. Matsusaka, N., D. Berg, W.E. Kollmer. 1985. In: Trace Elements in Man and Animals (C.F. Mills, I. Bremner, J.K. Chesters, eds.). Commonwealth Agricultural Bureaux, p. 404

2. Macapinlac, M.P., W.N. Pearson, W.J. Darby. 1966. In: Zinc Metabolism, (A.S. Prassad, ed.). Charles C. Thomas Publ. Springfield Ill. p. 142

3. Swenerton, H., L.S. Hurley. 1968. J. Nutr., 8.

4. Huber, A.M., S.N. Gershoff. 1970. J. Nutr., 949.

5. Bergman, B. 1970. Acta radiol. Ther. Phys. Biol. 9, 420.

6. Bergman, B., K.L. Wing. 1974. Acta physiol. scand., 451.

7. Hurley, L.S. 1969. Amer. J. Nutr. 22, 1332.

8. Hurley, L.S., S.J. Gowan, G. Milhaud. 1969. Proc. Soc. exp. Biol. (N.Y.) 130, 856.

9. Zipkin, J. 1973. J. Wiley u. Sons, NY. 453.

10. Morris, E.R., R. Ellis. 1980. J. Nutr. 110, 1037.

11. Brown, E.D., W. Chan, J.C. Smith. 1978. Proc. Soc. Exp. Biol. Med. 157, 211.

12. Harland, B.F., M.R.S. Fox, B.E. Fry. 1975. J. Nutr. 105, 1509.

13. Momcilovic, B., B. Belonje, A. Giroux, B.G. Shah. 1975. Nutr. Rep. Int. 12, 197.

14. Flanagan, P.R. 1984. J. Nutr. 114, 493.

15. Pallauf, J., M. Kirchgessner. 1972. Z. Tierphysiol. Tierernährg. u. Futtermittelkunde 30, 193.

16. Murray, E.J., H.H. Messer. 1981. J. Nutr. 111, 1641

17. Mc Lean, F.C., M.R. Urist. 1961. In: Bone, an introduction to the Physiology of skeletal tissues. University of Chigaco Press

18. Hurley, L.S., M.E. Gershwin, M. S. Golub. 1985. In: Trace Elements in Man and Animals, (C.F. Mills, I. Bremner, J.K. Chesters, eds.). Commonwealth Agricultural Bureaux, p. 197.

STUDIES ON IRON OVERLOAD: EFFECTS OF VITAMIN C AND DESFERRIOXAMINE TREATMENT

Z. T. COSSACK
Division of Nuclear biotechnique, Interuniversity Reactor Institute, Delft, Holland and Odense University Human & Clinical Nutrition Programme, Institute of Biochemistry, Odense, Denmark.

Introduction

In recent years the application of the extrinsic radioiron (1,2) have provided a clear picture of the factors affecting the absorption and excretion of iron. The effect of dietary ascorbic acid on the absorption of iron has been described in several studies (3,4,5). Megadoses of vitamin C may result in progressive accumulation of iron overload. In addition, some investigations have suggested that megadoses of vitamin C may result in redistribution of iron in tissues (3,4,6). It has been suggested that in iron overload, vitamin C administration could be potentially toxic, particularly on cardiac function (4,6,7). Clover et al. (8) indicated that a greater dose of iron ^{59}Fe was retained in the spleen of scorbutic, rather than in control, male guinea pigs and liver uptake of iron was greater than the spleen, but decreased in the scorbutic animals. Cook et al. (3) suggested that megadoses of vitamin C may induce some redistribution of body iron stores that would mask an actual increase in total body iron.

Desferrioxamine (DSF) is currently the most effective chelator available for the treatment of iron overload in man (4,7,9,10). It chelates ferric iron for its urinary excretion. Some investigators (9,11) have suggested that subcutaneous infusion of DSF may also cause a reduction in the gastrointestinal absorption of iron. However, their observation has not been fully elucidated.

The objectives of the present study were: 1) to study the effect of

Trace Element – Analytical Chemistry in Medicine and Biology, Vol. 4

ascorbic acid and Fe-overload on the distribution of orally administered dose of ^{59}Fe in iron-overload states, and 2) to study the effect of subcutaneous DSF treatment on the gastrointestinal absorption of iron ^{59}Fe in the mouse model.

Materials and Methods

Female mice of the Swiss strain, weighing 23-25 g, were divided into various treatment groups as shown in table I. All animals were randomly placed on wire-bottom cages, five mice per cage, and housed in a temperature-controlled (25°) environment, with 12 hour light-dark cycle. All animals were fed a casein based diet (Hope farms, Woerden, Holland) and were allowed to drink distilled-deionized water ad libitum. The experimental period lasted 5 weeks. Groups C, D, E and F were given iron dextran (5 mg Fe/mouse, intramuscular) five times during the first 3 weeks (total 25 mg/mouse). Groups B and D were given vitamin C (2 g ascorbic acid per liter of drinking water). Desferrioxamine (Ciba-Geigy Ltd., Basel, Switzerland) was given subcutaneously to group F for a period of 24 hours (12 hrs. prior to the ^{59}Fe-loading test and 12 hrs. following the oral dose of ^{59}Fe). Each mouse received 2 mg of desferrioxamine (in distilled deionized water) per 12 hrs. of infusion using a syringe-type infusion pump (Syringe-Driver-Type MS-16, Graseby Dynamics, England). ^{59}Iron-loading test was conducted as described; following 12 hours of starvation, mice were challenged with 0.28 µg ^{59}Fe per mouse. The iron was given as a solution of ferrous sulfate (0.2 ml/mouse) labelled with ^{59}Fe ferric chloride and was administered by intragastric intubation using a modified needle and syringe. Whole body counting (WBC) was done within one hour (zero-time) following the intubation using whole body counter provided with two NaI crystals. Mice were refed 6 hours following the zero-time measurement. Whole body counting was conducted again after 96 hours (in the vitamin-C exp) following the zero-time measurement. Corrections were made for background and radioactive decay. ^{59}Iron whole body retention (% WBR) was calculated using the equation:

$$\% \, WBR = \frac{\text{WBC at 96 hrs (corrected)}}{\text{WBC at zero-time}} \times 100$$

This equation expresses the net ^{59}Fe-retention as % of dose given by intubation. Mice were sacrificed immediately following the ^{59}Fe-loading test. Organs were isolated, counted for ^{59}Fe and wet weights were recorded. In the desferrioxamine experiment, mice were sacrificed 12 hours following intubation. Whole gastrointestinal tract was isolated and counted.

Table 1. Experimental design

	Vitamin C Exp.				Desferrioxamine Exp.	
Treatment/groups	A	B	C	D	E	F
Fe-overload	-	-	+	+	+	+
Ascorbic acid	-	+	-	+	-	-
Desferrioxamine	-	-	-	-	-	+

Results

Table 2 shows the retention of the oral dose of ^{59}Fe in the whole body and in the spleen, liver, kidneys, GIT and heart, and the distribution of the retained dose in those tissues. As shown, whole body retention of ^{59}Fe was increased significantly ($p<0.005$) when mice were supplemented with vitamin C (1.5 fold increase). This effect was observed in normal (B vs. A) and in Fe-overloaded mice (D vs. C). In addition, Fe-overload has resulted in a decrease in ^{59}Fe retention (73% decrease) regardless of vitamin C supplementation (C vs. A and D vs. B).

In normal mice (without Fe-overload), vitamin C supplementation has resulted in a significantly higher retention of ^{59}Fe in the spleen,

Table 2. Whole Body Retention and Distribution of Orally Administered ^{59}Iron

Group	Vitamin C Experiment				Desferrioxamine Experiment	
	A	B	C	D	E	F
As % of given oral dose:						
- whole body retention	3.70 ± 1.7^{a}	5.60 ± 1.4^{b}	1.0 ± 0.1^{c}	1.5 ± 0.6^{d}	-	-
% of control	100	151	27	40	-	-
- spleen	0.48 ± 0.14^{a}	0.71 ± 0.20^{b}	0.12 ± 0.04^{c}	0.10 ± 0.06^{c}	-	-
- liver	5.71 ± 2.7^{a}	6.29 ± 1.97^{b}	5.03 ± 1.19^{c}	7.25 ± 3.97^{d}	-	-
- kidneys	1.35 ± 0.74^{a}	2.27 ± 1.56^{b}	0.18 ± 0.06^{c}	0.20 ± 0.14^{c}	-	-
- G.I.T	1.57 ± 0.50^{a}	2.25 ± 0.95^{b}	0.64 ± 0.11^{c}	0.79 ± 0.39^{d}	0.82 ± 0.22^{a}	0.78 ± 0.19^{a}
- heart	0.59 ± 0.37^{a}	0.54 ± 0.5^{a}	0.05 ± 0.02^{b}	0.06 ± 0.03^{b}	-	-
Distribution (as % of retained dose):						
- spleen	1.44 ± 0.62^{a}	1.25 ± 0.21^{a}	1.01 ± 0.34^{b}	0.99 ± 0.59^{b}	-	-
- liver	15.4 ± 3.4^{a}	11.33 ± 2.62^{b}	42.17 ± 18^{c}	39.9 ± 19^{c}		
- kidneys	3.54 ± 0.57^{a}	3.71 ± 1.89^{a}	1.69 ± 0.61^{b}	1.59 ± 0.50^{b}	-	-
- G.I.T	4.66 ± 0.57^{a}	3.93 ± 0.75^{b}	5.91 ± 1.14^{c}	6.84 ± 2.1^{d}	-	-
- heart	0.53 ± 0.15^{a}	0.50 ± 0.19^{a}	0.41 ± 0.11^{b}	0.46 ± 0.26^{b}	-	-
Total wet wt. of tissue (mg)						
- spleen	106 ± 18^{a}	158 ± 35^{b}	121 ± 18^{c}	102 ± 23^{a}	-	-
- liver	1785 ± 300^{a}	1730 ± 85^{a}	2413 ± 188^{b}	2293 ± 304^{b}	-	-
- kidneys	606 ± 65^{a}	573 ± 63^{a}	510 ± 71^{b}	476 ± 53^{b}	-	-
- G.I.T	2332 ± 165^{a}	2677 ± 292^{a}	3024 ± 432^{b}	2692 ± 407^{b}	2590 ± 220	2493 ± 185
- heart	138 ± 20^{a}	131 ± 7^{a}	115 ± 12^{b}	109 ± 11^{b}	-	-

* a, b, c, d means, not having common superscripts are significantly different ($p<0.05$).

liver, kidneys, and GIT. However, there was no effect on the total retention in the heart (B vs. A). The effect of vitamin C supplementation, to Fe-overloaded mice, on the tissues retention of ^{59}Fe was not consistant: it resulted in increased retention in the liver ($p<0.01$) and GIT ($p<0.05$), and no change in the retention in the spleen, kidneys and heart.

Vitamin C supplementation did not affect the distribution of the retained dose of ^{59}Fe in the spleen, kidneys and heart in both Fe-loaded and non-loaded mice (D vs. C and B vs. A). However, lower % of the retained dose of ^{59}Fe was detected in the liver and GIT of vitamin C treated mice (without Fe-overload) (B vs. A). Iron-overload has resulted in a significant decrease in ^{59}Fe retained in the spleen ($p<0.05$), kidneys ($p<0.01$) and heart ($p<0.05$), and a drastic increase in the liver ($p<0.0001$) and GIT ($p<0.01$). As shown in table 2, the retention of ^{59}Fe in the GIT was not affected due to treatment with desferal in Fe-overloaded mice (F vs. E).

Discussion

The increase in the whole body retention of ^{59}Fe in response to vitamin C supplementation, as shown in the present study, is consistant with previous reports (3,4,5). We observed the same effect in both groups (with and without iron overload) and the effect of vitamin C was identical (1.5 times increase in ^{59}Fe whole body retention). In general, total wet weight of tissues except the spleen was not significantly affected by vitamin C supplementation (table 2). Thus, the increase in ^{59}Fe retention in those tissues, in response to vitamin C supplementation, may not be attributed to an increase in organ weight. We observed a greater dose of iron ^{59}Fe was retained in the liver in vitamin C treated mice. This is consistant with the finding of Clover et al (8) who used guinea pigs. However, their observation of an increased retention of ^{59}Fe in the spleen of scorbutic guinea pigs (as compared to controls) contradicts our finding in mice.

Our results indicate that the distribution of the retained dose of iron ^{59}Fe over the spleen, kidneys and heart was not affected by vitamin

C supplementation in mice with or without iron-overload (group D vs. C and B vs. A). However, in the liver and GIT of mice without iron-overload, the retention of ^{59}Fe (as % of retained dose) was significantly lower in response to vitamin C supplementation. The effect of iron over-load on the distribution of iron ^{59}Fe was not consistant: it decreased in the spleen, kidneys and heart and increased in the liver and GIT. The most drastic change was in the liver. Cook et al., (3) suggested that megadoses of vitamin C may induce some redistribution of body iron stores. Our data show that the liver and GIT are the compartments where major redistribution of iron occurs in response to vitamin C supplementation.

It has been reported (9) that subcutaneous infusion of desferrioxamine, in patients with a variety of iron-loading anemias, has resulted in a reduction in ^{59}Fe absorption. Those authors did not demonstrate whether subcutaneous DSF prevents Fe uptake by the GIT mucosal cells, prevents transfer of Fe from the cells to the plasma, or an increase in biliary iron excretion. However, they have shown that DSF promoted the re-excretion of absorbed iron in the urine which may be attributed to a rapid chelation of iron (by DSF) accompanied by a rapid urinary excretion. In the present study we observed no difference in the retention of iron ^{59}Fe, in the GIT, between the DSF treated and non-treated iron-overloaded mice which may indicate no effect of DSF treatment on the gastrointestinal absorption of iron. Further studies are needed to fully define the effect of DSF on Fe-absorption.

In conclusion, our results show that: 1) vitamin C supplementation exerted similar effects on iron absorption in mice, regardless of their iron status (controls vs. Fe-overloaded, 2) major redistribution of iron ^{59}Fe, in response to megadoses of vitamin C, occurred in the liver and GIT. Distribution was not affected in other organs, 3) iron-overload has resulted in redistribution of iron ^{59}Fe in spleen, liver, kidneys, GIT and heart, 4) treatment with subcutaneous desferrioxamine did not affect the gastrointestinal absorption of iron.

Acknowledgement:

This work was supported by the Interuniversity Reactor Institute Guest Investigator fellowship.

References

1. Hallberg, L., and E. Bjorn-Rasmussen. 1972. Scan. J. Hematol. 9, 193-199.

2. Cook, J.D., M. Layrisse, C. Martinez-Torres, R. Walker, E. Monsen and C.A. Finch. 1972. J. Clin. Ivest. 51, 805-810.

3. Cook, J.D., S.S. Watson, K.M. Simpson, D.A. Lipschitz and B.S. Skikne. 1984. Blood 3, 721-726.

4. Hussain, M.A.M., N. Green, D.M. Flynn and A.V. Hoffbrand. 1977. Lancet 1, 977-979.

5. Smith, C.H. and W.R. Bidblack. 1980. J. Nutr. 110, 1398-1408.

6. Wapnick, A.A. and S.R. Lynch. 1969. Brit. J. Hematol. 17, 563-568.

7. Conte, D., L. Brunelli, L. Ferrario, C. Mandelli, M. Quatrini, P. Velio and P.A. Bianchi. 1984. Acta. Hematol. 72, 117-120.

8. Clover, J.M., P.R. Jones, D.A. Greenman, R.E. Hughes and A. Jacobs. 1972. Brit. J. Exp. Path. 53, 295-300.

9. Pippard, M.J., J.T. Warner, S.T. Callender and D.J. Weatherall. 1977. Lancet, October 8, 737-739.

10. Editorial. 1984. Aust. NZ. J. Med. 14, 593-595.

11. Pippard, M.J., S.T. Callender and C.A. Finch. 1982. Blood 60, 288-294.

ENZYMES AS INDICATORS OF ZINC STATUS

George Everett and Jean Apgar
U. S. Plant, Soil and Nutrition Laboratory, Agricultural Research Service,
USDA, Tower Road, Ithaca, NY 14853, USA

Introduction

Although serum or plasma zinc (Zn) concentration is widely used as an indicator of Zn status, it is not as sensitive an indicator as is needed; and it is affected by factors other than Zn status (1, 2, 3). We have previously investigated acidic α-D-mannosidase (Mann) as a possible indicator of Zn status (4). In the current study we compared Mann with alkaline phosphatase (Alk Phos), an enzyme considered to be an indicator of Zn status (5) as well as with angiotensin-1-converting enzyme (ACE) and with 5'-nucleotidase (5'ND), both of which have been suggested to be indicators of Zn status (6, 7). Enzyme activities were measured in rats receiving a low Zn intake for 4 (Exp. 1; males) or 5 (Exp. 2; females) days. These brief periods of low Zn intake resulted in a mild deficiency which had less effect on growth and food intake than the 2 and 4 week periods that we have previously reported (4, 8). Because acidic Mann is a lysosomal enzyme, we have also investigated the effect of low Zn intake on the activity in plasma of three other lysosomal enzymes: acid phosphatase, N-acetyl-β-D-glucosaminidase and α-D-glucosidase.

Results

The effect of low Zn intake on the various enzyme activities is presented in Table 1. Mann activity was stimulated ($P<.01$) by the addition of Zn in vitro to a greater extent in plasma from -Zn than from +Zn rats. Basal activity was increased in -Zn

Trace Element – Analytical Chemistry in Medicine and Biology, Vol. 4

Table 1. Effect of Low Zinc Intake on Selected Plasma Enzymes

Enzyme	Exp. 1: Male (4 Day) +Zn	Exp. 1: Male (4 Day) -Zn	Exp. 2: Female (5 Day) +Zn	Exp. 2: Female (5 Day) -Zn
Body weight (g)	118±1[1]	102±1**	124±1	110±1**
Plasma Zn conc. (ug/ml)	1.7±.08	.53±.08**	1.6±.05	.40±.05**
Mann (nanomoles/ml/hr)	98±4	130±9*	188±12	140±10**
Increase with in vitro Zn	28±4	72±7**	59±4	134±6**
ACE (U/ml)[2]	17±.4	13±.5**		
Increase with in vitro Zn	4±.3	9±.5**		
ACE (micromoles/ml/hr)[3]			1.9±.06	2.0±.07
Increase with in vitro Zn			.01±.05	0.7±.11**
5'ND (U/L)	15±.4	9±.1**	13±.5	9±.2**
Alk Phos (Sigma U/ml)				
Am-Me-Prop buffer	7.5±.2	3.2±.1**	5.0±.2	3.8±.2**
Glycine buffer	8.4±.2	2.1±.1**	3.9±.2	2.3±.2**
Increase with in vitro Zn	7.7±.3	4.3±.2**	7.4±.4	7.4±.4
Acid Phos (Sigma U/ml)	2.8±.1	2.8±.1		
N-Acetyl-D-Glucosaminidase (μmoles/ml)	320±12	345±8		
α-Glucosidase (μmoles/ml)	348±7	377±8		

[1]Mean±SEM; 11-18 observations per mean.
[2]Colorimetric method (10).
[3]Fluorimetric method (11).

* Differs from +Zn, P<.05.
** Differs from +Zn, P<.01.

males (P<.05) but was decreased in -Zn females (P<.01).

Basal ACE activity was reduced (P<.01) in -Zn males but not in females. The reduction in the males was not as great as that reported by other investigators (6). Addition of Zn in vitro increased activity (P<.01) of ACE in both -Zn males and females to a much greater extent than in +Zn rats.

5'nucleotidase was reduced in plasma from both -Zn males and females (P<.01). Added Zn in vitro was not stimulatory with plasma from either -Zn or +Zn animals. Reduced activity in red cells and muscle of young male rats after 3-5 weeks of low Zn intake has been reported (9).

Alk Phos assayed in the presence of either of two buffers was reduced in -Zn males to a greater extent than in -Zn females. A similar reduction in Alk Phos activity in young male rats after four days of low Zn intake was reported by Roth and Kirchgessner (10). As these investigators also demonstrated (11), added Zn in vitro stimulated activity in the presence of glycine buffer. Zn was not stimulatory with 2-amino-2-methyl-1-propanol (Am-Me-Prop) as buffer, however. The stimulation by in vitro Zn in glycine buffer occurred in both +Zn and -Zn rats but was relatively greater in -Zn animals. Activity with added Zn in vitro was three times the basal activity in -Zn males and two times the basal activity in +Zn males. It was four times the basal activity in -Zn females and three times the basal activity in +Zn. Roth and Kirchgessner did not observe greater activation by in vitro Zn in Zn-deficient than in Zn-adequate rats (11).

Lysosomal enzymes other than Mann were assayed only in males and were not affected by low Zn intake. The higher basal activity of Mann in plasma from -Zn rats apparently was not due to release of lysosomal enzymes due to tissue catabolism because the activities of the three other lysosomal enzymes assayed were the same in plasma from -Zn and +Zn rats.

Discussion

A difference in the stimulatory effect of added Zn in vitro on Mann and ACE activity was apparent after 4 days of low Zn intake whether or not there was an effect on basal activity of the enzymes. The difference in Zn stimulation in vitro of these enzymes was also apparent in young male rats after 2 days of low Zn intake (data not shown). At that time there was no difference in the basal activities of either enzyme. The activity of 5'ND was also reduced at two days, and the activity of Alk Phos as well has been reported to be reduced by 2 days of low Zn intake (10). After 10 days of low Zn intake, however, by which time there was a marked effect on growth and feed intake, the activities of both 5'ND and Alk Phos were reduced in plasma of control rats with restricted-feed intakes (+ZnRI) as well as in that from -Zn rats (data not shown). Plasma 5'ND activity but not Alk Phos was reduced in +ZnRI rats in a previous experiment (8). 5'ND activity in muscle was also reduced in pair-fed controls as well as in -Zn rats after 5 weeks of treatment (9). As we have previously reported for Mann (4,8), reduced growth and/or feed intake did not affect the stimulation of ACE activity. ACE activity in plasma from rats with a low Zn intake for 10 days was markedly stimulated by Zn in vitro, but there was little stimulation of activity in plasma from either +Zn or +ZnRI rats. A restricted-feed intake control group was not included in the 4 and 5 days experiments because the difference in weights (Table 1) developed primarily in the last day before sampling. In the 10 day experiment reduced growth and/or feed intake increased activity of the three other lysosomal enzymes relative to activity in +Zn plasma, but the activities in -Zn plasma were not different from those in +Zn.

Because 5'ND is affected by reduced feed intake or growth and Alk Phos may be, Mann or ACE activity determined with and without added Zn in vitro seems to be more promising as an indicator of Zn status. The stimulatory effect of Zn in vitro on these enzymes occurred in animals in which the plasma Zn concentration was low, however. If the stimulatory effect of

Zn in vitro on Mann and ACE activity only occurs when plasma/serum Zn concentration is low, the enzyme assays will provide another but not necessarily a better way of determining Zn status than does plasma Zn concentration.

Methods

Approximately 100 g Sprague Dawley rats from our colony were given a low Zn (< 3 ppm) albumen-based diet unsupplemented (-Zn) or supplemented (+Zn) with Zn as acetate in drinking water (25 ppm). Heparinized blood was collected by heart puncture from anesthetized rats after 4 days of treatment in experiment 1 (males) and 5 days in experiment 2 (females). Aliquots of plasma were frozen until assayed. Mann was assayed by a colorimetric method (4); ACE, by either a colorimetric (12) or fluorimetric (13) method. Alk Phos was determined colorimetrically with reagents from Sigma kit 104-LL or with the kit buffer (2-amino-2-methyl-1-propanol) replaced by glycine buffer. Zn acetate was added to the incubation mixture in assays in which the stimulatory effect of in vitro Zn was studied. The concentration was 2 mM for Mann and 1 mM for Alk Phos and ACE. 5'nucleotidase was determined colorimetrically with Sigma kit 675. Acid phosphatase was determined with Sigma kit 104-AL; N-acetyl-β-D-glucosaminidase and α-D-glucosidase were determined as described by Tulsiani et al (14). Zn was determined by flame atomic absorption spectrophotometry. For Zn determinations, plasma was diluted 1:1 and compared with standards prepared in 12% sucrose to compensate for the increased viscosity of plasma samples. Means were compared by Student's unpaired t-test.

References

1. Solomons, N. W. 1979. Am. J. Clin. Nutr. 32, 856-871.

2. Abdulla,M. 1983. In: Zinc Deficiency in Human Subjects (A. S. Prasad, A. O. Cavdar, G. J. Brewer and P. J. Aggett, eds.). Alan R. Liss, New York. pp. 71-83.

3. Hambidge, K. M. 1981. Phil. Trans. Roy. Soc. Lond. B 294, 129-144.

4. Everett, G. and J. Apgar. 1980. In: Trace Element Analytical Chemistry in Medicine and Biology (P. Braetter and P. Schramel eds.). Walter de Gruyter, Berlin · New York. p. 491-501.

5. Luecke, R. W., M. E. Olman and B. U. Baltzer. 1968. J. Nutr. 94, 344-350.

6. White, C. L., I. Jacob, H. Dahlheim and J. Schorr. 1984. In: Fifth International Symposium on Trace Elements in Man and Animals. (In Press).

7. Pilz, R. B., R. C. Willis and J. E. Seegmiller. 1982. J. Biol. Chem. 257, 13544-13549.

8. Everett, G. and J. Apgar. Acta Pharmacol. Toxicol. In Press.

9. Prasad, A. S. and P. Rabbani. 1981. Trans. Assoc. Am. Physicians. 94, 314-321.

10. Roth, H.-P. and M. Kirchgessner. 1974. Z. Tierphysiol. Tierernaehrg u. Futtermittelkde. 32, 289-296.

11. Roth, H.-P. and M. Kirchgessner. 1979. Zbl. Vet. Med. A. 26, 835-840.

12. Lieberman, J. 1975. Am. J. Med. 59, 365-372.

13. Friedland, J. and E. Silverstein. 1976. Amer. J. Clin. Pathol. 66, 416-424.

14. Tulsiani, D. R. P., H. O. Buschiazzo, B. Tolbert and O. Touster. 1977. Arch. Biochem. Biophys. 81, 216-227.

SELENIUM STATUS PARAMETERS: RESPONSE TO A SIX-WEEK DIETARY MODIFICATION AND INTRA- AND INTERINDIVIDUAL VARIABILITY

A. van Faassen, J.M. Cardinaals " , P. van 't Veer,
Th. Ockhuizen

Department of Nutrition, TNO-CIVO Toxicology and Nutrition Institute, P.O. Box 360, 3700 AJ Zeist, The Netherlands

T. Muys

TNO-CIVO Food Analysis Institute, P.O. Box 360, 3700 AJ Zeist, The Netherlands

J.R.W. Woittiez

Energy Research Foundation ECN, Petten, The Netherlands

" present address: Notox, Hambakenwetering 31, 5231 DD 's-Hertogenbosch, The Netherlands

Introduction

The nutritional exposure of man to selenium (Se) can be assessed by combining dietary recall data with food composition data or by measuring biochemical parameters like the Se-concentration in blood fractions or toenail clippings and/or the activity of the Se-dependent enzyme glutathione peroxidase (1).
We had the opportunity to study various biochemical parameters during a dietary intervention trial in which meat and fish was replaced by meat or fish paste, leading to an increased Se-intake in the fish group. The data of the meat group could also be used to estimate the intraindividual and interindividual variability of the parameters. Both variabilities influence the accuracy in classifying subjects into levels and estimating correlations between the parameters (2).

Trace Element – Analytical Chemistry in Medicine and Biology, Vol. 4

The aim of this study was to determine two properties of Se-status parameters: 1) the response to modification of the Se-intake and 2) the intraindividual to interindividual variance ratio, which is indicated as individuality ratio r_i (3). We investigated the following parameters: the Se-concentrations ([Se]) in platelet-poor plasma (ppp), in red blood cells (rbc) and in clippings of the big toe nails (cbt) and the activity of Se-dependent glutathione peroxydase (GSH-Px) in ppp and in hemolysates (hemo).

Subjects and Methods

Twenty healthy young men (20-25 yrs) participated in a dietary intervention trial. After two weeks in which daily meat and fish were replaced by a meat paste, the volunteers were randomized into a group of ten continuing the consumption of the meat paste and another group of ten changing over to a fish (mackerel) paste. The Se-intake from the meat paste was 14 µg per day compared to 54 µg per day from the fish paste, calculated from the [Se] in the pastes measured by fluorometry at the TNO-CIVO Food Analysis Institute, Zeist. The difference in intake of 40 µg Se per day is about 50 % of the daily intake by Dutch adolescents(4). After overnight fasting and a 20-min rest in recumbent posture blood samples of 10 ml were collected by venapuncture in Venoject tubes containing Na-heparin at t= -2, 0, 3, 6 and 16 weeks. At t = -6 weeks blood samples were collected under non-standardized conditions. PPP was isolated by centrifuging at 10 °C for 5 min at 300 g followed by 10 min at 4500 g. Rbc were isolated by washing three times with 0.9 % (w/v) NaCl and centrifuging 10 min at 1200 g. Hemo are prepared by diluting 0.1 ml packed rbc with 50 volumes of aqua dest and centrifuging at 4 °C for 30 min at 2000 g. Cbt were collected during the year following the trial. At t = 6 weeks the left and right big toe nail were marked at the distal end of the lunula by making a groove in the nail with a razor blade and drawing a line in the groove by black ink.

At each clipping the distance between the black line and the distal end of the lunula was measured using a plastic ruler of Motala, kindly provided by Kinnarps Nederland B.V., Zutphen, The Netherlands. Date and distance were noted on the sampling sack (Minigrip Nederland B.V., Putten, the Netherlands) of each clipping. Linear growth rates of the big toe nails were calculated by regression analysis.

[Se] in ppp and rbc were measured by fluorometry (5). All reagents (analytical grade) were obtained from Merck, Darmstadt except 2,3-diamino-naphtalene which was supplied by Fluka AG (Buchs). The coefficients of variation measured by means of a control batch of ppp and rbc, were 6 and 5 % respectively.

[Se] in cbt were measured by non-destructive neutron activation analysis via the short-living Se^{77m} isotope after 20 s radiation in a flux of $5x10^{13}$ $cm^{-2}.s^{-1}$ using the fast rabbit system of the High Flux Reactor, situated at the Netherlands Energy Research Foundation ECN, Petten, The Netherlands (6). Before measurement the mark on the cbt was removed with a razor blade. The coefficient of variation of the analysis was 10 %.

The activity of GSH-Px in fresh ppp and hemo was measured by the coupled test system (7) in a Cobas-bio centrifugal analyzer. All reagents (analytical grade) were obtained from Merck, Darmstadt except NADPH, GSH-reductase (type III,1655 U/ml) and NaN_3 which were supplied by Boehringer, Mannheim, Sigma, St. Louis and BDH Chemicals, England respectively. Each assay mixture contained 0.1 mM H_2O_2, 0.8 mM GSH, 0.2 mM NADPH, 1U/ml GSH-reductase, 1.2 mM NaN_3, 5.0 mM Na_2EDTA and 50 mM K-phosphate buffer pH 7.4. The activity in ppp was measured at 37 °C. In hemo the activity was measured at 25 °C since at 37 °C the activity decreased in the course of the run. The coefficient of variation of the assay in ppp and hemo were 7 and 8 % respectively.

Differences in the Se-status parameters between the meat and the fish group were analyzed by applying analysis of variance to parameters of orthogonal polynomes in time (8). The [Se] in rbc at t = -6 and t = -2 weeks and the [Se] in ppp at t = - 2 weeks were taken into account as covariates in the analyses for [Se] in rbc and in ppp, respectively.

Intraindividual and interindividual variances were calculated by analysis of variance using a model comprising the effect of person and the residual intraindividual variation including variation due to day and analytical methodology.

Results

There were no statistically significant differences in the changes in Se-parameters between the meat and the fish group in any parameter (figure 1 and 2).

Figure 1. Mean Se-concentrations in ppp (□), rbc (○) and cbt (△) of the meat (open symbols) and fish (closed symbols) group. Vertical bars indicate ± s.e.m.

Figure 2. Mean activities of GSH-Px in ppp (□) and hemo (○) of the meat (open symbols) and fish (closed symbols) group. Vertical bars indicate ± s.e.m.

Including values at t = -6 and t = -2 weeks for the [Se] in rbc and at t = -2 for the [Se] in ppp did not effect the results. Therefore the variability in the Se-status parameters between t = 0, 3 and 6 weeks was interpreted as resulting from intraindividual variability during the six-week period and the variability between the 20 subjects from interindividual variability.
The mean (± s.d.) linear growth rate of the nail of the big toe, calculated from cbt of 19 subjects, was 0.062 (± 0.012) mm/day for the left one as well as for the right one. Because the nail is formed mainly at the matrix in front of the lunula (9,10) and since the lunula of the big toe nail is about 2 mm, the cbt with the mark will be formed about 4.5 weeks before t = 6 weeks, i.e. t = 1.5 weeks. The proximal and the distal ca. 2 mm clipping of the nail relative to the mark, had been formed 4.5 weeks before and after t = 1.5 respectively.
All relevant data were used for calculation of individuality ratios (table 1).

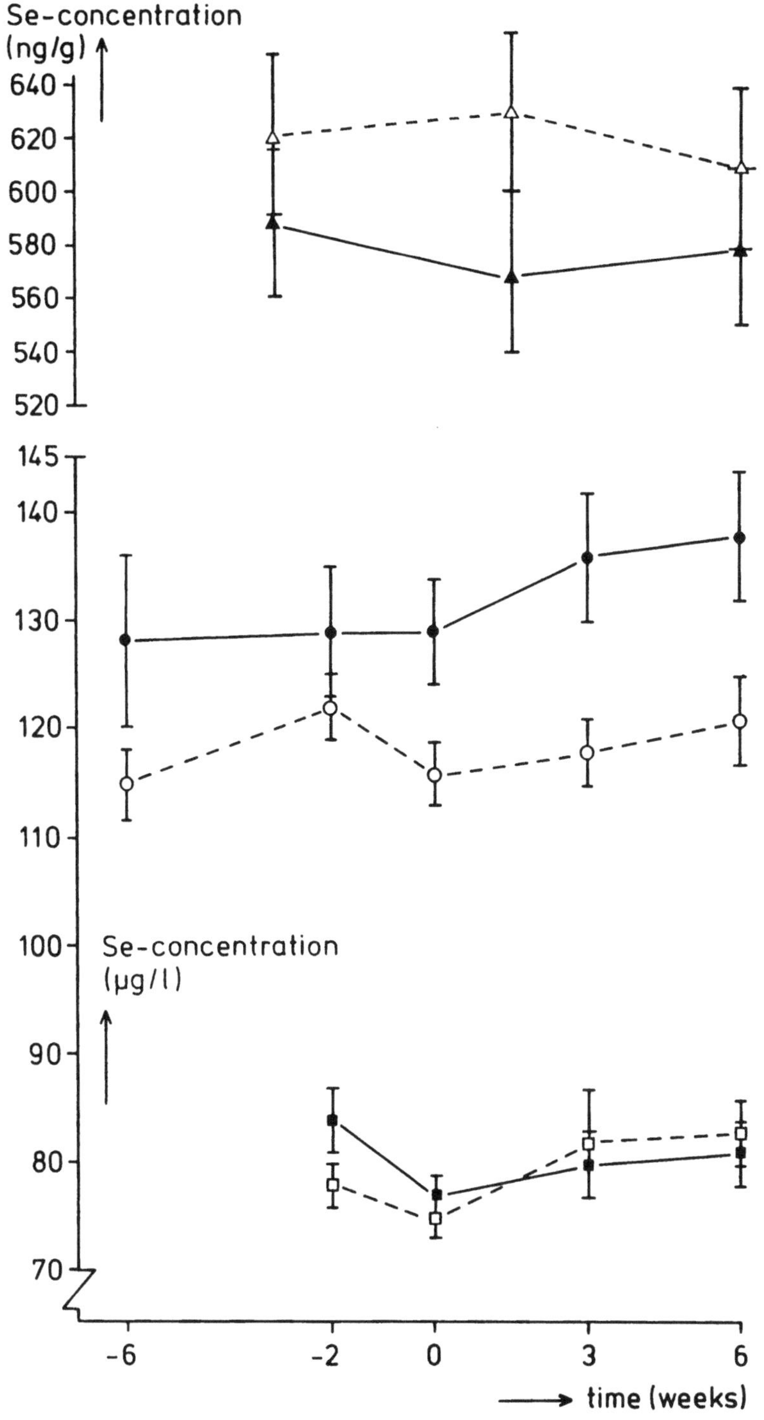
Se-concentration
(ng/g)
640
620
600
580
560
540
520
145
140
130
120
110
100
Se-concentration
(µg/l)
90
80
70
-6
-2
0
3
6
time (weeks)

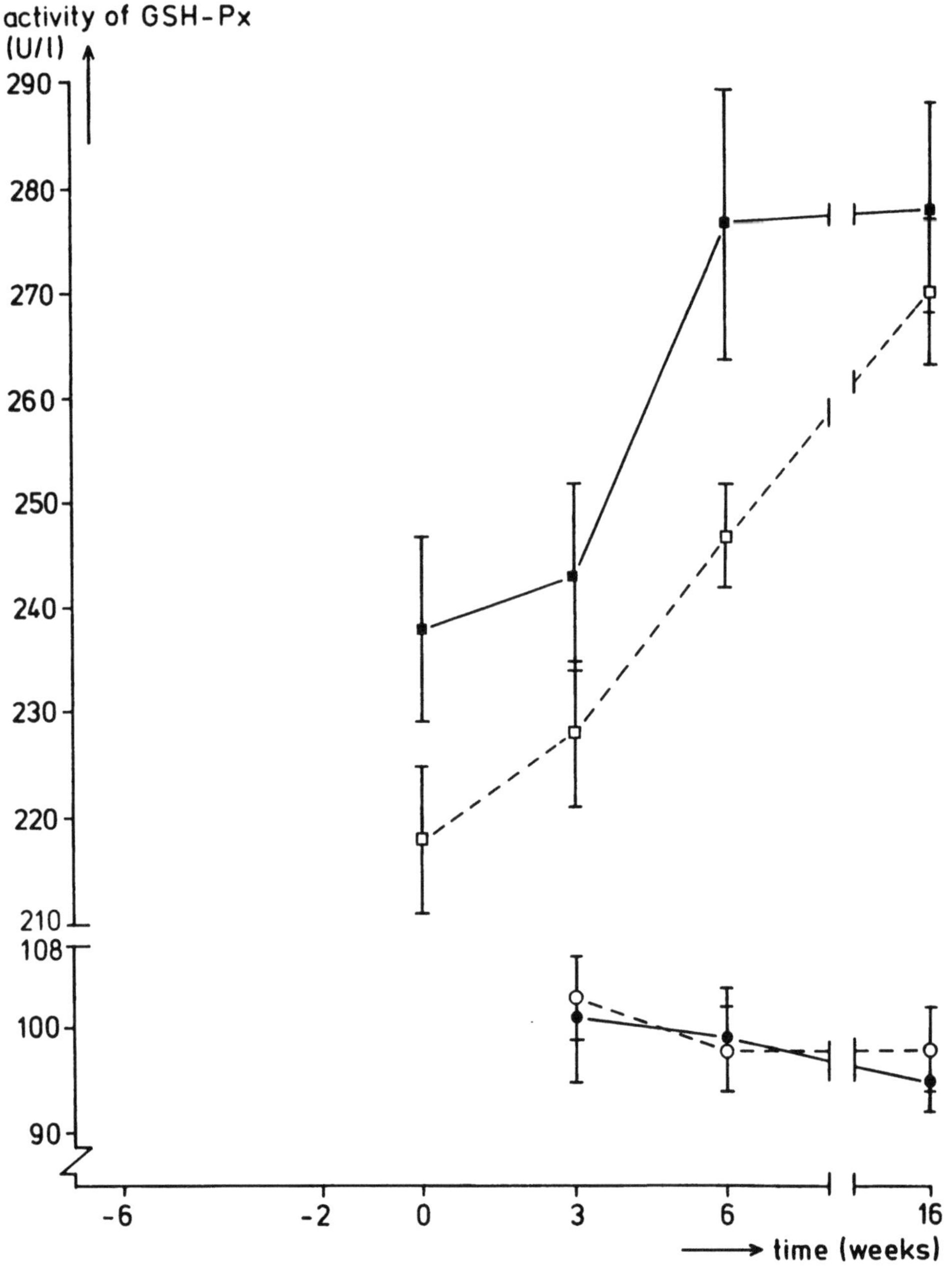
activity of GSH-Px
(U/l)
290
280
270
260
250
240
230
220
210
108
100
90
-6
-2
0
3
6
16
time (weeks)

Table 1 Individuality Ratios of Biochemical Se-status Parameters

parameter	mean value	coefficient of variation intra individual	inter individual	r_i *
[Se] ppp	79 μg/l	7.7	8.6	0.74
[Se] rbc	127 ng/g	4.3	12.2	0.12
[Se] cbt	600 ng/g	13.8	7.2	3.68
GSH-Px ppp	241 U/l	10.1	8.4	1.45
GSH-Px hemo	100 U/l	6.7	11.2	0.35

*defined as the intraindividual to interindividual variance ratio

The [Se] in cbt at t = 1.5 and 6 weeks are correlated with the [Se] in ppp and rbc at t = 3 and 6 weeks respectively, since there is considerable uncertainty in the times corresponding with the cbt. The correlations between the Se-status parameters are shown in table 2.

Table 2 Pearson Correlation Coefficients of Biochemical Se-status Parameters

Parameters	week			weighted* average
	t=0	t=3	t=6	
[Se] ppp / GSH-Px ppp	0.30	0.30	-0.06	0.08
[Se] ppp / [Se] rbc	0.44#	0.37	0.34	0.38#
[Se] rbc / GSH-Px hemo		0.11	0.28	0.20
[Se] rbc / GSH-Px ppp	0.24	0.22	0.25	0.24
[Se] cbt / [Se] ppp		0.78#	0.34	0.55#
[Se] cbt / [Se] rbc		0.20	0.17	0.15

* based on Fischer transformation 1/2 ln (1+r/1-r) weighted with the inverse of the variance 1/(n-3)
\# $p < 0.05$

Discussion and Conclusions

The higher intake of Se with the mackerel paste did not result in significantly higher values of biochemical parameters of the Se-status compared with the meat paste. For the [Se] and activity of GSH-Px in rbc and hemo respectively an increase was

not expected because rbc have a lifespan of 120 days, which implicates that after an increase in Se-intake during 6 weeks about one third of the rbc might have higher [Se] and GSH-Px activity. The significantly increase in the activity of GSH-Px in ppp in the meat as well as in the fish group may have been caused by an increased Se-intake in both groups since t = -2 weeks as compared to the subjects' usual diet. The [Se] and GSH-Px activity in ppp in the fish-group may not have increased more than in the meat group as a result of the higher content of the Se-antagonist Hg in the fish paste (0.02 mg/kg) than in Dutch meat in general (0.005-0.01 mg/kg) (4).
The estimated interindividual variability of the Se-parameters in ppp and rbc is higher than their intraindividual variability. The interindividual variability have been increased artificially as a result of the experimental difference in Se-intake. The response on the dietary modification may also have slightly increased the intraindividual variability. In addition the intraindividual variability is decreased due to the standardization of the Se-intake by replacing two main Se-sources meat and fish (4) by standardized meat or fish paste. However, the intake of the other important sources of Se, bread and milk products, with highly variable Se-concentrations (11), was not standardized. Thus although the intra- and interindividual variability may be slightly influenced by the dietary intervention, it will not have affected the individuality ratios substantially.
The higher intraindividual variability of the [Se] in cbt compared to the [Se] in ppp is probably caused by the higher analytical variability: 10 and 6 % respectively.
The higher intraindividual variability of the activity of GSH-Px in hemo compared to that of the [Se] in rbc is probably also a result of the higher coefficient of variation of the analytical method: 8 and 5 % respectively.
The [Se] and the activity of GSH-Px in rbc had the lowest individuality ratios. Therefore these parameters will classify persons with the highest degree of accuracy.

As expected, all correlations between the Se-status parameters appear to be positive. However, only correlations between the [Se] in ppp and in rbc and cbt reach statistical significance in this relatively small data set. The reason that no significant correlation between the [Se] in rbc and the activity of GSH-Px in hemo is observed, may be the inclusion of the activity of hemoglobin in the GSH-Px assay. The [Se] and the activity of GSH-Px in ppp may not be correlated since they responded to the dietary intervention in a different degree.
Summarizing, the increased intake of Se as a result of substituting meat by fish, did not significantly increase the Se-concentration and activity of glutathione peroxydase in platelet-poor plasma and red blood cells nor the Se-concentration in clippings of the big toe nails. Therefore, the (repeated) measurements of the Se-status parameters in the meat as well as the fish group could be used to estimate intraindividual to interindividual variance ratios and correlation coefficients. The parameters in red blood cells had the lowest variance ratios. Only the positive correlation between the Se-concentration in platelet-poor plasma and the Se-concentration in red blood cells and that between the Se-concentration in platelet-poor plasma and the Se-concentration in clippings of the big toe nails were statistically significant. We conclude that determining the Se-concentration and/or the activity of glutathione peroxydase in red blood cells will lead to a higher degree of accuracy in classifying persons within a population than measuring these parameters in platelet-poor plasma.

Acknowledgements

This study could be carried out benefiting from the activities of the International Working Party "Fish against thrombosis" - coordinator dr. G. Hornstra, Department of Biochemistry, Limburg University, P.O. Box 616, 6200 MD Maastricht, The Netherlands. Their investigations were supported by the Netherlands Heart Foundation. For the measurements of the Se-parameters in the trial reported here - coordinator E.J. van der Beek, M.D., Department of Nutrition, TNO-CIVO Toxicology and Nutrition Institute - financial support was obtained from the Koningin Wilhelmina Fonds (Netherlands Cancer Foundation) (grant CIVO 84-4).

References

1. Levander, O.A. 1985. Considerations on the assessment of selenium status. Fed.Proc. 44, 2579.

2. Liu, K., J. Stamler, A. Dyer, J. McKeever, P. Mc Keever. 1978. Statistical methods to assess and minimize role of intra-individual variability in obscuring relationships between dietary lipids and serum cholesterol. J.Chronic Dis. 31, 399.

3. Leissing, N., R. Izzo, H. Sargent. 1985. Variance estimates and individuality ratios of 25 serum constituents in beagles. Clin.Chem. 31, 83.

4. Vos, R.H. de, W. van Dokkum, P.D.A. Olthof, J.K. Quirijns, T. Muys, J.M. van der Poll. 1984. Pesticides and other chemical residues in Dutch total diet samples (June 1976-July 1978). Fd.Chem.Toxic. 22, 11.

5. Koh, T.S., T.H. Benson. 1983. Metals and other elements; critical reappraisal of fluorometric method for determination of selenium in biological materials. J.Assoc.Off.Anal.Chem. 66, 918.

6. Woittiez, J.R.W. 1985. Analysis of selenium in environmental and biological samples by neutron activation. In: Proc.Int.Symp. on Nucl.Anal.Chem., Halifax, Nova Scotia, Canada (in press).

7. Flohe, L., W.A. Guntzler. 1984. Assays of glutathione peroxidase. Meth.Enzymol. 105, 114.

8. Rowell, J.G., D.E. Walters. 1976. Analysing data with repeated observations on each experimental unit. J.Agric.Sci., Camb. 87, 423.

9. Runne, U., C.E. Orfanos. 1982. The human nail. Karger Gazette 44/45, 6.

10. Hopps, H.C. 1977. The biological bases for using hair and nail for analyses of trace elements. Sci.Total Environ. 7, 71

11. Bos, P.M.J., R.J. Egger, P. van 't Veer, A. van Faassen, M. Wedel, R.J.J. Hermus. 1985. Possible sources of variation in selenium content of whole-meal bread, white bread, and cow's milk in the Netherlands. Preliminary results. In: Strategies in food quality assurance: analytical, industrial and legal aspects. Proc. third European Conference on Food Chemistry, March 26-29, 1985, Antwerpen, België (W. Baltes et al., eds.), III-23.

ZINC AND CADMIUM CONCENTRATIONS IN LIVER AND PLACENTAS OF RATS: ARE THEY INTERRELATED?

W. Hazelhoff Roelfzema, A.M. Roelofsen, J.H.J. Copius Peereboom-Stegeman

Laboratory for Histology and Cell Biology, University of Amsterdam
Academic Medical Centre, 1105 AZ Amsterdam, The Netherlands

R.F.M. Herber

Coronel Laboratory for Occupational and Environmental Health, University of Amsterdam, Academic Medical Centre, 1105 AZ Amsterdam, The Netherlands

SUMMARY

Cadmium and zinc levels were determined in maternal livers and placentas from rats after exposure to cadmium during pregnancy. In the liver, both cadmium and zinc concentrations increased after exposure to cadmium. These increases were dose-dependent and related to the period of exposure. Cadmium and zinc concentrations were strongly interrelated (r=0.95). In the placenta the cadmium concentration increased after exposure to cadmium. In contrast to the situation in the liver, the zinc concentration in the placenta was not affected by exposure to cadmium.

INTRODUCTION

Zinc is an essential trace element. In Western countries zinc deficiency is rare. There may be, however, some groups at risk for zinc deficiency, caused by unbalanced diets (elderly people, students). In pregnant women, low serum zinc concentrations are correlated with risks to mother and infant, viz. abnormal deliveries, inefficient labour, atonic bleeding, abnormal gestation periods, immaturity or dysmaturity of infants, congenital malformations and missed abortions (1). In people living in industrial areas or working in cadmium plants, the cadmium burden can be enhanced severely

Trace Element – Analytical Chemistry in Medicine and Biology, Vol. 4

(2). As cadmium is known to disturb zinc-dependent metabolism, exposure to this heavy metal might increase the risk for zinc deficiency. Zinc-dependent metabolic processes are affected by cadmium in several ways. Cadmium might inactivate enzymes by replacing zinc (3,4). In the placenta this might cause decreased transport of zinc and other essential metals to the fetus (5). As the placenta can be considered as a liver for the fetus (6), similarly a cadmium-dependent increase of the zinc concentration might be expected in the placenta. However, it is known that in fetuses from cadmium exposed animals the zinc concentration is decreased (5). As the placenta consists mainly of fetal tissue, a decreased zinc concentration might be expected as well. Increased cadmium concentrations, together with decreased zinc concentrations are found indeed in human placentas, delivered in winter (7).
To investigate the relationship between the cadmium burden and the zinc status of the placenta, cadmium and zinc concentrations were determined in the liver and placentas from pregnant rats after parenteral exposure to cadmium during pregnancy. Rather low doses were used to relate the results to exposure in humans. The dose-effect relation was studied. Furthermore, the effect of cadmium on zinc levels was investigated in several stages of pregnancy.

MATERIALS AND METHODS

Virgin female Wistar rats, approximately 200 g, were mated with males (5:1) during at least 4 hours. Mating was confirmed by the presence of sperm in vaginal smears (day 0). The rats were housed individually in macrolon cages under standard laboratory conditions ($21^{o} \pm 1^{o}C$; light/dark schedule 12/12 hrs, rat chow (Hope farms, Woerden, The Netherlands) and tap water freely available). From day 0 onwards the rats were injected daily subcutaneously with $CdCl_2$ in saline, in a quantity of 2 ml/kg. Control animals received equal amounts of saline.
Two types of experiments were performed:
1. A dose-effect experiment (exp. 1): Five groups of rats were injected daily up to and including day 18 (starting on day 0) with different doses. The used doses were 0, 1.1, 2.2, 4.4 and 8.8 µmol Cd/kg/day. On day 19 the animals were sacrificed.

2. A time-response experiment (exp. 2): The animals were injected daily with saline or 4.4 µmol Cd/kg/day, starting on day 0, up to and including day 13, 15, 17, and 19, respectively. On day 14, 16, 18, and 20, respectively, the animals were sacrificed.
In both experiments, 24 hours after the last injection, placentas and maternal livers were removed and blood was collected after nembutal anaesthesia (48 mg/kg i.p. after light ether anaesthesia). Before use, all instruments and tubes were thoroughly cleaned in 1M HNO_3 during one night, rinsed 5 times in demineralized water and allowed to dry. All samples were weighed, stored in polyethene tubes at -40°C, freeze-dried during 72 hours (freeze dryer Breda Scientific LY-5-FM, The Netherlands) and homogenized using a coarse quartz bar after determination of the dry weight. The freeze-dried samples were stored in polyethene tubes at 20°C in dark.
Two livers and placentas per group were used for comparing cadmium and zinc concentrations in exp. 1. In exp. 2, two livers and four placentas were used per gestational day. Cadmium concentrations in blood were determined by electrothermal atomization atomic absorption spectrometry (ETA-AAS; Varian A.A6).

Determination of cadmium in blood:

Instrumentation: A Varian Techtron A.A6 spectrometer was used in conjunction with a Varian CRA-60 atomizer, a Varian ASD53 dispenser, a custom-made pyrometer (8) and a custom-made background monitor system (9).

Experimental conditions:

analyte line	: λ=228.8 nm
band width	: λ=0.5 nm
lamp	: cadmium hollow cathode (Varian); lamp current I= 3 mA, hydrogen continuum background
atomizer	: pyrolytically coated mini-Massman
sample volume	: 5 µl
sheating gas	: nitrogen, 3.9 l/min.
drying	: 15 s 80°C
ashing	: 15 s 350°C
atomizing	: 1.5 s 1500°C.

Cadmium and zinc determinations in tissues were performed by solid sampling ETA-AAS with the Zeeman effect background correction technique (Grün SM1) with reference to standard materials (bovine liver NBS 1577a; oyster tissue NBS 1566, obtained by the National

Bureau of Standards, USA). The accuracy obtained by this technique was found to be satisfying previously (10).

Minimum detection limits: cadmium (λ=228.8) 0.005 ng
cadmium (λ=326.1) 0.2 ng
zinc (λ=307.6) 0.5 ng

Precision : 95% confidence interval width ≤10%.

The concentrations were calculated in µg/g w.w.

Solid sampling technique:

Instrumentation: A Grün SM 1 (FRG) spectrometer was used in conjunction with a Sartorius 4503 MP6 (FRG) microbalance and a Goerz Metrawatt SE 120 (FRG) double channel stripchart recorder.

Experimental conditions:

analyte line : cadmium: λ=228.8 nm, λ=326.1 nm
zinc λ=307.6 nm.

lamp : cadmium ENF (Grün); heating current I=0.40 A (both top and bottom); lamp current 5.8.
zinc ENF (Grün); heating current I=0.80 A (both top and bottom); lamp current 5.0.

atomizer : graphite oven (Massman type) and graphite vessel (L'vov platform).

time constant : τ=0.3 s

sheating gas : nitrogen, 0.15 l/min.

		Cadmium		Zinc	
		time	adjusted temperature	time	adjusted temperature
heating step	1	45 s	500°C	45 s	500°C
	2	15-90 s	500°C	15-90 s	500°C
	3	7-10 s	2000°C	7-15 s	2000°C

RESULTS

The internal doses of cadmium - as cadmium in blood - are presented in tables 1 and 2. The concentration of cadmium in blood is related to the dose (r=0.99; table 1). The period of exposure does not influence the cadmium concentration in blood (table 2).
In the *liver*, the cadmium concentrations are strongly related to the dose (r=0.99). After 19 days exposure to 4.4 µmol Cd/kg, the cadmium concentration was increased up to 85 µg/g w.w. (fig. 1a,

table 1). A slight increase of the cadmium concentration is observed between 14 and 20 days of exposure. This increase is rather low, due to the weight gain of the liver during pregnancy (fig. 1b table 2).

In both experiments, the zinc concentration too is increasing with the dose of cadmium (r=0.97) or the period of cadmium exposure (fig. 1a,b; table 1,2). In control animals, the zinc concentration in the liver remains constant during gestation, notwithstanding the weight gain of the liver. Calculated from both experiments, a strong relationship between the cadmium and the zinc concentration in the liver is observed (r=0.95; fig. 1c).

In the *placenta* too, the cadmium concentration is strongly related to the dose (r=0.98; fig. 2a, table 1). After 19 days of exposure to 4.4 µmol Cd/kg, the cadmium concentration was increased up to 2.48 µg/g w.w. In exp. 2, a strong increase of the cadmium concentration is observed due to the period of exposure (r=0.90), notwithstanding the weight gain of the placenta during gestation (fig. 2b, table 2).

Fig. 1 Cadmium and zinc concentrations in the liver after exposure to cadmium. ● Cd in exposed animals, ■ Zn in exposed animals, ▲ Zn in control animals.
a: after 19 days exposure to different doses (exp. 1)
—●—: CdL= 3.2 + 1.66 x dosis. r=0.99.
-■-: ZnL= 31 + 9.2 x dosis. r=0.97.
b: after exposure up to day 14, 16, 18 or 20 (exp. 2).
—●—: CdL= 19 + 2.85 x day of pregnancy. r=0.69.
-■-: ZnL= 45 + 2.03 x day of pregnancy in exposed animals r=0.50.
···▲··: ZnL= 18 + 0.34 x day of pregnancy in control animals r= 0.45.
c: correlation between cadmium and zinc concentrations.
——: ZnL= 29 + 0.62 x CdL. r=0.95.

Fig. 2 Cadmium and zinc concentrations in the placenta after exposure to cadmium. ● Cd in exposed animals, ■ Zn in exposed animals, ▲ Zn in control animals.
a: after 19 days exposure to different doses (exp. 1).
—●—: CdP= 0.049 + 0.53 x dosis. r=0.98.
-■-: ZnP= 8.0 + 0.0064 x dosis. r=0.25.
b: after exposure up to day 14, 16, 18 or 20 (exp. 2).
—●—: CdP= -5.50 + 0.47 x day of pregnancy. r=0.90.
c: no correlation between zinc and cadmium concentrations in the placenta.
ZnP= 10.3 - 0.14 x CdP. r=-0.10.

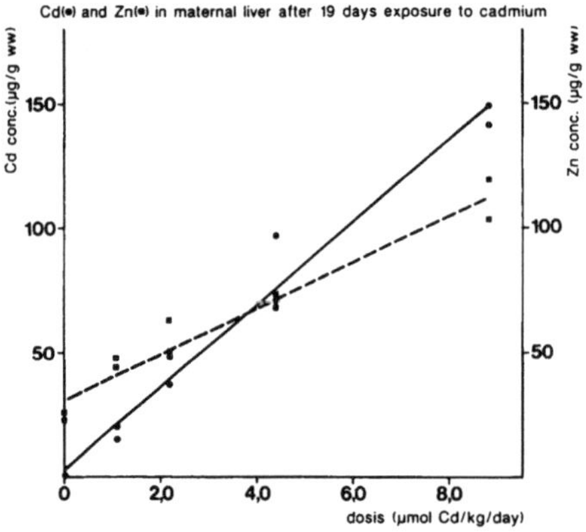

Fig. 1a

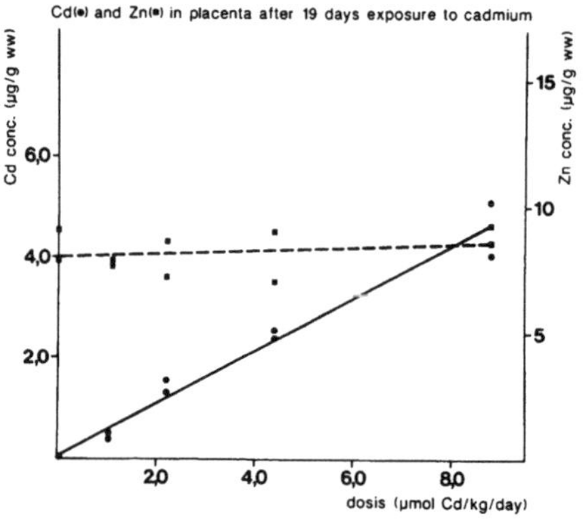

Fig. 2a

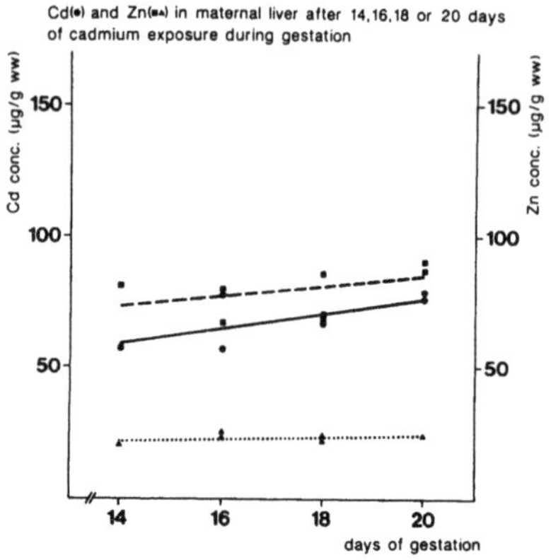

Fig. 1b

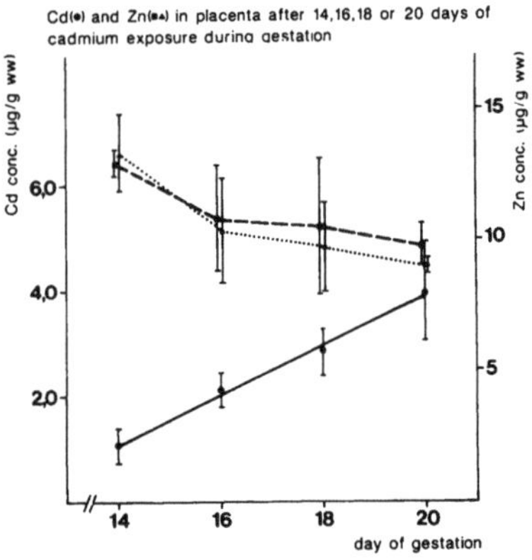

Fig. 2b

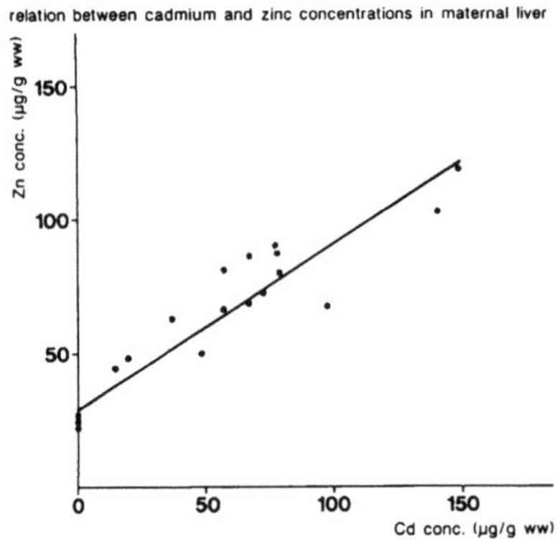

Fig. 1c

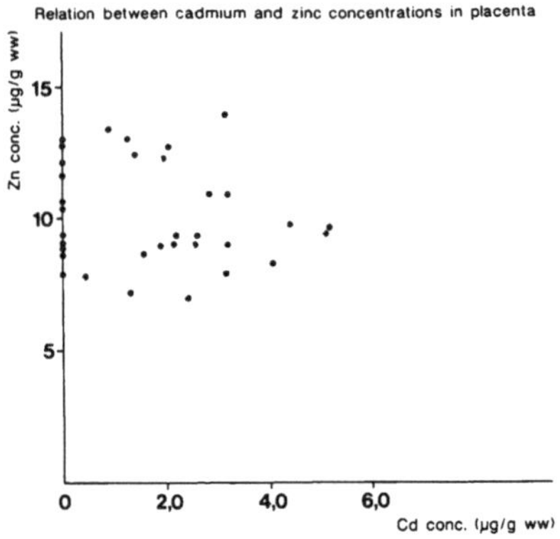

Fig. 2c

In exp. 1, the zinc concentration of the placenta is not influenced by exposure to cadmium: 7.2 - 9.4 µg/g w.w. in control animals and all doses (fig. 2a, table 1).
In exp. 2, the zinc concentrations of placentas from both control and cadmium exposed rats decrease during gestation from 13 µg/g on day 14 down to 9.4 µg/g on day 20 (fig. 2b, table 2). Thus, the zinc concentration of the placenta is not influenced by exposure to cadmium and no correlation exists between the zinc and the cadmium concentrations (fig. 2c).

Table 1. Cadmium and Zinc Concentrations in Liver and Placenta after 14 days s.c. Exposure to Different Doses of Cadmium (exp. 1).
Values as obtained from individual animals.

dosis (µmol/kg/day)	blood Cd (µg/ml)	liver Cd (µg/g w.w.)	liver Zn (µg/g w.w.)	placenta Cd (µg/g w.w.)	placenta Zn (µg/g w.w.)
0	0,0011	0.013	24	< 0.002	9.4
0	± 0,0002	0.012	24	< 0.002	7.9
1.1	0,13	20	48	0.42	7.8
1.1	± 0,023	15	44	0.44	7.8
2.2	0,34 ± 0,082	37	63	1.3	7.2
2.2		49	50	1.5	8.6
4.4	0,58 ± 0,029	72	73	2.6	9.0
4.4		98	68	2.4	7.0
8.8	0,98 ± 0,014	149	119	5.1	9.4
8.8		141	103	4.1	8.3

Table 2. Cadmium and Zinc Concentrations (μg/g w.w.) in Liver, Placenta and Blood (μg/ml) after 14, 16, 18 or 20 days Exposure to Cadmium (4.4 μmol/kg s.c.) (exp. 2).

gestiational day	Blood	Liver					Placenta			
			Controls		Cd-exposed		Controls		Cd-exposed	
	Cd (μg/ml)	mean liver weight (g)	Cd	Zn	Cd	Zn	Cd	Zn	Cd	Zn
14	0.28 ±0.028 (n=3)	12.3±0.95 (n=5)	< 0.003 (n=2)	21 (n=1)	57 (n=1)	81 (n=1)	<0.003 (n=1)	13.3±1.5 (n=3)	1.1±0.3 (n=4)	12.9±0.5 (n=3)
16	0.29 ±0.025 (n=4)	13.6±0.73 (n=5)	<0.003 (n=2)	24±2 (n=2)	67±1 (n=2)	73±9 (n=2)	<0.003 (n=1)	10.3±2.0 (n=4)	2.1±0.3 (n=4)	10.8±2.0 (n=4)
18	0.33 ±0.009 (n=4)	14.4±1.86 (n=4)	<0.003 (n=2)	24±1 (n=2)	67±1 (n=2)	78±12 (n=2)	<0.003 (n=2)	9.7±1.7 (n=4)	2.8±0.5 (n=4)	10.5±2.6 (n=4)
20	0.28 ±0.033 (n=3)	14.6±1.22 (n=3)	<0.003 (n=2)	24 (n=1)	77±1 (n=2)	89±2 (n=2)	<0.003 (n=1)	9.0±0.1 (n=2)	4.0±1.0 (n=4)	9.8±0.8 (n=4)

DISCUSSION

In this study with pregnant rats, cadmium was accumulated in the liver and in the placenta. The liver accumulated about 30 times as much cadmium as the placenta. The values for cadmium in blood, obtained from exp. 1, using the same dosis as in exp. 2 (4.4 µmol/kg) did not correspond very well with the values obtained from exp. 2. This is contributed to the use of a different batch of animals. On the contrary, the values for the cadmium concentrations in tissues from exp. 1 (collected on day 19) agreed rather well with the values obtained from exp. 2, using the same dose and comparing with the 18th or 20th day of gestation.
In both liver and placenta a strong dose related increase of the cadmium concentration was established. Moreover, a relation between the length of exposure and the cadmium burden was observed. The zinc concentrations in these two organs reacted rather different: In the *liver*, the zinc concentration increased both related to the dose and to the period of exposure. A strong relationship between the cadmium and zinc concentration is obvious. In contrast, in the *placenta* the zinc concentration is not affected by exposure to cadmium: In exp. 1 the zinc concentration remains constant, irrespective of the dose of cadmium. In exp. 2, in control as well as in cadmium exposed rats the zinc concentration in the placenta is decreasing, not influenced by cadmium exposure.
Increase of the zinc concentration in the liver after oral cadmium exposure was observed by Pond and Walker (5) and by Sasser et al. (11). Sasser et al. (11) observed normal zinc concentrations in the placenta of cadmium exposed rats on day 21. The high zinc concentration in these studies were accompanied by high levels of metallothionein, whereas in the placenta the metallothionein concentration did not seem to increase considerably. Our study indicates that the zinc concentration in the placenta of rats, parenterally exposed to cadmium, remains within the levels obtained from control rats, as observed from day 14 up to day 20.
Summarizing, the zinc concentration in the placenta is not influenced by cadmium exposure, in contrast to the situation in the liver (where the zinc concentration is increased) or the situation in the remaining fetal tissues (where the zinc concentration is decreased (5). Although the total zinc concentration in the placen-

ta is not affected, probably the zinc is replaced by cadmium in transport proteins, thus influencing the zinc status of the fetus. Further research is required to investigate in what respect the zinc level in the fetus is changed. It is still an unsolved problem why cadmium and zinc concentrations in liver show a strong relationship, whereas in the placenta they are not interrelated.

Acknowledgement

This research was financially supported by the Netherlands Technology Foundation (STW).

REFERENCES

1. Jameson, S. 1978. Acta Med.Scand. 593 (suppl.), 1-89.
2. Herber, R.F.M., M.A. Verschoor, A.A.E. Wibowo. In: Environmental toxin reviews, vol. Cadmium (O. Hutzinger and S.H. Safe, eds.). Springer Verlag, Heidelberg. In press.
3. Vallee, B.L., D.D. Ulmer. 1972. Ann.Rev.Biochem. 41, 91-128.
4. Pool, M.L. 1981. Toxicol.Environ.Chem.Rev. 4, 179-203.
5. Pond W.G., E.F. Walker jr. 1975. Proc.Soc.Exp.Biol.Med. 148, 665-668.
6. Villee, C.A. 1962. Am.J.Obst.Gynec. 84, 1684-1694.
7. Dawson, E.B., H.A. Croft, R.R. Clark, W.J. McGanity. 1968. Am. J.Obst.Gynec. 102, 354-361.
8. Herber, R.F.M., H.J. Pieters, A.M. Roelofsen, W. van Deijck. 1984. Spectrochimica Acta 39B, 397-405.
9. Herber, R.F.M., H.J. Pieters. Int. Lab. In press.
10. Herber, R.F.M., A.M. Roelofsen, W. Hazelhoff Roelfzema, J.H.J. Copius Peereboom-Stegeman. 1985. Fres.Z.Anal.Chem. 322, 743-746.
11. Sasser, L.B., B.J. Kelman, A.A. Levin, R.K. Miller. 1985. Toxicol.Appl. Pharmacol. 80, 299-307.

PLATELET AND ERYTHROCYTE KINETICS STUDIED WITH SELENIUM-74

J. Kiem, G. Koslowski, H. Weese, L.E. Feinendegen
Institute of Medicine, Nuclear Research Center Jülich GmbH,
D-5170 Jülich, F.R. Germany

Introduction

In 1978 we reported that there is no human cell with a selenium concentration as high as in the platelet (1). This fact still holds true after detection of systematic errors which affected these early data of 1978 (2) so that the normal selenium concentration in human platelets is - on wet weight basis - in the order of 500 ng/g (3). Since platelets are known to be most important in the process of hemostasis and thrombosis we had put forward the hypothesis that they are the link between selenium deficiency and the increased deathrate from cardiovascular diseases thereby observed (1).

Several experiments were performed to examine this concept (3). They showed that

- platelets tend to retain their selenium firmer during selenium depletion than do other blood components;
- the activity of the selenoenzyme glutathione peroxidase and - if double-peak distribution is assumed - the selenium concentration are decreased in platelets of about 50% of patients with acute myocardial infarction;
- a high selenium concentration seems to have an anti-aggregating effect for platelets.

Since no information is available in the literature regarding the metabolic pathway of selenium in platelets we used the stable isotope ^{74}Se to study and compare platelet and erythrocyte kinetics.

Trace Element – Analytical Chemistry in Medicine and Biology, Vol. 4

Materials and Methods

A normal human male subject aged 36 years received orally 1 mg of ^{74}Se (Rohstoff-Einfuhr GmbH, F.R.G.) as $Na_2{}^{74}SeO_3$ in water. Starting from 2 days before until 18 days after 10 ml samples of citrated blood were taken daily, and in intervals of 4-21 days for further 7 months. Duplicate samples of platelets, erythrocytes and plasma were obtained and analysed for ^{74}Se by instrumental neutron activation analysis together with other nuclides such as ^{58}Fe and ^{64}Zn as described elsewhere (2, 3). One of the duplicate samples was irradiated for 6 hours, the other for 48 hours. Due to good correspondence the mean values of the duplicates were used for further calculations. Concentrations were based on wet weight after correction for trapped plasma, counting rates were used without such corrections.

Results and Discussion

Fig. 1 shows the platelet/plasma concentration ratios T/P of ^{74}Se. The abscissa gives logarithmically the time in days. As seen on the left T/P is 6-7 before taking ^{74}Se. Since in this case ^{74}Se is proportional to total Se platelets contained 6-7 times more Se than plasma. This agrees well with our previous observations and underlines once more the extraordinary high concentration of selenium in platelets (3).

After intake of ^{74}Se T/P decreases by a factor of about two and remains nearly constant for 4 days. Therefore about 50% of selenium seems to be freely exchanged in platelets and to enter these cells directly from blood plasma when supplied as selenite.

It is known from the literature (4) that megakaryocytopoiesis lasts 4-5 days so that it would take about 4 days until the first platelets derived from ^{74}Se-labeled early precursor cells appear in the blood stream. In fact a rapid threefold increase of T/P can be observed after day 4. Therefore the remaining about 50% of platelet selenium - normally approximately 250 ng/g - seems to follow the bone marrow pathway when supplied as selenite.

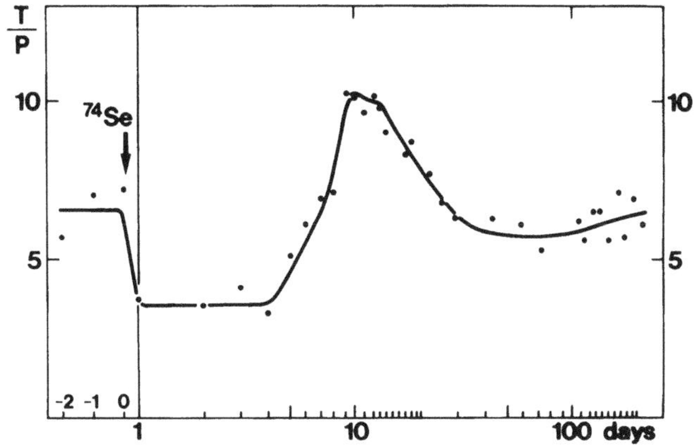

Fig. 1: Ratio of platelet (T) to plasma (P) concentration of ^{74}Se

The literature reports a normal platelet life span of 10 days (4). In fact 10 days after the first raise on day 4 T/P starts to fall on day 14. Therefore the selenium which is incorporated into platelets via their early precursor cells in the bone marrow is firmly bound and leaves platelets only after their death. From day 30 onwards T/P becomes again equilibrated at the level of 6-7 observed before intake of ^{74}Se.

From these data we can not predict whether selenium follows similar pathways when supplied in other chemical forms. In this context it is also interesting to note that ^{75}Se-methionine has since long time been used as in vivo label to study platelet kinetics (4). Perhaps in these studies the selenium pathway is used rather than the methionine pathway as generally believed. Thereby selenium may be separated from methionine by the liver.

Fig. 2 shows the data for erythrocytes (E). After intake of ^{74}Se E/P decreases by a factor of about five. Therefore - if supplied as selenite - about 20% of selenium seems to be incorporated into erythrocytes directly from blood plasma, the remaining 80% via the bone marrow pathway.

According to the literature normal erythrocytes have a life span

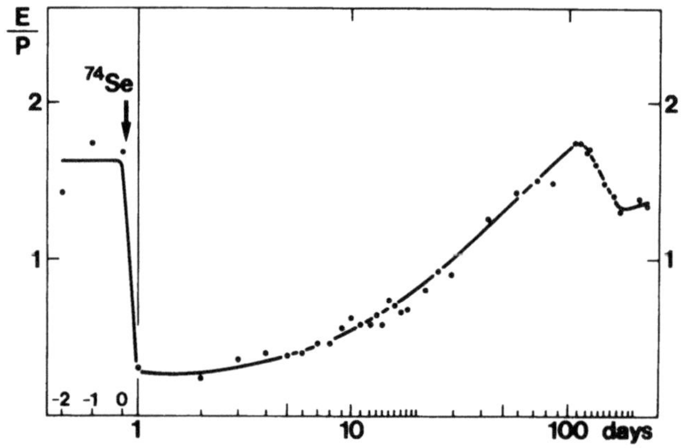

Fig. 2: Ratio of erythrocyte (E) to plasma (P) concentration of ^{74}Se

of 120 days (5). In fact E/P declines after about 120 days according to the death of the ^{74}Se-labeled erythrocytes which first entered the blood stream from the bone marrow. This proves that like for platelets bone marrow incorporation of selenium in erythrocytes is irreversible. Therefore oral administration of $Na_2{}^{74}SeO_3$ can be used as in vivo label to measure platelet and erythrocyte life span.

In order to simplify the technique we divided the counting rates of ^{75}Se and ^{65}Zn (Se/Zn) for platelets (T) and of ^{75}Se and ^{59}Fe (Se/Fe) for erythrocytes (E) irrespective of neutron flux and sample weight. Fig. 3 and 4 show that the critical marks of 14 days for platelets and 120 days for erythrocytes can easily be detected. This simple test circumvents any cell manipulation typical for in vitro labeling techniques. It promises to be useful in patients where platelet and/or erythrocyte kinetics have to be studied, but application of radiotracers is disadvantageous, e.g. in obstetrics and pediatrics.

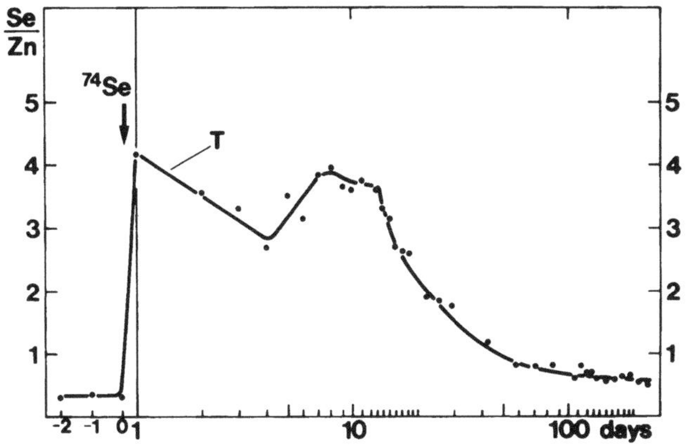

Fig. 3: Ratio of ^{75}Se to ^{65}Zn counting rate (Se/Zn) in platelets

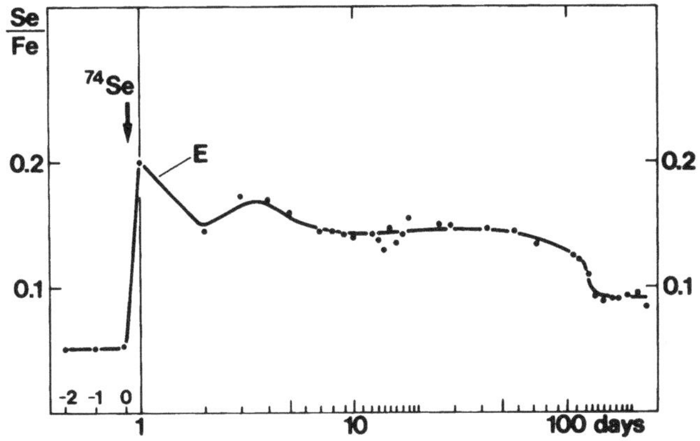

Fig. 4: Ratio of ^{75}Se to ^{59}Fe counting rate (Se/Fe) in erythrocytes

Conclusions

If supplied as selenite about 50% (80%) of selenium seems to be irreversibly incorporated into platelets (erythrocytes) via the bone marrow pathway. Therefore the stable tracer $Na_2{}^{74}SeO_3$ can be used to measure platelet and/or erythrocyte kinetics.

References

1. Kiem, J., G.V. Iyengar, H. Borberg, K. Kasperek, M. Siegers, L.E. Feinendegen and R. Gross. 1979. In: Proc. Nuclear Activation Techniques in the Life Sciences 1978 (IAEA-SM-227/53). IAEA, Vienna, pp. 143-164.

2. Kiem, J., H. Weese and L.E. Feinendegen. 1984. Fresenius Z. Anal. Chem. 317, 438-440.

3. Kiem, J. and L.E. Feinendegen. 1984. In: Trace Element Analytical Chemistry in Medicine and Biology, Vol. 3 (P. Brätter and P. Schramel, eds.). Walter de Gruyter, Berlin - New York, pp. 339-345.

4. Aster, R.H.. 1971. In: Platelet Kinetics (J.M. Paulus, R.H. Aster and H. Breny, eds.). North-Holland Publishing Company, Amsterdam - London, pp. 3-23.

5. Kaboth, W. and H. Begemann. 1971. In: Physiologie des Menschen, Band 5 (O.H. Gauer, K. Kramer and R. Jung, eds.). Urban und Schwarzenberg, München - Berlin - Wien, pp. 17-18.

PARTICIPATION OF METALLOTHIONEIN AND CEREBRAL STRUCTURES IN IRON HOMEOSTASIS OF ANEMIC RATS

B. Ribas, M.A. Brenes, F.J.De Pascual, J. Del Rio, M.I. Sánchez-Reus
Institute of Biochemistry, CSIC-Complutense University, Faculty of Pharmacy, 28040-Madrid, Spain.

Introduction

The low turnover of cadmium (1) is connected with its ability of protein binding and principally with the metalloproteins, a fact which explains its importance in Toxicology. It is accumulated in tissues inhibiting the biochemical function of specific proteins which affect growth and development (2). Cadmium induces the synthesis of metallothioneins (3,4) through the feed back mechanism, and its molecular isoforms are of great significance in Occupational Medicine.

The biological behaviour of cadmium as protein marker made it possible to detect the miroin in rat brain (5,6), with similar physico-chemical characteristics as the metallothionein (3),and to establish the tight correlation between transferrin and metallothionein, and their both increment in iron deficient rats and posthemorrhagic anemia (7). The hypothesis that metallothionein could participate in iron absorption in the intestinal mucosa as well as in the implication of the brain, in iron homeostasis, constitutes the object of this work.

The participation of the brain in iron homeostasis was not yet studied, transferrin and ferritin are the main proteins involved in iron metabolism and a number of proteins are considered to advance the knowledge of iron absorption (8-12). The unbalanced iron homeostasis is tightly connected with mental diseases (13-15) some of which, with unknown ethiology, are empirically remedied with iron therapy (16,17).

Trace Element – Analytical Chemistry in Medicine and Biology, Vol. 4

Material and Methods

Label of duodenal metallothionein with ^{59}Fe in posthemorrhagic anemia: a volume of 0.5 ml of 5 µCi of $^{59}FeCl_3$ (Amersham, U.K.), pH 2 was administered through intragastric sonda to male Wistar rats of ca. 200 g with posthemorrhagic anemia, and normal rats as controls. Radioactivity of duodenum in cpm/g fresh tissue for this experiment and for the next ones is measured with BETAmatic (Kontron) or Autogamma 5110 (Packard) detectors, 10 min after the radionuclide administration, and the routine physiological and chemical techniques.

Label of brain metallothionein (miroin (5,6)) with intracerebroventricular administration of ^{59}Fe and ^{109}Cd in iron deficient rats and normal ones as controls: a volume of 2 µl with 2 µCi of $^{59}FeCl_3$ or $^{109}CdCl_2$, is applied to each lateral ventricle of anemic rats with iron deficient diet. The animals which received iron are killed two hours later, and those which received cadmium, four hours later in order of their kinetics (18,19). The main cerebral structures signalized afterwards are dissected and radioactivity evaluated as before expressed. Anemic condition is controled by the determination of hemoglobinemia, hematocrit and proteinemia. Total iron binding capacity, Lowry et al., Zahler and Clelland and Mancini are applied as routine analyses as expressed elsewhere (7).

Kinetics of incorporation of ^{59}Fe in brain is studied with experimental models as iron deficient, iron loaded and normal rats: 5 µCi $^{59}FeCl_3$ in 0.5 ml volume is administered intraperitoneally, and allow to incorporate during a period of 45 min. Iron deficient animals are fed with iron deficient diet, and to iron loaded ones 1 mg Fe, chemical form $FeSO_4$, is injected intraperitoneally once a day during 3 days before the radionuclide administration.

Results

Posthemorrhagic anemia shows a marked incorporation of ^{59}Fe in rat duodenum 10 min after the intragastric administration of the radionuclide in relation to the controls as it is expressed in

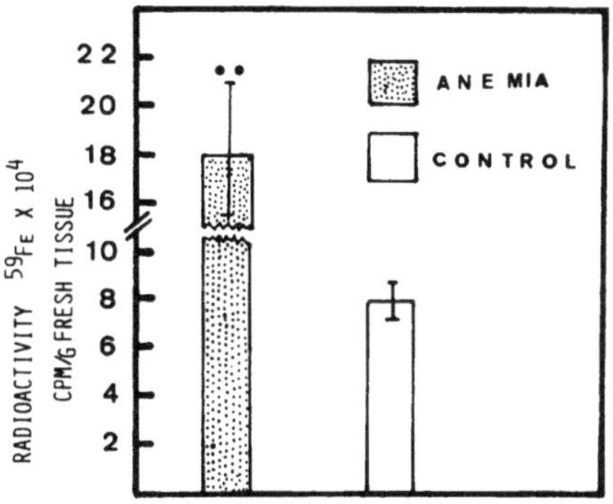

Figure 1. Incorporation of ^{59}Fe in duodenum of post-hemorrhagic anemia and control rats, 10 min after the intragastric administration of 0.5 ml with 5 μCi of $^{59}FeCl_3$ pH 2

figure 1. The results of the administration of 4 μCi of $^{59}FeCl_3$, through the intracerebroventricular route to the rat groups submitted to iron deficient diet are shown in figure 2, and the coeficient between anemic and control rats are expressed in figure 3. Both figures represent the iron incorporation in the main cerebral structures of anemic and normal rats.

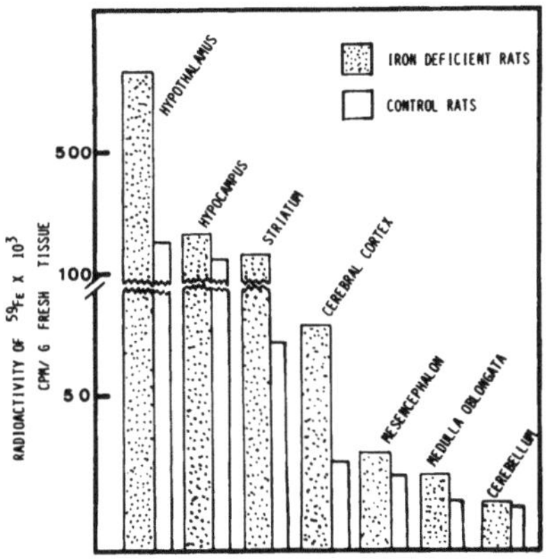

Figure 2. Incorporation of ^{59}Fe in cerebral structures of iron deficient and normal rats, after the intracerebroventricular application of a total of 4 μCi $^{59}FeCl_3$.

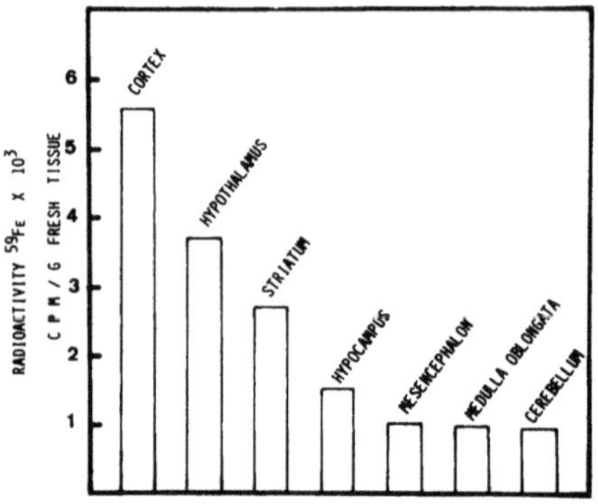

Figure 3. Anemic/control ratios of the main cerebral structures, of the rats expressed in figure 2, administered with the iron isotope.

Figure 4 and 5, represent the comparative results after the administration of the cadmium radionuclide used in this experiment. In both figures 3 and 5, the cerebral cortex shows the higher ratio anemic/control iron and cadmium accumulation.

Figure 6 shows the incorporation of ^{59}Fe in the encephalon of the two experimental models, iron loaded and iron deficient rats compared to the normal ones, after the intraperitoneal administration of $^{59}FeCl_3$.

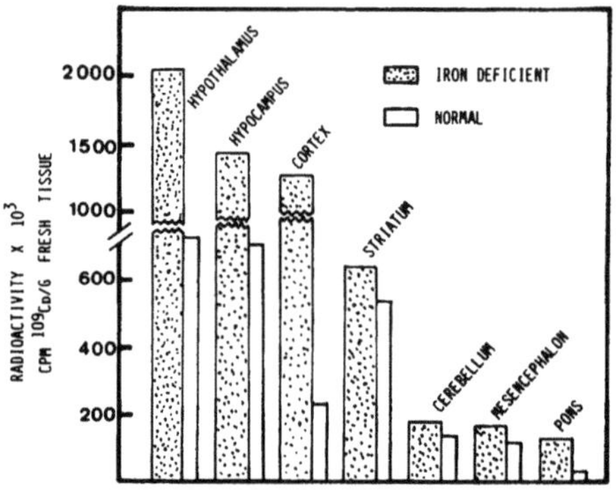

Figure 4. Incorporation of ^{109}Cd in cerebral structures of iron deficient and normal rats, after the intracerebroventricular application of a total of 4 μCi $^{109}CdCl_2$.

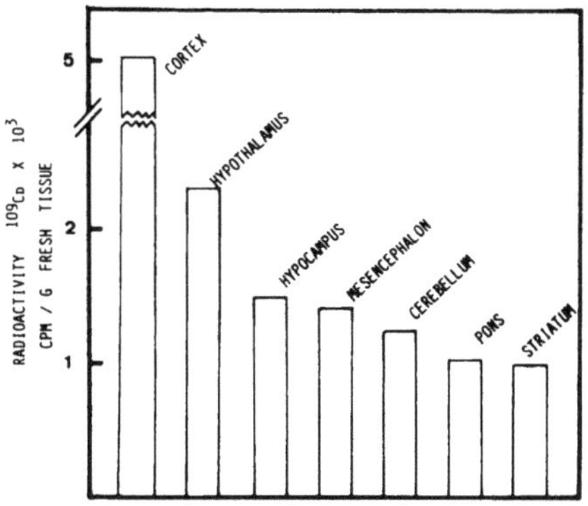

Figure 5. Anemic/control ratios of the main cerebral structures of the rats represented in figure 4, administered with the cadmium radioisotope.

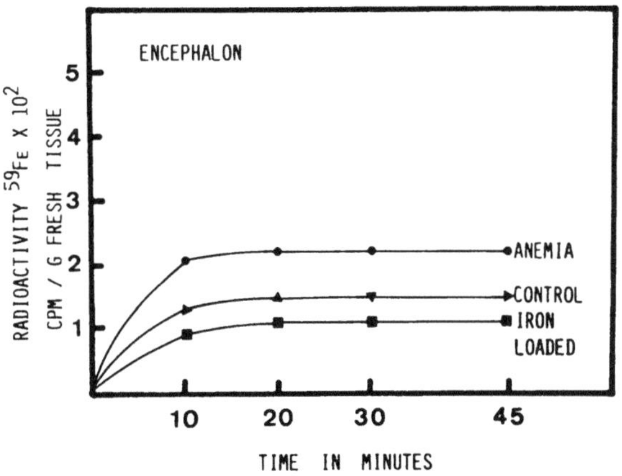

Figure 6. Kinetics of incorporation of ^{59}Fe in encephalon of normal rats as controls, compared to iron deficient animals submitted to iron deficient diet, and to iron loaded ones with 1 mg Fe in chemical form of $FeSO_4$/rat/day, during 3 days by intraperitoneal route. Each animal is injected with 5 μCi $^{59}FeCl_3$ through the intraperitoneal way and mean values are obtained with 5 to 8 animals.

Discussion

Metallothionein (MT) function is not yet known, it is studied in several biologic systems (3,20), and the recent works on anemia (7) lead us to express the hypothesis of its participation in iron homeostasis. In this work, after the similar incorporation profile between cadmium and iron in cerebral structures, there is proposed the binding of brain MTs (5,6) with both cations.

The metallothionein induction in the intestinal mucosa and brain after the same stimulus, the anemia, led us to propose the participation of the MT in both, the iron absorption by the intestinal mucosa, and brain involvement in iron homeostasis. These results help to consider the share of brain biochemistry in unresolved different human pathological and neurological Syndromes with unknown ethiology, remedied with empirical iron therapy (13-17). In this sense it could be proposed: Restless legs and Plummer-Vinson (Kelly-Patterson) Syndromes.

Since 1937, iron homeostasis is basically observed in the intestinal absorption (21), and other data completed the concept of "mucosal barrier" (22-24). Recent investigations bringt forward new parameters on iron homeostasis concerning new proteins (25-29), which could be connected with the catalytic activity of the MT as a biological chelate sequestering and releasing zinc and copper. The compartmental steady-state between transferrin and ferritin was studied time ago and also recently (10,24,30). Others studied the signification of iron corporal content and iron status in the diet consumed 1-3 days before its absorption (31-32).

Interesting interrelationships between iron status, ferritin and EEG response, are connected with iron pathology (15-33). According to those results, cerebral cortex is one of the main cerebral structures with higher ^{59}Fe and ^{109}Cd incorporation ratio, with biochemical signification at the molecular level in these experiments. The limbic system has a high iron incorporation according to the observations of zinc and other metal content (34,35), and with its attractive endocrine and neurochemical biochemistry. It is surprising how quickly iron crosses the blood brain barrier in both directions.

New data are contributing with the better knowledge of iron homeostasis, suggesting specific transport mechanisms in the bone marrow (36) and transferrin receptors in rat brain (37). The data exposed here could suggest for the brain, a feed back mechanism for iron absorption and homeostasis, similar to the well established for other hormonal axes.

References

1. Cotzias, G.C., D.C. Borg, K.B. Selleck: Am. J. Physiol., 201, 927-930 (1961)

2. Bondia S, A. Jaudi, B. Ribas, A. Santos Ruiz, M.I. Sanchez Reus: 1980. In: In: Trace Element Anal. Chem. Med. Biol. 1980, vol 1, (P. Brätter, P. Schramel, eds.). Walter de Gruyter Co., Berlin-New York, pp 37-46.

3. Kägi, J.H.R. and Nordberg M., eds., in: Metallothionein, 1st Internat. Meeting on Metallothionein, 1979, Zürich, Switzerland, Birkhauser Verlag, Basel-Boston-Stuttgart.

4. Bondia S, B. Ribas, A.M. de la Torre, A. Santos Ruiz, vol 12, 1980, In: Adv. Physiol. Sci., Nutrition, Digestion, Metabolism, (T. Gati, L.G. Szollar, Gy. Ungvary, eds.), Pergamon Press-Akademiai Kiadó, pp 71-75.

5. Ribas B., An. Med., 67, 941-953 (1981).

6. Ribas B, In: Trace Element Anal. Chem. Med. Biol., vol 2, 1983, (P. Brätter, P. Schramel, eds.), Walter de Gruyter Co., Berlin-New York, pp 585-597.

7. Iniesta M.P., C. Rubio, B. Ribas, 2nd Intern. Meeting on Metallothionein, 1985, August, Zürich, Switzerland, p 66.

8. Forth W, 1983, In: Intestinale Resorption von Eisen und Chemisch Verwandten Metallen, Dünndarm, Handbuch der Inneren Medizin 5. Auflage, 3, Teil A, (W.F. Caspary, ed.), Springer Verlag, Berlin-Heidelberg-New York.

9. O'Donell M.W.,T.M. Cox: Biochem. J., 202, 107-115 (1982).

10. Savin M.A., J.D. Cook: Blood, 56, 1029-1035 (1980).

11. Johnson G., P. Jacobs, L.R. Purves: J. Clin. Invest., 71, 1467-1476 (1983).

12. Osterloh K., R.J. Simpson, K. Raja, S. Snape, T.J. Peters: Biochem. Soc. Trans., 13, 763-764 (1985).

13. Strong E.K.: 1916, In: Health Comm., Publication 3, Rockefeller Foundation p 121.

14. Burman D.: Clin. Hematol., 2, 257-268 (1973).

15. Tucker D.M., R.A. Swenson, H.H. Sandstead, In: Chapter 8, Neurobiology of the Trace Elements, vol 1, (1983), pp 269-291.

16. Pollit E., R.L. Leibel: J. Pediatr., 88, 372-381 (1976).

17. Lozoff B, G. Brittenham, F.E. Viteri, J.J. Urrutia: 1982. In: Iron Deficiency, Brain Biochemistry and Behaviour, (E. Pollit, R.L. Leibel, eds.), Raven Press, New York.

18. De la Torre A.M., S.M. Katarneh, M.I. Perez Escolano, M.A. Arranz, B. Ribas: 1984, vol 3, In: Trace Element Anal. Chem. Med. Biol., (P. Brätter, P. Schramel, eds.), Walter de Gruyter Co., Berlin-New York, pp 649-656.

19. Arranz M.A., M.I. Perez Escolano, M.C. Saiz Vadillo, I. Basagoiti I., S.M. Katarneh, B. Ribas, R. Cadórniga: Rev. Esp. Fisiol., 40, 365-370 (1984).

20. Metallothionein, 2nd Intern. Meeting on Metallothionein, Department of Biochemistry University Zürich, August 1985, Switzerland, Abstracts Book, pp 1-79.

21. McCance R.A., E.M. Widdowson: The Lancet, 233, 680-684 (1937).

22. Hahn P.F., W.F. Bale, J.F. Ross, W.M. Balfaur, G.H. Whipple: J. Exptl. Med., 78, 169-188 (1943).

23. Granick S.: Chem. Rev., 38, 379-403 (1946).

24. Wheby M. S., V.H. Crosby: Blood, 22, 416-425 (1963).

25. Blanc B., H. Isliker.: Helv. Physiol. Pharmacol. Acta, 23, 145-148 (1965).

26. Pearson W.N., M.B. Reich: J. Nutr., 99, 137-140 (1969).

27. Huebers H., E. Huebers, W. Forth, W. Rummel: Life Sci., 10, Part I, 1141-1148 (1971).

28. Worwood M., A. Jacobs: Life Sci., 10, Part I, 1363-1373 (1971).

29. Boulard M., M. Selin, Y.C. Najean: Proc. Soc. Exptl. Biol. Med., 139, 1379-1384 (1972).

30. Nathanson M.H., G.D. McLaren, G.M. Saidel: Computers Biomed. Res.,17, 55-70 (1984).

31. Fairweather-Tait S.J., A.J.A. Wright: Brit. J. Nutr., 51, 185-191 (1984).

32. Fairweather-Tait S.J., A.J.A. Wright: 2nd Workshop on Carcinogenic Mutagenic Metal Compounds, January 1986, Villars-sur-Ollon, Switzerland, Abstracts Book, Comm. Nr 9.

33. Tucker D.M., H.H. Sandstead. 1982. In: Iron Deficiency: Brain Biochemistry and Behaviour, (E. Polllit, R.L. Leibel, eds.) Raven Press New York, 161-181.

34. Hock A., U. Demmel, S. Schicha, K. Kasperek, L.E. Feinendegen: Brain, 98, 49-64 (1975).

35. Sourkes T.L., 1982, In: Iron Deficiency, Brain Biochemistry and Behaviour, (E. Pollit, R.L. Leibel, eds.), Raven Press New York.

36. Soda R., M. Tavassoli: J. Ultrastruc. Res., 88, 18-29 (1984).

37. Hill J.M., R. Michael, R.J. Weber, C.B. Pert: Proc. Natl. Acad. Sci. USA, 82, 4553-4557 (1985).

REGRESSION MODEL FOR THE THYMIC GLAND MORPHOLOGY BY CADMIUM AND ZINKIN INDUCTION BY ZINC.

B. Ribas, F.J. de Pascual, J. del Río, M.I. Sánchez-Reus.

Institut of Biochemistry, CSIC-Complutense University, Faculty of Pharmacy
28040-Madrid, Spain

Introduction

Toxic heavy metals and organic compounds in drugs abuse are considered as chemical ligands inducing molecular processes as the receptor synthesis, increment of its concentration through the possible blockade and feed-back mecanism, and alteration of its molecular dinamic. Cadmium and zinc, both bivalent metals, affect in a different sense numerous biological processes (1,2), the understanding of metal homeostasis, is not complete and metallothionein (MT) is involved in such processes (3). MT is identified as the major intracellular metal-binding protein in a variety of normal and pathological tissues (3,4).

Thymus involvement in metalothionein, metal homeostasis and immune-response is more scarcely known (5,6), nevertheless zinc is localized in the epithelial cells of the thymic gland (7).

The decrement of the immune capacity, lung infections and neoplasias that appear in the cadmium treated animals (8,9) could be connected with the physiology and molecular biochemistry of thymus, moreover of the thymus regression factor, which is most likely activated by cadmium or other toxics through genetical activation, rather than through homeostatic imbalance. Zinc MT is also characterized in the thymus gland (6) and cadmium induces two proteins (10), nevertheless the MT polymorphism has been recently stablished in other systemic organs in six isometalothioneins (11). The object of this work was considered after the observation of the opposite morphological changes of thymus gland with the cadmium and zinc treatment wich was undertaken in the hystological and biochemical routes.

Trace Element – Analytical Chemistry in Medicine and Biology, Vol. 4

Material and Methods

Animals and treatments.- Experiments are repeated with three groups of male Wistar rats with an average weight about 100 ± 20 g. In each case next treatments were followed: 1.- Control group fed with standard food and water "ad libitum", and injected with 0.1 ml of normal saline serum. 2.- A second group submitted to intraperitoneal injection of 0.1 ml of $ZnCl_2$/100 g. weight and fed like before. 3.- A third group submited to intraperitoneal injection of 0.1 ml of $CdCl_2$/100 g. weight, following the same feeding conditions than the forementioned groups.
All solutions were obtained with Merck's chemical products, using 1 mg of each element in destilled and sterile water, lasting the treatment four weeks, and three animals from each group were sacrified every week.
Samples Processing.- The analysis process was based in the hystological planimetry of the slides, obtained through cardiac perfussion with heparinized normal serum and Bouin's fixing solution, so that a non-deformed and perfectly fixed organic image could be obtained.
Afterwards, organs were mantained for 24 hours in the forementioned fixator and dehydrated in increasing strenghted alcohols, including the pieces in Merck's parafine after bathing them in toluene. The hystological slides were about 4 µm thick and staining was realized mixing either Pappenheim's panoptic method or hematoxyline-eosine dichromic one. Planimetry was made projecting the slides in a well-known density millimetred paper with a photograph enlarger, drawing then exactly the cortical and medullar profiles as well as the interlobulillar partitions, with a posterior image confirmation throughout an exhaustive microscopical scanning.
Seven slides from every gland were used, employing a cephalic one, a medium-cephalic one, three medial, one medium-caudal and a caudal one. In every case we tried to use the same organographic regions with identical spatial orientation.
The statistical method has been designed by the authors, using for it an Apple IIe personal computer, and it is based in the partial and total areas, the frecuency of cortex/medulla ratio, (C/M) and lobulilles number.
Every week of the four the experiment lasted, the thymus of some of the rats treated with zinc and cadmium - as well as the controls - were weighed and cut with scissors in four columns of PBS (Phosphate Buffer Saline 0.1 M, Ph 7.4 in NaCl 0.8%) and their cells isolated by centrifugation of 1400 rpm. 10^7 cells/ml PBS were marked with 250 µCi ^{125}I and 5 µl lacto-peroxidase (0.1 mg/ml PBS) and the proteins extracted with Nonidet P-40

0.4% and 80,000 cpm submitted to polyacrylamide gel electrophoresis and autoradiography in AR.10 Kodak films.

Results

In figure one are shown the drawned planimetric images of the hystological cuts from the rats thymus treated during 21 days with zinc and cadmium through intraperitoneal route. Thymusses were removed and processed for their planimetric processing as signalized in Material and Methods.

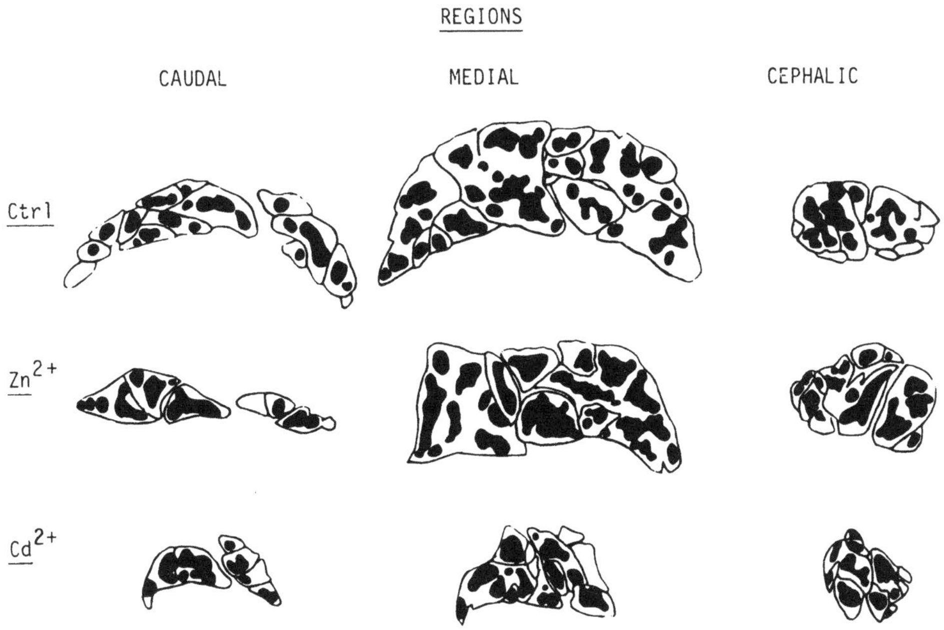

Figure 1.- Morphological changes of thymus, after the intraperitoneal administration for 21 days of 1mg Zn or Cd/kg/day compared to the controls.The thymus figures are obtained by drawing their planimetric projections.

Next page shows, in table 1,the data obtained by the statistical analysis of the thymus images, refered to the areas and number of lobulilles.In table 2 are shown the frecuency of the ratio C/M, being very remarkable the first data referred to cadmium in the third week.

Table...1

After the statistical analysis, results (briefly represented) referred to areas and lobulillar ratios were as folows

	C.	M.	T.A.	C/M	M.L.A.	N.L./S
			Starting Week			
Ctrl	1159	341	1500	3.4	69.4	6.2
			First Week			
Ctrl	2401	729	3130	3.3	66.6	6.7
Zn	1577	658	2235	2.4	31.5	10.1
Cd	2135	627	2762	3.4	65.8	6.0
			Second Week			
Ctrl	2709	907	3616	2.9	106.3	4.9
Zn	1979	1099	3078	1.8	57.0	9.0
Cd	2046	853	2899	2.4	46.0	9.0
			Third Week			
Ctrl	3125	1419	4544	2.2	56.8	11.4
Zn	2306	1357	3663	1.7	66.0	8.7
Cd	1030	716	1746	1.4	26.5	9.4

NOTE: C=Cortical Area M=Medullar Area T.A.=Total Area M.L.A.=Medium Lobulillar Area N.L./S= # lobulilles/Slide

Table...2

The calculation of the ratio frecuency C/M offers the next result

Week	Ratio →	0-1	1-2	2-3	3-4	4-5	5-6	6-7
0	Ctrl	0	4	12	13	4	1	0
	Ctrl	0	6	10	11	6	2	0
I	Zn	0	39	20	9	1	1	1
	Cd	0	9	16	5	6	1	1
	Ctrl	0	4	11	10	2	1	1
II	Zn	0	34	18	4	2	0	0
	Cd	0	11	19	13	3	2	0
	Ctrl	0	20	29	7	3	3	1
III	Zn	0	5	9	12	13	6	2
	Cd	12	27	13	3	1	0	0

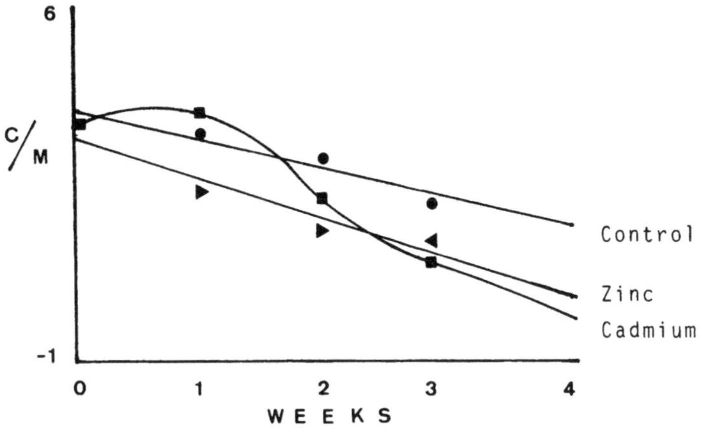

Figure 2. Graphic representation of the ratios cortex/medulla in function of time of the three groups of animals: treated with zinc and cadmium, compared to the controls.

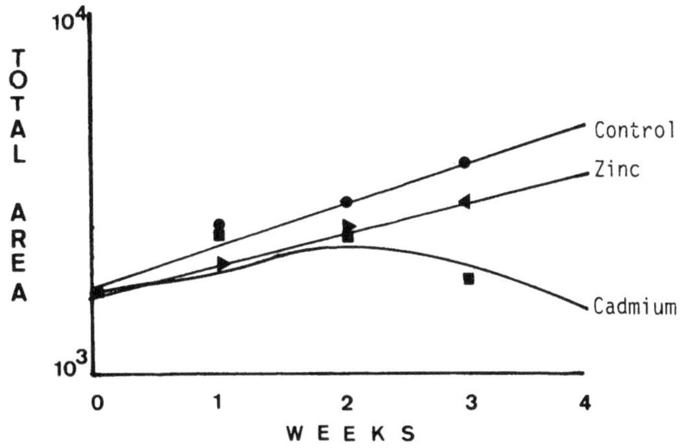

Figure 3. Evolution of the thymus total areas during the period of treatment with zinc and cadmium, compared with the control animals as expressed in Material and Methods.

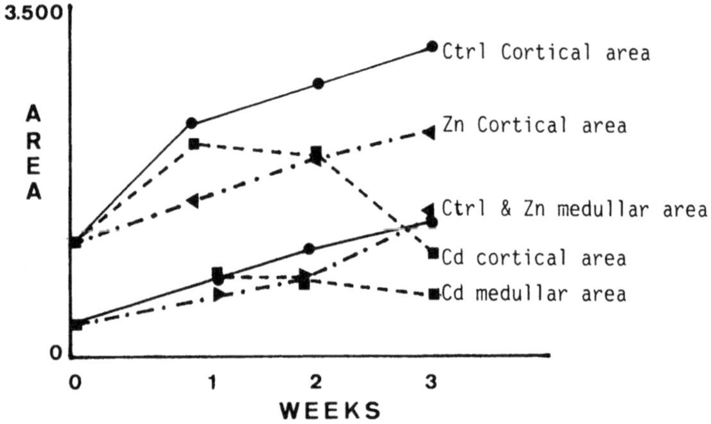

Figure 4. Cortical and medullar variations suffered by thymus along the three weeks treatment lasted. Notice the parallelism showed by cortical and medullar areas of control and zinc and the very different behavior of cadmium both areas. This shows that all variations suffered by cadmium treated animals are exlusively due to cortical regression.

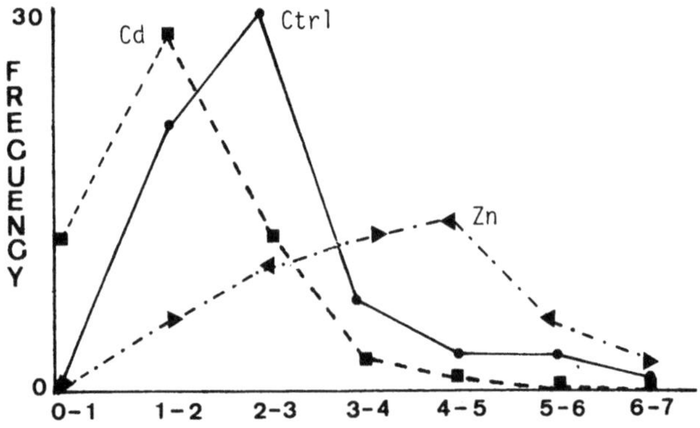

Figure 5. Frecuencies of the ratio C/M in the third week. Notice that only cadmium presents a high frecuency of a higher medullar area than cortical one. Fact that never occurs neither in control nor with zinc treatment, which shows a very a high statistic mode (That means, quite a bigger cortical area than a medullar one).

The functions obtained for the graphic representations in figure 2 are, for cadmium: $Y = (\sin(X) \times 0.40 + 1.49)^2$ $(r = 0.99)$

for zinc: $Y = -0.56\ X + 3.75$ $(r = -0.94)$

for control $Y = -0.39\ X + 3.95$ $(r = 0.97)$

The functions obtained for the graphic representations in figure 3 are, for cadmium: $Y = 1/(\cos(X)) \times 1.79.10^{-4} + 5$ $(r = 0.74)$

for zinc: $Y = 733.2\ X + 7.86$ $(r = 0.99)$

for control: $Y = 961.8\ X + 793$ $(r = 0.97)$

In the case of cadmium, in both graphic representations, the regression have been made using the curvilinear one, due to the special behaviour that this element provokes in the evolution of the physiological parameters of the thymus.

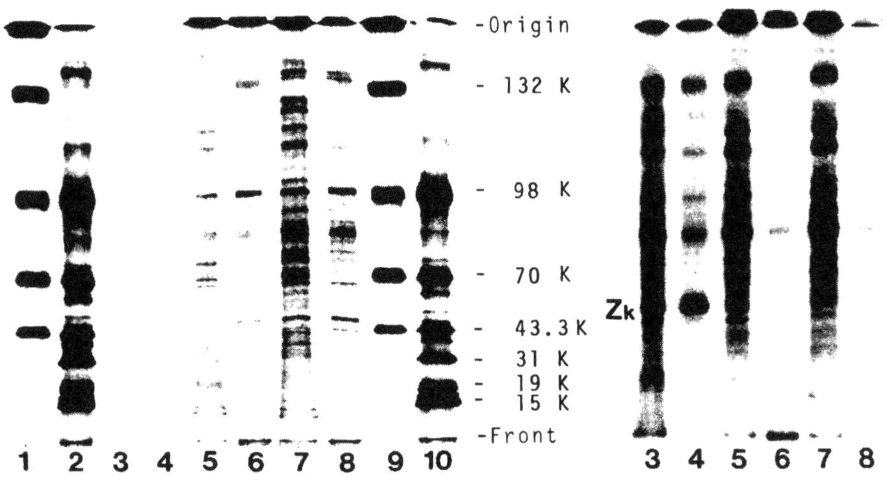

Figure 6. SDS-Polyacrylamide gel electrophoresis and autoradiographic patterns of proteins from thymus cells marked with ^{125}I and lactoperoxydase, extracted with Nonidet P-40 0.4 % and submitting 80 000 cpm to 20 % T gel. Molecular weight markers: 1,2,9 an 10 on the both sides; $ZnSO_4$ treatment: 3=pellet and 4=supernatant; $CdSO_4$ treatment: 5=pellet and 6=supernatant; and controls: 7=pellet and 8=supernatant. Zinkin is marked with a **Zk** which appears with a Mr$\simeq$ 45 Kilodaltons.

Discussion

From the table 1, it is deduced that, although the higher is the total tissue area in control group, the lower is the C/M ratio, which means a higher cortical growth. This ratio is inverted in the case of zinc treated animals, whom presents a higher C/M ratio as the total area gets bigger. In the case of cadmium, it can be observed an obvious decrement in the total area exclusively due to the cortical area, meaning a strong decrement of lymphocytes production, remaining its germinal zone almost vestigial This fact could be in connexion to the human anergic condition and alterations induced by cadmium, of the defense mechanisms, in which the thymus plays an important role (5,8,9). It is remarkable the increment in the medial lobulillar area induced by zinc, essential trace element localized in the thymic celullar reticulum (7). Thymulin, a zinc containing hormone, is found in thymic epithelial cells (12) and in its dense granules and vacuoles (7), but such cells are found in both, cortical and medullary areas of the thymus (7). Nevertheless, cadmium induces a great regression in its cortical area, evidences a decrement of this parameter, which mechanism is not known.

It is showed also in figure 1,the difference between cadmium and control rats, for while medium lobulillar area descends in control animals, and remarkably for cadmium treated rats. In the case of zinc, it could be suggested an increment of the physiological function of the thymus.

The second table expresses the fact that cadmium causes a decrement in cortical area (figure 2), that while control and zinc treated animals maintain a C/M ratio near to 1 or 2, in the third week this relation is inverted, so that numerous lobulilles may be found in which their medullar area is bigger than the cortical one. As showed that the medullar size remains constant (table 1), it is obvious that it is due to a decrement of the cortical area. This changes may suppose a strong inactivation of the lymphopoïetic capacity of the thymus. The decrement of the immune capacity, lung infections and neoplasias, that appear in cadmium treated animals (8,9), could be in connexion with thymus physiology and the thymus regression factor, which is morewell activated by cadmium through an unknown mechanism.

Zinc metallothionein is also characterized in the thymus gland (6), cadmium induces two proteins (10), and in systemic organs its polymorphism is established to six isometallothioneins (11). Here it is showed the susceptibility of the thymus gland to be inhibited by cadmium; and to induce a protein to be characterized, the zinkin, by zinc administration. The nonapeptide thymulin, a zinc containing hormone, present in the cytoplasmic vacuoles of thymic epithelial cells (12), seems to have a lower Mr than zinkin, which appear to have in

SDS-polyacrylamide gel electrophoresis an approx. Mr of 45,000. On the other hand, zinkin seems to be a membrane protein. Research is doing in order to establish the metal content, its exact Mr and relationships with the metallothionein. The physiological behaviour of thymus show different capacity than the liver, where cadmium induces specific and inespecific proteins, like the isometallothioneins and increment its weight, in spite of the pathological changes already well established. Regression and activation capacities for cadmium and zinc respectively, are observed on the growth of the rat thymic gland.

References

1. Ribas B., M. Dean, W. Schmidt: Zlb. Arbeitsmed., 4, 81-83 (1977).

2. Bondia S., A. Jaudi, B. Ribas, A. Santos Ruiz, M.I. Sanchez Reus. Trace Element Anal. Chem. Med. Biol., vol 1, 1980, (P. Brätter, P. Schramel, eds.), Walter de Gruyter & Co., Berlin-New York, pp 37-46.

3. Christie N.T., M. Costa: Biol. Trace Element Res., 6, 139-158 (1984)

4. Kägi J.H.R., Nordberg M. eds., Metallothionein, Experientia, 34, 19-371, Suppl. (1979).

5. Blakley B.R.: Canad. J. Compar. Med., 49, 104-108 (1985).

6. Olafson R.W.: Canad. J. Biochem. Cell Biol., 63, 91-106 (1985).

7. Nabarra B., S. Halpern, D. Kaiserlian, M. Dardenne: Cell Tissue Res., 238, 209-212 (1984).

8. Boisset M., Cl. Boudène: Toxicol. Applied Pharmacol., 57, 335-345 (1981).

9. Gardner D.E., F.J. Miller, J.W. Illing, J.M. Kirtz: Bull. Europ. Physiopathol. Resp., 13, 157-174 (1977).

10. Maytin E.V., D.A. Young: J. Biol. Chem., 258, 12718-12722 (1983).

11. Hunziker P.E., J.H.R. Kägi: Biochem J., 231, 375-382 (1985).

12. Schmitt D., J.C. Monier, M. Dardenne, J.M. Pleau, J.F. Bach: Thymus, 4, 221-231 (1982).

Session IV

TRACE ELEMENT ANALYSIS OF BODY FLUIDS AND TISSUE SAMPLES

Chairman:

G. Tölg

Poster Reports:

G. Knapp

G. Tölg

APPLICATIONS OF ELECTROANALYTICAL FLOW ANALYSIS IN THE TRACE ELEMENT DETERMINATION OF BIOLOGICAL MATERIALS

Wolfgang Frenzel, Peter Brätter

Spurenelemente in Gesundheit und Ernährung, Hahn-Meitner-Institut, Glienicker Str. 100, D-1000 Berlin 39

Introduction

There is a rapidly expanding appreciation of the significance of trace elements in the field of biomedical research [1]. The recognition of potential toxic effects as well as the essential role of trace elements in human metabolism has tremendously increased the need for trace element determinations of environmental and biological samples. This demand has been met with the development of suitable analytical techniques highlightening the usefulness of neutron activation analysis (NAA) [2,3], flameless atomic absorption spectrometry (AAS) [4,5], and inductively coupled atomic emission spectrometry (ICP-AES) [6,7].

Electroanalytical methods such as potentiometry, polarography, voltammetry, amperometry and others are often reported to offer outstanding features [8-13] which in many cases cannot be obtained on using the other methods mentioned. However, with regard to the analysis of biological samples the use of electroanalytical methods is not wide spread in clinical or other laboratories involved in trace element research. This discrepancy between the potential feasibility and the actual application of electrochemical methods in the analysis of biological samples can be attributed to the following reasons:

I. Susceptibility to interferences.
   Since all electrochemical events occur at the respective electrode surfaces the response is prone to adsorption phenomena. Additionally, mutual elemental interferences are observed in

Trace Element – Analytical Chemistry in Medicine and Biology, Vol. 4

real samples which are often more pronounced than in other analytical methods.

II. Tedious sample pretreatment.
Electroanalytical methods are principally capable of measuring activities only. Hence, the elements under consideration have to be brought into a uniform chemical state by means of sample manipulation or digestion procedures in order to determine their respective concentrations. Digestions, however, are laborious, timeconsuming and include a high risk of contamination.

III. Lack of suitable and commercially available instrumentation.
Manufacturers proof of the feasibility is often limited to simple applications. Reduction to practice in clinical laboratories requires the tiresome elaboration for special applications and immense effort in testing the procedures developed. Additionally, the high number of samples accruing in clinical laboratories require at least partial automation which is difficult to produce with electroanalytical equipment.

IV. The need for skilled operators.
Although usually the instrumentation required is much simpler than for other analytical methods, trouble shooting is complicated due to the various possible failures. Sometimes "surprises" emerge just as things are starting to go well which call for trained staff.

V. Tradition and conservatism.
In the world of the clinical laboratory, people are especially busy with routine work and evade new techniques because the status quo will be upset. Prejudices may also play a role.

There is no doubt, that these points are related and the general acceptance of electroanalytical methods presupposes improvements with respect to selectivity, sensitivity and accuracy. Moreover, the extension of electroanalysis into the clinical environment demands at least partially automation with minimum maintenance and observation.

At the Hahn-Meitner-Institut we looked out to revitalize old techniques and to develop new ones with emphasis towards their potential application for routine trace element analysis of biological samples. Major aims were to overcome some of the aforementioned drawbacks, simplify the analytical procedure, and make these methods more reliable. One of the keys to success has been the incorporation of electroanalytical sensors into flow-systems. This will be demonstrated in detail for the direct potentiometric determination of fluoride and simultaneous determination of cadmium, lead and copper by means of electrochemical stripping analysis. Further support of the usefulness of electrochemical flow analysis will be given by examples from the relevant literature.

Flow Injection Potentiometry

Ion Sensitive Electrodes (ISE) are readily applied in clinical chemistry for the determination of Na^{+}, K^{+}, Ca^{2+}, F^{-}, for the analysis of blood gas and for sensing enzymatic activity [8,9,14-16]. Automatic analyzers are available for some of these elements and have been proved to fulfil routine requirements. However, the determination of fluoride for instance lacks sensitivity due to the limited linear dynamic range and the detection limits reported are around 20 - 100 µg l^{-1} F^{-} [17,18]. Additionally the slow electrode response at low fluoride levels requires at least 10 min/analysis. The recently introduced Flow Injection Analysis (FIA) [19,20] has been widely recognized as a powerful tool in many areas of analytical chemistry and was also applied to ISE measurements [21-23]. The technique is based on the injection of a sample into a non-segmented carrier stream which transports the sample to the flow-through detector cell (Fig. 1). The electrode response follows the transient concentration profile of the sample, the peak height of which is a measure of the sample concentration. The improvements obtained by the adaption of FIA for the direct potentiometric determination of fluoride are published [24]. They may be summarized as follows: Increased detection limits (≈ 1 ppb), high sample throughput (up to 180/h), short start-up time (usually 10 min),

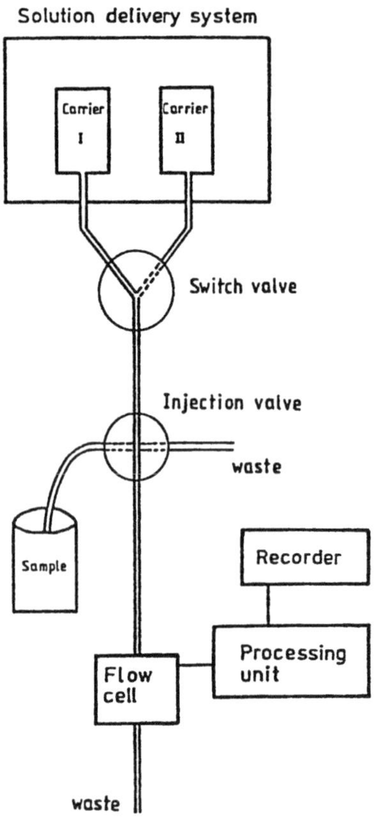

Fig.1: Basic Setup for Electrochemical Flow Through Analysis.

ease of operation and minimum maintenance. Furthermore, contamination problems are restricted to the sampling and preconcentration step since the system is self-controlled with respect to carry-over and blank levels. A careful investigation of the influence of the composition of the carrier and the ion strength adjustment buffer (TISAB) which is recommended for fluoride determinations [17] turned out the necessity of optimizing these solutions according to the particular kind of sample [24].

Urine could be successfully analyzed using a citrate buffer with a sample throughput of 40/h as is shown in Fig. 2. Repetitive injections of standards and samples confirm the high precision of the method. The accuracy was checked by recovery experiments and the analysis of NBS standard reference material (NBS 2671a). In Table 1 the results of 5 urine samples and the NBS standard are given, respectively.

Table 1: Results of Fluoride Recovery in the Analysis of Urine and the Determination of Fluoride in NBS Reference Material.

Sample	fluoride added [$\mu g \cdot l^{-1}$]	fluoride found [$\mu g \cdot l^{-1}$]	recovery [%]
U 1/1 (diluted 1:10)	--	206	--
	100	299	97.7
	200	398	98.0
U 1/2	--	335	--
	100	419	96.3
	200	541	101.1
	500	809	96.9
U 4/1	--	285	--
	20	305	100.0
	50	319	95.2
	100	356	92.5
U 5/1 (diluted 1:5)	--	211	--
	20	230	99.6
	50	276	105.8
	100	291	93.6

		certified value [$mg \cdot l^{-1}$]	fluoride found (n=5) [$mg \cdot l^{-1}$]
NBS 2671a	elevated level	5.7 ± 0.3	5.2 ± 0.1
	low level	0.55 ± 0.03	0.48 ± 0.04

Recently we started to use the same set-up for the determination of iodide which turned out to give excellent results and improved detection limits [25]. In Table 2 a survey of the relevant literature concerning Flow Injection Potentiometry for biomedical applications is presented.

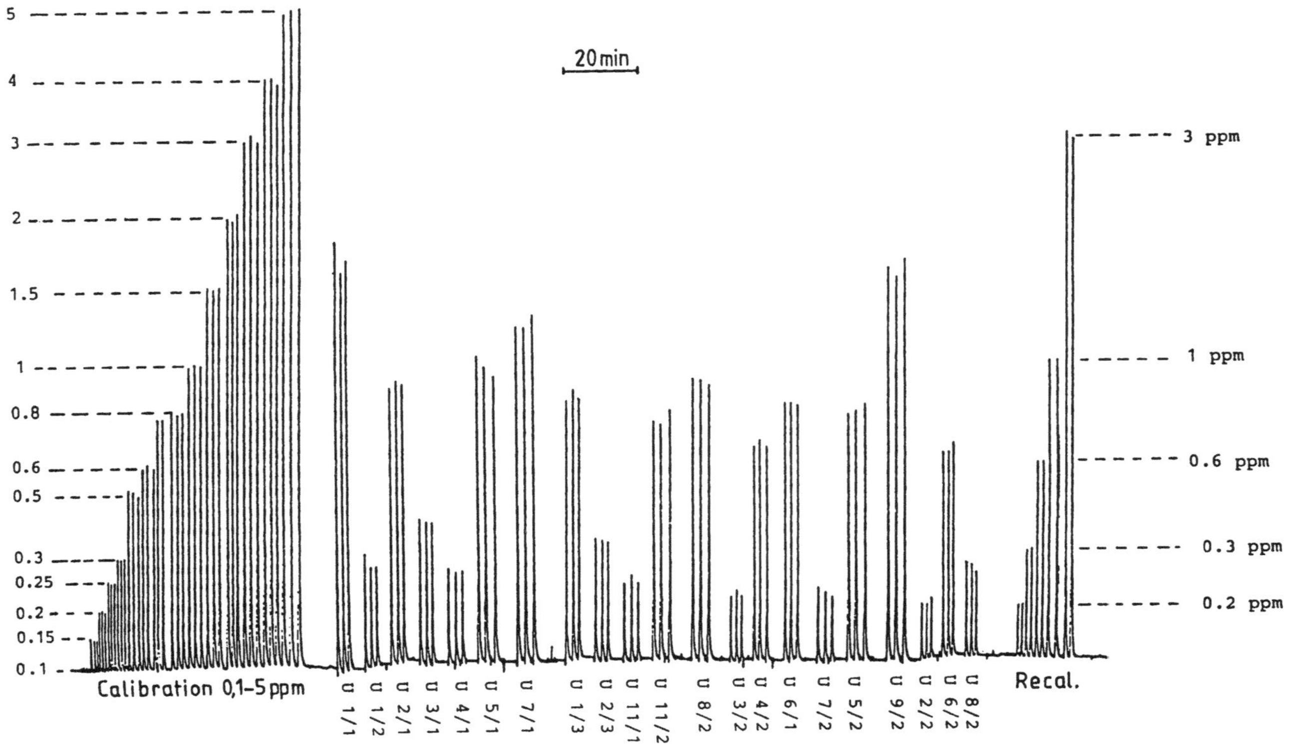

Fig.2: Fast Sequential Determination of Fluoride in Urine by Flow Injection Potentiometry
Original recorder trace of successive injection of standars and urine samples which are mixid with citrate ionic strength buffer in a 1:1 ratio.

Table 2: Literature Survey on the Application of Flow Injection Potentiometry for the Determination of Biomedical Relevant Parameters.

Element	Ref.	Sample	Remarks
Ca^{2+}	26/28	Serum	in-vitro monitoring micro samples
K^{+}	27/29 34	Serum, Blood	high sample throughput; chemfet simple instrumentation;
Na^{+}	27/34	Serum, Blood	high sample throughput; chemfet
pH	26/28 34	Serum, Blood	in-vitro monitoring; capillary electrodes; chemfet
NH_3	30	Blood, Serum, Plasma	gas sensitive electrode
Urea	31/32	Serum	indirect via pH change [31], enzyme reactor NH_3 electrode [32]
F^{-}	24/33	Bone, Urine	improved detection limit, high sample throughput
Cl^{-}	34	Blood	high sample throughput, reduced interference
CO_2	35	Whole Blood	gas-sensitive electrode, heterogenous samples; chemfet
Protein	36	Standards	indirect via Cu-electrode
Creatinin	37	Urine, Whole Blood Plasma	indirect via NH_3-semiconductive electrode

Flow Injection Electrochemical Stripping Analysis

Trace elements in body fluids, tissue and other kinds of biological samples are presented at concentrations ranging from the subnanomolar to the micromolar level. Hence, the sensitivity and dynamic range of the respective analytical method is an important factor. Electrochemical stripping techniques such as Anodic Stripping Voltammetry (ASV) [13,38,39] and Potentiometric Stripping Analysis (PSA) [40,41] are among the most sensitive methods available and are frequently used for the determination of heavy metals in biological samples [e.g. 42-44]. A survey of the current literature indicates a good agreement between the results obtained with stripping techniques and other methods [e.g. 45-47]. However, as pointed out before the information regarding the trace element content can only be obtained after suitable sample treatment. Solid samples (e.g. tissue, bone, teeth, food) must be digested. In the case of body fluids and beverages the addition of metal releasing agents or just sample dilution prior to analysis is discussed [46,48-50]. However, background interferences and spurious signals due to residual organic matter most likely appear and significantly reduce the detection limits and reliability of the methods. Therefore, sample pretreatment procedures for the application of electrochemical techniques are often more laborious and increase the contamination risk. Additionally, electroanalytical stripping techniques suffer from inherent interferences with overlapping peaks, intermetallic compounds, electrode deterioration, and adsorption phenomena being the most important problems.

The performance of electrochemical stripping in FIA [e.g. 51,52] largely avoids the aforementioned problems because easy matrix exchange between the deposition and stripping step can be applied. Thus strong acid samples (wet digests) can be directly injected without interference from the hydrogen reduction current and deoxygenation of the sample is unnecessary [52-55]. Further advantages can be obtained by optimizing the carrier with respect to sensitivity and selectivity. Being a closed system contamination problems are reduced to the sampling and pretreatment steps.

The flow system used is basically identical with that shown in Fig. 1. The carrier solution which is used for preplating a mercury film on a glassy-carbon substrate also serves as the stripping medium. It contains 10 mg l^{-1} Hg^{2+} in 1 mol/l KNO_3 /0.05 mol/l HNO_3 and was deaerated prior to analysis. A wall-jet type flow-through electrode was applied and connected to a PAR 174 polarographic analyser or PSA-computer [40], respectively. In Fig. 3a linear scan voltammograms obtained after successive injection of 10 µg l^{-1} Cd^{2+} and Pb^{2+} in 1 mol/l HNO_3, respectively, are shown. Obviously, there is no interference due to the high sample acidity. Signals obtained on running a calibration in the 10 - 200 µg l^{-1} range are shown in Fig. 3b. The figure indicates a detection limit around 2 µg l^{-1} which may be improved by the use of larger sample volumes (longer deposition times) and/or the choice of a more sensitive stripping mode (e.g. Differential Pulse, Square Wave [56]).

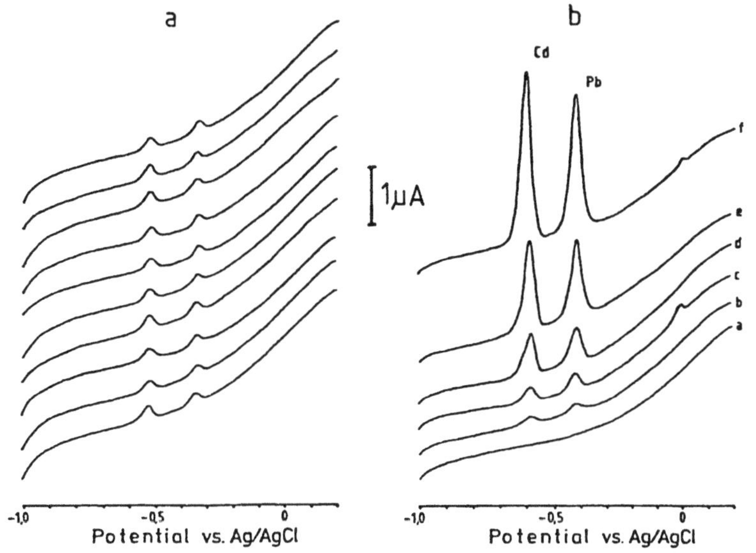

Fig.3: Flow Injection Anodic Stripping Voltammetry for the Determination of Lead and Cadmium.
Sample volume 60 µl; Flow-rate 0.4 ml/min.

A) Repetitive injections of a 10 µg/l standard
B) Calibration with increasing concentrations
a,blank; b,10; c,20; d,50; e,100; f,200 µg/l.

The precision of the method is excellent in all determinations and rarely exceeds 5 % relative standard deviation. The accuracy achieved is also good as can be seen from the results of the analysis of the standard reference materials (Table 3).

Table 3: Accuracy of Flow Injection Stripping Analysis Obtained in Analysing Reference Materials and by Comparison with other Methods.

A Cadmium Determination in Horse Kidney compared with other Methods [57]

method	mean content ppm (dry mass)	standard deviation
FI-ASV	192.1	6.6
FIPSA	189.9	9.3
Zeeman AAS	195.3	4.6
ICP-AES	180.8	2.3
ENAA	188.0	14

B Copper and Lead Values obtained for NBS-Orchard Leaves (n = 5)

element	certified value ppm	found FIPSA	FI-ASV
Cu	12 ± 1	n.d.	13 ± 3
Pb	45 ± 3	41 ± 4	39 ± 3

C Lead in IAEA Animal Bone H5 Reference Material (ppm,dry mass)

range given	mean of accepted values	found by FI-ASV
1.1 - 21.3	2.97 ± 1.08	2.48 ± 0.31

*ENAA = Epithermal Neutron Activation Analysis

Conclusions and prospects

The foregoing has evidently demonstrated that the performance of electrochemical analysis in flow systems can be beneficial for the trace element determination in biological matrices. Many problems involved in ISE-mea-surements and stripping analysis are substantially minimized and the general analysis procedure is greatly simplified. Finally, flow methods are found to enhance the possibilities of automation.

The potential of electrochemical analysis for trace element speciation which is recognized to be of utmost importance in biomedical research [58] is also worthy of mention and has been reported

[59]. It is again the flow approach which offers exciting new possibilities in coupling electroanalytical with for instance spectrometric detectors in a single line FIAsystem [60]. Thus, simultaneous information regarding the free and total amount of trace elements becomes available.

It is hoped that our work contributes towards the support of the acceptance of electroanalytical methods in the field of biomedical research and will help to put an end to the restricting reasons mentioned in the introduction of the present paper.

Acknowledgement

We like to thank Frank Chisela and Philip Gardiner for stimulating discussions and critical comments.

References

1. Trace elements in medicine, ed. H. Zumkley, Vol.1, 1984.
2. K. Heydorn, "Neutron Activation Analysis for Clinical Trace Element Research", Vol.1+2, Boca Raton, Fl., CRC Press, 1984.
3. P. Brätter, Radiochim. Acta, 34 (1983) 85.
4. B. Welz, "Atomabsorptionsspektrometrie", Verlag Chemie, Weinheim 1983.
5. B. Welz, Ärztl. Lab., 25 (1979) 168.
6. M. Thompson, J.N. Walsh, " A Handbook of ICP Spectrometry", Blackie and Sons Ltd., Glasgow-London, 1983.
7. R.L. Dahlquist, J.W. Knoll, Appl. Spectrosc., 32 (1978) 1.
8. K. Cammann, "Das Arbeiten mit Ionenselektiven Elektroden", Springer-Verlag, Berlin 1977.
9. R.N. Khuri, in "Ion Selective Electrodes", ed. R.A. Durst, NBS Spec. Publ. 314, Washington, 1969, p.287.
10. B. Fleet, R.D. Joe, in "Advances in Voltammetric Techniques", Vol. 4, ed. L.S. Bark, Chem. Soc., London, 1976, 1.
11. J.D. Czaban, Anal. Chem., 57 (1985) 345A.
12. J. Koryta (ed.), "Medical and Biological Applications of Electrochemical Devices", Wiley, Chichester, 1980.
13. J. Wang, "Stripping Analysis", Verlag Chemie, Weinheim 1985.
14. G.J. Moody, J.D.R. Thomas, Ion-Sel. Electrode Rev.,1 (1979) 187.
15. D.N. Band, I. Treasure, in "Ion Selective Electrode Methodology", Vol.2., ed. A.K. Covington, Boka Raton, CRC Press, Fl., 1979, p. 41.
16. C. Fuchs, in Ref. 12, p. 93.
17. Instruction Manual for the Fluoride Electrode, Orion Research Inc., 1977.
18. D.C. Cowell, Med. Lab. Technol., 32 (1975) 73.
19. J. Růžička, E.H. Hansen, "Flow Injection Analysis", Wiley, New York,1981.
20. C. Riley, B.F. Rocks, R.A. Sherwood, Talanta, 31 (1984) 879.
21. J. Růžička, E.H. Hansen, NBS Spec. Publ. 519, 1979, p. 501.

22. E. Pungor, K. Tóth, A. Hrab'czy-Páll, Trends in Anal. Chem., 3 (1984) 28.
23. E. Pungor, Z. Fehér, G. Nagy, K. Tóth, G. Horvai, M. Gratzl, Anal. Chim. Acta, 109 (1979) 1.
24. W. Frenzel, P. Brätter, Anal. Chim. Acta, in press.
25. W. Frenzel, presented at "Electroanalysis na hEireham" conference, Dublin, 1986.
26. E.H. Hansen, J. Růžička, A.K. Ghose, Anal. Chim. Acta, 100 (1978) 151.
27. J. Růžička, E.H. Hansen, E.A.G. Zagatto, Anal. Chim. Acta., 88 (1977) 1.
28. J. Růžička, A. Ramsing, Scand. J. Clin. Lab. Invest. Suppl., 42 (1982) 35.
29. M.E. Meyerhoff, P.M. Korvach, J. Chem. Educ., 60 (1983) 766.
30. M.E. Meyerhoff, Y.M. Fraticelli, Anal. Lett., 14 (1981) 415.
31. J. Růžička, E.H. Hansen, A.K. Ghose, H.H. Mottola, Anal. Chim., 51 (1979) 199.
32. L. Gorton, L. Ögren, Anal. Chim. Acta, 130 (1981) 45.
33. W. Frenzel, D. Gawlik, to be published.
34. H. Müller, Anal. Chem. Symp. Ser., 8 (1981) 279.
35. J.J. Harrow, J. Janata, Anal.Chim.Acta, 174(1985) pp.115,123.
36. M.L. Hitchman, F.W.M. Nyasula, Anal. Chim. Acta, 173(1985)337.
37. F. Winquist, J. Lundström, B. Danielsson, Anal.Chem.,58 (1986) 145.
38. E. Barendrecht, "Stripping Voltammetry", Electroanal. Chem., Vol. 2, Marcel Dekker Inc., New York, 1967, p. 53.
39. T.M. Florence, J. Electroanal. Chem., 168 (1984) 207.
40. D. Jagner, Trends in Anal. Chem., 2 (1983) 53.
41. S. Jaya, T.P. Rao, Rev. Anal. Chem., 6 (1982) 343.
42. J. Wang, J. Electroanal. Chem., 139 (1982) 225.
43. J.P. Franke, R.A. deZeeuw, Arch. Toxicol., 37 (1976) 47.
44. S.B. Adeloju, A.M. Bond, M.H. Briggs, Anal. Chem., 57 (1985) 1386.
45. M. Stoeppler, C. Mohl, P. Ostapczuk, M. Goedde, M. Roth, E. Waidmann, Fresenius Z. Anal. Chem., 317 (1984) 486.
46. D. Jagner, M. Josefson, S. Westerlund, K. Årén, Anal. Chem., 53 (1981) 1406.
47. F.W. Sundermann, S.S. Brown, M. Stoeppler, D.B. Tonks, in "IUPAC Collaborative Interlaboratorative Studies in Chemical Analysis", H. Egan, T.S. West (eds.) Pergamon, 1982.
48. S. Dogan, G. Nembrini, W. Haerdi, Anal. Chim. Acta, 130 (1981) 385.
49. T.R. Copeland, J.H. Christie, R.A. Osteryoung, R.K. Skogerboe, Anal.Chem. 45 (1972) 2171.
50. G. Morell, G. Giridhar, Clin. Chem., 22 (1976) 221.
51. J.A. Wise, W.R. Heinemann, P.T. Kissinger, Anal. Chim. Acta., 172 (1985) 1.
52. W. Frenzel, P. Brätter, Anal. Chim. Acta., 179 (1986) 389.
53. J. Wang, B.A. Freiha, Anal. Chem., 57 (1985) 1176.
54. D. Jagner, K. Årén, Anal. Chim. Acta, 141 (1982) 157.
55. E.B. Buchanan, D.D. Soleta, Talanta, 29 (1982) 207.
56. R.A. Osteryoung, J. Osteryoung, Phil. Trans. R. Lond., A 302 (1981) 315.
57. F. Chisela, Thesis, Technical University Berlin, 1986.
58. K. Cammann, Fresenius Z. Anal. Chem., 320 (1985) 429.
59. P.H.E. Gardiner, Topics in Current Chemistry, 1986, in press.
60. W. Frenzel, in preparation.

OPTIMIZATION OF A DIRECT PROCEDURE FOR THE DETERMINATION OF SELENIUM IN PLASMA AND ERYTHROCYTES USING ZEEMAN EFFECT ATOMIC ABSORPTION SPECTROSCOPY.

J. Nève, S. Chamart and L. Molle
Laboratory of Analytical Chemistry and Toxicology, Institute of Pharmacy, Free University of Brussels, Campus Plaine 205-1, B-1050 Brussels, Belgium.

Introduction

Determination of selenium (Se) in biological fluids generally needs long and complicated procedures, sometimes liable to be affected by systematic analytical errors(1). Trying to find a more rapid and direct technique has therefore been the preoccupation of many analysts during these last few years. Saeed, Thomassen and Langmyhr were the first to report satisfactory results for serum Se with a direct graphite furnace atomic absorption spectroscopic(AAS) procedure : they recommended nickel (II) as a matrix modifier and used deuterium arc for background correction (2). Reliability of the technique was confirmed by some authors (3,4). However, Welz and coworkers (5) recently investigated the thermal stabilization of three Se compounds (-II, IV and VI) in the presence of several modifiers, and established a clear difference in the thermal behavior of these species when atomized with nickel (II). This was considered as a possible source of error when undigested serum samples are analyzed in the furnace and that Se (IV) is used to prepare the reference solution (5). Copper (II), mixed with magnesium (II) was found more acceptable for the stabilization of all considered Se compounds.

A second difficulty in the direct determination of Se was the occurence of spectral interferences at the Se resonance line of 196.0 nm, that were not completely compensated by the classical deuterium arc background corrector.Non specific signals are indeed generated from matrices containing appreciable amounts of iron and phosphate, such as whole blood or seminal plasma (5-9). To some authors, the interference could be clearly demonstrated for serum Se as well (8,10). One of the best alternative to avoid this problem was to carry out the determination with Zeeman effect background correction (8,10).

In a recent study (11), we critically investigated the analytical performances of a direct AAS procedure, previously developped by Welz and coworkers (12). Se was stabilized with a Cu-Mg mixed modifier and the Zeeman effect applied

Trace Element – Analytical Chemistry in Medicine and Biology, Vol. 4

for background correction. In order to better destroy the organic matrix, oxygen was flowed in the furnace during a low temperature pretreatment step (11). Accuracy of this technique was established by the analysis of two serum reference materials and by comparaison with an independent technique (13,14). In the present work, we describe further steps in the optimization of the procedure. The advantages of three matrix modifiers are first critically examined. Then, the interferences due to iron and phosphate are studied at concentrations relevant to biological analysis. Finally, the methods for calibration of the spectrophometer are reported and some practical problems are reviewed.

Experimental

Apparatus

Perkin-Elmer spectrophotometers, model 3030 (deuterium correction) and Zeeman 3030, with an HGA-600 graphite furnace programmer, an AS-60 autosampler, and a PR-100 printer. Se - EDL lamp 6W. Slit width 0.7nm. Wavelength 196.0 nm. Pyrolytically coated graphite tubes with inserted L'vov platform. Argon and oxygen as purge and alternate gas, respectively. Corrected atomic absorbance and background absorbance were calculated and displayed by the instrument. Evaluation was exclusively done with integrated absorbance signals, A.s (integration time 3.0 s). Typical furnace conditions are given in table 1.

Table 1 : time-temperature program for the determination of Se in serum.

Step	1	2	3	4	5	6	7
T° setting, °C	130	450*	450*	1250*	2200	2600	20
Ramp time, s	1	10	1	1	0	1	1
Hold time, s	30	30**	20**	10**	3	5	5
Argon flow, ml/min	300	-	300	300	0	300	300
Oxygen flow, ml/min	-	300	-	-	-	-	-

* Conditions for a new graphite tube, when running water is used to cool the furnace. With more than 150 firings, conditions become 500, 500 and 1350, respectively.

** For erythrocytes or whole blood Se, time conditions are 50, 40 and 20, respectively.

Biological samples

Lyophilized human serum, Seronorm, batch 103 and 105, from Nycomed (formerly Nyegaard and Co, Oslo). Se content was recommended on the basis of a IUPAC interlaboratory trial : 93 ± 7 and 91 ± 6 ng Se/ml, respectively (15). Plasma and erythrocytes from Belgian adults : venous blood samples were collected in acid washed tubes containing heparin; plasma and buffy coat were removed after centrifugation; red cells were washed twice with an isotonic NaCl solution, then resuspended in this solution (dilution 1+1).

Reagents and standard solutions.

All reagents were of analytical grade. Two stock Se solutions were prepared (25 mg Se/l) and compared before use : one with black elemental Se, dissolved in 2 ml HNO_3 65%, then in bidistilled water; the other with Na_2SeO_3. 5 H_2O dissolved in diluted HNO_3 (2ml HNO_3 65%/l water). Working Se solutions were obtained by dilution in 0.1 N HCl (1000 ng Se/ml). These solutions were stable for several months at room temperature. Mixed Cu-Mg modifier contains 0.05 g Cu, 0.1 g Mg (as nitrates), and 0.015 g triton X-100 in 100 ml HNO_3 0.2 % (m/v) (12). Ni-modifier contains 0.25 g Ni (as nitrate) and 0.015 g triton X-100 in 100 ml HNO_3 0.2 %. Pd-Mg modifier contains 0.13 g Pd (as chloride), 0.10 g Mg (as nitrate) and 0.015 g triton X-100 in 100 ml HNO_3 0.2 %.

Procedure

The sample (0.25-0.30 ml) was diluted 1 + 3 with the modifier solution (premix), and 15 µl aliquots were atomized. The standard addition procedure was adopted for calibration : one blank and two sample solutions containing 30 ngSe and 60 ng Se (0.03 and 0.06 ml of the working Se solution) were simultaneously prepared.

Results and Discussion

Thermal stabilization of different Se compounds

Thermal pretreatment curves for different Se species - Na_2 SeO_3 in aqueous solution and in a serum matrix, D-L-selenomethionine in aqueous solution and native serum Se (dilution of sernorm) - were established in the presence of three matrix modifiers : Ni, Cu-Mg and Mg-Pd. The Zeeman effect was applied for background correction.

Conditions of Figure 1 A were those routinely used by a majority of authors: Ni as the modifier and wall atomization in pyrolytically coated graphite tubes (2-4). In accordance with these investigators, we found that the temperature could be raised up to 1000°C without losing Se as Na_2 SeO_3 . When the element was present as D-L-selenomethionine, higher temperatures could be reached (1200°C) suggesting that the thermal behavior is greatly dependent on the chemical form of the Se species. This was further demonstrated by the investigation of optimal conditions with serum Se. At temperatures less than 1050-1100°C, we were unable to obtain consistent results with the biological fluid because of a strong disturbance of the Zeeman signal, probably due to high background absorption. A similar, but less intense disturbance was also observed in the same conditions with a deuterium background corrector. However, reproducible Se recoveries and more classical absorbance profiles, with a reduced background level, were established for higher pretreatment temperatures i.e. in the range 1050-1400°C. Serum Se therefore appeared as the most stable of investigated Se species. Surprisingly, Na_2 SeO_3 mixed with serum Se behaved exactly as this last species,

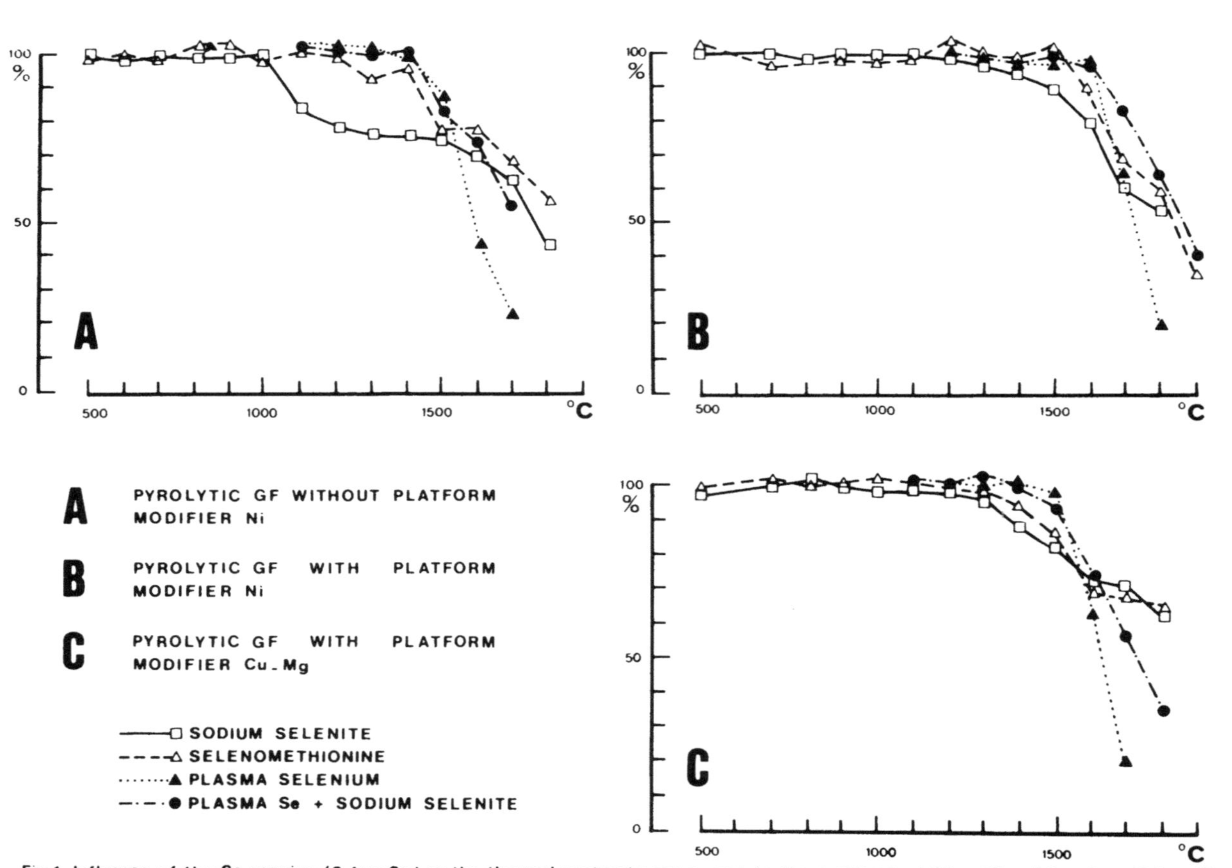

Fig.1. Influence of the Se species (0.4 ng Se) on the thermal pretreatment curves in the presence of 26 μg Ni or 5.3 μg Cu + 10.5 μg Mg. Integrated absorbance signals are expressed in % of highest registered values.

suggesting that the inorganic form is more completely stabilized in a biological matrix.

The comparaison of figures 1 A and 1 B illustrates the effects of atomization off a L'vov pyrolytic graphite platform, in thermal equilibrium conditions. By applying a set of experimental conditions proposed by Slavin and coworkers (16), and described in the "Stabilized temperature platform furnace (STPF)" concept, it is indeed expected that the temperature in the graphite tube is stabilized before the element to be determined is atomized. The main improvements that we observed consisted in a significant increase in sensitivity (11), and in a more comparable behavior of the investigated Se species in the range 1200-1400°C : thermal pretreatment curves were moved towards higher temperatures and the differences between the Se species were less pronounced than in figure 1 A. In opposition to the results of Welz and coworkers (5), we did not find a decrease in the integrated absorbance signals of $Na_2 SeO_3$ at temperatures between 500 and 1000°C. However, we were able to confirm the satisfactory results obtained by these authors with a Cu-Mg mixed matrix modifier (5) : although the element was typically lost at temperatures slightly lower than those registered with Ni (approx. 100°C), the overall behavior of all Se compounds was very similar in the range 1100-1250°C (Figure 1 C).

As far as the third modifier was concerned, Mg-Pd, the thermal pretreatment curves were quite comparable with those observed with Ni (results not shown).

Study of interferences

In a preliminar report on the direct determination of Se in serum reference materials, we established a better accuracy using Zeeman effect for background correction instead of the classical deuterium corrector (11). At that moment, no attempt was made to find out the origin of the discrepancy, but it was suggested that the last corrector was not efficient enough to tackle with some interferences from elements of the biological matrix. It had indeed been reported that compounds such as iron and phosphate may create compensation problems with the deuterium arc source due to spectral interferences (2, 6, 7-9, 17, 18). Iron has several absorbing lines so close to the primary 196.0 nm Se line that they fall within the spectral bandpass : during atomization, these lines absorb radiation from the continuum source but not from the analyte source, and the absorption signal is overcorrected (2, 6, 9, 17-19). Phosphate may generate interferences due to the presence of fine structures in the broad-band background signals : inaccurate correction is caused by the difference in the background measured by the continuum source and by the analyte radiation source (6, 7, 9, 17-19).

Figure 2 summarizes the results of our investigations on the correction of

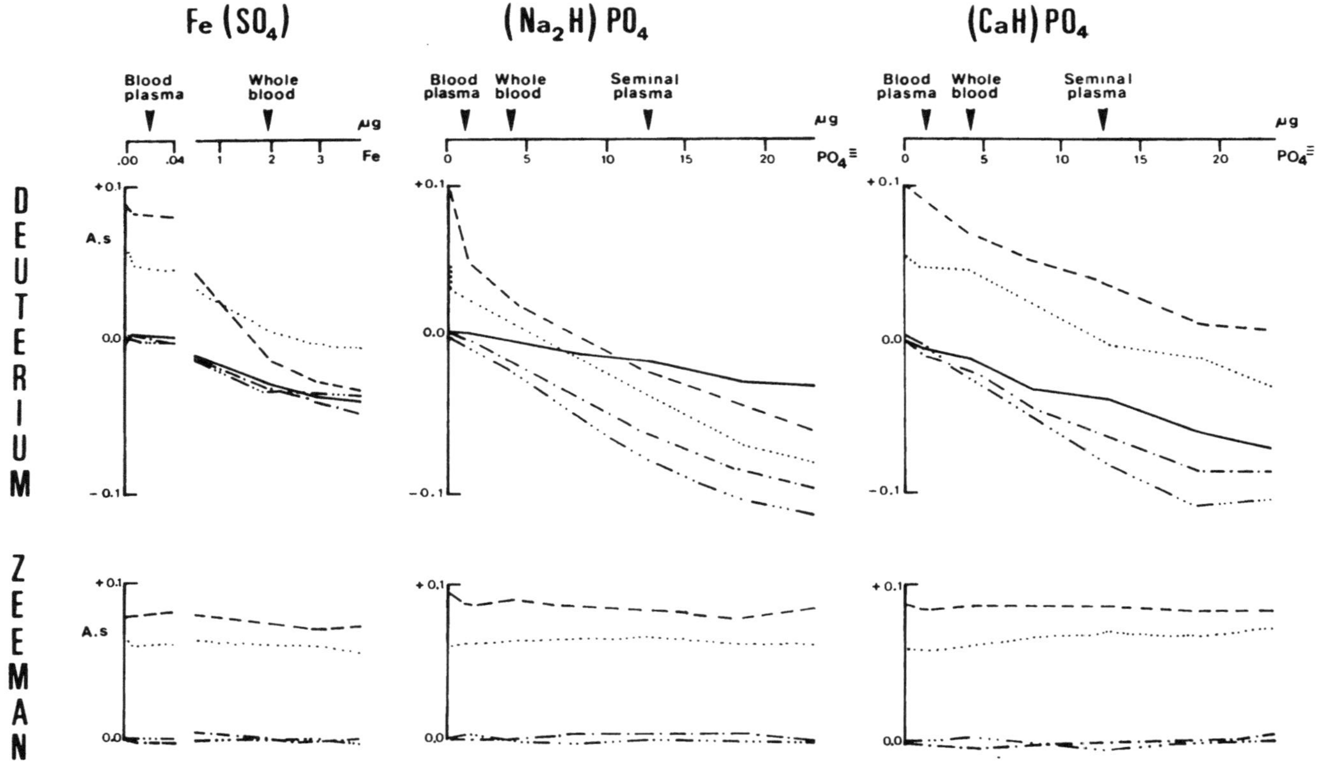

Fig.2. Integrated absorbances at the 196.0 nm Se line in the presence of various amounts of interfering element (IE):

– – – – Se + Cu–Mg .... Se+Ni ———— IE alone – . – . – IE+Cu–Mg —...—... IE+Ni

these interferences by the classical deuterium system and with the Zeeman effect. When solutions containing iron or phosphate were atomized in the absence of Se, negative absorption signals were registered at the 196.0 nm Se line when deuterium correction was applied : this clearly indicated overcompensation for the interfering background signal (6, 17, 18). In the present conditions, negative signals were first observed with 0.4 µg Fe (as sulfate), 4.1 µg PO_4 (as NaH_2PO_4) and 1.0 µg PO_4 (as CaH PO_4) in the furnace. Subsequent extent of the interference was dependent on the concentration of the interfering species. Contrary to the findings of several authors (2, 6, 8), the addition of modifiers such as Ni (26 µg) or Cu-Mg (5.3-10.5 µg) did not result in the suppression of uncorrectable non specific signals. It is possible that such a correction depends on the sample matrix and the modifier concentration (6, 9).

When solutions containing Se (0.4 ng in the furnace) were analyzed in the presence of a stabilizing metal, overcompensation signals were observed at lower concentrations of interfering species : 0.04 µg Fe and 1.0 µg PO_4 (both salts). Interferences were therefore clearly registered in solutions containing iron and phosphate ions at concentrations similar to those of human blood plasma (Figure 2). Such an observation confirms the results of Welz and coworkers (10). The phenomenon was so large for higher concentrations, corresponding to those of human whole blood or seminal plasma, that no reliable data for Se could be obtained.

The use of a Zeeman instrument was really the best alternative to avoid these interferences (5, 8,16-18). In these conditions, negative signals were never registered in the investigated concentration range of interfering species (Figure 2). A slight modification in Se sensitivity was only observed in some instances with increasing amounts of interfering compounds.

Calibration of the spectrophotometer and analytical performances

The influence of the modifiers on the accuracy of the determination of Se in two serum reference materials was studied using Zeeman effect for background correction. Results of table 2 show the excellent performances of all stabilizing metals when calibration was performed with the standard addition procedure. Only the Mg-Pd modifier could not be premixed with serum as a precipitation occured in these conditions : it had to be added on the top of the sample after dispensing it on the platform (no premix). On the contrary, the use of aqueous solutions (working Se solution : 15, 30, 45 and 60 ng Se/ml) for calibration delivered inaccurate results. This indicates that the behavior of involved Se species remained significantly different despite a careful optimization of analytical conditions.

Table 2 : Se (ng/ml) in two serum reference materials by Zeeman corrected GFAAS ($x \pm SD$, n).

	Analyte addition procedure		Calibration with aqueous standards
	Ser. 103	Ser. 105	Ser. 105
Modifier Ni	-	92 ± 8(15)	113 ± 9(7)
Cu-Mg	92 ± 7(35)	93 ± 7(32)	80 ± 5(8)
		92 ± 5(7)*	
Mg-Pd	-	92 ± 8(9)*	-
Recommended content	93 ± 7(100)	91 ± 6(200)	91 ± 6(200)

* Samples were not premixed with the modifier.

We tested a more rapid calibration procedure based on the calculation of Se levels in unknown samples from data previously collected on samples of a similar fluid mixed with two standards (standard addition autocalibration procedure). Reliability of the procedure was demonstrated by the comparaison with the results obtained by standard addition procedure ($r = 0.998$, for 14 different serum samples). As accuracy of this procedure could be influenced by the progressive loss in sensitivity resulting from tube aging, we examined the factors liable to reduce the rate of this evolution. The introduction of oxygen in graphite tube during a low temperature pretreatment step improved reproductibility of the determination and tube lifetime (11). A decrease in the oxygen content of the sheath gas had a similar effect (11). Finally, we had to select the lowest pretreatment temperatures in the predetermined optimal range (table 1). In this respect, a clear influence of the cooling system on the actual furnace temperature was documented : the pretreatment temperatures established when running water was used to cool the furnace (water temperature 15°C) were indeed systematically higher than those established with a commercial continuous cooling system (40°C). In the range 1100-1500°C, the difference amounted to 200°C !. In optimized conditions, pyrolytically coated graphite tubes could tolerate 350-400 firings. No significant loss in sensitivity could be detected during approximatively 50 consecutive firings.

Performances of the modifiers were finally compared by calculating the "characteristic mass", m_o, which is the Se quantity (in pg) in the furnace that produces an integrated absorbance signal equivalent to 0.0044 A.s. This parameter is a measure of the sensitivity when the furnace technique is used. Results of table 3 show that the highest sensitivity was measured with Mg-Pd and Cu-Mg modifiers.Somewhat poorer performances obtained with Ni suggest that this agent was less efficient to deal with interferences affecting atomization efficiency of serum Se. Our data are in good agreement whit those of literature reporting a spread for m_o values from 20 to 30 pg/0.0044 A.s.

Table 3 : Se m_o data (new pyrolytic graphite furnace with platform).

Se species	Modifier	Pretreat. T°, °C	m_o
Aq. $Na_2 SeO_3$	Ni	500-1350	28-33
	Cu-Mg	500-1250	26-32
Serum Se	Ni	1250-1450	30-38
	Cu-Mg	1150-1350	25-29
	Mg-Pd	1250-1450	18-26

Conclusions and Practical Applications

Direct determination of Se in biological fluids by GFAAS remains a delicate task. This study does not demonstrate considerable differences between three matrix modifiers as analytical performances are concerned. Far more important is the efficiency of the background corrector system, even in matrices containing low levels of potentially interfering elements, such as in blood plasma. Most adequate conditions are the presence of a Cu-Mg modifier and the use of the Zeeman effect for background correction. The method was routinely applied for the direct determination of the element in biological samples such as blood and seminal plasma, synovial fluid, whole blood or erythrocytes. For this last fluid, time conditions during ashing had to be increased (table 1) in order to better eliminate larger amounts of carbonaceous residues. Careful optimization of analytical conditions taking in mind factors liable to affect sensitivity and reproductibility of the determination, and comparaison of some results with those of a method of established accuracy (1, 11, 14) were other indispensable steps in the development of this direct AAS procedure.

Acknowledgements

Our thanks are expressed to Mrs. A. Harmegnies and C. Sommereyns for technical assistance, and to the "Ministères de l'Emploi, du Travail et du Budget, Belgium" (contract CST 30090); to Mr. B. Welz, G. Schlemmer and W. Salvin, Perkin-Elmer Co for helpful discussions; to Mr. W. Stelzer, C. Devillez and P. Scutyzer, Perkin-Elmer Belgium (Ets Van der Heyden), for the opportunity to use the PE 3030 spectrophotometer; and to Mrs. De Greef, Nycomed Belgium, for kind supply of scronorm samples. Financial support of the "Fonds de la Recherche Scientifique Médicale", Belgium (credits 9.4518.84 and 1.5.419.86 F) is also gratefully acknowledged.

References

1. Nève, J., Hanocq, M., Molle, L., Lefebvre, G., The Analyst, 107, 934-952 (1982)

2. Saeed, K., Thomassen, Y., Langmyhr, F., Anal. Chim. Acta, 110, 285-289 (1979)

3. Oster, O., Prellwitz, W., Clin. Chim. Acta, 124, 277-291 (1982).

4. Alfthan, G., Kumpulainen, J., Anal. Chim. Acta, 140, 221-227 (1982).

5. Welz, B., Schlemmer, G., Voellkopf, U., Spectrochim. Acta 39 B, 501-510, (1984).

6. Saeed, K., Thomassen, Y., Anal. Chim. Acta, 130, 281-287 (1981).

7. Saeed, K., Thomassen, Y., Anal. Chim. Acta, 143, 223-228 (1982).

8. Carnrick, G., Manning, D., Salvin, W., The Analyst, 108, 1297-1312 (1983).

9. Bauslaugh, J., Radziuk, B., Saeed, K., Thomassen, Y., Anal. Chim. Acta, 165, 149-157 (1984).

10. Welz, B., Melcher, M., Schlemmer, G., Fresenius, Z, Anal. Chem., 316; 271-276 (1983).

11. Nève, J., Molle, L., Acta Pharmacol. Toxicol., in press.

12. Welz, B., Melcher, M., Schlemmer, G., Trace Element Analytical Chemistry in Medecine and Biology, Vol. III. Bratter, P., Schramel, P., eds. W. de Gruyter, Berlin 1984, pp 207-215.

13. Nève, J., Hanocq, M., Molle, L., Anal. Chim. Acta, 115, 133-145 (1980).

14. Welz, B., Melcher, M., Nève, J., Anal. Chim. Acta, 165, 131-140 (1984).

15. Ihnat, M., et al., Acta Pharmacol. Toxicol., in press.

16. Slavin, W., Manning, D., Carnrick, G., At. Spectrosc., 2, 137-145 (1981).

17. Manning, F., Myers, S., Slavin, W., Anal. Chem., 52, 741-746 (1980).

18. Wibetoe, G., Langmyhr, F., Anal. Chim. Acta, 165, 87-96 (1984).

STUDY OF THE SPECIFIC ASPECTS OF THE ANALYTICAL DETERMINATION OF THE SELENIUM CONCENTRATION IN WHOLE BLOOD OF NEW-BORNS AND YOUNG CHILDREN

M.A. Deschuytere and H. Deelstra

University of Antwerp, Department of Pharmaceutical Sciences, Universiteitsplein 1, B-2610 Wilrijk, Belgium

Introduction

The increasing interest in selenium as a trace element, has led to the development of a great variety of analytical procedures for its determination. Nevertheless, the accurate determination of selenium using a graphite furnace atomic absorption spectrometer (GFAAS) still remains an intriguing subject in selenium research. Among the problems arising with this technique are spectral interferences as well as volatilisation losses. The volatility of selenium was found to be reduced by thermal stabilization with nickel. This matrix modifier enables higher ashing temperatures to be used, without losses of selenium. In this way the unspecific absorption of organic matter and salts is minimized. The spectral interferences originating from important quantities of iron and phosphorous species on the selenium 196.0 nm and 204.0 nm resonance lines are are probably the origin of the exclusion of the direct deuterium-arc corrected electrothermal AAS determination of selenium in whole blood and acid-digested biological tissues. For these determination procedures, a Zeeman effect based instrument should therefore be recommended.

The purpose of this work , however was to obtain a method for

Trace Element – Analytical Chemistry in Medicine and Biology, Vol. 4

the selenium analysis of small volumes (150 ul) of whole blood, using a GFAAS with deuterium background corrector. Digestion followed by extraction had to be omitted, since these procedures require at least 1 ml of sample (1,2). The use of a molybdenum coated graphite furnace with L'vov platform was responsible for the disappearance of the iron interference, which is the major interfering compound by the direct determination of selenium in whole blood.

The technique developed in this work, was then applied for the analysis of the selenium concentration in venous and capillary blood samples of young children.

Experimental

Apparatus and Reagents

The measuring apparatus set up consisted of a Perkin-Elmer model 4000 atomic absorption spectrometer, equipped with a deuterium continuum background corrector, a selenium electrodeless discharge lamp and a HGA-500 furnace with an AS-40 autosampler.

Commercial available 1000 ug/ml standard solution of selenium (Titrisol, Merck) was used. The working standards of 0, 10, 20 and 30 ppb selenium in a nickel solution, were prepared daily.

A 1 % (v/v) solution of Triton-X-100 was used to dilute the samples.

Molybdenum Coating

The molybdenum coating was applied to avoid negative peaks as a result of the iron present in whole blood. When the molybdenum coating was carried out as described by Manning and Slavin (3), although definetely lower, the negative absorption peak did not

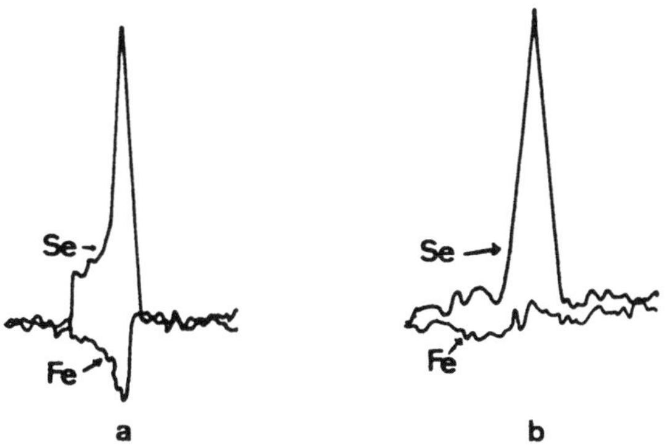

Fig. 1. Se and Fe absorption on the 196 nm resonance line of Se. The program setting of the graphite furnace is as described in Table II.
a) Successive injection of 15 ul of a Fe and Se solution into a non-coated graphite tube with L'vov platform
b) Successive injection of 15 ul of a Fe and Se solution into a molybdenum-coated graphite tube and L'vov platform

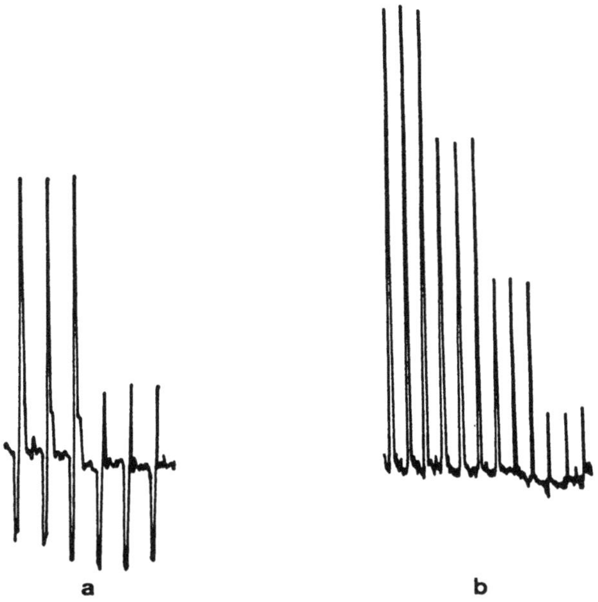

Fig. 2. Standard addition of a whole blood sample. Dilution of the sample is performed as described in sample preparation.
a) injection of 15 ul sample volume into a non-coated grahite tube with L'vov platform
b) injection of 15 ul of sample volume into a molybdenum-coated graphite tube and L'vov platform

completely disappear. Molybdenum coating with higher concentrations of ammoniummolybdate, or more injections of the coating solution, resulted in a decrease of the sensitivity, without lessening the negative peaks. Coating with lower concentrations, and repeated injections, resulted in a complete disappearance of the negative peaks (fig. 1 - 2). The program setting of the graphite furnace was the same as described by Manning and Slavin (3) and is given in table I.

As a side advance of the coating procedure, the background absorption caused by the presence of nickel in the sample solution, almost completely disappeared.

Table I. Program Setting of the Graphite Furnace for the Molybdenum Coating Procedure

Step	t_{ramp}(s)	t_{hold}(s)	Temperature (°C)	Int.Flow (ml/min)
1	10	10	110	300
2	15	15	550	300
3	9	9	2500	0

4 injections of 35 ul of an approximately 5% molybdate solution were carried out

Nickel Concentration

The Ni concentration used was the same as described by Alftan and Kumpulainen for their Se determination in plasma (4). Higher Ni concentrations were tested, resulting in higher sensitivities, for which however, significant blank measurements were responsible. Lower Ni concentrations resulted in losses of Se.

Preliminary mixing of the Ni solution with blood causes precipitation after almost 15 minutes; this is responsible for significant losses of Se, as was also described by Huguet et al. (5). Therefore the direct addition of the Ni solution into the graphite tube was checked. However this method resulted in lower sensitivity, possibly because of a less efficient mixing of the sample with the Ni solution.

Sample Preparation

Samples were prepared just before measurement.

100 ul of whole blood are diluted with an equal volume of a 1 % Triton X-100 solution. 40 ul of this dilution are added to 500 ul of the working standard (containing 0.25% Ni, and 0, 10, 20 or 30 ppb Se). The solution is mixed and brought into a plastic sample cup, used for the autosampler. The Se concentration of each cup is determined by three replicated injections of 15 ul into the graphite furnace, which is operated under the conditions given in table II.

The standard addition plot is calculated with the linear regression program of a desk calculator.

A reference whole blood sample was prepared by pooling several human blood samples, and dividing the pool in 1 ml portions that were stored at -40 C.

Table II. Program Setting of the Graphite Furnace for the Selenium Determination

Step	t_{ramp}(s)	t_{hold}(s)	Temperature (°C)	Int. Flow (ml/min)
1	10	10	110	300
2	5	5	250	300
3	5	20	1200	300
4	0	4	2200	0
5	2	2	2600	300

Injection volume = 15 ul

Results

The within-day variation of the described method amounts to

4.5 %, the day-to-day variation amounts to 5.3% (n=10) for a mean Se concentration in whole blood of 122 ppb. The detection limit is 0.02 ng Se injected into the graphite furnace. The recovery for the addition of 20 ng and 60 ng of Se to 1 ml of whole blood amounts to 100.2% and 103.1% respectively.

Table III shows the Se concentration of the reference whole blood pool, as determined by GFAAS, Hydride Generation AAS and a fluorimetric method.

The results of the analysis of venous and capillary blood samples of 5 children are shown in table IV.

Table III. Mean Se Concentration of the Reference Whole Blood Pool as Determined by Different Analytical Techniques

Determination Method	Number of Determinations	Mean Se Conc. (SD) (ppb)
GFAAS	10	123 (6)
Fluorimetry (6)	3	121 (5)
HGAAS (6)	5	116 (12)

Table IV. Se Concentration in Venous and Capillary Blood of 5 Children (between 1 and 4 years old)

Child nr	Se Conc (ppb) Venous Blood	Se Conc (ppb) Capillary Blood
1	106	102
2	59	54
3	65	63
4	78	85
5	99	99

Discussion

The direct determination of Se in whole blood by means of a GFAAS with deuterium background corrector is hindered by the presence of iron. Iron absorbs radiation of the deuterium continuum, but does not absorb radiation of the Se lamp. This results in a negative or false correction applied to the absorption signal of Se (7). Hilderbrand states that the Se absorption signal occurs later than the background absorption of Fe (8). However, as can be seen in fig. 1, when no molybdenum coated graphite furnace and L'vov platform are used, the negative peak of Fe appears almost simultaneously with the Se absorption peak, under the conditions used.

However when the graphite furnace and L'vov platform are coated with molybdate, as described before, no Fe interference was observed; also the background absorption, as a consequence of the presence of Ni in the sample solution, disappeared. The reason of this phenomenon may be explained by the properties of the molybdenum carbide that is probably formed by the coating procedure. Molybdenum carbide belongs to the interstitial carbides, which show an electron band structure and other characteristic properties that are not fundamentally different from the ones present in the pure metal (9). Hence, the disappearance of the Fe interference could possibly be explained by the formation of strongly metal-metal bounded compounds.

To conclude, in cases of determination of Se in small volumes of whole blood (100 ul), using a GFAAS equiped with a deuterium arc background corrector, we strongly recommend to employ a molybdenum coated graphite tube.

This technique was applied for the determination of Se in capillary and venous blood samples of little children. Only five venous and capillary blood samples were analyzed so far, hence no definite conclusions can be drawn. However, since for these samples no significant difference between the Se concentration in venous and capillary blood was found, these results do arise the supposition that 100 ul of capillary blood would be sufficient to determine the Se status of a child.

Acknowledgement

M.A. Deschuytere is a research assistent at the Belgian National Fund for Scientific research.

References

1) Tulley, R. T. and H.P.Lehmann.1982. Flameless atomic absorption spectrometry of selenium in whole blood, Clin. Chem.28/7, 1448-1450.

2) Oster, O. and W. Prellwitz.1982.A methodological comparison of hydride generation and carbon furnace atomic absorption spectroscopy for the determination of Se in serum, Clin. chem. Acta 124, 277-291.

3) Manning, D. C. and W. Slavin.1978. Determination of lead in a chloride matrix with the grapite furnace, Anal. Chem. 50/9, 1234-1239.

4) Alftan, G. and J. Kumpulainen.1982.Determination of selenium in small volumes of blood plasma and serum by electrothermal atomic absorption spectrometry, Anal. Chem. Acta 140, 221-227.

5) Huguet, C.,Ph. Chappuis, J. Peynet, F. Thuillier, A. Legrand, and F. Rousselet. Serum selenium assay by atomic absorption spectrometry: Its value in the laboratory diagnosis of non-obstructive cardiomyopathies, Ann. Biol. Clin. 41, 277-283.

6) Verlinden, M..1981.Determination of Selenium by Atomic Absorption Spectrometry, doctorate's thesis.

7) Manning, D. C..1978.Spectral interferences in graphite furnace atomic absorption spectroscopy. I. The Determination of selenium in an iron matrix. At. Abs. Newsl.17/5,107-108.

8) Hilderbrand, D. C..1983..Flameless atomic absorption determination of selenium in human blood. Reply to comments, At. Spectrosc. 4/4, 164.

9) Cotton, F. A. and G. Wilkison.1972. Advanced Inorganic Chemistry: a Comprehensive Treatise, third edition. Wiley Interscience.

TRACE ELEMENT DISTRIBUTIONS IN HUMAN LIVER LOBULES DETERMINED WITH MICRO-PIXE

D. Heck

Kernforschungszentrum Karlsruhe GmbH, Institut für Kernphysik III, D-7500 Karlsruhe, Federal Republic of Germany

A. Ochs

Medizinische Universitätsklinik, D-7800 Freiburg, Federal Republic of Germany

A. Klempnow, K.P. Maier

Medizinische Klinik, Fachbereich Gastroenterologie, Städtische Krankenanstalten, D-7300 Esslingen, Federal Republic of Germany

C. Kratt

Institut für Pathologie, Städtische Krankenanstalten, D-7300 Esslingen, Federal Republic of Germany

Introduction

Many diseases of the liver are accompanied with changes of the trace element (TE) concentration in the blood serum. For example patients suffering from hemosiderosis and hemochromatosis show an increased concentration of serum iron. Similar findings exist for the liver tissue itself, where these deficiencies also produce an elevated Fe-level. But for a detailed study of the degeneracy of the metabolic processes, it is important to localize the TE changes within the liver lobules; these lobules with a typical diameter of 1 mm form the functional units of the liver tissue. Therefore an analytical method is needed, which is sensitive enough to determine the TE's at a concentration of µg/g and additionally shows a spatial resolution of the magnitude of the hepatocytes (liver cells, about 12 µm diameter), to give the TE distribution within the lobule. These both requirements are combined in the method of proton induced X-ray emission (PIXE) [1,2] with micrometer focussing of the exciting proton beam [3].

Trace Element – Analytical Chemistry in Medicine and Biology, Vol. 4

Sample Preparation and Irradiation

For the experiments human liver samples were taken from patients during surgical cholecystectomies or by needle biopsy. Immediately after the sampling, the liver pieces - normally less than 1 cm^3 - were snap frozen to the temperature of liquid Nitrogen. The frozen sample blocks were cut with a cryomicrotome into slices of 4 - 8 µm thickness, transferred directly onto Formvar^R backing film of 20 - 50 $\mu g/cm^2$ thickness without melting and freeze dried for the irradiation. By this procedure any migration of the elements across the sample slice is prevented. It can be monitored by the distribution of those TE's which occur in ionic form with high mobility (K, Ca, Cl). Their vanishing concentration gradient hints onto a melting of the sample before drying. Similar homogenizing effects we could not observe for TE's like Fe, Cu, and Zn, which seem to stay localized also in imperfectly treated samples. Each second slice was mounted onto a microscopic glass slide and stained for histological examinations. As the unstained slices on the backing films show very poor contrast in an optical microscope, we compared them with the neighbouring stained slices to select regions which seemed interesting for the TE analysis.

The irradiations were performed with the ion microprobe [4] installed at the 3.75 MV Van de Graaff accelerator of the Karlsruhe nuclear research center. Protons of 3 MeV energy are collimated and focussed to form a particle beam of 3 - 5 µm diameter with currents of 600 - 1000 pA. At the point of impact on the sample surface, the protons induce the emission of X-rays, which are registered in a Si(Li) semiconductor detector. The energy distribution of the X-rays reflects the local elemental composition. To derive the TE concentration from the observed X-ray yield, the production probability [5], the slowing down of the protons [6] and the X-ray self absorption [7] within the sample are taken into account. The simultaneous detection of the protons elastically backscattered from the matrix elements C, N, and O yields the bombarded tissue mass - typical 0.1 pg - which is used to normalize the TE amount to dry weight (DW) concentrations [8]. Absolute concentration values are attained by comparison with a reference foil of known thickness and composition [8].

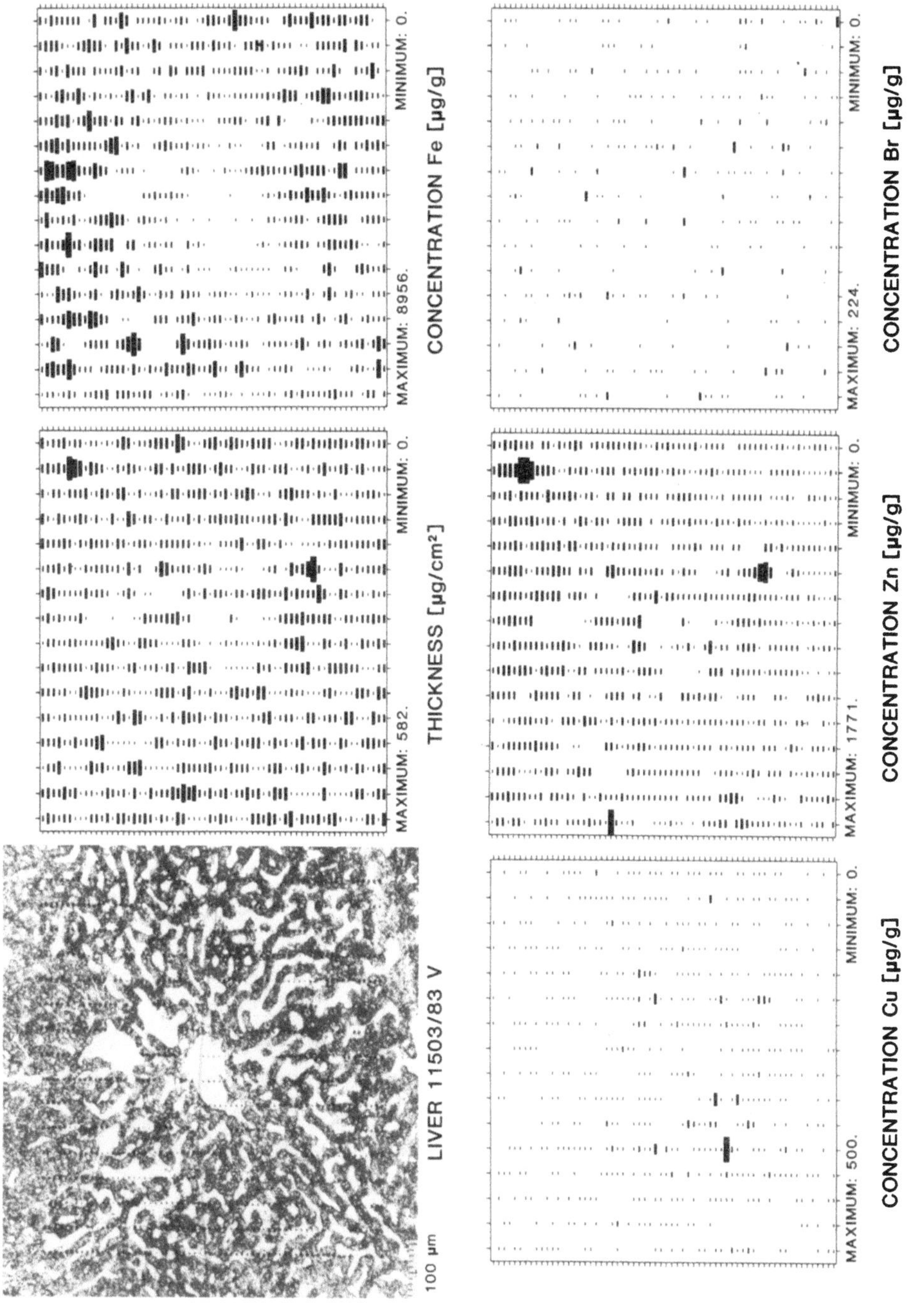

Fig. 1 Trace element distributions within a normal human liver.

The proton beam is swept across the sample surface in a meandrous path, thus covering an area of 0.6 x 0.6 mm^2, in which the TE distributions are determined. The fast scanning speed of 1 cm/sec reduces the heating of the liver samples. Despite these means the beam impact cracks the chemical bonding by which hydrogen is released out of the carbon hydrides, while most of the TE's remain in a frame of carbon. This carbonization along the trace of the beam is visible on the samples (cf. figs. 1 to 3).

Results

Healthy livers. In a first series of measurements [9] we established the reproducibility of the method. On five consecutive microtome slices of one liver sample we identified the same lobule for irradiation. The analysis results of all five measurements are compared and the averaged TE concentrations scatter from the mean value by max. 25% for Fe, Cu, and Zn; for Br with much poorer statistics the deviation reaches 40%. Considerable fluctuations from patient to patient are observed. For hepatocytes of 30 healthy patients, we found TE concentrations in the following ranges: Fe from 200 to 1500 µg/g DW, Cu from 20 to 50 µg/g DW, Zn from 200 to 700 µg/g DW, and Br from 4 to 15 µg/g DW. Comparing these values with literature [10] we find an agreement within the confidence limits. Our values may also be compared with recent results [11] for fresh matter taking into account the average water content of the liver tissue of 72% [10]. The slightly higher concentration of Zn and lower concentration of Br of our measurements is explained by the selection of the irradiation zones: We favoured the lobular centres, neglecting the periphery, where the portal area and the connecting tissue is localized. The latter tissue structures generally show a slightly lower content of Zn, while there the Br concentrations reach values up to 50 µg/g DW. Fig. 1 shows an example of a scan across the lobule of a healthy human liver. For the displayed TE's we detect no pronounced concentration gradient within the lobule.

Cirrhotic livers. Based upon 12 investigated liver samples with alcohol toxic, post-hepatitic or primary biliary cirrhosis (PBC) we derive common features of the TE concentration patterns for Fe, Cu and Br.

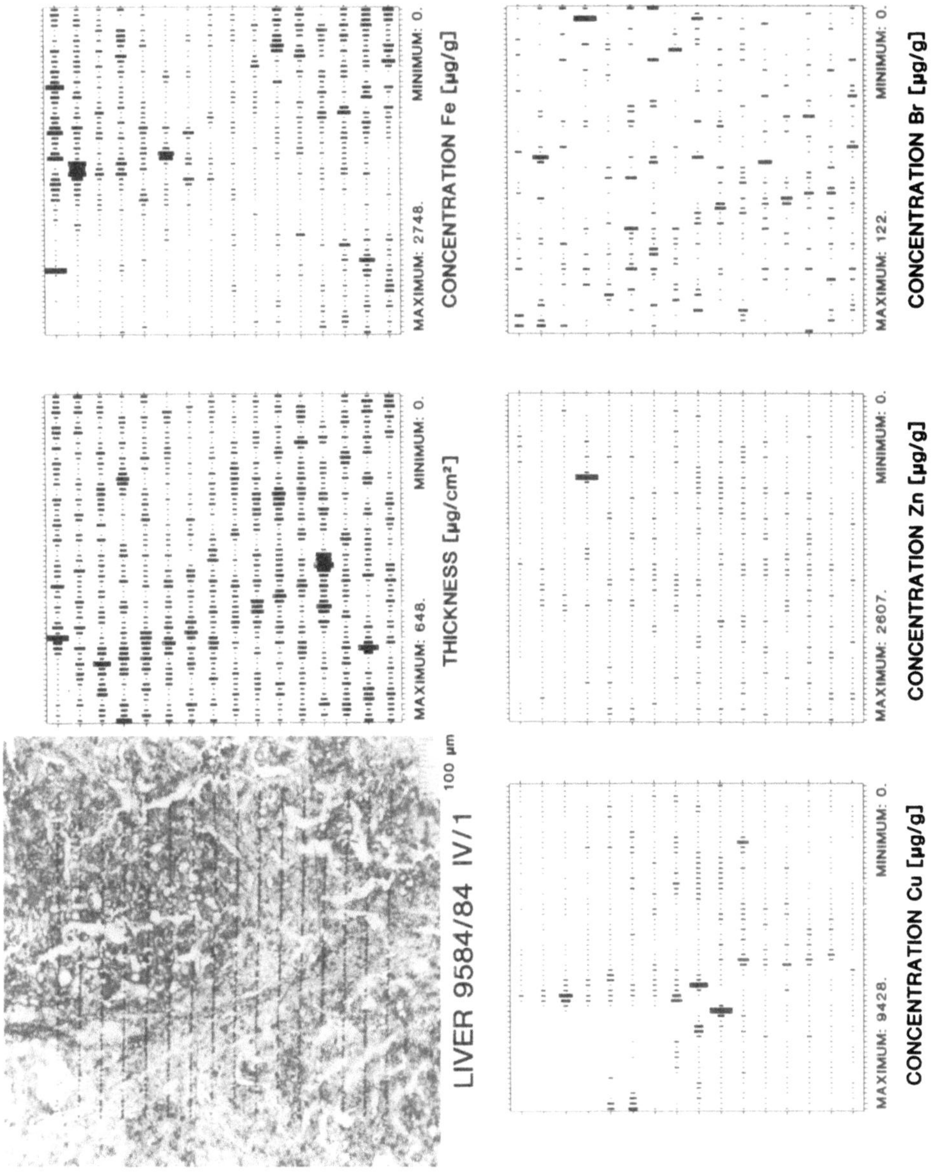

Fig. 2 Trace element distributions within a cirrhotic human liver (unknown etiology, presumably PBC).

Most striking is the rise of the Cu-concentration up to values well above 1000 µg/g DW. In the case of a cirrhosis with unknown etiology (presumably PBC), which is shown in fig. 2, we find the Cu-content going up closely to 10000 µg/g. The points of these Cu-accumulations are strictly localized to the areas of active fibrotic degeneracy of the parenchyma. This is in accordance with the observations of Watt et al. [12], who was able to localize the Cu-deposites within the hepatocytes, which line the fibrosed part of a PBC liver. In the adjacent fibroticly altered portal tracts, the Cu-content is not or only slightly increased and reaches values of 30 - 500 µg/g varying from patient to patient and with the stage of the cirrhosis. Similar findings apply for the parenchyma, but the absolute Cu-concentrations lie with 40 - 800 µg/g DW mostly above the values of the fibroticly degenerated areas.

The fibrotic portions are easily recognized in most TE patterns. In fig. 2 a pronounced fibrotic string leads from the upper left corner to the middle of the lower margin. A less evident branch divides the right half of fig. 2, ending just above the middle of the right margin. In fig. 3 of an alcohol toxic cirrhosis with mild inflammatory activity, a broad fibroticly transformed region separates a lobule with its center close to the lower left corner from the parenchyma at the upper right side, parts of which are destroyed during the preparation procedure. Indications onto these fibrotic strings are: a) a different thickness, which results from the freeze-drying process at regions with different content of water, b) the Br-concentration is significantly elevated compared with the content in the parenchyma, and c) Fe is depleted out from the fibrotic part of the liver. The latter finding is also observed in neutron activation analysis of cirrhotic liver pieces [13].

Summary

These results demonstrate that with the micro-PIXE method the TE distributions in the micro range can be determined to give hints onto the enzymatic reactions; by macroscopic TE determinations these detailed structures are averaged out.

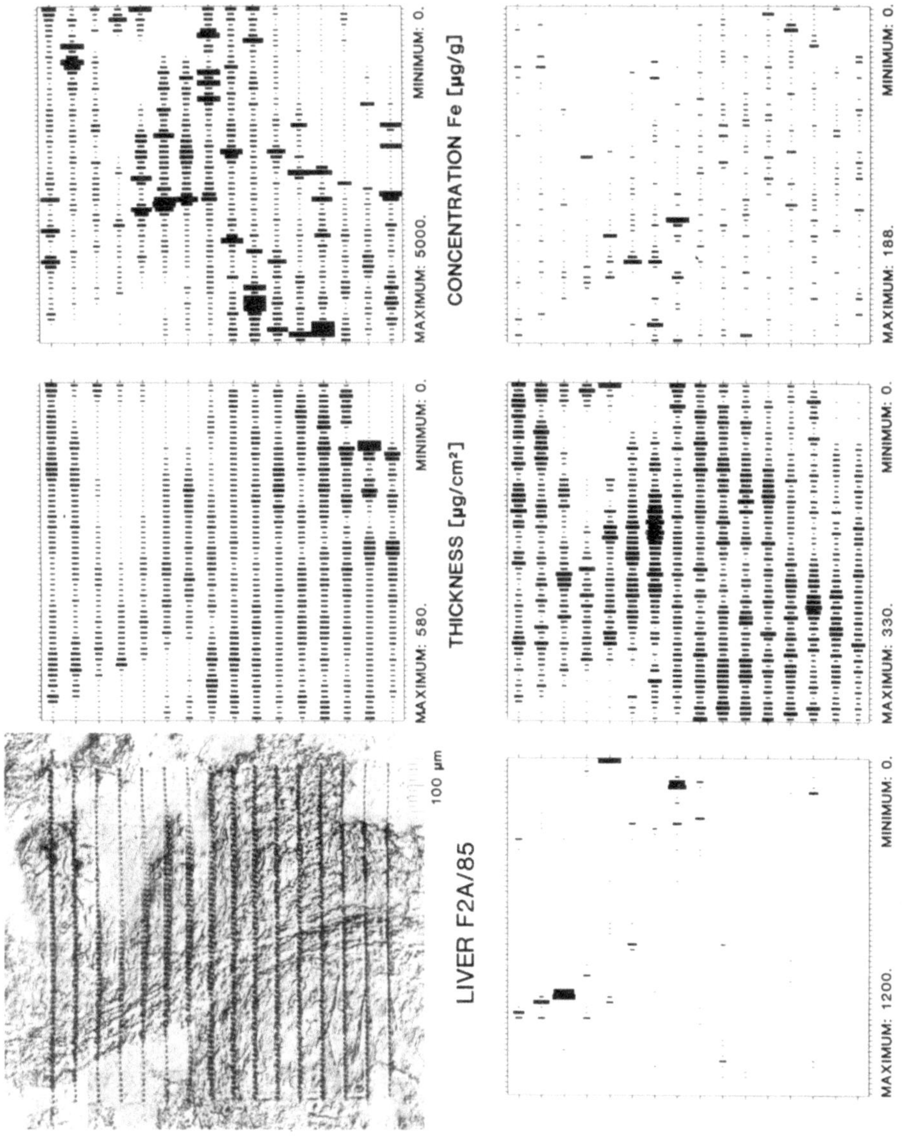

Fig. 3 Trace element distributions within a cirrhotic (alcohol toxic) human liver.

References

1. Johansson, T.B., R. Akselsson, S.A.E. Johansson. 1970. X-ray analysis: Elemental trace alanysis at the 10^{-12} g level. Nucl. Inst. Meth. 84, 141-143.

2. Folkmann, F.. 1975. Analytical use of ion-induced X-rays. Journal of Physics E8, 429-444.

3. Cookson, J.A., A.T.G. Ferguson, F.D. Pilling. 1972. Proton microbeams, their production and use. J. Radioanal. Chem. 12, 39-52.

4. Heck, D.. 1985. The Karlsruhe ion microprobe setup and its applications. Atomkernenergie - Kerntechnik 46, 187-192.

5. Johansson, S.A.E., T.B. Johansson. 1976. Analytical application of particle induced X-ray emission. Nucl. Instr. Meth. 137, 473-516.

6. Northcliffe, L.C., R.F. Schilling. 1970. Range and stopping power tables for heavy ions. Nuclear Data Tables A7, 233-463.

7. Veigele, WM.J.. 1973. Photon cross sections from 0.1 keV to 1 MeV for all elements Z = 1 to Z = 94. Atomic Data Tables 5, 51-111.

8. Heck, D., E. Rokita. 1984. Local matrix mass thickness determination in scanned micro-PIXE by proton backscattering. Nucl. Instr. Meth. in Phys. Research B3, 259-262.

9. Ochs, A., D. Heck, C. Kratt, G. Schneider, K.P. Maier. 1984. Protoneninduzierte Röntgenemission zur Spurenelementanalyse in menschlichen Lebergewebsproben. Verhandl. Deutsch. Gesellschaft innere Medizin 90, 1531-1534.

10. Iyengar, G.V., W.E. Kollmer, H.J.M. Bowen. 1978. The Elemental Composition of Human Tissues and Body Fluids. Verlag Chemie, Weinheim. New York.

11. Zeissler, R.. 1983. In: The Pilot National Environmental Specimen Bank (Zeissler, R., S.H. Harrison, S.A. Wiese eds.). National Bureau of Standards special publication 656, pp. 81-89,

12. Watt, F., G.W. Grime, J. Takacs, D.J.T. Vaux. 1984. The Oxford scanning proton microprobe: a medical diagnostic application. Nucl. Instr. Meth. in Phys. Research B3, 599-605.

13. Kostić, K., R. Ristanović, V. Obradović, M. Djodjević, R. Drasković. 1980. In: Trace Element Analytical Chemistry in Medicine and Biology (Brätter, P., P. Schramel eds.). Walter de Gruyter, Berlin. New York. pp. 601-610.

Direct determination of selenium and rubidium in plasma and blood using energy dispersive X ray fluorescence (EDXRF). Relation between rubidium and selenium levels and diseases.

F. Rastegar, E. A. Maier, R. Heimburger and M. J. F. Leroy
Laboratoire de Chimie Minérale, Unité Associée 405 du C.N.R.S. - Ecole Nationale Supérieure de Chimie de Strasbourg - Strasbourg (France)

C. Christophe
Savima - 1, Quai des Bateliers - Strasbourg (France)

C. Ruch
Siemens AG, Gerätewerk Karlsruhe (Bundesrepublik Deutschland)

We are developing a research program using energy dispersive X-ray fluorescence (EDXRF) for the determination of trace elements in biological samples (1,2,3). Two features may justify the choice of this technique:

- microsampling
- low detection limits associated with a multielemental analysis.

The present report deals with selenium and rubidium determination in plasma and whole blood. Selenium has been extensively studied (5) and EDXRF will be compared to the other techniques for the determination of this element. Rubidium is detected at very low level and the aim of this work is, after a compilation of the data, to bring information regarding the possible role of rubidium. Schwarz (4) ten years ago, has suggested that rubidium could be essential and this question is still pending due to the lack of appropriate experiments and also of consistent values obtained independently with various techniques.

Trace Element – Analytical Chemistry in Medicine and Biology, Vol. 4

Lombeck (6) reported a concentration of 12 ± 3 μg/ml of rubidium in blood whereas, Schulten (7) using field desorption mass spectrometry proposed 0.167 μg/ml Rb in plasma.

In this paper we studied a selected population (990 individuals) for the determination of selenium and rubidium in plasma and blood. Then a comparaison of patients affected by various diseases has been done.

Equipment

The energy-dispersive apparatus we used was a prototype developed primarily for analysis of small samples, to attain detection limits in the microgram per liter range. It has been described previously (2). In this study, a molybdenum filter, 150 μm thick, eliminated the x-ray continuum. (Bremsstrahlung). For all measurements the Mo x-ray tube was set at 40.0 kV and 25.0 mA. Results did not differ between measurements with air in the x-ray path or under reduced pressure. Accumulation of the data from an x-ray irradiation took 500 s.

Sample Preparation and Assay

All samples were diluted with doubly-distilled water that contained two internal standards : yttrium 4,50 μg/mL, and vanadium 100 ug/mL (Figure 1).

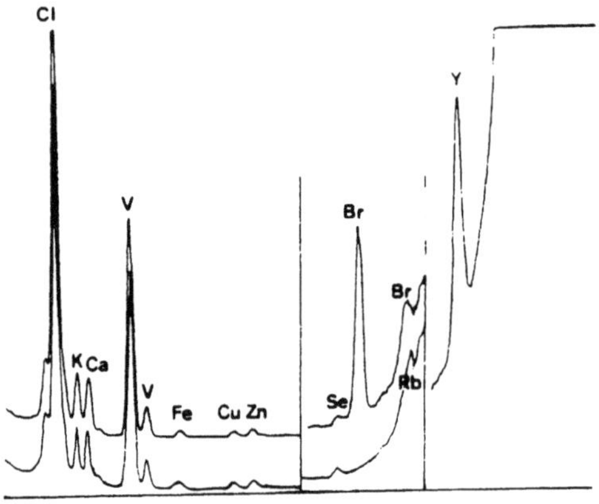

upper : Spectrum of plasma with internal standards.

lower : Same spectrum with Br deduction

Figure 1

The use of internal standards minimizes errors due to instrument fluctuations, geometry variation in sample density, and pipetting. The water had been previously analyzed and found to contain no trace elements. The standards were added as NH_4VO_3 and $Y(NO_3)_3$ (both Merck "pro analysi" grade). In this study we pipetted 25 µL of the diluted sample onto a 4-µm-thick polypropylene film suspended on a Teflon ring and dried this at room temperature, protected from pollution with dust.

Results and Discussion

Analytical Procedure

In developing the analytical procedure we used a commercial control-serum preparation (Seronorm, batch 161; Nyegaard § Co, Oslo, Norway).
Internal standards : Biological samples subjected to the quasi-monochromatic Mo x-ray irradiation exhibit a low, flat background continuum between 2 and 13 keV. The program used (Super Ml, Tracor Northern) performs both automatic elemental identification of energy dispersive x-ray spectra and calculation of peak intensities corrected for background and normalizing to yttrium. The simultaneous use of two internal standards (e.g., V and Y) provides a valuable control, both of the analysis and of the matrix influence.

Samples

Plasma were diluted 1:1 with the internal standards solution whereas whole blood samples were diluted 1:5. To the internal standards solution used for dilution of whole blood samples we added PEG (polyethylene glycol 20,000) for obtaining an homogeneous drying and avoiding fregmentation of the obtained deposit.

Calibration curves

The calibration curves, obtained by the method of known additions, were determined by linear regression of the readings vs the added concentrations, such that the sum of the deviations of the individual readings from the straight calibration line was set equal to zero while keeping the sum of the squares of the deviations at a minimum. Figure 2 shows the calibration curves for Rb and Se in plasma, and Rb in whole blood.

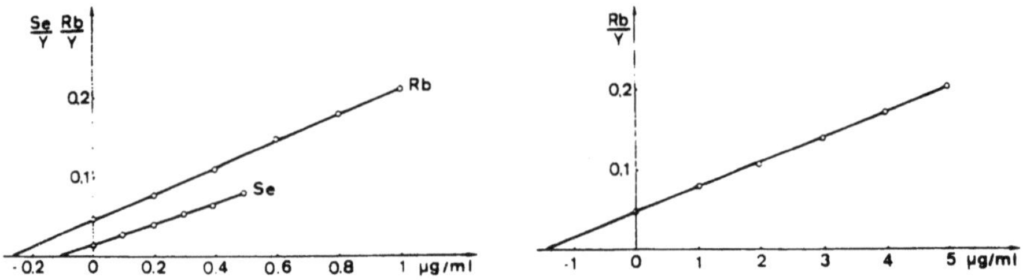

plasma whole blood

Figure 2 : Calibration straight line by the method of known additions

In all cases correlation coefficients for the fitted values were good. The reproductibility of the calibration points is also remarkably good (Table 1).

Precision and reproductibility

To evaluate the precision of the apparatus we counted the same sample five times, keeping all conditions constant. To assess reproductibility, we analysed five different aliquots of the same plasma and whole blood. Table 2 lists the results and the detection limits in both media.

PRECISION AND REPRODUCTIBILITY FOR Se AND Rb IN PLASMA AND BLOOD

Table 1

PLASMA						BLOOD		
Se			Rb			Rb		
µg/ml	precision %	reproductibility %	µg/ml	precision %	reproductibility %	µg/ml	precision %	reproductibility %
0,112	11,20	11,84	0,275	12,20	13,20	1,51	5,20	7,90
0,212	6,05	3,87	0,475	10,20	11,30	2,51	4,90	5,39
0,312	8,20	10,90	0,675	5,20	4,05	3,51	3,10	3,10
0,412	4,10	2,64	0,875	2,50	1,49	4,51	2,10	1,89
0,512	3,20	2,86	1,075	3,74	4,20	5,51	2,50	3,20
0,612	4,50	5,07	1,275	3,70	4,10	6,51	2,80	3,80

Distribution of the Concentration

With the appropriate coefficient, the "Super ML" program gives the results directly in nanograms per milliliter. Figure 3 shows the distribution of the concentrations for selenium in plasma and for rubidium in blood and plasma. The representative population for calculation of the mean-values and the standard deviations is about 990. This population is taken from an analytical biological laboratory, it has been chosen among patients without established disease and therefore considered healthy.
The distribution of selenium is quite narrow and fits the values quoted by Tölg (5) in reviewing other techniques.

In addition the direct excitation of the sample improves the detection limit of Se Compared to total reflection EDXRF.

For rubidium the values for plasma are in good agreement with those of Schulten (7) obtained with FD MS. In blood we find 1.57 + 0.43 µg/ml with 72.8% of the population within 1 SD.

PLASMATIC SELENIUM | PLASMATIC RUBIDIUM | BLOOD RUBIDIUM

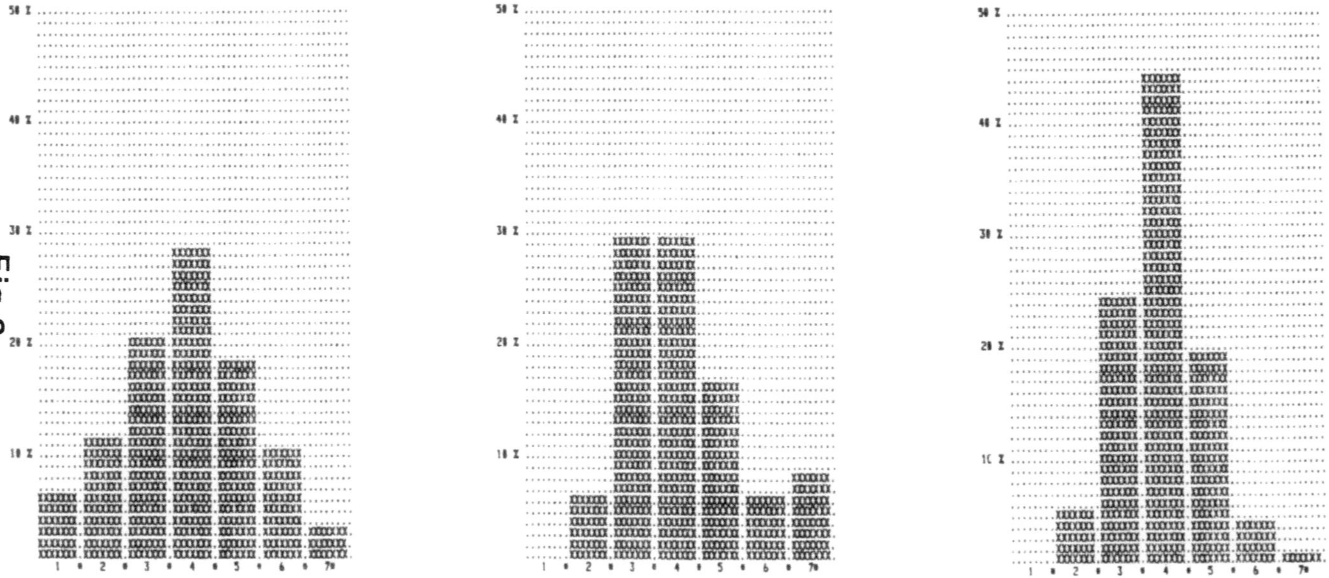

Fig 3

PLASMATIC SELENIUM

Representative population	: 995
Mean µg/ml	: 0.0831
Standard deviation	: 0.015
% For 1 Sd	: 63.9447
% For 2 Sd	: 98.7437

PLASMATIC RUBIDIUM

Representative population	: 994
Mean µg/ml	: 0.1547
Standard deviation	: 0.0379
% for 1 Sd	: 66.0377
% for 2 Sd	: 95.9742

BLOOD RUBIDIUM

Representative population	: 984
Mean µg/ml	: 1.5668
Standard deviation	: 0.4311
% for 1 Sd	: 72.8659
% for 2 Sd	: 94.6138

Distribution of Se and Rb in relation with diseases

A classification of 28 diseases using trace elements levels in plasma and blood for Ca, Fe, Cu, Zn, Se, Br, Rb, Mg and I (tetraiodo tyroxine) has been done. When taking increase as + and decrease as - for each element, none of the disease presents the same trend when considering the previous trace elements. Table 2 gives results for some Rb and Se.

Table 2

	Plasma		blood
	Se	Rb	Rb
Cancer	- -	+ + +	+
Infections	- -	+	+ +
Diabetes	+ + +		+ +
Hypertension	+	-	+
Hypotension		+ + +	- -
Multiple sclerosis	- -	- -	- -
Hyperlipidemia	- - -	-	+

Selenium

The results obtained for selenium are quite consistent with those of the litterature which proves that EDXRF is a reliable technique with the advantage of avoiding any pretreatment of the sample.

The variations of Se concentration in plasma for the various considered diseases are significative and agree with previously published data (5).

Rubidium

Considering Rb some dues indicating an essential role for rubidium are

- a repartition of Rb between plasma and erythrocytes which suggests a regulation.
- a specific modification of rubidium levels with respect to diseases.

In a parallel study concerning rubidium in blood for a selected population

we found that rubidium has always a narrow repartition but the mean concentration seems to be dependent on diet and sex (8).
The most interesting feature is the correlation coefficient of 0.8 between K^+ concentration and Rb^+ concentration in plasma or blood. Schulten (7) found a correlation coefficient of 0.55 using the same approach. Since most of the studies in the litterature of the role of potassium use ^{83}Rb as a tracer, the question of a simultaneous role of potassium and rubidium in membrane processes is pending and we believe that studies involving simultaneous determination of K and Rb may answer the possible essentiality of rubidium. This question appears to be important in view of the variations of the rubidium concentrations we observed for different diseases. In the case of multiple sclerosis a depletion of rubidium is observed for plasma and blood. This would support Schulten's results (7) of a depletion of rubidium in urine from multiple sclerosis patients.
For hypertension and hypotension it would be most interesting to confirm the behavior of rubidium which trends are opposite for the two diseases and also between plasma and blood. This result, although very interesting, needs to be confirmed using various methods of investigations and strict medical protocole to certify the diagnosis and to select a homogeneous population.

Concerning selenium our results and the analytical procedure previously described will be applied to the determination of this element in relation with diseases. But, more interesting, the multielemental analysis obtained will be used to determine the interactions between trace elements. We are presently studying the effect of cis-platinum treatment of cancers on the levels of trace elements in plasma and blood in function of the type of cancer and the type of treatment(26).

BIBLIOGRAPHY

1) C. RUCH, F. RASTEGAR, R. HEIMBURGER, E.A. MAIER and M.J.F. LEROY J. Anal. Chem. (1985), 57, 1691 - 1694.

2) E.A. MAIER, F. RASTEGAR, R. HEIMBURGER, C. RUCH, A. PELLETIER, A. MAIER and M.J.F. LEROY.
Clin. Chem. (1985), 31, 551 - 555.

3) F. RASTEGAR, E.A. MAIER, R. HEIMBURGER, C. CHRISTOPHE, C. RUCH and M.J.F. LEROY.
Clin. Chem. (1984), 30, 1300 - 1303.

4) K. SCHWARZ, in essentiality versus toxicity of metals.
Clinical Chemistry and Chemical Toxicity of Metals
S.S. BROWN (Ed), North Holland Biomedical Press, Amsterdam (1977) p.p. 3-32.

5) G. TÖLG, in trace element analytical chemistry in medecine and biology
P. BRÄTTER and P. SCHRAMEL (Eds), W. de GRUYTER, Berlin, (1984) p.p. 95-125.

6) I. LOMBECK, K. KASPEREK, L.E. FEINENDEGEN and H.J. BREMER.
Biol. Trace Element Res. (1980), 2, 193.

7) H.R. SCHULTEN, P.B. MONKHOUSE, C. ACHENBACH and R. ZISKOVEN.
Experientia (1983), 39, 736.

8) G. LAPITAJS, P. DE BIEVRE, M.J.F. LEROY, W. MULLER to be published.

9) D.E. RICHARDS, J.C. ELLORY and I.M. GLYNN
Biochimica et Biophysica Acta, 648 (1981) 284-286.

10) J.S. FLIER, P. USHER and M. DELUISE
Diabetes (1981), 30, 975.

11) I.M. GLYNN and D.E. RICHARDS
J. Physiol. (1982), 330, 17 - 43.

12) S.J.D. KARLISH and W.D. STEIN
J. Physiol. (1982), 328, 295-316.

13) S.J.D. KARLISH and W.D. STEIN
J. Physiol. (1982), 328, 317 - 331.

14) S.J.D. KARLISH and W.R. LIEB and W.D. STEIN
J. Physiol. (1982), 328, 333 - 350

15) N.A. ISMAIL, E.S.M. ELDENSHARY, L.A. IDAHL, P. LINDSTRÖM, J. SEHLIN and I.B. TÄLJEDAL.
Acta Phys. Scand. (1983), 118, 167 - 174.

16) S.D. SETH, M.P GUPTA, A.H. KAUR and A.K. BASU
Indian J. Exp. (1981), 19, 782 784

17) H.M. RHEE
Naunyn- Schmiedeberg's Arch. Pharmacol. (1982), 318, 344 - 348.

18) Y. IMAIZUMI and M. WATANABE
J. Physiol. (1981), 316, 33 - 46.

19) G. POCOCK
Molecular Pharmacology, (1982), 23, 681 - 697.

20) T. TAMAGAWA and J.C. HENQUIN
Am. J. Physiol. (1983), 244, E 245 - E 2452.

21) G. SCHMIDT, H. EIBL and W. KNOLL
J. Membrane Biol. (1982), 70, 147 - 155.

22) H. AOSHIMA, D.J. CASH and G.P. HESS
J. Am. Chem. Soc. (1981), 20, 3467 - 3474.

23) S. VYAS and R.M. MARCHBANKS
J. Neurochem. (1981), 37, 1467 - 1474.

24) J.G. KELLY, R. McC. Mc MILLEN and D.D. Mc DEVITT
Br. J. Clin. Pharmac. (1983), 15, 49 - 53.

25) T.J. HOUGEN and W.F. FRIEDMAN
Am. J. Physiol. (1982), 243, H 517 - H 522

26) M.L. SARGENTINI, E.A. MAIER, R. HEIMBURGER and M.J.F. LEROY
to be Published.

DETERMINATION OF CADMIUM AND LEAD IN ANIMAL LIVER AND FREEZE DRIED BLOOD BY GRAPHITE FURNACE ZEEMAN ATOMIC ABSORPTION SPECTROMETRY AND SOLID SAMPLING

W. Erler, R. Lehmann, U. Voellkopf
Bodenseewerk Perkin-Elmer & Co GmbH
Postfach 1120, D-7770 Überlingen

Introduction

Much of the work during the early days of atomic spectroscopy was concerned with solid samples. The use of arc and spark exitition is inherently a technique to analyse solid samples. With the growing importance of flame atomic absorption spectrometry, however, the analysis of dissolved samples became more favourable for a big variety of different materials. Unlike flame atomic absorption spectrometry, graphite furnace atomic absorption offers the possibility to analyse solid samples directly without preceding digestion. A number of analysts have reported about their attempts to perform the direct analysis of solid samples with the graphite furnace. Though the method not always proved to be successful.

Solid sampling offers several advantages in comparison with the analysis of samples after wet digestion: No time consuming digestion procedures or separation; dilution or preconcentration steps are applied. This reduces the risk of contamination by chemicals, solvents or beakers. Furthermore the analysis of microsamples becomes possible. Thus the local dis-tribution of trace elements within a sample can be determined.

In the past failures in the analysis of solid samples, as well as in the analysis of solutions often were caused by complexity of the processes occurring in the graphite furnace, and the lack of knowledge about these processes. Also the instrumentation did not offer the possibility to recognize and overcome interferences as modern spectrometers do. With the development of the Stabilized Temperature Platform Furnace Concept (STPF) the introduction of AA-Spectrometers with fast electronics, undistorted graphic display of atomization peaks and effective background correction, solid sampling technique became a reliable tool besides the conventional analysis of dissolved samples.

Trace Element – Analytical Chemistry in Medicine and Biology, Vol. 4

Instrumental

- Description of the cup-in-tube technique.

The solid sampling accessory must meet different requirements: it should, of course, be able to allow the determination of trace elements in solids, but doing this, it should not effect neither the analytical flexibility nor the reliability of the instrument.

The new solid sampling device comprises:

-- a specially slotted graphite tube

-- a graphite cup which is inserted into the tube (fig. 1).

Both are made of high density graphite, coated with pyrolytic graphite.

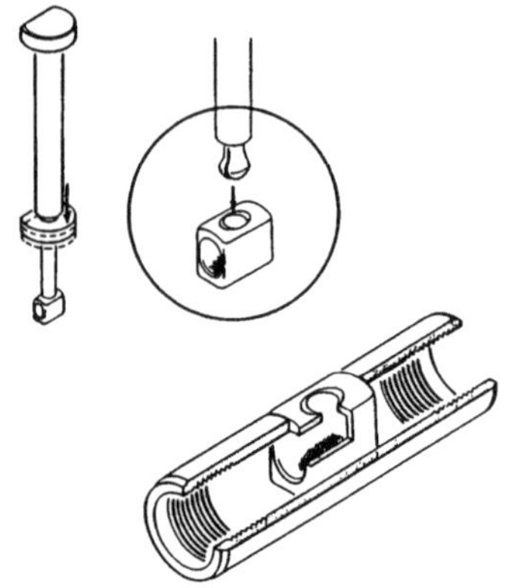

Fig. 1: The solid sampling cup and its handling with the special tweezer.

The shape and dimensions are designed to provide an easy insertion and removal of the cup. The large radiation beam aperture prevents vignetting. The addition of liquids by the autosampler through the pipetting hole is possible like into a normal graphite tube. The low mass of the cup (approx. 160 mg) gives good tare to sample weight ratios.

The determinations were carried out with an Perkin-Elmer Zeeman 3030 System and an AS-60 autosampler. An AD-6 autobalance was used for weighing the solid samples.

Figure 2 shows the heating characteristics of the inner tube wall, the surface of the L'vov platform and the solid sampling cup. The measurement has been

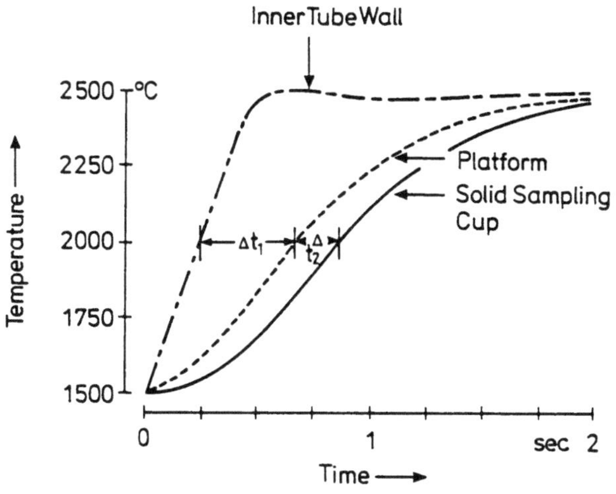

Fig. 2: Heating characteristics of the inner wall of a pyrolytically coated graphite tube, a L'vov platform and the inner surface of a solid sampling cup. Maximum power heating from 1500 to 2500 °C.
Δt = time delay to reach the same temperature.

described earlier (1, 2). Of analytical interest is the delayed heating of the cup, which makes the atomization take place under conditions which are close to the thermal equilibrium of the furnace. The same effect has been used very succesfully with the L'vov platform in order to reduce or even to eliminate nonspectral interferences. This means that the concept of the Stabilized Temperature Platform Furnace is transferred straightforward to a solid sampling technique. Also other benefits of the STPF-concept like matrix modification, peak area evaluation, maximum power heating and calibration against aqueous reference solutions can be used without difficulties.

- Setting of the Instrument

-- Choice of the wavelength:

In this work the primary lines, i.e. 283.3 nm for lead and 228.8 nm for cadmium were chosen. In many samples, however, it is preferable to choose other lines in order to reduce the sensitivity. Other possibilities to influence the sensitivity of the determination are the variation of the gas flow during the atomization step and in some extent the choice of an appropriate sample weight. The sample weight for most samples is between 0.1 and 1 mg.

-- The programming of the graphite furnace is shown in Table 1. Typical for many analyses with solid sampling is the high gas flow rate during the atomization step. This measure frequently must be taken in order to reduce the sensitivity. Also an extended pretreatment phase may be helpful to destroy the matrix of the sample as complete as possible.

Table 1: Furnace parameters of the HGA 600 for the determination of lead in solid samples of biological origin.

Step	Furnace Temperature (C)	Time (s) Ramp	Hold	Internal Gasflow (ml/min)	Read
1	100	5	20	300	
2	150	15	20	300	
3	800	30	20	300	
4	2100	0	2	100	X
5	2650	1	3	300	
6	20	1	10	300	

Calibration

Langmyhr (3) pointed out that the matrix effects the rate of atomization and thus the shape of the peak. As already observed in the analysis of liquids, the direct determination of samples against aqueous reference solutions may lead to wrong results. The use of the L'vov Platform and peak area integration, however, made the direct determination possible and method of addition unnecessary. Experience with the solid sampling cup device has shown, that with this accessory in many cases it is possible also when applying this solid sampling technique, to calibrate the instrument simply with aqueous reference solutions.

As this method will not work in any case, the analysis of solid samples often is carried out against reference materials. In this work calibration was performed with diluted, slightly acidified (0.2 % nitric acid) aqueous reference solutions.

Sample Treatment

Also the solid sampling technique does not completely eliminate sample (pre-) treatment. The main difference to measurements after wet digestion is, that all

chemical sample treatment must take place inside the graphite furnace. To perform an effective sample treatment inside the furnace, the technique of matrix modification was transferred to solid sampling.
The following mixture has proved to be an efficient matrix modifier for cadmium and lead in samples of biological origin:

2 mL	HNO_3	(2 % in water)
2 mL	$(NH_4)_2HPO_4$	(4 % in water)
20 µL	$Mg(NO_3)_2$	(10 % in water)
20 µL	Triton X-100	(1 % in water)

10 µL of this mixed modifier were pipetted to samples and aqueous reference solutions.

Signal Tracings

Figure 3 shows highly time-resolved signals for cadmium in reference solution and matter of biologic origin. Obviously the rate of atomization is different in solutions and solid samples. Peak height evaluation would lead to wrong results. Using peak area integration, however, it is possible to compare both signals directly to each other. The signals also show that background attenuation is moderate under the applied conditions.

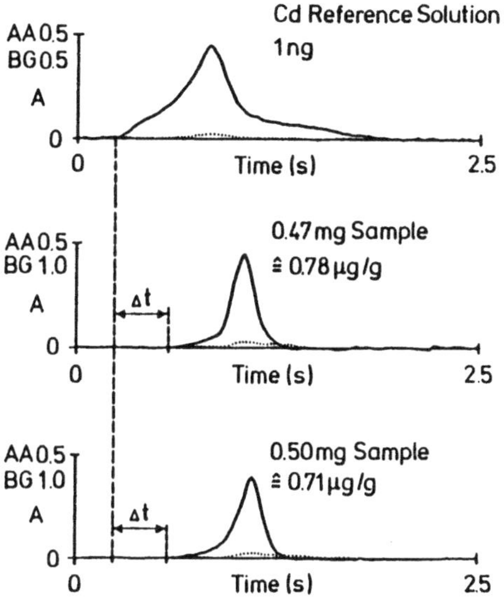

Fig. 3: Highly resolved signals for cadmium in an aqueous reference solution and in matter of biological origin. Typical is the time delay Δt for the beginning of atomization between solution and solid.

Results

The element concentrations found by applying the described technique are summarized in Table 2 and, if possible, compared with certified values or values which had been obtained by determination after wet digestion. The results were calculated from 5 to 10 single measurements. The relative standard deviation was between 4 and 14 %. In comparison to standard deviations which are obtained in determinations from decomposition solutions (2 to 3 %), the results of the solid sampling technique do not seem to be very reliable. Comparing such determination, it must not be forgotten, that the reproducibility of measurements of one solution may be better, but that sample pretreatment during decomposition and dilution incorporates the threat of contamination and losses, which will lead to wrong results. For highest accuracy it still will be advantageous to determine trace elements from digestion solutions. For screening purposes, however, and if a fast analysis is needed, the method of solid sampling will be a suitable tool.

Table 2: Contents of lead and cadmium in liver and blood samples.

Sample	Content in ug/g			
	Pb		Cd	
	solid sampl.	comparative or certified value	solid sampl.	comparative or certified value
NBS 1577	0.35	0.34	-	-
NBS 1577a	0.14	0.14	0.42	0.44
Fresh liver	3.30	2.90	-	-
Fresh liver	0.27	0.26	0.41	0.46
Fresh liver	0.09	0.11	0.51	0.45
Dried liver	0.70	0.75	1.07	0.99
Dried liver	3.34	3.0	1.06	0.93
Dried blood	4.27	4.33	-	-
Dried blood	0.19	0.19	-	-

Application of the cup-in-tube technique to other materials.

Besides the biological materials described in this work, already other samples could be analyzed successfully by the cup-in-tube technique. This will be of special interest for materials which are difficult to decompose by acid digestion, like synthetic polymers.

Table 3: Other materials which have been successfully analyzed by the cup-in-tube technique.

Samples	Analytes
Biological Material	
- hay, grass, leaves, spinach, etc.	Cd, Cr, Cu, Pb
- bones, hair	Cd, Cr, Cu, Pb
- mussles, lobster, plaice, oysters	Cd, Cr, Cu, Mn, Pb, Se
- tobacco	Cd, Pb
- insects	As, Cd, Pb
- human tissue (liver, etc.)	Cd, Pb
Soil Samples (IAEA-SRM)	As, Cr, Cu, Pb
Papers	Cd, Pb
Plastics	
- polyester	Mn, Rb, Zn
- Nylon (polyamides)	Cu, Mn, Rb
- others	Cd, Cu, Mn, Pb
Gun shot Residues	
- cloth material and skin	Sb

Background - is ZEEMAN-Effect BE-Correction essential?

The cup-in-tube technique was developed on AA-spectrometers with AC ZEEMAN-effect background correction. Further experience has shown, however, that not in all cases must background become so severe that ZEEMAN-effect background correction is essential. It can be stated therefore that there are many samples which can be analyzed without any difficulty and danger of getting wrong results with D_2-corrected AA-spectrometers as well. A prerequirement of course is, that the samples are volatilized under conditions close to the thermal equilibrium, and that the sample pretreatment inside the tube could be achieved to a sufficient degree.

HANGING MERCURY DROP ELECTROCHEMICAL PRECONCENTRATION AND SEPARATION TECHNIQUE FOR INDUCTIVELY COUPLED PLASMA WITH ELECTROTHERMAL VAPORIZATION

Henryk Matusiewicz

Technical University of Poznań, Department of Analytical Chemistry, 60-965 Poznań, Poland

Introduction

It has become quite common to apply electrochemical deposition as a pre-concentration and separation method for instrumental determinations of metals in complex matrices such as biological fluids, waters, food and mineral digests, and in general, in any samples in which there is a high content of interfering matrix elements. Controlled potential electrolysis performed on hanging mercury drop electrode (HMDE) is one of the main analytical method, which uses the electrolytic preconcentration and separation. This technique has previously proved succesful for complicated matrices (1).

Electrothermal vaporization (ETV) devices have proven to be useful for introduction of a microvolume of analyte into the inductively coupled plasma (ICP) for multielement analysis, and several applications have recently been reported (2). Adequate detection power, compared with graphite furnace atomic absorption spectrometry (GFAAS), neutron activation analysis (NAA) or anodic stripping voltametry (ASV) van be achieved using ETV for sample introduction, but a number of matrix interferences arises from the vaporization/atomization transport and plasma effects which have yet to be elucidated fully (3).

Combining two analytical techniques: HMDE and ETV-ICP with their advantages should be well suited for the determination of trace and ultratrace elements that are soluble in mercury. Also, the method should provide a good simultaneous separation-preconcentration step prior to ETV-ICP analysis.

Recently, electrolytic reduction of copper ions onto a HMDE was reported as a sampling technique for direct sample insertion device coupled with ICP (4). The method was not tested on "real" samples and investigations were with very limited information.

In addition, the detection limits were not as good as could have been expected.

The object of this work was to examine whether ETV-ICP with HMDE technique could be applied for the determination of seven elements in biological standard reference materials. Certified reference materials were chosen mainly to determine accuracy and validate proposed analytical method (5).

Experimental

Instrumentation

The ETV-ICP instrumentation has been fully described previously (3).

Princenton Applied Research (PAR) Model 384 polarographic analyzer, equipped with Model 303 static mercury dropping electrode and Model 305 stirrer were used. The working electrode was a HMDE with a surface area of 1.3 mm^2 and a weight of 2.2 mg. Reproducible stirring of 700 rpm of the solution was controlled by microprocessor. Cell solutions were deaerated for 5 min with highly purified nitrogen.

Reagents

Distilled, deionized water was used. The mercury was a triple-distilled (Bethlehem Apparatus Corp.). Standard solutions were prepared daily from 1000 μg/ml ICP reference stock standards of metals (Spex Industries, Inc.). Pyrolytically coated graphite cuvettes with a crater toward one end were used. Sulfuric and nitric acids (J.T. Baker Chem. Co. "Ultrex") for dry ash were used.

Sample digestion procedure

1 g of NBS Bovine Liver 1577, Rice Flour 1568 and 5 ml of NIOSH-NBS Urine were ashed using the procedure described by Satzger et al. (6).

Procedure

Transfer 1 ml sample solution to the Teflon cell containing 4 ml pre-electrolyzed HOAc/NaOAc buffer-electrolyte. Transfer Teflon stirring bar to cell and clamp cell in place over stirring motor. Purge solution with oxygen-free nitrogen saturated with electrolyte for 5 min. The mercury drop was formed automatically after deaeration period. Set electrodeposition potential at -1.5 V;

deposition time 10 - 20 min with stirring; equilibration time 45 s. All measurements were made at 25° C with Ag/AgCl reference electrode. During the equilibration time, lower the cell, rinse mercury drop with distilled water, and transfer mercury drop directly to a preburned graphite cuvette with a crater. Then volatilize completely the mercury on a hot plate with a graphite support by about 20 min evaporation at 140° C prior to ETV-ICP excitation. The residue was evaporated and excited in ETV-ICP. Duplicate exposures were made of each standard and sample. The content of each element was established from the calibration curves and by method of additions of µl volumes of mixed standard solution to sample cell.

Results

Preconcentration - Separation Step

In order to get as many metals as possible deposited into the HMDE, the applied potential chosen was -1.5 V vs. Ag/AgCl reference electrode. At that potential, trace metals such as Ag, Al, Au, Bi, Cd, Cr, Co, Cu, Ga, Ge, In, Fe, Mn, Mo, Ni, Pb, Pd, Pt, Sb, Sn, Tl and Zn can be simultaneously deposited.

For a given deposition period, the sensitivity of the method depends on the rate of the mass transport to the electrode during electrolysis. For the present work, a stirring rate of 700 rpm was used, which was the highest available in this instrument.

Obviously, the reproducibility of the drop area considerably affects the reproducibility of the analytical measurement in the second analytical step: ETV-ICP technique. Reproducibility of the drop area was 4.1 % RSD.

Despite matrix interferences in ETV-ICP (3) careful rinsing of the mercury drop was necessary to avoid transferring of adhered solution with mercury to the crater of graphite cuvette. Measured sodium concentration, by ETV-ICP, after rinsing the drop was in the range of 0.5 to 1.0 µg/ml (without rinsing in the range of 100 to 200 µg/ml).

Elemental volatility study was not performed because under evaporation heating conditions at 140° C it was unlikely to loss trace metals, but to vaporize excess amount of mercury was absolutely necessary which otherwise can cause erroneous high blanks observed for trace metals, possibly as a result of spectral interferences and plasma characteristics changes.

Electrodeposition Time

The electrolysis time was varied from 5 to 45 min. The sensitivity was increased by prolonging the electrolysis time, but nonlinearily. The dependence of detection limits for 8 elements on electrodeposition time are given in Table 1. An improvement in the trace elements detection limits by a 5 to 38 % is possible after excruding extra drops of mercury, because of amalgam diffusion process into the glass capillary.

Table 1. Detection Limits (ng/ml) for 5 to 45 min Electrolysis

Element	Electrodeposition time (min)					
	5	10	15	20	30	45
Cd	12.8	5.6	3.7	3.2	2.9	2.4
Co	7.5	3.3	2.4	2.2	1.9	1.8
Cu	8.1	6.1	4.7	4.1	3.3	2.8
Mn	7.7	4.4	3.6	3.1	2.7	2.4
Ni	21.5	14.4	9.5	8.0	7.4	6.9
Pb	15.9	10.9	7.4	6.0	4.4	3.4
Sb	18.9	11.3	8.5	6.7	5.2	3.8
Zn	22.1	9.3	7.5	6.6	5.6	4.8

Parameter Selection for ETV-ICP

The ICP outer and intermediate argon gas flow rates, the heating rate, the outer argon flow around the graphite cuvette, carrier argon flow rates, vaporization temperature, viewing height and rf power were kept constant for all elements investigated based upon earlier results (3).

Analytical Parameters

The detection limits, calculated as previously described (3) were obtained for 8 elements using the electrochemical deposition for ETV-ICP combination for a 10 min time and are shown in Table 2. The values are presented in terms of both concentration (ng/ml) and analyte mass (pg) to simplify comparisons. Since no detection limits obtained by similar technique are available, the results are compared only with those obtained by ETV-ICP (3) and GFAAS (7).

Table 2. Comparison of Detection Limits Obtained with HMDE-ETV-ICP technique, ETV-ICP and GFAAS

Element	Wavelength (nm)	Detection limit (3σ)+ Concentration (ng/ml)	RSDø (%)	Literature (3) (ETV-ICP) (ng/ml)	(pg)	(7) (GFAAS) (ng/ml)§	(pg)≠
Cd	228.8"	5.6	9.6	3	9	0.03	1.5
Co	228.6	3.3	2.2	4	12	1.2	60
Cu	324.7	6.1	4.0	13	39	0.9	45
Mn	257.6	4.4	2.9	0.7	2.1	0.2	10
Ni	231.6"	14.4	9.8	20	60	3	150
Pb	220.3	10.9	7.2	12	36	0.6	30
Sb	217.6	11.3	8.6			1.2	60
Zn	213.8"	9.3	6.2	2	6	0.02	1

+The limit of detection corresponds to the concentration giving a signal equivalent to three times standard deviation of the background.

øRelative standard deviation of six measurements from a single sample preparation.

§On the basis of a consumed sample volume of 50 μl.

≠My calculation based on reported limit of determination (10σ).

"Used in the second order on the polychromator.

Simultaneous or off-peak background measurements for each element were not applied because of the very small sample amount available, therefore, the background values measured at the analytical wavelength for the blank were employed as background correction with all samples. It should be noted that no blanks were found for all elements, except for Zn where blank was found to slightly vary. Sporadically, erroneously high blanks were observed for Ag and Al proved to be also troublesome. When related to the detection limits for ETV-ICP (3) with the same equipment improved concentration detection limits, except for Mn, Cd and Zn, were obtained with the present method, as a result of preconcentration-deposition technique applied.

The precision of the present technique is dependent on many experimental parameters which are related to the electrodeposition, electrothermal vaporization and excitation steps. The measured precision, expressed as relative standard deviation (RSD) by a series of separate experiments (6 replicate measurements of 8 trace elements at 15 times their detection limits) was 2 - 10 %.

The biological standard reference materials analysis results obtained by calibrating with synthetic aqueous standard solutions and simultaneous-element standard additions to biological materials after decomposition in sulphuric acid are essentially identical. Therefore, for routine measurements of biological material samples, the analysis data could be based on calibration using multielement synthetic standard solutions.

Practical Analytical Applications to Biological Materials

Having assessed the analytical quality of the method, a preliminary study of the practical utility and accuracy of the electrodeposition technique was tested on the reference values. Three NBS standard reference materials: SRM 1577 Bovine Liver, SRM 1568 Rice Flour and NIOSH-NBS Freeze-Dried Urine were analyzed with the method described. Because the NBS reference materials are certified, these materials were used to verify the overall performance of the analytical technique. The analytical results are presented in Table 3. In virtually all cases results obtained showed good agreement with the NBS certified values. This suggests the validity of the proposed procedure, therefore, the utilization of the electrodeposition preconcentration-separation scheme was effective in eliminating significant matrix interference effects.

Table 3. Analytical Results for the NBS Standards[+]

Element	SRM 1577 Bovine Liver		SRM 1568 Rice Flour		NIOSH-NBS Urine	
	Concentration					
	found ($\mu g/g$)	certified ($\mu g/g$)	found ($\mu g/g$)	certified ($\mu g/g$)	found ($\mu g/ml$)	certified ($\mu g/ml$)
Cd	0.35±0.03	0.27±0.04	§	0.029±0.004	§	∅
Co	0.25±0.01	0.18″	§	0.02±0.01	§	∅
Cu		193±10		2.2±0.3		0.030±0.005
Mn	9.6±0.3	10.3±1.0	20.8±0.6	20.1±0.4	0.1±0.003	∅
Ni	§	∅	§	0.16″	0.97±0.10	1.00±0.11
Pb	0.51±0.04	0.34±0.08	1.77±0.13	∅	§	∅
Sb	§	0.005″	§	∅	§	∅
Zn	120±7.4	130±13	17.7±1.1	19.4±1.0	1.53±0.10	∅

[+]Single analysis throughout; results given as $\mu g/g$ of dry weight of Bovine Liver and Rice Flour samples; duplicate determinations.

″Uncertified values.

∅Not reported.

§Below limit of detection or not detected.

Conclusions

The present study indicate that the HMDE-ETV-ICP technique can considerably extend the analytical capabilities of the ICP method by combining sufficient preconcentration-separation with multielement analysis. Thus, using this combined method, it was possible to directly determine trace metals in a real samples such as biological materials. Electrochemical preconcentration-deposition time of 10 min have been found satisfactory. For lower trace element concentrations this time could be expanded to even 1 hour.

Acknowledgement

Experimental part of this work was done at the EARC, US FDA, Cincinnati, U.S.A., and the author gratefully acknowledge financial support.

References

1. Matusiewicz, H. 1985. Analyst 110, 1481.
2. Matusiewicz, H. 1986. J. Anal. At. Spectrom. 1 (in press)
3. Matusiewicz, H., Fricke, F.L., Barnes, R.M. 1986. J. Anal. At. Spectrom. 1 (in press)
4. Habib, M.M., Salin, E.D. 1985. Anal. Chem. 57, 2055.
5. Sutarno, R., Steger, H.F. 1985. Talanta 32, 439.
6. Satzger, R.D., Clow, C.S., Bonin, E., Fricke, F.L. 1982. J. Assoc. Off. Anal. Chem. 65, 987.
7. Koch, O.G., LaFleur, P.D., Morrison, G.H., Jackwerth, A., Townshend, A., Tölg, G. 1982. Pure Appl. Chem. 54, 1565.

Session V

SIGNIFICANCE OF TRACE ELEMENTS IN MEDICINE

Plenary Lectures:

G.N. Schrauzer

C.J.M. van Tiggelen

Chairman:

R.J. Shamberger

W. Fresenius

G.N. Schrauzer

Poster Reports:

G.N. Schrauzer

C.J.M. van Tiggelen

TRACE ELEMENTS IN CANCER DIAGNOSIS AND THERAPY. A REVIEW

G.N.Schrauzer
University of California at San Diego
La Jolla, California, U.S.A.

Introduction

Although sophisticated instrumentation is available for the rapid simultaneous determination of multiple elements in biological materials, the new analytical technology is still largely a research tool and is not yet widely used in clinical practice. In part, this is because the development of the analytical instruments is still proceeding at a rapid pace, creating uncertainty as to which technique will ultimately prove to be the most reliable and cost-effective. More seriously, however, the value of mineral analyses is not yet generally accepted; many clinicians feel, for example, that much of the information gathered is ephemeral or only of theoretical interest. Indeed, prior to the practical implementation of any clinical test, a large body of empirical information on patients and controls must be collected, processed and documented. This is particularly difficult to achieve with cancer patients, as these form an unusually heterogeneous group, differing greatly from one to another with respect to site, type of tumor, stage of disease, treatment history, general health status etc. However, although the data is still incomplete, some generalizations are now possible which should prove that knowledge of the mineral status of cancer patients is of considerable diagnostic and prognostic significance.

Mineral Status of Cancer Patients

Most of the available information on the trace element status of cancer patients was accumulated with data obtained by means of standard AA- or related spectroscopic techniques. Thus, our discussion will initially focus only on Na, K, Ca, Mg, Cu, Fe, Zn, Li and Se. Most interpretations offered are furthermore based on the results of serum analyses. It should be kept in mind that serum values of some elements may remain normal irrespective of significant losses in tissues, bone or liver. As serum only reflects the extracellular space, the analysis of isolated erythrocytes and platelets has

Trace Element – Analytical Chemistry in Medicine and Biology, Vol. 4

been recommended as a means of obtaining information on the status of the intracellular compartment.

Cancer patients typically exhibit subnormal serum levels of sodium, potassium, magnesium, iron, zinc, while calcium is usually elevated, especially if osseous metastases are present(1). Serum copper may be elevated, normal or low, depending on the nature and stage of malignant disease(2).

Elevated copper levels are indicative of inflammatory responses of the host organism to the tumor but may also be unrelated to the malignant process. They are often seen in patients with rapidly growing tumors, especially in cases of early cancer of the lung, in patients with melanoma, Hodgkin's Disease, carcinoma of the stomach, or with primary or secondary osteosarcoma.

In female breast cancer patients, copper is also often elevated, the copper levels correlate with elevated ferritin- and CEA levels(3).

Elevated copper levels are observed in patients with infiltrating carcinoma of the liver but also in patients with acute infections, benign acute and chronic diseases of the liver, lupus erythematosus, as well as during pregnancy and in women on steroidal contraceptives.

Since serum copper is usually low in advanced orterminal patients, periodic monitoring of serum copper is useful for patient follow-up. It should be noted, however, that serum copper may also remain entirely normal in some patients in spite of progressing malignant disease. This occurs if blood group- and tumor cell membrane antigens are similar and is associated with a poor prognosis(4).

Serum zinc is subnormal in most actively growing cases of cancer of the lung. Since zinc deficits may lead to lymphopenia, anorexia, taste and smell dysfunction, it would be logical to recommend extradietary supplementation. However, although no reports of adverse reactions to such supplementation have been documented, it is felt that caution is necessary in the light of animal experiments which demonstrated that supplemental zinc abolished the cancer-protecting effects of Se and stimulated tumor growth(5). In contrast, zinc-deficient animals showed diminished tumor growth rates and improved survival(6).

Serum iron levels are low in virtually all cancer patients. Iron is essential for cell growth and is normally delivered to cells by the serum-iron binding protein transferrin. Some tumor cells produce transferin-like proteins of their own. If large, such tumors may withdraw sufficient amounts of iron to cause a systemic Fe deficit. However, since the anemia of cancer patients is usually caused by a tumor-induced disturbance of iron utilization in the reticulo-endothelial system(1), iron supplementation is contraindicated unless proven to be due to nutritional deficiency. On the other hand, it is well known that chronic iron deficiency, in conjunction with other nutritional deficiencies, previously gave rise to Plummer-Vinson syndrome, notably in Swedish women. This condition is associated, inter alia, with esophageal varices and increased risk of cancer of the upper gastrointestinal tract. However, iron excess may also increase the susceptibility to cancer development because excess iron interferes with host defense mechanisms against malignant cells. Further research is necessary to establish specific guidelines for the management of the abnormal iron metabolism in cancer patients. In special cases, controlling the transport of iron to cancer cells could inhibit tumor growth and/or increase the effectiveness of cancer chemotherapy. The complex relationships of iron to neoplasia has been the subject of a recent review(7).

Low levels of sodium and potassium are frequently encountered in cancer patients and may have numerous causes. Here it need only be mentioned that abnormal sodium levels may be associated with disturbances of fluid retention or excretion which in turn may interfere with therapy, and that severe potassium deficiency may produce acute cardiac distress, muscular weakness, paralysis and respiratory failure. Deficiencies are corrected by administering potassium chloride or potassium phosphate, with appropriate precautions against induction of hyperkalemia.

Although Delbet(8) more than 40 years ago suggested, on the basis of epidemiological considerations and animal studies, that Mg deficiency increases the cancer susceptibility and may retard tumor growth, studies by other authors subsequently showed that Mg supplementation in animals with transplanted tumors may accelerate tumor growth. Thus, interest in Mg as a possible cancer-protecting agent waned until about 1950, when Heinemann(9), Nieper(10) and other workers reaffirmed the earlier epidemiological findings. Other studies demonstrated that low cancer rates are observed in certain Mg-rich regions of Egypt(11-12), while the extremely high stomach cancer rates in certain regions of the Ukraine were associated with low Mg availability(13). In Mg-deficient areas of Poland, a higher incidence of cattle leukemia was observed(14). In animal experiments, young rats fed Mg deficient diets were found to develop thymic tumors classified as malignant lymphoma or lymphosarcoma(15,16). The incidence of malignant lymphomas was observed only in severely Mg deficient animals; neonatal rats were particularly vulnerable when Mg-deficient(15-17). In acutely Mg deficient rats and calves increases of circulating leukocytes have been observed(18). In Mg-deficient rats, myeloid leukemia developed(19,20). Bone abnormalities intermediate between true neoplasia and connective tissue hyperplasia were also seen(21). Numerous enzymes, including ATP-ase and alkaline phosphatase require Mg as cofactor. The element is also a constituent of the nucleic acids and believed to be responsible for the maintenance of its structural stability.

Magnesium deficiency may result from inadequate intake (energy-protein malnutrition), impaired absorption, or increased excretion due to the use of diuretics. Hypomagnesia may also be coupled with hormonal imbalances (hyperaldosteronism, hyperthyroidism, diabetic acidosis) and hypercalcemia. Correction of Mg deficiency is readily achieved by extradietary supplementation of magnesium salts. Although, as was mentioned above, low serum magnesium levels are frequently observed in cancer patients, this finding is not consistent and does not necessarily indicate a lack of Mg or a need for supplementation. Low serum Mg levels may actually be caused by the high Mg demand of rapidly growing tumors. In some instances, cyclic variations in plasma and erythrocyte Mg levels of cancer patients have been observed; the shortest intervals between peak levels occurred in patients with highly malignant tumors(22). This may suffice to illustrate the complexities of Mg metabolism in cancer patients and the need for further research. The subject is treated in greater detail in an excellent review by Seelig(23).

Elevations of serum Ca are particularly pronounced in patients with bone metastases, in leukemias, lymphomas and myelomas. Control of hypercalcemias involves the administration of prednisone and diuretics (e.g. furosemide), together with other supportive measures, but the description of these therapeutic details is outside the scope of this review.

According to Heger(24), subnormal serum levels of lithium are frequently observed in cancer cases. This first member of the group of alkali elements is believed to be essential; goats rendered Li-deficient demonstrated diminished weight gains; in female goats, reduced fertility was observed(25). Since relatively large doses of lithium appear to be beneficial in the control of the manic phase of manic-depressive psychosis, lithium is believed to be required for the maintenance of emotional stability. DeVeaugh-Geiss, however, urged critical scrutiny in the use of this agent in psychiatric practice(26). Since undoubtedly not all of the physiologic functions of lithium have been elucidated, existing deficiencies should be cautiously corrected, where necessary.

The Special Case of Selenium

Selenium is presently receiving world-wide attention by cancer researchers, epidemiologists and the general public. While animal experiments continue to confirm its anticarcinogenic properties, new epidemiological studies reaffirm its protective role against various types of cancer in humans. Since the voluminous literature on this subject cannot be covered here, readers desirous of more complete information are referred to recent reviews and monographs(27-30).

While most authors see the main value of selenium in cancer prevention, its role in the diagnosis and therapy of cancer must by no means be ignored.

Although several studies have shown that cancer patients usually exhibit lower blood- or serum selenium levels than healthy controls or patients with nonmalignant diseases, some apparent exceptions to this rule have also been reported. Thus, in an often quoted study conducted in New Zealand, the differences between cases and controls was vanishingly small(31). This result is not surprizing, however, as New Zealand is a low-Se country and the dietary intakes of Se are generally low. As blood- and serum Se levels are mainly determined by the dietary intakes, it is obvious that differences between cases and controls should also be small.

In contrast, the blood-Se levels of Japanese women with and without breast cancer demonstrated significant differences. The average Se levels in samples of whole blood obtained from 79 Japanese women with newly diagnosed breast cancer was 0.195 ± 0.059 µg/mL as compared to 0.285 ± 0.032 µg/mL of the healthy controls(32).

In healthy subjects, blood Se levels reflect the dietary Se intakes which are a function of the types and Se-contents of the foods ingested. Individual blood Se levels thus are not only determined by regional variations of Se in foods but are also indicators of life-style and dietary preferences. Since the typical diet in the Western industrialized countries is not only lower in Se but also higher in sugar, animal proteins and fats, the combined effects of its adoption by the Japanese are likely to increase the breast cancer risk. The traditional Japanese diet provides abundant amounts of Se, 250-300 µg/day, about twice the amount typically obtained by American or European adults, and is significantly lower in sugar, fats and animal protein(32).

It is of interest that patients with benign fibrocystic breast disease, a preneoplastic condition associated with a 3 fold higher breast cancer risk, also exhibited lower blood Se levels than the healthy controls.

Blood Se levels of American women with breast cancer and benign fibrocystic breast disease were also lower than those of healthy controls. The differences between cases and controls were smaller, however, and the blood Se levels of 0.191 ± 0.023 µg/mL of the healthy American women, whose breast cancer risk is 4-5 times greater than that of Japanese women, were similar to those of the Japanese breast cancer patients.

Low blood or serum Se levels in cancer patients indicate diminished antioxidant protection and may render patients more vulnerable to damage of the liver and the heart, especially when undergoing cytotoxic chemotherapy. Adriamycin (doxorubicin), for example, causes peroxidative damage to cardiac mitochondria which is aggravated in selenium deficiency(33). In a recent study, over 50% of the patients with cancer exhibited depressed RBC GSH-Px activity, which correlated significantly with age and weight loss(34); supplementation with 200 ug of selenomethionine i.v. prevented further declines of GSH-Px activity. Large scale clinical trials should be conducted to test the effects of Se supplementation in patients undergoing chemotherapy or radiation therapy; based on initial observations with smaller groups of patients, Se has generaly benign effects(35). This is supported by recent animal experiments. While selenium by itself has previously been shown to inhibit the growth of certain well-known animal tumor cells, e.g. Ehrlich ascites or L-1210 leukemia cells (36,37), it was found to be ineffective in rats against transplanted prostatic adenocarcinoma cells (type RPMI 1640) at levels of 6 ppm in form of selenite in the drinking water(38). However, selenite clearly reduced the toxicity of MGBG (methylglyoxal-bis-guanylhydrazone), one of the most toxic polyamine synthesis inhibitors, without lowering its antitumor activity(38). Selenium similarly reduced the toxic side effects in animals treated with cis-platinum without diminishing its therapeutic effects(39). It is likely, therefore, that selenium compounds will in future find applications as adjuvants in cytotoxic human cancer chemotherapy.

The Roles of Selenium Antagonists

Many of the toxic heavy metals (Hg, Ag, Au, Cd, Pb, Sn, As, etc.), and under appropriate conditions, even some of the essential trace elements (e.g. Zn), are Se-antagonistic and are known or may be suspected to abolish the cancer-protecting effects of selenium. Accordingly, the chronic exposure even to low, so-called "no-effect-" levels of such elements may increase the cancer risk. As some of the toxic elements in addition are mutagenic, carcinogenic or act as tumor growth factors, the routine screening for possible heavy metal exposures is important for healthy subjects and cancer patients alike. With the modern analytical instrumentation now available, rapid simultaneous determinations of the majority of the toxic elements in biological materials is now possible. For example, a laboratory using vacuum induction-coupled argon-plasma quantometer (VIPAC-) spectrometry in the U.S. routinely provides results for 29 elements (Ca, Mg, Na, K, Cu, Zn, Fe, Mn, Cr, Co, Li, Mo, P, Se, Si, V, S, Sr, Ba, B, Au, Ag, Sn, Sb, W, Zr, Pb, As, Hg, Cd, Al, Ni, Be). As some of the elements of interest cannot be determined in blood or serum, samples of hair can analyzed instead. (Hair analysis results must be interpreted with caution because they may in part reflect external exposures; the use of samples of pubic hair or of toenail clippings is advisable).

To assess the current metal burden of Americans a market-basket study was conducted in 1973 by the FDA(40). According to this study, 15-20 year-old American males typically ingest 60.4 µg of Pb, 51.2 µg of Cd, 2.89 µg of Hg, 10.1 µg of As and 149 µg of selenium per day. The combined, food-only intakes of the toxic metals thus correspond to at least 1.0 µmol/day. Considering that only 2.0 µmols of Se are ingested during the same period, of which not all is bioavailable, it may be expected that a significant fraction of the total Se is being inactivated through interactions with heavy metals. This problem is likely to become more serious at higher metal exposures, which are easily attained e.g. through environmental or occupational dust inhalation, water contamination, etc., and thus could lead to a significant reduction of the physiological activity of Se. While little quantitative information is available concerning the bioavailability of Se in human organs and tissues, studies of animal feedstuffs revealed that e.g. only 22.4% of the total Se in tuna meal, 15.1% of the Se in meat and bonemeal and only 8.5% of the Se in fish meal is bioavailable(41). In fishmeal, Se is known to be rendered unavailable through complexation with mercury. The origin of the low bioavailability of Se in the other feedstuffs is not known but could be due to complexation to other metals. It is reasonable to assume, therefore, that a significant percentage of the ingested Se is used up to detoxify heavy metals. If the dietary Se intakes are low in the first place, the continued intakes of Se-antagonistic heavy metals could ultimately produce latent or acute Se deficiency syndromes. That this is a realistic proposition follows from comparisons of the Se, Pb, Hg, Cd, and As contents of samples of hair of healthy American Children and adults(42). In most cases, the combined molar equivalents of the toxic heavy metals exceeded the amounts of Se; molar ratios of toxic metals to Se of 10:1 - 50:1 were not uncommon. Under normal dietary conditions and in the absence of occupational exposures, the effects of Se inactivation by heavy metals may not be clinically noticeable except by an increased suceptibility to cancer.

Experiments with laboratory animals have demonstrated that low-level exposures to some of the Se-antagonistic elements suffice to abolish the cancer-protecting effects of selenium(43). In this regard, Pb and As are of particular interest. As human exposures to Pb have increased by several orders of magnitude since prehistoric times (44), it is possible that the natural resistance of early man to cancer gradually diminished due to increased Pb exposure. Arsenic, on the other hand, is a recognized environmental hazard and cancer risk factor in certain areas of Taiwan(45) and Argentina(46). It is also an occupational carcinogen. Recent studies of Gerhardsson et al.(47) suggest a protective effect of selenium on lung cancer in Swedish smelter workers: Lung cancer appeared in workers with the lowest lung tissue Se levels and the highest heavy metal/Se quotient. Ya-Jun Chu et al(48), similarly demonstrated in 1984 that blood Se concentrations of residents in a high lung cancer region of China were lower than in a region with low lung cancer incidence. According to this study, workers frequently exposed to As exhibited lower blood Se contents, the lowest Se values were observed in the lung cancer patients. In our own studies, a significant inverse correlation was found between the blood Se levels of healthy subjects and the age-corrected male lung cancer mortalities in different countries (49).

In the past, the apparent carcinogenic effects of arsenic have been the subject of much discussion. The problem is still not solved as it is difficult to induce cancer in experimental animals solely by exposure to arsenicals. In our own experiments, the exposure of MMTV-infected female weanling mice led to rapid weight gains and produced adult animals of apparently good health. However, the animals showed no resistance against the tumorigenic effects of MMTV and developed rapidly growing mammary tumors early in life; even multiple mammary tumors appeared, a normally rare event(50). Arsenite interferes with the transport and metabolism of Se but does not appreciably inhibit GSH-Px activity(51). Selenium, a powerful inhibitor of many enzymes involved in cell-division and growth, presumably exerts is cancer-protecting effects by modulating the rate of cell division(52). Based on recent experiments with liver cells, Hoekstra et al defined Se as an antiproliferative agent(53). In accord with these views, Se could prevent the metabolic activation of cells invaded by tumorviruses or chemical carcinogens without preventing normal mitotic activity. In cells subjected to the modulating influence of Se, errors of DNA replication leading to malignant transformation could thus be prevented.

Although different Se-antagonists elicit different responses, at appropriate levels all appear to have the ability to abolish the cancer--protecting effects of Se.

As epidemiological evidence clearly implicates arsenic, cadmium, nickel and chromium as etiological factors in human cancer(54), cancer patients should also be screened for possible exposures to these elements before treatment is instituted.

Some Neglected and Emerging Elements in Cancer Research

In addition to the abovementioned elements several other relatively common trace elements may play a role in cancer etiology and therapy, but interest in them is currently low or just emerging:

Iodine, for example, is considered by Eskin(55) to be an essential factor in the prevention of cellular changes of breast tissue that may lead to malignant transformation. The epidemiological evidence here is controversial as high breast cancer rates in some regions correlate with low iodine intakes, while in others the opposite seems true.

Manganese, an essential element, is of interest beause of its anticarcinogenic effects in animal tumor model systems(56). Mn intakes in different populations vary considerable; they are comparatively low in the U.S. and high in regions Asia where Mn-rich tea is regularly consumed. A possible association of Mn deficiency with cancer susceptibility in Finland has been suggested(57).

Silicon, an essential element(58), is of interest because it is needed for oesteogenesis and the formation of cartilage and connective tissue. A study of its effect on osteolytic metastases could produce new therapeutic leads.

Molybdenum deficiency has been implicated in the etiology of cancer of the esophagus(59). In rats exposed to N-nitrososarcosine, 2 ppm of Mo in the drinking water reduced the incidence of tumors in the esophagus and forstomach(60). It also inhibited mammary carcinogenesis induced by N-methyl-N-nitrosourea(61), but increased the incidence of lung adenoma in mice(62). In consideration of its role in the enzyme xanthine oxidase,

Heger(24) recommends the administration of Mo supplements as a means of treatment of intestinal dysbacteria in tumor patients.

Chromium, though carcinogenic in the form of chromate, is an essential element in the +3 oxidation state; deficiencies are associated with impaired glucose tolerance, elevated cholesterol and blood lipids(63). Exposure of MMTV-infected female mice to 5 and 10 ppm Cr (as the acetate) in the drinking water, had no effect on the incidence of mammary tumors. However, fewer tumors were observed in the groups exposed to 30 ppm of Cr(64).

Vanadium is possibly an essential element. In the +4 oxidation state it is mutagenic in certain bacterial test systems(65), but presently is not considered to be carcinogenic. Its role as a potential anticarcinogen should be investigated.

While barium and strontium are unlikely to conquer prominent positions in cancer therapy, they could exhibit Ca-antagonistic properties and might exhibit interesting effects in malignant osteogenic processes. For related reasons, studies of the effects of fluoride on osseous metastases should be studied.

Germanium was for some time used in Germany in the treatment of pulmonary tuberculosis prior to the availability of effective chemotherapeutic agents. Germanium appears to be particularly popular in Japan, where books on the apparent anticarcinogenic effects of certain organogermanium derivatives have been published(66). "Spirogermanium", 2-[3-(dimethylamino)propyl]-8,8-diethyl-2-aza-8-germaspiro[4,5]decane, is a patented antineoplastic germanium compound(67). Last but not least, a variety of organotin compounds have been found to possess antitumor activity. Di- and triorganotin derivatives of Amino Acids, for example, have Antileukemia P-388 activity(68).

Concluding Remarks

It cannot be denied that the introduction of the essential mineral elements further complicates an already difficult area of medical activity, especially since their functions are also intimately coupled with those of vitamins and other nutrients. However, their neglect may have serious consequences for the patient. While mineral analyses are helpful for the assessment of the mineral status, this review has hopefully made it abundantly clear that detected deficiencies in cancer patients need not, and in some cases must not, be corrected by supplementation. Periodic mineral analyses also have considerable value in the monitoring of treated patients. While blood or serum analyses are adequate for this purpose, hair analyses may sometimes be more informative, although their interpretation is not always straightforward. Such analyses may reveal the presence of toxic elements which are increasingly recognized to play important roles in the development and growth of tumors. They may also indicate changes in the disease process in response to therapy that may not be detectable by other methods. However, because of the nonspecific nature of the results, such analyses can only provide valid conclusions in conjunction with other tests. For example, abnormal hair Na/K ratios and concentrations almost unfailingly indicate an active disease process (not necessarily malignant). Similarly, cancer patients also show subnormal hair Fe levels, which obviously can have many causes. In patients in remission, mineral element profiles often normalize. However, even hopelessly ill patients with large, rapidly growing metastases may show essentially normal hair mineral profiles, especially if their tumors elicit no antigenic responses. In this case such analyses may be used to confirm the infaust prognosis.

Further research is evidently necessary before mechanistic details of the mineral metabolism of tumor patients is fully understood. It is already apparent, however, that such studies could provide new approaches to therapy. Thus, in animal experiments, the artificial induction of nutritional zinc deficiency was found to have effects comparable to cytostatic chemotherapy.

Last but not least, mineral analyses may in future become important in cancer risk assessments of healthy subjects and ultimately point the way toward successful prevention.

References

1. Windstosser, K., Die Summationsdiagnostik auf Karzinom und Praekanzerose. 2nd Ed., Vol. 1, Verl. f. Medizin Dr.Ewald Fischer, Heidelberg 1982, p. 69.
2. Bayer, W., in: Mineralstoffwechsel und Abwehrsystem, (Eds. Schmidt, K., and Bayer W.), Verl. f. Medizin Dr. Ewald Fischer, Heidelberg 1982, pp. 17-32.
3. Bayer, W., in "Mineralstoffwechsel beim Tumorpatienten" (Eds., Schmidt K. and W. Bayer), Verl. f. Medizin Dr. Ewald Fischer, Heidelberg 1984, pp. 47-58.
4. Nieper, H. A., op. loc cit. ref. 3, pp. 71-78.
5. Schrauzer, G.N, and White, D.A., Bioninorg. Chem. 6, 265-270 (1976).
6. Pories, W.J., DeWys, W.D., Flynn, A.M., Mansour, E.G., and Strain, W.H., Advan. Exp. Med. and Biol. 91, 243-257 (1978).
7. Weinberg, E. D., Biol. Trace El. Res. 3, 55-80 (1981).
8. Delbet, P., "Politique Preventive du Cancer", Editions Denoel, Paris, 1944.
9. Heinemann, H., Hippokrates 24 , 744-746 (1953).
10. Nieper, H.A., Med. Welt 154, 271, 1282 (1960).
11. Schrumpf-Bierron, Bull. Acad. Med. Paris 105 , 818-821 (1931).
12. Bablet, J., and Bader, H., Bull. Assoc. France d'Etude Cancer 21 , 570-573 (1932).
13. Bazikinian, K.L., First Internatl. Symposium on Magnesium, Vittel, France, ed. Durlach, J., p.593.
14. Aleksandrowicz, J., and Skotnicki, A. B., "Leukemia Ecology", publ. by U.S.Dept. of Commerce, Natl. Techn. Inf. Service, Springfield, VA 22161, pp. 141-146.
15. Jasmin, G., Biol. Trace El. Res. 1, 217-228 (1979).
16. Bois, B., Sandborn, E. B., And Messier, P. E., Cancer Res. 29, 763-775 (1969).
17. Averdunk, R., Bippus, P.H., Gunther, T., and Merker, H.J., J. Cancer Res. Clin. Oncol. 104, 63-73 (1982).
18. Larvor P., Girard, M., Brochart, M., Parodi, A., and Sevestre, J., Ann. Biol. Anim. Bioch. Biophys. 4, 345-355 (1964).
19. Battifora, H. A., McCreary, P.A., Hahneman, B.M., Laing, G.H., and Hass, G.M., Arch. Path. 122 , 610-618 (1968).
20. Battifora, H., Proc. 1st Int. Symp. on Magnesium, Vittel, France, ed Durlach, J., 501-516 (1971).
21. Hunt, B.J., and Belanger, L.F., Calc. Tiss. Res. 9 , 17-25 (1972).
22. Collery P., Coudoux P., and Geoffroy H., Proc. 18th Ann. Meeting Amer. Coll. of Nutr., Houston, June 1977, pp. 219-231.
23. Seelig, M.S., Biol Trace El. Res. 1 , 273-297 (1979).
24. Heger, H., in Mineralstoffwechsel beim Tumorpatienten Vol., eds. Schmidt K. and Bayer W., Verl. f. Medizin Dr.Ewald Fischer, Heidelberg, 1984, pp. 59-70.
25. Anke M., Hennig A., Groppel B., and Kronemann H., Proc. Arbeitstagung Mengen- u. Spurenelemente,Leipzig 1981, pp. 217-239 (1981).
26. DeVeaugh-Geiss, Med. Hypothes. 4, 521-530 (1977).
27. Shamberger, R.J., Biochemistry of Selenium, Plenum Press, New York, 1983.
28. Schrauzer, G.N., Selen, Neuere Entw. a. d. Biol., Bichem. u. Medizin, Verl. f. Medizin Dr.Ewald Fischer, Heidelberg, 1983.

29. Schrauzer, G.N., in "Mineralstoffwechsel beim Tumorpatienten", Vol. 3, eds Schmidt K., and Bayer, W., Verl. f. Medizin Dr.Ewald Fischer, 984, 39-46.
30. Schrauzer G.N., in "Vitamins, Nutrition and Cancer, ed. Prasad, K.N., Karger, Basel, 1984, pp 240-250.
31. Robinson M.F., Godfrey P.J., Thomson, C.D., Rea H.M., and van Rij A.M., Am. J. Clin. Nutr. 32, 1477-1485 (1979).
32. Schrauzer G.N., Molenaar, T., Mead, S., Kuehn, H., Yamamoto, H., and Araki, E., Jpn. J. Cancer Res. (Gann) 76, 374-377 (1985).
33. Doroshow, J.H., Locker G.Y., and Myers, C.E., J.Clin.Invest. 65, 128 135 (1980).
34. Batist, G., Katki, A.G., Ferrans, V.J., and Myers, C.E., J. Amer. Coll. Toxicol. 5, 87-93 (1986).
35. Donaldson, R, quoted in ref. 29, loc cit.
36. Greeder, G.A., and Milner J.A., Science(Washington) 209, 825-826 (1980).
37. Milner J.A., and Hsu, C.Y., Canc.Res. 41, 1652-1656 (1981).
38. Kuehn, K., Dunzendorfer, U., Whitmore, W.F., and Schrauzer, G.N., Biol.Trace El. Res. 8, 237-250 (1985).
39. Berry, J.P. et al. Cancer Res. 44, 2864-2868 (1984).
40. Mahaffey, K.R., Corneliussen, P.E., Jelinek, C.F., and Fiorino, J.A., Environm. Hlth. Perspect. 12, 63-69 (1975).
41. Cantor, A.H., Moorhead, P.D., Musser, M.A., in "Selenium in Biomedicine", eds. Spallholz, J.E., Martin, J.L. Ganther, H.E., Avi Publ. Co. Westport, Conn., 1981, pp. 192-202.
42. Unpublished results.
43. Schrauzer, G.N., Kuehn, K., and Hamm, D., Biol.Trace El. Res. 3, 185-196 (1981).
44. Patterson, C.C., Arch. Environ. Health 11, 344-363 (1965).
45. Chen, C.J., Chuang, Y.C., Lin, T.M., and Wu, H.Y., Cancer Res. 45, 5895-5899 (1985).
46. Astolfi,E., Maccagno, A., Fernandez, J.C.Garcia, Vaccaro,R., and Stimola,R., Biol.Trace El. Res. 3, 133-144 (1981).
47. Gerhardsson, L., Brune, D., Nordberg, I.G.F., and Wester, P.O., Brit. J. Industr. Med. 1985, 617-626.
48. Chu, Y.J., Liu, Q.Y., Hou, C., and Yu, S.Y., Biol.Trace El. Res. 6, 133-137 (1984).
49. Unpublished results.
50. Schrauzer, G.N., White, D.A., McGinness, J.E., Schneider, C.J., and Bell, L.J., Bioinorg. Chem. 9, 245-253 (1978).
51. Little, C., and O'Brien, P.J., Biochem. biophys. Res. Commun. 31 , 134-150 (1968).
52. G.N.Schrauzer, Advan. Nutr. Res., ed. Draper, H.H.,Vol. 2, 219-244 (1979).
53. LeBoeuf, R.A., Laishes, B.A., and Hoekstra, W.G., Cancer Res. 45, 5489-5495 (1985).
54. Costa, M., Metal Carcinogenesis Testing. The Humana Press Inc., Clifton, N.J. 1980, pp. 25-40.
55. Eskin, B.A., Biol. Trace El. Res. 5, 399-412 (1983).
56. Yamane, Y., and Sakai, K., Jpn. J. Cancer Res.(Gann), 64, 563-573 (1973).
57. Marjanen H., and Soini, S., Ann Agric. Fenn., 11, 391-411 (1972).

58. Carlisle, E.M., in "Biochemistry of the Essential Ultratrace Elements", ed. Frieden, E., Plenum Press, New York, 1984, pp. 257-291.
59. Yang, C.S., Cancer Res. 40, 2633-2644 (1980).
60. Luo, X.M., Wei, H.J., Hu, G.G., Shang, A.L., Liu, Y.Y., Lu, S.M., and Yang, S.P., Fed. Proc. 40, 928 (1981).
61. Wei, H.J., Luo, X.M., Sproat, H.F., and Yang, S.P., Fed. Proc. 42, 670 (1983).
62. Stoner, G.D, Shimkin, M.B., Troxell, M.C., Thompson, T.L., and Terry, L.S., Cancer Res. 36, 1744-1747 (1976).
63. Borel, J.S, and Anderson, R.A., in "Biochemistry of the Essential Ultratrace Elements" ed Frieden, E., Plenum Press, New York, 1984, pp. 175-199.
64. Unpublished observations.
65. Nishioka, H., Mutation Res. 31, 185-189 (1975).
66. Asai, K., Organic Germanium, Japan Publications Inc., Tokyo, 1980, ISBN 0-97040-474-1.
67 Schein, P.S., et al., Cancer Treat. Rep. 64 , 1051 (1980).
68. Cardarelli, N.F., "Tin as a Vital Nutrient: Implications in Cancer Prophylaxis and other Physiol. Processes", CRC-Press, Boca Raton, Florida, 1986.

TRACE ELEMENTS, MICROSOMAL LIVER ENZYME INDUCTION AND AMINO-ACIDS: A NEW PERSPECTIVE IN BIOLOGICAL PSYCHIATRY.

Cees J.M.van Tiggelen, M.D.
Consultant Geriatrician
Dandenong Psychiatric Centre,
Dandenong, 3175-Victoria, AUSTRALIA

Introduction

Nutritional status has always drawn attention in psychiatry: several mental disorders have been associated with changes in nutritional status, whilst in many organic mental disorders nutritional deficiencies have been described, not always adequately discriminating between cause and effect.
In particular the nutritional effects of long-term consumption of medication such as phenylhydantoin (PHT) and drugs such as alcohol have been set out in detail: effects on bioavailability of vitamins, of trace elements and of aminoacids have been described and have been associated with the development of varying degrees of neuro-myelo-encephalopathy. (1, and ref. there).
It seems plausible that at least some of these nutritional effects are caused by the capacity of e.g. PHT and alcohol to induce microsomal liver enzymes.
Recently, hard evidence has been presented (2), associating neurotoxic effects of industrial chemicals with the process of microsomal liver enzyme induction. Overviewing the literature, mounting evidence can be found , indicating that microsomal liver enzyme induction plays a role in the development of certain neuro-psychiatric disorders, especially when taken into account that such a process has considerable influence on the levels of amino-acids.
Over the last years the role of changed patterns of aminograms has attracted a lot of attention in neurobiology (3-14), indicating that

Trace Element – Analytical Chemistry in Medicine and Biology, Vol. 4

figure 1

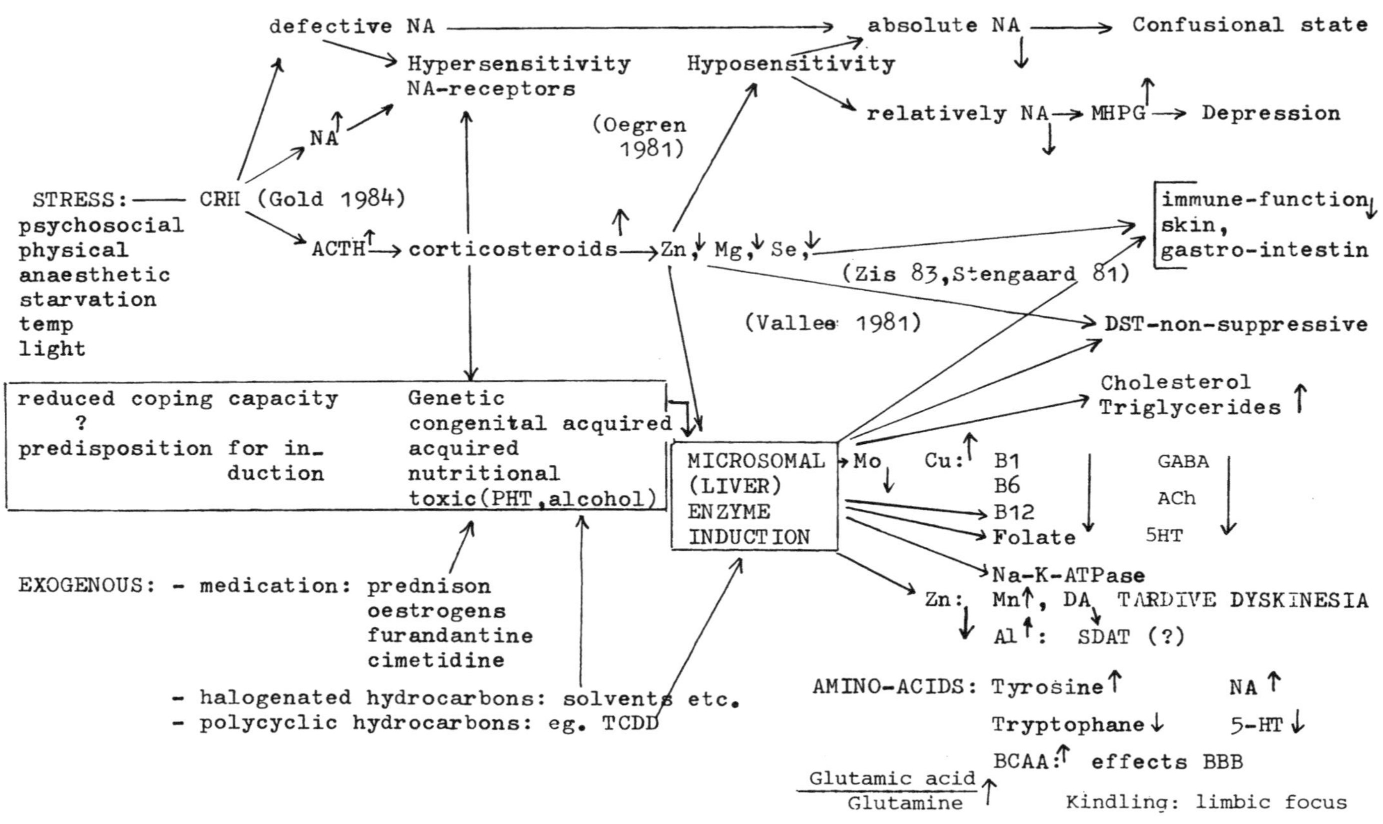
defective NA
absolute NA
Confusional state
Hypersensitivity
NA-receptors
Hyposensitivity
relatively NA
MHPG
Depression
NA
(Oegren
1981)
STRESS:
psychosocial
physical
anaesthetic
starvation
temp
light
CRH (Gold 1984)
ACTH
corticosteroids
Zn, Mg, Se,
immune-function
skin,
gastro-intestin
(Zis 83,Stengaard 81)
(Vallee 1981)
DST-non-suppressive
Cholesterol
Triglycerides
reduced coping capacity
?
predisposition for in_
duction
Genetic
congenital acquired
acquired
nutritional
toxic(PHT,alcohol)
MICROSOMAL
(LIVER)
ENZYME
INDUCTION
Mo
Cu:
B1
B6
B12
Folate
GABA
ACh
5HT
Na-K-ATPase
Zn:
Mn, DA
TARDIVE DYSKINESIA
Al:
SDAT (?)
EXOGENOUS: - medication: prednison
oestrogens
furandantine
cimetidine
- halogenated hydrocarbons: solvents etc.
- polycyclic hydrocarbons: eg. TCDD
AMINO-ACIDS: Tyrosine
NA
Tryptophane
5-HT
BCAA:
effects BBB
Glutamic acid
Glutamine
Kindling: limbic focus

the focus of attention in neurobiological research is shifting from the black box with neutrotransmittors, the brain, to the equally black box of the liver, which produces the precursors for the neurotransmittors: the amino-acids.
It is far too early to say that Hippocrates was absolutely right, when he localised 'melancholia' in the black bile of a sick liver, but there are some indications that he may have been on the right track after all.
It is against this background that I have explored the possible relation between trace-element status and amino-acid patterns: assuming that in two ways microsomal liver enzyme induction plays a role.
On the one hand, it has been demonstrated that microsomal liver enzyme induction influences biological availability of trace-elements, whilst on the other hand is is very well known that trace-element status has an influence on the same process: it is e.g. demonstrated that zinc deficiency facilitates the process of enzyme induction.
My base-line hypothesis is that primary or secondary trace-element abnormalities facilitate - through the process of microsomal liver enzyme induction - the development of abnormal concentrations of amino-acids, resulting in neuro-psychiatric disorder.
I will describe a brief case-history, which was basically the origin of this hypothesis, followed by a report on findings in selected patient-groups, examining in a small group of patients with a neurasthenic-depressive syndrome the aminograms before and after nutritional supplementation with mainly trace-elements.
The flow-chart in figure 1 describes schematically the hypothesis.

Case history

Mr.T., a 58 year-old man, was referred in May 1984 for psycho-geriatric assessment of his suspected incipient pre-senile dementia of the Alzheimer type. Since the age of 10 he had been suffering from post-traumatic epilepsy and had received long-term anti-epileptic medication, mainly barbiturates initally, later on phenylhydantoïn (for 26 years) and valproate (10 years).

The presenting problem was a rapidly increasing amnestic syndrome and severe behavioural disturbances such as indecent exposure, which had resulted in criminal charges.
Neurological examination showed the presence of autonomous and peripheral sensoro-motor neuropathy, confirmed in the electromyogram. Cerebellar dysfunction was found, as well as truncal ataxia. Primitive reflexes showed presence of subcortical encephalopathy, no signs or symptoms of cortical encephalopathy. He was diagnosed as a man with anti-epileptic medication induced neuro-myelo-encephalopathy, presenting as neuropathy, amnestic syndrome and organic personality change.
I assumed that the anti-epileptic induced microsomal liver enzyme induction had resulted in reduced biological availability of several nutrients such as the demonstrated reduced levels of vitamin B 12 in cerebrospinal fluid and the increased ratio serum-Copper/serum-Zinc.
He was treated with oral zinc-medication, 50 mg. elemental zinc TDS and oral vitamin B 12, 250 mcg, TDS from May 1984 till March 1985. In March 1985 he commenced on the Recommended Daily Allowance trace-element solution with added vitamin B 12 (10 ml. per day, containing 30 mg. Zinc, 0.2 mg. Chromium, 0.1 mg. Selenium, 0.3 mg. Molybdenum, 0.2 mg. Iodine and 100 mcg. Cyanocobalamine).
His anti-epileptic medication remained unchanged till June 1985, when he was re-assessed. At that time his EEG and EMG had improved considerably. His biochemistry had normalised considerably. Clinically he improved to a major extent, reported disappearance of his symptoms of neuropathy and improvement of his memory.
On re-examination in March 1986 his neurological condition revealed no abnormalities, his biochemistry was normal, his clinical status was perfect: after 20 years of being an invalid pensioner, he had decided to start looking for a full-time job.
For details of his biochemical condition, see table 1.

Table 1

Biochemical data Mr.T.:

	Cu	Zn	Cu/Zn	Glutamic Acid	Glutamine	Gla/Glu	R-Tryp	R-Tyr	B 12	DST
May 84:										
Serum	19.4	10.5	1.8	248	157	1.57	0.10	0.14	675	Non-suppressive
CSF:	--	--	--	11	174				7	
Aug+84:	23.6	16.2	1.5							
March 85:	22.2	17.1	1.3							
June 85: serum	19.7	17.1	1.2	137	1061	0.12	0.07	0.17	831	----
CSF:	--	--	--	-	916				115	
March 86:	15.8	11.8	1.3	113	560	0.20	0.12	0.15	450	suppressive

Legenda: Cu and Zn in micromol/l,

amino-acids in micromol/l, free amino-acid

B 12 in nanogram/l.

Gla/Glu: ratio glutamic acid/glutamine

R-Tryp: ratio Tryptophane/ (Valine + Leucine + Isoleucine + Phenylalanine + Tyrosine)

R-Tyr: ratio Tyrosine / (Valine + Leucine + Isoleucine + Phenylalanine + Tryptophane)

Clinical data

Intrigued by these unexpected findings, we have been able to assess in 108 patients, presenting with varying degrees of depressive, neurasthenic and psycho-somatic syndromes, the aminograms and the ratio se-Cu/se-Zn. The support of Unipath.Lab. Melbourne was highly appreciated in providing facilities for testing amino-acids and trace-elements by AAS and HPLC amino-acid analysis.
Out of this group of 108 patients we selected the groups of patients with pre-senile dementia, with long-term phenylhydantoin medication and the group with a history of chronic alcoholism.
At the time of preparation of this paper there were 10 patients amongst the total number of patients, who for mainly legal reasons had a repeat-aminogram after 4-6 months treatment with the trace-element solution: these patients were all patients who had been presented originally with a neurasthenic-depressive syndrome or an organic affective syndrome, presumably caused by exposure to potentially neurotoxic chemicals: these patients were involved in legal action or liability questions.
These 10 patients are described resp. as pre-RDA and post-RDA group.
The values of the control-group were provided by the pathology laborotory: the controls are twenty healthy athletes, who underwent a periodical medical examination. Unfortunately. in this group no copper- and zinc-levels were tested.
The statistical work-up was done by the computer-department of the Gippsland Institute of Adult Education (Mr.Dowsley), with the help pf the department of statistics. With the two sample t-test the different groups were compared with the control-group regarding the ratio glutamic acid/glutamine.
One way analysis of variance was also used for the glutamic acid/ glutamine ratio.

Results

table 2:

	ratio Cu/Zn $\pm$ S.D.		ratio GLA/Glu $\pm$ S.D.		R-Tryp. $\pm$ S.D.	
Patients (80)	1.2	0.4	0.25	0.18	0.10	0.03
Controls (20)	--	--	0.24	0.04	0.12	0.01
Pre-RDA (10)	1.4	0.3	0.49	0.41	0.10	0.02
Post-RDA (10)	1.2	0.3	0.21	0.10	0.09	0.03
Alcoholics (10)	1.4	0.3	0.51	0.18	0.09	0.02
Dilantin (5)	1.5	0.4	1.00	0.42	0.10	0.01
Alzheimer (3)	1.3	0.2	0.54	0.10	0.09	0.03

Two sample t-test ratio Gla/Glu compared with controls:

-patient-group:	T = - 0,57	P = 0.57	DF 97.7
-pre RDA:	T = - 1.97	P = 0.080	DF 9.1
-post RDA:	T = 0.74	P = 0.48	DF 10.5
-alcohol:	T = - 4.72	P = 0.000	DF 9.5
-dilantin:	T = - 4.09	P = 0.015	DF 4.0
-Alzheimer:	T = - 5.36	P = 0.33	DF 2.1

Figure 2:

One way analysis of variance ratio Glutamic acid / Glutamine:

Source	DF	SS	MS	F
Factor	6	3.7509	0.6251	15.47
Error	131	5.2930	0.0404	
Total	137	9.0439		

```
                                     Individual 95 pct CI's for mean
                                     based on pooled stdev
Level         N    Mean      Stdev   -------+---------+--------+-------
Patients     80  0.2548    0.1839      (x-)
Controls     20  0.2420    0.0417     (--x--)
Pre-RDA      10  0.4967    0.4074           (---x---)
Post-RDA     10  0.2174    0.1015    (---x---)
Alcoholics   10  0.5119    0.1784            (---x---)
Dilantin      5  1.0051    0.4168                       (---x----)
Alzheimer     3  0.5487    0.0978          (------x-------)
                                     -------+----------+--------+-------
                                          0.30      0.60     0.90

Pooled STDEV = 0.2010
```

The results indicate that regarding ratio Glutamic acid/Glutamine the group pre-RDA is different from the control-group and different from the patient-group, but similar to the alcohol, dilantin and small Alzheimer group.
The results also indicate that the pre-RDA group, (before treatment different from the control group), after treatment with RDA does not differ from the control group.
It does not seem warranted to compare the pre-RDA group and the post-RDA group with the two-sample t-test: however, notwithstanding the small numbers and the considerable standard deviation, there seems to be sufficient reason to state that the treatment with RDA has resulted in 'normalisation' of amino-acid levels as expressed in ratio glutamic acid/glutamine.

Discussion

The above reported results, all be it the preliminary results on small numbers, are not in disagreement with the basis-hypothesis, that manipulation of trace-element status has an effect on amino acid status. To what extent supplementation with trace-elements in conditions associated with known microsomal liver enzyme induction has a normalising effect on abnormal aminograms is presently under investigation: we are in the process of following up the aminograms in the patients reported above in the alcoholics, dilantin and Alzheimer group.
Clinical improvement in these patients suggests a therapeutic effect of the trace-element solution, but we are awaiting the hard biochemical data. If and when in further work the basis hypothesis can be confirmed, it will mean that further support is given to the hypothesis that microsomal liver enzyme induction plays a major role in the development of (organic) mental disorder.
In my opinion this could have far reaching therapeutic and preventative implications in the management of psychiatric disorders.
It can be argued then, that in major depressive disorder with 'vital symptoms' microsomal liver enzyme induction is involved, correctable

with orthodox anti-depressive medication in a far lower dose, if supported by nutritional supplementation: it could well be, and I will be happy to discuss this during question-time, that a non-suppressive dexamethasone suppression test is an indicator of microsomal liver enzyme induction combined with an absolute or relative zinc and/or molybdenum deficiency.

It can be argued, that development of tardive dyskinesia due to psychotropic medication, can be prevented by providing simultaneously supplementation, especially extra Molybdenum in order to restore the integrity of steroid-receptors, in particular oestrogen-receptor stability (15).

It can be argued that trace-element supplementation is essential in the treatment of post-partum-depression, again related to impaired steroid-receptor function associated with zinc and molybdenum deficiency.

Another area which will require intensive research is the area of children with minimal brain dysfunction; I have the preliminary clinical impression, that in these children a congenitally acquired predisposition to microsomal liver enzyme induction plays a major role in the development of the neuro-psychiatric problems: correction and/or prevention of actual microsomal liver enzyme induction under stress-conditions may well be very important in management.

Conclusion

The presented data strongly suggest an important role for trace-element status in the maintenance of amino-acid homeostatis. It becomes increasingly clear that amino-acid imbalance is significantly associated with neuro-psychiatric disorders, either as a cause or as an effect.

It seems that the process of microsomal liver enzyme induction is instrumental in the development of amino-acid abnormalities.

Considerable evidence is available linking microsomal liver enzyme induction with changed biological availability of many nutrients, including some trace-elements.

From a teleological point of view it can be argued that conditions of chronic inescapable stress conditions, through the mechanism of corticosteroid-induced enzyme induction, result in the development of a self-protective reaction: the reduction of biological availability of nutrients to the central nervous system is a plausible mechanism.
I suggest that investigations into the changes of nutritional status of hibernating animals during the process of going into hibernation and coming out of it could provide relevant basic information on the biochemistry of stress-related disorders.
Summarised, the proposed interaction between trace-elements, microsomal liver enzyme induction and amino-acid homeostatis deserves more attention: this could play a significant role in our understanding of the development of symptoms in neuro-psychiatric conditions, in particular in relation to stress-related disorders.
It could well be that we begin to realise that some patients with stress-related disorder manifest their pathology in an individual way, due to genetic predispositions: the clinical manifestation of stress-induced microsomal liver enzyme induction may show up in some people as a major depressive disorder, in other people as an immune-disorder: as development of cancer, as an auto-immune disease, as an allergy, as asthma or as a skin-disorder.
In another group the stress-related enzyme induction will result in a raised cholesterol and/or triglyceride level, resulting in cardio-vascular pathology.
It will be interesting to find out in further research to what extent the empirical approach of nutritional medicine, orthomolecular psychiatry, naturopathy and homeopathy has in fact a scientific base.
It will also be interesting to find out to what extent man-made environmental pollution in the broadest sense of the word increases the nutritional requirements, conditional for physical and mental well-being.
Above all, against the background of the emphasis in this paper on the liver and its function, it will be fascinating to find out whether Hippocrates, with his emphasis on the black bile in patients with melancholia, was right after all.......

Would it not make us feel humble, when once more it could be demonstrated scientifically that ancient knowledge, gained by visionary intuiton and guiding mankind throughout the process of evolution, was spot-on right!

I acknowledge gratefully the help of Mrs.de Bever and Mrs.Anderson, librarians, and of Mrs.Gielen in preparing the manuscript.
The support of Unipath Pathology Laboratory, Melbourne, was indispensable.

References

1. Abou-Donia M.B., Lapadula D.M., Campbell G., Timmons P.R.: The synergism of n- Hexane-induced Nuerotoxicity by Methyl Isobutyl Ketone following subchronic (90 days) inhalation in hens: induction of hepatic microsomal cytochrome P-450. Toxicol.Appl.Pharmacol.: 81, 1-16 (1985)

2. van Tiggelen C.J.M.: Disturbed Behaviour in the Elderly. Australian Family Physician: 15, 107-118 (1986)

3. Schwarcz R., Meldrum B.: Excitatory aminoacid antagonists provide a therapeutic approach to neurological disorders. The Lancet, ii, 140-143 (1985)

4. Simon R.P., Swan J.H., Griffiths T., Meldrum B.S.: Blockade of N-methyl-D-Aspartate receptors may protect against Ischemic Damage in the brain. Science, 226, 850-852 (1984)

5. Smith C.C.T., Bowen D.M., Francis P.T., Snowden J.S., Neary D.: Putative amino acid transmittors in lumbar cerebrospinal fluid of patients with histologically verified Alzheimer's dementia. J. Neurol. Neurosurg. Psychiatry 48, 469-471 (1985)

6. Branchey L., Branchey M., Worner T.M., Zucker D., Shaw S., Lieber C.S.: Association between amino acid alterations and hallucinations in alcoholic patients. Biol.Psychiatry 20, 1167-1173 (1985)

7. Watanable A., Takei N., Higashi T., Shiota T., Nakatsukasa H., Fujiwara M., Sakata T., Nagashima H.: Glutamic acid and glutamine levels in serum and cerebrospinal fluid in hepatic encephalopathy. Biochemical Medicine 32, 225-231 (1984)

8. Shiota T., Nakatsukasa H., Fujiwara M. et al: Plasma amino acid imbalance in alcoholic liver cirrhosis. Biochemical Medicine 32, 181-188 (1984)

9. Branchey L., Branchey M., Shaw S., Lieber C.S.: Depression, suicide and aggression in alcoholics and their relationship to plasma amino acids. Psychiatry Research 12, 219-226 (1984)

10. Pratt J.A., Jenner P., Johnson A.L., Shorvon S.D., Reynolds E.H.: Anticonvulsant drugs alter plasma tryptophan concentrations in epileptic patients: implications for antiepilectic action and mental function. J.Neurol. Neurosurg.Psychiatry 47, 1131-1133 (1984)

11. Engelsen B.A., Fosse V.M., Myrseth E., Fonnum F.: Elevated concentrations of glutamate and aspartate in human ventricular cerebrospinal fluid during episodes of increased CSF pressure and clinical signs of impaired brain circulation. Neuroscience Letters 62, 97-102 (1985)

12. Möller S.E., de Beurs P., Timmerman L., Tan B.K., Leynse-Ybema H.J., Cohen-Stuart M.H., Hõpfner Petersen H.E.: Plasma tryptophan and tyrosine ratios to competing amino acids in relation to anti-depressant response to citalopram and maprotiline. Psychopharmacology 88, 96-100 (1986)

13. Eriksson T., Carlsson A.: Adrenergic influence on rat plasma concentrations of tyrosine and tryptophan. Life Sciences 30, 1465-1472 (1982)

14. Voog L., Eriksson T.: Toluene-induced decrease in rat plasma concentrations of tyrosine and tryptophan. Acta Pharmacol et Toxicol 54, 151-153 (1984)

15. Atrache V., Ratajczak T., Senafi S., Hähnel R.: Purification of the Molybdate-stabilized 9-10 S Estradiol receptor from calf uterus. J.Biological Chemistry 260 (10), 5936-5941 (1985)

ALUMINIUM IN THE MAINTENANCE DIET OF RATS

D. Gawlik, P. Brätter, P. Gardiner. W. Gatschke
Hahn-Meitner-Institut für Kernforschung Berlin

W. Meyer-Sabellek
Klinikum Steglitz der Freien Universität Berlin

Introduction

In previous investigations on the incorporation of aluminium by nephrectomized rats we noticed that the maintenance diet used for the control animals contained up to 250 $mg \cdot kg^{-1}$ aluminium. This is approximately the fivefold content of the average european and north american human diet (1).

Consequently we investigated several charges of rat maintenance diets from the two german manufacturers Altromin and Sniff and found that our first observation was no exception. All measured values were between 190 and approximately 270 $mg \cdot kg^{-1}$.

By a measurement of the different components of one charge of the often used maintenance diet Altromin 1320 it became apparent that the constituents made very different contributions to the total aluminium content (Fig. 1). In this diet only two of the main components, namely soybeans and fish meal, contribute considerably to the total element content. The minor components green meal and a substance specified as $CaCO_3$ together however contribute to 60 % despite their low weight portion of together 3.5 %.

Recent improvements in analytical instrumentation have now made it possible to study effects of small doses of aluminium in human and animal diet on the mineral and trace element metabolism (1,2).

In the present study the following questions were posed:
1) Does the known depression of the absorption of phosphorus and fluorine occur at the concentrations mentioned above?

2) Is there an influence on the metabolism of the essential trace

Trace Element – Analytical Chemistry in Medicine and Biology, Vol. 4

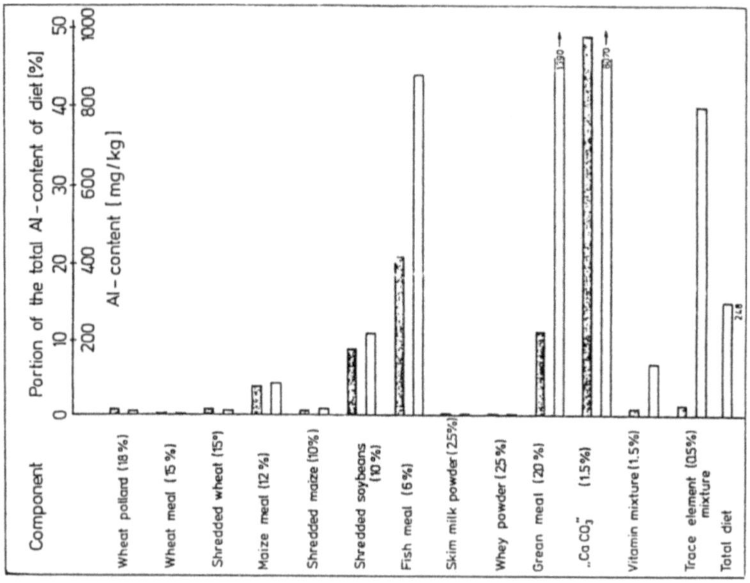

Fig. 1 : Aluminium content (empty bars) of the various components in the maintenance diet Altromin 1320 and their relative contribution to the total content (filled bars).

elements iron, zinc and selenium and of rubidium at these concentrations?

3) Is there an accumulation of aluminium in bone or brain at these dietary levels?

To answer these questions we fed rats over a period of five months with diets of different aluminium content between 7 and 920 $mg \cdot kg^{-1}$. We then measured the changes in the content of the above mentioned elements in a number of tissues, in blood components as well as some physiological parameters.

Experimental

Two diets from the manufacturer Altromin were used (Fig. 2). The normal maintenance diet 1320 with an aluminium content of 190 $mg \cdot kg^{-1}$ served as control. To the diet C 1000 which contained originally 7.3 $mg \cdot kg^{-1}$ two different amounts of the element were added leading to concentrations of 189 and 916 $mg \cdot kg^{-1}$ aluminium respectively. The level of 189 corresponded thereby to the element

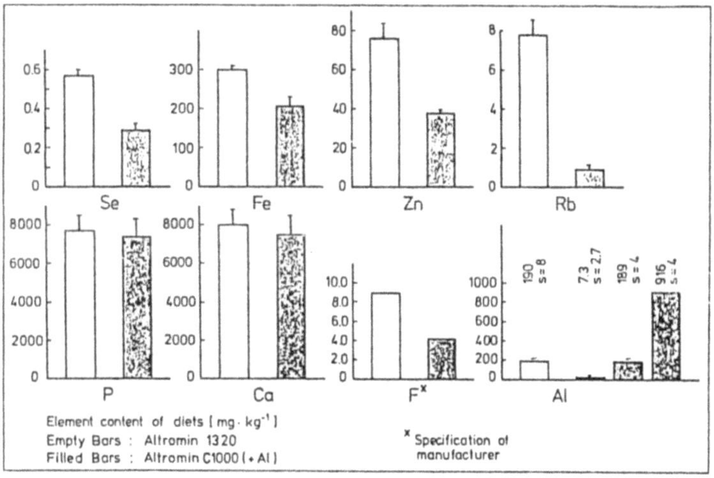

Fig. 2: Element contents of the maintenance diet Altromin 1320 used for the controls (empty bars) and of the experimental diet C 1000 (filled bars).

content of the maintenance diet 1320. The aluminium was given as aluminium hydroxyde gel, a compound which is used as active agent in many german phosphate binders and antacids as Alugel $^{(R)}$ and was delivered by the company Giulini.

The trace element contents of the two diets are slightly different. The contents in the diet C 1000 tend to be lower, but in both diets the levels are in the recommended range and in accordance with the declaration of the manufacturer.

Fourty-five female Wistar rats (Strain WIST (SPF 67 Han.) Mura) aged three months entered the experiment (Fig. 3). Up to this time they had been fed the maintenance diet Altromin 1320 and the control group of 10 rats received the same diet further. For the experiment three groups, each consisting of ten rats were formed, receiving the above described diet with 7.3, 189 and 916 $mg \cdot kg^{-1}$ aluminium. After two and five months experimental time five rats were killed after narcotization with Nembutal$^{(R)}$ by bleeding through heart puncture.

Element determination: (Fig. 4)

The trace elements selenium, iron, Zinc and rubidium were determined after freeze drying in blood components, in soft tissue, brain and bone and in the diets by means of instrumental neutron

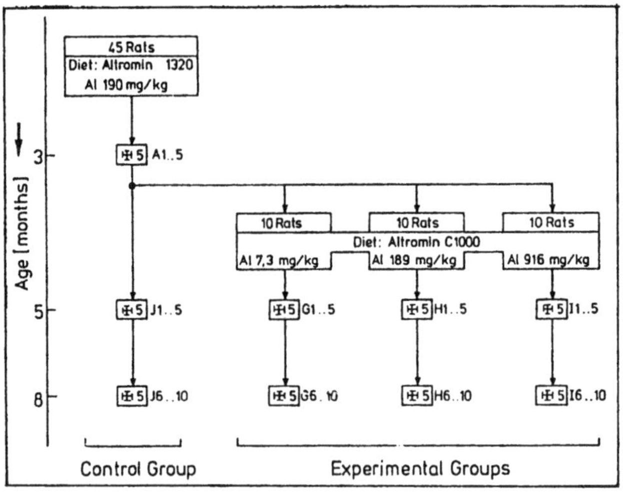

Fig. 3: Schedule of the experiment

Sample	Elements	Method
Blood components Soft tissues and Cartilage	Se, Fe, Zn, Rb	INAA (L)
Brain	Se, Fe, Zn, Rb Al	INAA (L) GFAAS
Bone	Se, Fe, Zn Ca, P, F Al	INAA (L) INAA (S) GFAAS
Diet	Se, Fe, Zn, Rb Ca, P, Al	INAA (L) INAA (S)

INAA (L): Neutron activation analysis. Irrad. time 5d
INAA (s): Neutron activation analysis. Irrad. time 20s
GFAAS: Graphite furnace atomic absorption spectrometry

Fig 4: Methods of element determination

activation analysis (INAA) after five days irradiation time (3). Calcium, fluorine and phosphorus in bone (defatted, dry) were detected by the same method, but after only 20 seconds irradiation (4.5). Additionally we measured after wet digestion in HNO_3 aluminium in brain and bone with graphite furnace atomic absorption spectrometry (6) and in the diet with INAA.

There are still some problems concerning the analysis of aluminium and fluorine in bone. The analytical situation is best characterized by the results of an intercomparison experiment of the IAEA from 1982 with the animal bone standard (Fig. 5). Just three laboratories took part in the fluorine analysis two of them offered a value of about 450 $mg \cdot kg^{-1}$.

The aluminium analysis of eleven laboratories resulted in values of smaller than 5 $mg \cdot kg^{-1}$ up to about 30000 with an overall mean of the accepted laboratory averages of 101 $mg \cdot kg^{-1}$. We don't think that this scatter can be explained by inhomogeneities of distribution as it is the case for cobalt in this standard.

In the meantime we have confirmed our values for the fluorine and aluminium content by independent methods.

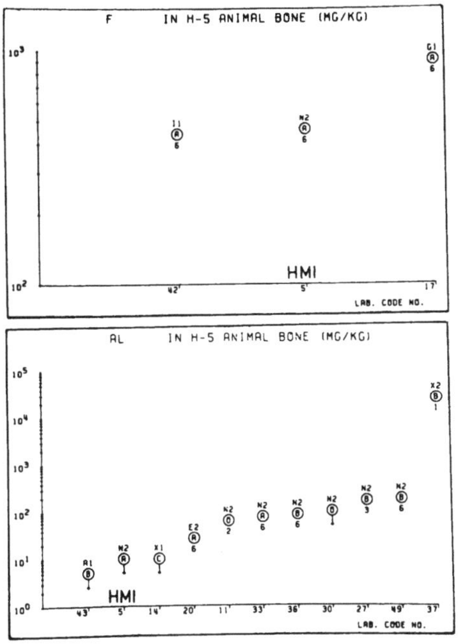

IAEA Intercomparison 1982 **Top: Fluorine**
Animal Bone H-5 **Bottom: Aluminium**

Fig. 5: Results of the determination of the elements fluorine and bone in the IAEA intercomparison experiment animal bone standard H-5 (1982)

In the case of fluorine Dr. Frenzel of our laboratory after wet digestion using ion-sensitive electrodes obtained a value of 459 $mg \cdot kg^{-1}$ (s = 18; n = 5; LQD = 40) which agrees very well with the 461 $mg \cdot kg^{-1}$ (s = 24, n = 5; LQD = 50) from neutron activation analysis.

Concerning aluminium we measured 2.6 $mg \cdot kg^{-1}$ (s = 0,33; n = 5; LQD = 0,20) with graphite furnace atomic absorption spectrometry. For this element we had been below our detection limit of 10 $mg \cdot kg^{-1}$ with neutron activation analysis after short time irradiation.

It seems that work has to be done to achieve certified values for these two elements in the animal bone standard.

Other determinations:
A number of blood data were received by routine clinical methods: plasma urea, total proteine, relative dry mass, alcaline phosphatase, potassium, calcium, magnesium and the erythrocyte and reticulocyte number.

Results and discussion

Over the five months experimental time the ad libitum fed rats developed normally. The three experimental groups did not differ between each other or compared to the controls with respect to food and water consumption which were between 12.4 and 13.2 g/d and between 18.1 and 18.7 ml/d respectively. They showed no sign of clinical abnormalities. The normal plasma urea content of between 38.7 and 42.1 mg/dl confirms an intact renal function.

We observed a slight decrease in the plasma total proteine content in the high aluminium group (Fig. 8). The implication of this will be discussed later in connection with the plasma zinc and selenium values.

Concerning the detectable element changes: the elements zinc, selenium, iron, rubidium, fluorine and phosphorus were measured in the groups with five animals with a standard deviation of roughly 10 %. Approximately half of the standard deviation is due to biological scattering. This results in detectable changes of between 15 % on the 95 4 level of probability and 32 % on the 99.9 % level.

In figure 6 the changes of element content caused by dietary aluminium are summarized. The numbers signify the relative differences of the group with the highest dietary aluminium content compared to this with the lowest:
There are no changes in phosphorus, iron and rubidium but changes in the order of -20 % in plasma selenium and plasma zinc with an additional selenium decrease in a number of tissues and a drastic decrease in bone fluorine.

Samples	Elements						
Plasma	(Se) -17%	Fe	(Zn) -22%	K	P	Ca	Mg
Erythrocytes	Se	Fe	Zn	Rb			
Liver	Se	Fe	Zn	Rb			
Muscle (biceps femoris)	(Se) -14%	Fe	Zn	Rb			
Cartilage (xiphoid)	(Se) -13%	Fe	Zn	Rb			
Brain (cerebrum)	Se	Fe	Zn	Rb	Al		
Brain (cerebellum)	(Se) -19%	Fe	Zn	Rb			
Brain (rest)	(Se) -6%	Fe	Zn	Rb			
Bone (femur epiphysis)	Se	Fe	Zn	Al	(F) -57%	P	Ca
Bone (femur diaphysis)	Se	Fe	Zn	Al	(F) -53%	P	Ca

◯ Differences ($P > 99\%$) after 5 months time of experiment. (Dietary aluminium 916mg·kg^{-1} compared with 7.3mg·kg^{-1}).

Fig. 6: Summary of the observed differences in element content between the groups with dietary aluminium content of 916 mg·kg^{-1} and 7.3 mg·kg^{-1} after 5 months experimental time.

Phosphorus

The diet with the highest aluminium content had a phosphorus/aluminium ratio of 8.2. Even at this value we find no decrease in plasma inorganic phosphorus. The absence of hypophosphataemia is supported by the fact that there was no histological sign of osteomalacia.

Iron

No differences in the content of blood components and tissues became apparent. Neither we find indications of disturbed erythropoiesis.

The influence of oral aluminium on the iron metabolism is discussed controversially.Two investigators found a decrease in the absorption of iron (7,8) while two others were not able to confirm this result (1,9). Recently a modulation of the aluminium absorption by the iron absorption was proposed, a kind of common pathway of metal absorption (10).

Severe aluminium intoxication seems indeed to be connected with various types of anaemia. But this had been produced in rats by heavy loading via intravenous or intraperitoneal injections (11).

Rubidium

Additionally we had included rubidium into the group of trace elements investigated. The reason was that in earlier studies we had found a total body depletion of that element in renal insufficient men and rats. But oral aluminium does not change the contents of rubidium in any of the tissues under investigation.

Aluminium

Owing to high blank values caused by the digestion procedure we were not able to measure aluminium in the brain samples.

In table 1 the values obtained for the aluminium content of the femur epiphysis are summarized. Generally they are low compared to those given in the literature and are subject to high standard deviations.

An increase of the element contents with age is deductible from the significant differences between the rats of 5 and 8 months age in the control group and in the high aluminium group, respectively.

At 8 months age there is a trend to higher bone contents with higher dietary aluminium in the experimental groups, but this is not significant.

Age (months)	Altromin 1320	Altromin C 1000 (+Al)		
	$190\ mg \cdot kg^{-1}$	$7.3\ mg \cdot kg^{-1}$	$189\ mg \cdot kg^{-1}$	$916\ mg \cdot kg^{-1}$
3	210 s=20 (10%)			
5	220^{+} s=80 (36%)	390 s=180 (46%)	520 s=290 (56%)	470^{++} s=90 (19%)
8	560^{+} s=260 (46%)	470 s=210 (45%)	790 s=360 (46%)	890^{++} s=220 (25%)

$^{+}P > 95\ \%$ $^{++}P > 99\ \%$

Table 1: Aluminium content ($\mu g \cdot kg^{-1}$) of the proximal epiphysis of the femur (defatted, dry) in rats (n=5) fed with diets containing different amounts of aluminium.

Fluorine (Fig. 7)

The influence of oral aluminium on fluorine absorption is well known (12). We will not discuss the development of the fluorine loss out of the bone which becomes apparent in the picture, but restrict ourselves to the bottom part where the fluorine contents after 5 months experimental time are given for different dietary aluminium. There is a difference of 41 % in fluorine bone content between the rats receiving 189 $mg \cdot kg^{-1}$ and those receiving only 7.3 $mg \cdot kg^{-1}$. This may not be important with respect to the fluorine metabolism but may cause secondary interaction effects.

Selenium and zinc (Fig. 8)

After an experimental time of five months we noticed a decrease in the plasma content of selenium and zinc in the group with medium and high aluminium intake compared to that with the lowest value of 7.3 $mg \cdot kg^{-1}$. These differences were not detectable after only two months time of feeding. In the case of selenium there are additional decreases of the same order in some tissues as muscle, carti-

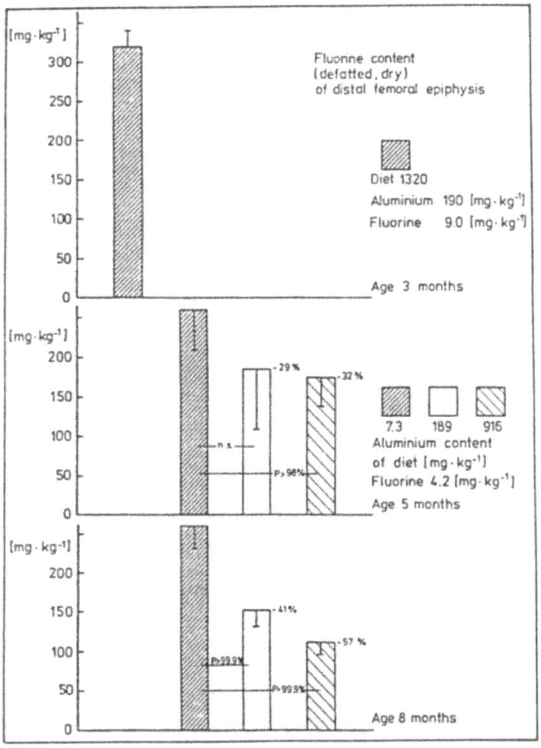

Fig. 7: Fluorine content of distal femural epiphysis in rats receiving a diet with different aluminium content

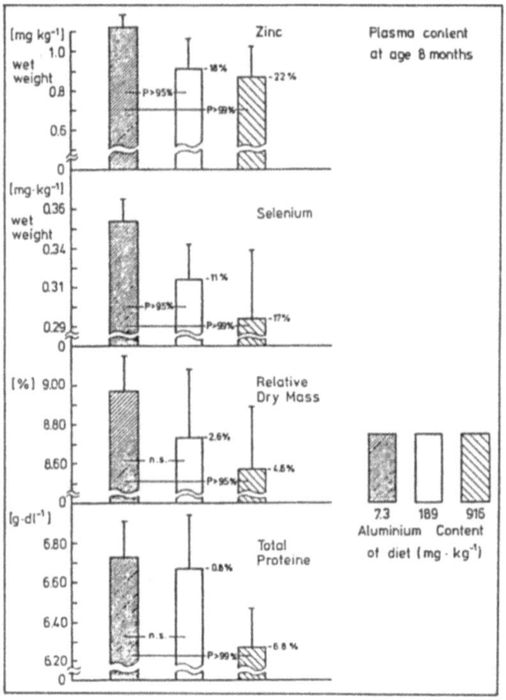

Fig. 8: Zinc and selenium content of blood plasma and plasma dry mass and total proteine in rats receiving a diet with different aluminium content after 5 months experimental time.

lage and brain. This finding does not allow to conclude a direct interaction of aluminium with the metabolism of selenium and zinc but may be due to secondary effects as fluorine decrease or changes in gastrointestinal pH. A part of these effects can be deduced from plasma dilution as the reduction in relative plasma dry mass and total proteine shows.

An interaction between zinc and aluminium in the sense that suboptimal levels of dietary zinc increase the aluminium accumulation in the brain is described by Wenk et al. for rats (13).

Conclusions

It seems to be useful, during experiments with rats, in which the fluorine, selenium or zinc metabolism is essential, to control the aluminium levels of the diet.

We believe that it is desirable to ask the manufacturers of rat diets to include the content of this element into the specifications of the products. It follows from the measured element distribution among the components of the diet that a reduction of the total aluminium content is possible by simply replacing a few constituents.

Concerning the effects of aluminium medication as phosphorus binder or antacid it should be kept in mind that if aluminium hydroxyde is replaced by other phosphate binders,as it is often proposed,the bone fluorine content will presumably rise faster as it does under aluminium medication. Even in renal insufficient patients receiving an aluminium dose 1.5 g/d, we found a doubling of the fluorine content in iliac crest biopsies within approximately 4 years.

More experiments are needed, to clarify the influence of long time supply of small doses of oral aluminium on the zinc and selenium metabolism.

Acknowledgement

The authors are grateful to Mr. J. Franke from the Institute of Nuclear Medicine, Klinikum Steglitz, Freie Universität Berlin for his skilful performance of the animal experiment and to Mrs. G. Röhlicke for her help in performing the neutron activation analysis.

References

1. Greger, I. L., M. J. Baier. 1983. Am. J. Clin. Nutr. 38, 411.
2. Weberg, R., A. Berstad, J. Aaseth, J. A. Falch. 1985. Scand. J. Gastroenterol. 20(6), 741.
3. Behne, D., H. Jürgensen. 1978. J. Radioanal. Chem. 42, 447.
4. Behne, D., P. Brätter, D. Gawlik, C. Keller, J. Möller, U. Rösick. 1976. In: Medizinische Physik, Vol. 2 (W. J. Lorenz ed.) Alfred Hüthig, Heidelberg. pp. 239-252.
5. Gatschke W., D. Gawlik, D. Kraft. 1980. J. Clin. Chem. Clin. Biochem. 18, 403.
6. Gardiner, P. E., M. Stoeppler, H. W. Nürnberg. 1985. Analyst 110, 611.
7. Rastogi, S. P., F. Padilla, C. M. Boyd. 1975. Proc. Am. Soc. Nephrol. 8, 21.

8. Skirne, B. S., S. R. Lynde, J. D. Cook. 1981. Gastroenterology 81, 1068.

9. Blumberg, A. 1977. Schweizer Med. Wochenschr. 107, 1064.

10. Cannata, J. B., C. Suarez, V. Cuesta, R. Rodriguez Roza, M. T. Allende, J. Herrera, J. Perez Llanderal. 1985. Proc. Eur. Dial. Transpl. Assoc. Eur. Ren. Assoc. 21, 354.

11. Drüeke, T. B., B. Lacour, M. Touam, J. P. Jucquel, J. J. Plachot, G. Cournot-Witmer, P. Galle. 1986. Kidney Intern., 29(18), 45.

12. Spencer, H., L. Kramer, C. Norris, E. Wiatrowski. 1980. Clin. Pharmacol. Theor., 28(4), 529.

13. Wenk, G. L., K. L. Stemmer. 1983. Brain Res., 288, 393.

LASER MICROPROBE MASS ANALYSIS (LAMMA) OF PARATHYROID GLANDS FROM DIALYSIS PATIENTS

A.H. Verbueken[1], F.L. Van de Vyver[2], E.J. Nouwen[3], F. Roels[4], M.E. De Broe[3] and R.E. Van Grieken[1]

Departments of Chemistry[1], Medicine[2] and Nephrology-Hypertension[3], University of Antwerp (U.I.A.), B-2610 Wilrijk-Antwerp; Department of Human Anatomy[4], Free University of Brussels (V.U.B.), B-1090 Brussels, Belgium.

Introduction

Aluminum intoxication is a severe threat in patients on chronic dialysis (CD). In these individuals, aluminum accumulates in the body, mainly in bone and liver. It may lead to bone disease (vitamin-D-resistant osteomalacia) and to damage of the brain (dialysis encephalopathy) (1-3). The preferential uptake of aluminum by the parathyroid glands of CD patients has also been reported (4). In rats loaded with aluminum, the parathyroid glands contained significantly more aluminum per unit mass than did thyroid tissue or cervical muscle (5). Circumstantial evidence has been provided that aluminum interferes with parathyroid hormone (PTH) secretion (6). In patients with aluminum-induced osteomalacia (AIO), low serum PTH levels may be the result of either high serum calcium or a direct action of aluminum on the parathyroid glands. It has indeed been shown in vitro that high concentrations of aluminum suppress PTH secretion (7, 8).
Furthermore, parathyroid gland function appears to play an important role in the development of AIO. In fact, sustained hyperparathyroidism seems to protect bone mineralization against the adverse effects of aluminum since the development of typical AIO has often been observed following parathyroidectomy (9). Also, aluminum shows a different distribution throughout the bone in patients with severe osteitis fibrosa as opposed to patients with AIO, even if bulk bone aluminum levels are comparable. Microanalytical studies have demonstrated that in bone specimens showing osteitis fibrosa the aluminum is more evenly distributed and the Aluminon staining usually is negative, while in AIO aluminum accumulates at the osteoid/calcified-bone boundary and the Aluminon staining is positive (10-12).

The purpose of the present study was to disclose the localization of aluminum (and other elements) in human parathyroid glands from CD patients with significant aluminum accumulation. The microanalytical method used for the subcellular localization was laser microprobe mass analysis (LAMMA) which was

Trace Element – Analytical Chemistry in Medicine and Biology, Vol. 4

preferred to other techniques (13), because of its simultaneous multiple-element detection combined with high sensitivity and favorable spatial resolution of analysis (1-2 μm) (14). LAMMA was applied to microtome sections of parathyroid gland tissue obtained during partial parathyroidectomy in two CD patients with secondary hyperparathyroidism. Putative sources of aluminum were aluminum-contaminated dialysate as well as prolonged oral administration of aluminum-containing phosphate binders.

Experimental

Tissue specimens :

Parathyroid gland tissue was obtained at parathyroidectomy from two CD patients with secondary hyperparathyroidism. From Patient 1 (M.V.) a thyroid gland sample was obtained at the same time.

Specimen preparation :

Common procedures for tissue fixation using glutaraldehyde and osmication as routinely performed in electron microscopy (EM) were applied for the thyroid and parathyroid gland samples. Dehydration was achieved in stages with acetone followed by embedding in epoxy resin (Epon). After block polymerization, sections of the desired thickness (0.2 μm) were cut using glass knives on a Reichert ultramicrotome and mounted on labelled copper EM-grids without any supporting film to avoid additional mass interferences from organic ions in LAMMA. From the parathyroid gland samples ultrathin sections (70 nm) have also been prepared and stained with alcoholic uranyl acetate and lead citrate to improve the ultrastructural information provided by the electron micrographs. The histological sections for LAMMA were left unstained. Parathyroid tissue from Patient 2 (W.V.) had in addition been stained for acid phosphatase by the lead precipitation method to differentiate lysosomal organelles at the EM-level (15, 16).

Electron microscopy :

The histological samples were examined by transmission electron microscopy (TEM) with a Zeiss EM 109 electron microscope. TEM-electron micrographs were taken before and after LAMMA analysis to enable the identification of the analyzed subcellular fragments and to measure the evaporated area left by the laser impact in LAMMA (Fig. 1).

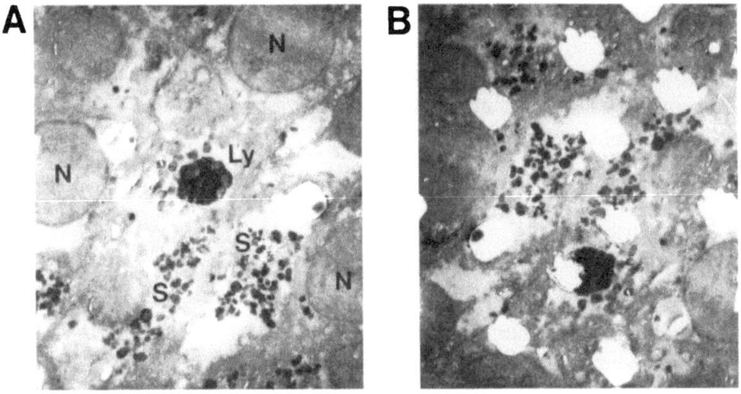

Fig. 1. Electron micrographs (TEM) of a thin unstained section of human parathyroid gland (Patient 1, thickness 0.2 μm) taken before (A) and after (B) perforation by the laser beam. It demonstrates the spatial resolution of analysis : the diameter of the laser-induced holes is about 2 μm in this example (N=nucleus, Ly=lysosome, S=secretory granules).

Laser microprobe mass analysis (LAMMA) (14) :

The laser-induced mass spectra were obtained with the LAMMA-500^{R} instrument of Leybold-Heraeus (Köln, F.R. Germany). The histological section is viewed with a binocular light microscope. The LAMMA system is equipped with two lasers : the first is a low-power red He-Ne laser which is focused to a micrometer-sized spot onto the section to indicate the subcellular region to be analyzed. Collinear with this pilot laser is a frequency-quadrupled (265 nm) Nd:YAG laser which provides a high-power pulse (Q-switched); it vaporizes and ionizes a selected area of a few μm^2 of the thin section (Fig. 1). The created positive or negative ions are extracted and separated in a time-of-flight mass spectrometer and thus produce a complete mass spectrum for each laser pulse. The technique was applied to unstained tissue sections and combined with EM to identify the analyzed subcellular structures. LAMMA was also used for the simultaneous demonstration of aluminum and acid phosphatase activity (presence of lead). Semi-quantitative determination of local elemental concentrations is possible by comparison with suitable standard samples (17).

Bulk analysis :

Bulk analysis of the tissue samples was carried out for aluminum and iron by means of electrothermal atomic absorption spectrometry (ETAAS) following procedures outlined elsewhere (18).

Results and Discussion

Bulk concentrations of aluminum and iron as measured by ETAAS were elevated for both the parathyroid and thyroid glands (Patient 1 : 2.4 µg/g Al, 4220 µg/g Fe for fresh parathyroid tissue, and 1.4 µg/g Al, 350 µg/g Fe for the thyroid sample; Patient 2 : 4.5 µg/g Al, 375 µg/g Fe for the parathyroid glands). Thus, severe iron accumulation in the parathyroid glands was observed as well. Transiliac bone biopsy of Patient 1 showed osteitis fibrosa, abundant iron in the bone marrow and a negative histochemical staining for aluminum (16).

Ultrastructural study of parathyroid gland histology revealed a striking abundance of electron-dense granules in the cytoplasm aggregated near the nuclei (probably secretory granules). This feature was in particular expressed for Patient 1 (Figs. 1, 2).

The microanalytical study by means of LAMMA demonstrated the presence of aluminum in lysosomal bodies of the parathyroid cells (see for example Fig. 3). The lysosomal nature was verified on the acid phosphatase stained tissue which, in LAMMA, provided Al^+ signals simultaneously with extreme Pb^+; the remaining background of the latter, however, precluded further differentiation. In addition, for both CD patients, aluminum (and variable iron) were localized in the smaller and electron-dense granular structures, which were considered to be secretory granules, accumulated in groups near the nuclei (Fig. 4). LAMMA of the cytoplasm showed the presence of iron (Patient 1) but no aluminum. A direct comparison of the ion signals with those recorded from suitable reference standards under identical instrumental conditions suggested local concentrations of aluminum in the range of 100 to 300 ppm (Patient 1) (16, 19).

In the thyroid sample (Patient 1) aluminum and iron were concomitantly localized within lysosomes of follicular epithelial cells (Fig. 5).

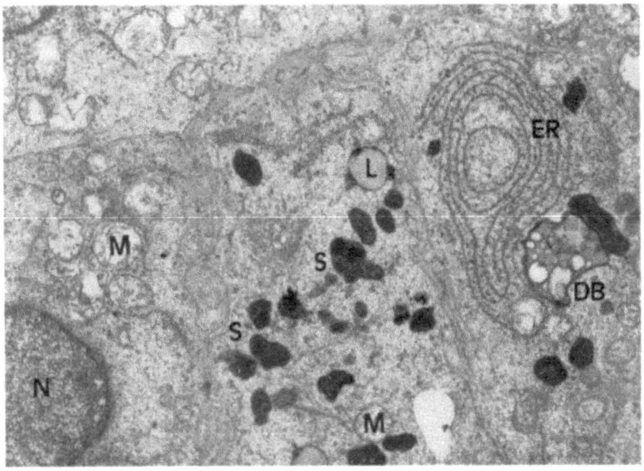

Fig. 2. Electron micrograph (TEM) of an active chief cell from parathyroid gland of Patient 1 (ultrathin section stained with heavy metals, original magnification 10,000 X). Numerous "storage" secretory granules (S), large lamellar and circular profiles of endoplasmic reticulum (ER) are present in cytoplasm (N=nucleus, M=mitochondria, L=lipid body, DB= dense body containing translucent areas of different sizes).

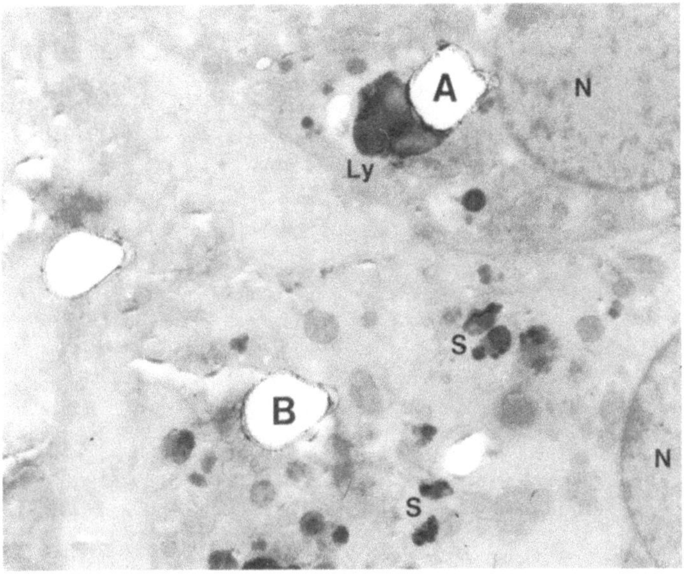

Fig.3. Electron micrograph (TEM) of a chief cell from the parathyroid gland after LAMMA (Patient 2, unstained 0.2 μm-thick section, OsO_4-postfixation, original magnification 10,000 X). Next to the nucleus (N) a rather large and electron-dense organelle (probably a lysosome, Ly) has been analyzed. The cytoplasm also contains scattered secretory granules (S). The LAMMA spectrum derived from region A displayed the presence of aluminum, while in region B no aluminum was detected.

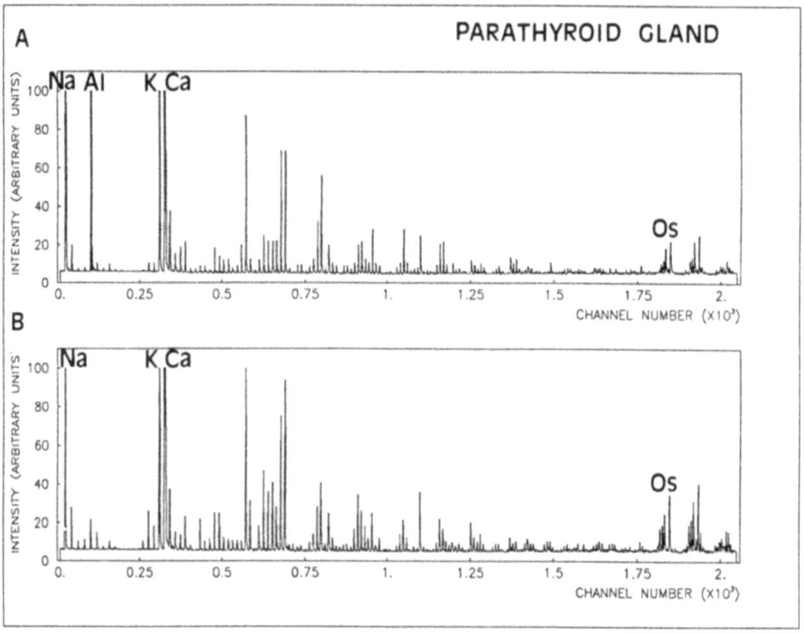

Fig. 4. Positive-ion LAMMA spectra from a parathyroid gland cell (Patient 2). Spectrum A taken from the smaller electron-dense granules within the cytoplasm (secretory granules) clearly demonstrates the presence of aluminum. Other spectra display a concomitant localization of iron (in particular for Patient 1). Spectrum B taken from a nearby region without any visible granules does not show aluminum. The osmium signals result from the postfixation procedure.

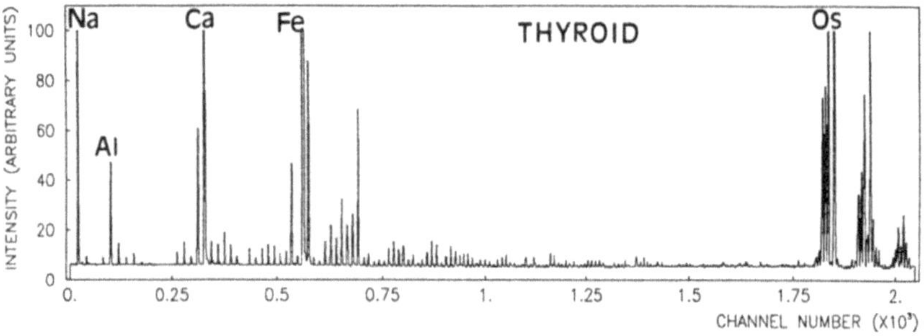

Fig. 5. Positive-ion LAMMA spectrum of a thyroid lysosome (Patient 1, follicular epithelial cell). A concomitant localization of aluminum and iron is demonstrated. The osmium signals result from the postfixation procedure.

It is known that extensive ultrastructural alterations of the parathyroid tissue occur in response to a stimulatory or inhibitory action (e.g. calcium concentration). Also, acute inhibition after a period of stimulation leads to an accumulation of secretory granules in the cells, suggesting that the inhibition of secretion preceeds the inhibition of synthesis of the parathyroid hormone (20). The presence of aluminum in the secretory granules suggests a direct interference of aluminum with PTH secretion.

It is important to note that in cases where electron probe X-ray microanalysis and histochemical staining (13) were negative, LAMMA succeeded to localize aluminum in the human parathyroid tissues, despite rather low bulk concentrations. The relevance of these findings has to be further investigated with regard to the relation between the ultrastructural localization of aluminum (and iron) and aluminum-induced toxicity.

Acknowledgements

This work was supported in part by the Belgian Ministry of Science Policy through research grants 80-85/10 and 82-87/47. A. Verbueken acknowledges the Belgian National Foundation for Scientific Research (N.F.W.O.) for a position as Senior Research Assistant. The skillful secretarial work of Erik Snelders is highly appreciated.

References

1. Alfrey, A.C., Hegg, A., Craswell, P., Am. J. Clin. Nutr. 33, 1509 (1980).

2. King, S.W., Savory, J., Wills, M.R., CRC Crit. Rev. Clin. Lab. Sci. 14, 1 (1981).

3. Parkinson, I.S., Ward, M.K., Kerr, D.N.S., J. Clin. Pathol. 34, 1285 (1981).

4. Mendes, V., Jorgetti, V., Nemeth, J., et al., Néphrologie 5, 65 (1984).

5. Cann, C.E., Prussin, S.G., Gordan, G.S., J. Clin. Endocrinol. Metab. 49, 543 (1979).

6. O'Hare, J.A., Murnaghan, D.J., Am. J. Med. 77, 229 (1984).

7. Morrissey, J., Rothstein, M., Mayor, G., Slatopolsky, E., Kidney Int. 23, 699 (1983).

8. Morrissey, J., Slatopolsky, E., Kidney Int. 29, S41 (1986).

9. Andress, D.L., Ott, S.M., Maloney, N.A., Sherrard, D.J., N. Engl. J. Med. 312, 468 (1985).

10. Cournot-Witmer, G., Zingraff, J., Plachot, J.J., et al., Kidney Int. 20, 375 (1981).

11. Verbueken, A.H., Van de Vyver, F.L., Van Grieken, R.E., Paulus, G.J., Visser, W.J., D'Haese, P., De Broe, M.E., Clin. Chem. 30, 763 (1984).

12. Verbueken, A.H., Van de Vyver, F.L., Paulus, G.J., Visser, W.J., Verpooten, G.A., De Broe, M.E., Van Grieken, R.E., in Trace Element Analytical Chemistry in Medicine and Biology, Vol. 3, P. Brätter and P. Schramel, Eds., Walter de Gruyter, Berlin, p. 375 (1984).

13. Verbueken, A.H., Van de Vyver, F.L., Van Grieken, R.E., De Broe, M.E., Clin. Nephrol. 24, S58 (1985).

14. Verbueken, A.H., Bruynseels, F.J., Van Grieken, R.E., Biomed. Mass Spectrom. 12, 438 (1985).

15. Roels, F., Paulus, G., De Broe, M.E., Path. Res. Pract. 179, 230 (1984).

16. Verbueken, A.H., Van de Vyver, F.L., Visser, W.J., Roels, F., Van Grieken, R.E., De Broe, M.E., Biol. Trace Elem. Res., in press.

17. Verbueken, A.H., Van Grieken, R.E., Paulus, G.J., de Bruijn, W.C., Anal. Chem. 56, 1362 (1984).

18. D'Haese, P.C., Van de Vyver, F.L., de Wolff, F.A., De Broe, M.E., Clin. Chem. 31, 24 (1985).

19. Verbueken, A.H., Van de Vyver, F.L., Visser, W.J., De Broe, M.E., Van Grieken, R.E., Adv. Mass Spectrom., in press.

20. Oldham, S.B., Fischer, J.A., Capen, C.C., Sizemore, G.W., Arnaud, C.D., Am. J. Med. 50, 650 (1971).

SPECIATION STUDY OF CHROMIUM DURING PERITONEAL DIALYSIS

B. Wallaeys, R. Cornelis

Laboratory of Analytical Chemistry, Rijksuniversiteit Gent
Proeftuinstraat 86, B-9000 Gent (Belgium)

F. Belpaire

Heymans Institute of Pharmacology, University Hospital
De Pintelaan 185, B-9000 Gent (Belgium)

N. Lameire

Renal Division, University Hospital
De Pintelaan 185, B-9000 Gent (Belgium)

Introduction

A previous study (1) on trace elements in serum of patients on continuous ambulatory peritoneal dialysis (CAPD) revealed an extremely high serum Cr level ± S.D. of 4.30 ± 1.92 $ng.ml^{-1}$ compared to the normal level ± S.D. of 0.16 ± 0.08 $ng.ml^{-1}$ (2). This deviation of normality stems from the very high Cr content of the dialysate : 1.22 ± 0.47 $ng.ml^{-1}$. As such an exposure may cause deleterious effects, an attempt was made to shed some light on the Cr species involved in the transport from the dialysate to the blood serum. The subsequent protein-binding of Cr was investigated as well.

Behaviour of Inorganic Elements on GPC Gels

It is well known (3,4,5,6,7,8) that metal ions may be retarded during Gel Permeation Chromatography (GPC) on highly cross-linked gels. As the size of the pores normally exceeds that of the hydrated metal by a several orders of magnitude, another mechanism than exclusion has to be accounted for. To our knowledge, the behaviour of metal ions on gels with a lower percentage of cross-linkage, as used for protein separations, has not yet been described in the literature. Therefore

Trace Element – Analytical Chemistry in Medicine and Biology, Vol. 4

we investigated the eventuality of adsorption of a series of metal ions on Sephacryl S200, Sephadex G75 and Biogel P100 which are envisaged to use for protein separation.

Methods and materials

The elutions were studied with the aid of radiotracers. Five ml of a solution containing ^{24}Na, ^{75}Se, ^{86}Rb, ^{65}Zn, ^{124}Sb and ^{134}Cs (buffer : 0.1 M NH_4Ac, pH = 7.2) with carrier concentrations < 50 $ng.ml^{-1}$ was brought on top of a Sephacryl S200 column (L = 1m, Φ = 26 mm, V_0 = 150.4 ± 1.2 ml (n=5)). The activity was eluted with 0.1 M NH_4Ac at a flow-rate of 1.10 $ml.min^{-1}$ and the fractions (10 ml) were counted for γ-radioactivity on a Ge-Li detector.

Separate columns (V_{tot} = 20 ml) were used to investigate the adsorption behaviour of Cr ($^{51}Cr(III)$) as well as $^{51}Cr(VI)$) in two buffer systems (10 mM TRIS and 0.1 M NH_4Ac) on three gels for protein separation : Sephacryl S200, Sephadex G75 and Biogel P100.

These elutions were subsequently followed by elutions with normal human serum, to investigate whether adsorbed substances became bound to serum proteins.

Results and discussion

The elution characteristics for the elements Na, Zn, Se, Rb, Sb and Cs are given in table 1. As expected all the elements, except Zn however, eluted in the same zone at the end of the GPC spectrum, indicating that no interaction of the ion with the chromatographic medium occurs. The different behaviour of Zn reveals that this element becomes completely adsorbed on the column. A subsequent run with human serum on this contaminated column led to the binding of Zn with serum proteins, as shown in fig. 1. The ^{65}Zn activity appeared to be bound to proteins in the albumin-transferrin regio. Overlooking such a phenomenon may have led to artifactual conclusions in Zn speciation studies.

Table 1 : Elution volumes and recovery after a GPC run of Na, Zn, Se, Rb, Sb and Cs on Sephacryl S200 in 0.1 M NH_4Ac

Radioactive label	Elution volume of peak maximum (ml)	% Recovery
^{24}Na	440 - 450	100
^{65}Zn	/	0
^{75}Se	440 - 450	94
^{86}Rb	440 - 450	96
^{124}Sb	430 - 440	93
^{134}Cs	440 - 450	101

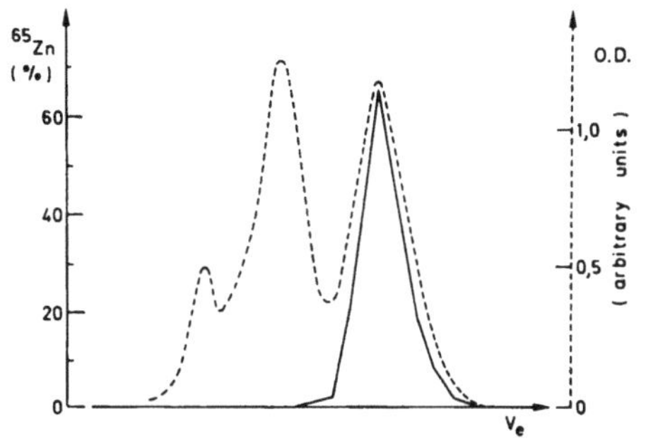

Fig. 1 : Elution of human serum on a ^{65}Zn contaminated Sephacryl S200 column

The results for elution of either $^{51}Cr(III)$ and $^{51}Cr(VI)$ on three permeation gels are summarized in table 2. The main issue is that Cr(III) was totally adsorbed on each of the 3 GPC columns. The outcome did not differ when using TRIS or NH_4Ac eluens. The adsorption of Cr(III) may tentatively be explained by the formation of hydrolysis products in the absence of complexing agents. Elution of normal human serum on the ^{51}Cr contaminated columns did not give rise to decontamination as was the case for ^{65}Zn. Therefore GPC may nevertheless offer a suitable tool for studying speciation of Cr in plasma, as any residual Cr present on the column appears to stay there unaffected.

Table 2 : Behaviour of Cr(III) and Cr(VI) on three GPC gels

Cr species	Sephacryl S200	Sephadex G75	Biogel P100
$^{51}Cr(III)$	adsorbed	adsorbed	adsorbed
$^{51}Cr(VI)$	adsorbed	not adsorbed	weakly adsorbed

Chemical Form of Cr in Dialysate

Several investigators have demonstrated that the metabolism and hence the beneficial or toxicological properties of Cr are highly dependent upon its oxidation state (9,10,11).

The presence of Cr in dialysate is probably due to a mere contamination by the ingredients the fluid is made of and by the processing in stainless steel utensils. In order to gain information on the chemical form of Cr in dialysate, we separately investigated the stability of both Cr(III) and Cr(VI) in this medium (12). It appeared that Cr(VI) was quickly reduced to its trivalent form, due to the presence of lactate. Such a reduction is described in the literature (13) to be extremely slow at pH = 7.4, but as both redox systems are pH dependent, the reduction of Cr(VI) seems more likely at pH = 5.5 (the pH of the dialysate) :

$$E_0' \ (HCrO_4^-/Cr(III),\ pH = 5.5,\ T = 25^\circ C) = 0.44$$

$$E_0' \ (\text{pyruvate/lactate},\ pH = 5.5,\ T = 25^\circ C) = -\ 0.10$$

On the other hand Cr(III) spiked in a dialysate bag appeared to be stable and dialyzable, even after two months storage.

Availability of Cr during Peritoneal Dialysis

One of the impressive features of the CAPD treatment is the very rapid almost complete correction for acidosis, by diffusion of dialysate lactate to the blood and release of bicarbonate from the body into the peritoneal cavity (14). Such an exchange results in a rapid pH increase of the dialysate, which subsequently may give rise to hydrolysis of the dialysate Cr, which then becomes a non-dialyzable compound. Prior to any peritoneal dialysis experiment, we therefore investigated the possible hydrolysis formation of Cr in different buffering agents, such as lactate, bicarbonate, phosphate ... (15). We found that Cr existed as a dialyzable molecule in the presence of lactate and bicarbonate, and not of phosphate. 'In vitro' titration of dialysate with 0.1 M NaOH revealed that the pH exerts no effect on the dialyzability of Cr. Possibly, hydrolysis of Cr is prevented by complexation with lactate or bicarbonate, which may explain the availability of Cr(III) throughout the dwell period of a peritoneal dialysis session.

Protein Binding of Cr during Peritoneal Dialysis

One of the most intriguing findings about the Cr transport is its diffusion from the dialysate to the blood against a concentration gradient. As mentioned in the introduction, the Cr content in the serum of the CAPD patients amounts to 4.30 $ng.ml^{-1}$ versus 1.22 $ng.ml^{-1}$ in fresh dialysate. It therefore seems plausible that Cr binds to serum constituents and hence is transported to different target organs. Therefore we investigated the binding of Cr to proteins during peritoneal dialysis in rabbits.

Uptake of Cr during Acute Peritoneal Dialysis in Rabbits

Methods

Male rabbits were anesthetized with oral urethane (1.25 $g.kg^{-1}$). Both arteria and vena femoralis were catheterized. A constant saline infusion was given (15 $ml.h^{-1}$) and blood pressure was followed during the course of the experiment. A peritoneal catheter was introduced in the body and $^{51}Cr(III)$ labelled dialysate (ca. 20 $ng.ml^{-1}$) was instilled (75 $ml.kg^{-1}$ body weight). The rabbits were heparinized (Liquémine). During the dwell period blood samples were taken and counted for ^{51}Cr. Dialysate was sampled as well. Blood was centrifuged and plasma samples were transferred to a Sephacryl S200 column and eluted with 0.1 M NH_4Ac. The U.V.-spectrum was recorded and the fractions were collected and counted for ^{51}Cr.

Results

A typical disappearance curve of ^{51}Cr from dialysate and subsequent uptake in plasma is shown in fig. 2A en B. The data could be fitted fairly well to a decreasing exponential curve in the case of dialysate and to a saturation function in the case of plasma.

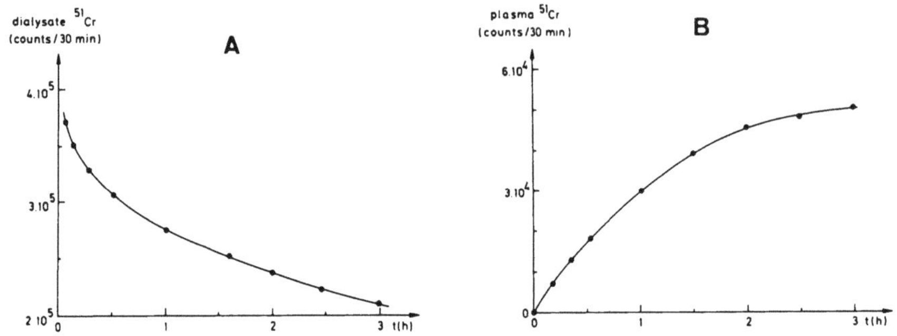

Fig. 2 : ^{51}Cr disappearance in dialysate (A) and uptake in plasma (B) versus dwell time

The gradual uptake in the protein regio of the rabbit plasma can be followed in fig. 3. Obviously Cr, transported as Cr lactate complex, was quickly transferred to plasma proteins with a moleculair weight between 70000 and 90000. The next step consisted in finding out to which specific protein the Cr is bound to. The shoulder peak occurring at t = 1.5 h and t = 2.5 h is Cr still bound to lactate.

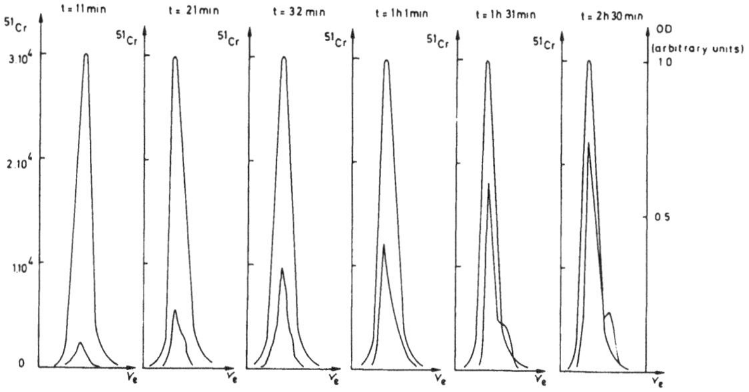

Fig. 3 : Sequential increase of ^{51}Cr in the albumin-transferrin regio during an acute peritoneal dialysis session in rabbits

Discrimination of the Specific Cr Binding Proteins

Nephelometric measurement of albumin and transferrin concentrations in the fractions after GPC of plasma on Sephacryl S200 revealed that both proteins eluted in exactly the same fractions, and hence direct determination of the relationship between Cr and albumin and Cr and transferrin, as suggested by Cochran et al. (16) is not possible. Therefore a separation on DEAE-TRISACRYL was necessary.

Methods and materials

Similar to the description of methods in the previous chapter, 5 ml $^{51}Cr(III)$ 'in vivo' labelled rabbit plasma was transferred to a column (L = 1 m, Φ = 26 mm) packed with Sephacryl S200. Elution was performed with 0.1 M NH_4Ac at a flow rate of 72 $ml.h^{-1}$. Fractions were collected and counted for ^{51}Cr radioactivity. Protein concentration of albumin and transferrin was measured by laser nephelometric detection. The fractions containing both proteins were pooled and lyophilized . The residual substance was dissolved in 10 mM TRIS and eluted in a saline gradient (ca. 2.5 $mM.cm^{-1}$) on DEAE-TRISACRYL. U.V. absorbance was measured, and the fractions were counted for ^{51}Cr radioactivity.

Results and discussion

Fig. 4 A represents typical U.V. and ^{51}Cr activity spectra of labelled plasma. The outcome of the elution on DEAE-TRISACRYL is shown in fig. 4 B. The latter elution clearly points to the fact that Cr is mainly bound by transferrin. These results are in agreement with the findings by Hopkins et al. (17), who suggested that 70% to 90% of the $^{51}Cr(III)$ administered to rats was bound by the β-globulin. On the other hand, Venezia et al. (18) recently found the Cr binding to be non-specific. Their finding seems rather questionable because the elution they performed on Sephadex G200 is unsatisfactory in separating the different proteins of interest.

Binding of Cr with Transferrin

Specific binding of Cr to human transferrin was measured by incubating a standard solution of substantially iron free, human transferrin (Sigma) ($5.2.10^{-3}$ mM) with a series of ^{51}Cr labelled solutions with increasing Cr concentration, dissolved in 1.39% HCO_3^- buffer (pH = 7.2). The protein bound Cr was determined after ultrafiltration through centriflo cones (Amicon) at ca. 1000 RPM.

Results and discussion

The amount of ultrafiltrated ^{51}Cr served to calculate the quantity of protein-bound Cr. The data were subsequently treated to be plotted as a Scatchard plot. The outcome is shown in fig. 5. This experiment revealed that at least 2 binding sites were available for Cr binding. The equilibrium constant determined from the plot was ca. $10^{9}.mol^{-1}$. This value lies much lower than the equilibrium constant for Fe (ca. $10^{24}.mol^{-1}$), and is also below the binding constant for Al (ca. $10^{13}.mol^{-1}$) as investigated by Cochran et al. (16).

Conclusions

During CAPD treatment, Cr is absorbed from the dialysate, accounting for high Cr levels in the serum of those patients. The existance of Cr(VI) compounds in the dialysate can be ruled out due to the presence of lactate. Cr(III), which normally would hydrolyse at physiological pH, is complexed as Cr lactate and hence exists as a dialyzable compound.
Acute peritoneal dialysis was applied to rabbits, with the aid of $^{51}Cr(III)$ labelled dialysate. GPC of the plasma revealed a gradual uptake of ^{51}Cr in the albumin-transferrin regio all over the dwell period. Binding of Cr to transferrin was demonstrated by gradient elution on DEAE-TRISACRYL. Incubation of solutions of increasing Cr concentration with transferrin, followed by the assessment of the amount of protein-bound Cr, allowed to calculate an approximate association constant of $10^{9}.mol^{-1}$. Since transferrin is usually saturated for 30 % only with Fe, this protein should indeed be considered as a potential binding site for Cr. Its much lower association constant in comparison to that for Fe practically precludes the possibility that Cr would affect the iron transport in serum.

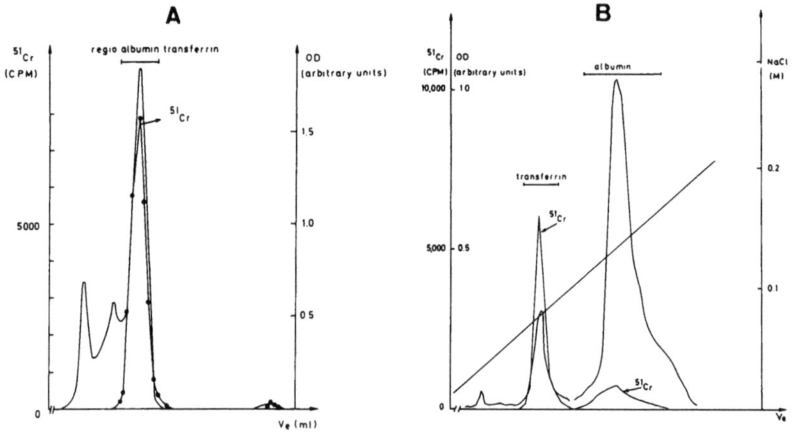

Fig. 4 : U.V. and ^{51}Cr spectrum of plasma GPC on Sephacryl S200 (A) and elution of the albumin-transferrin regio on DEAE-TRISACRYL in a saline gradient (B)

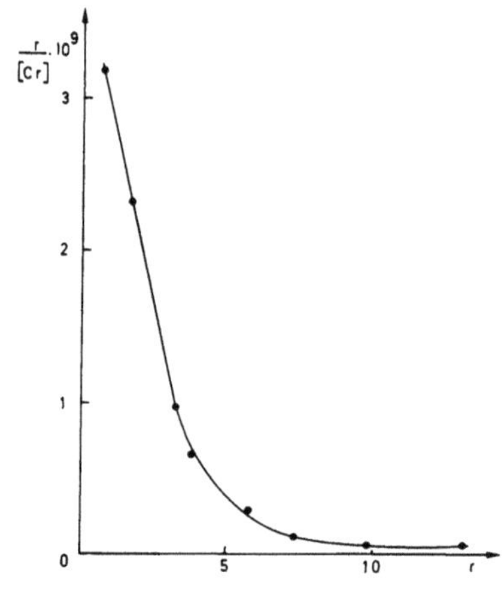

Fig. 5 : Scatchard plot of Cr binding data on human transferrin, in the presence of HCO_3^-

$$\text{with} \quad r = \frac{\text{moles Cr bound to protein}}{\text{total moles protein}}$$

[Cr] = free Cr concentration

References

1. Wallaeys, B., R. Cornelis, L. Mees, J. Hoste, N. Lameire. 1984. Atomkernenergie-Kerntechnik, Suppl. to 44, 624-630.

2. Versieck, J., J. Hoste, F. Barbier, H. Steyaert, J. De Rudder, H. Michels. 1978. Clin.Chem. 24, 303-308.

3. Pesek, J.J., R.L. Pecsok. 1972. Anal.Chem. 44, 620-622.

4. Ortner, H.M., H. Dalmonego. 1974. J.Chrom. 89, 287-294.

5. Ortner, H.M., H. Krainer, H. Dalmonego. 1973. J.Chrom. 82, 249-261.

6. Ortner, H.M.. 1975. J.Chrom. 107, 335-339.

7. Ortner, H.M., H. Dalmonego. 1975. J.Chrom. 107, 341-349.

8. Ortner, H.M., O. Packer. 1972. J.Chrom. 71, 55-65.

9. Langård, S.. 1982. In : Biological and Environmental Aspects of Chromium (S. Langård, ed.). Elsevier Biomedical Press, Amsterdam. New York. Oxford. pp. 149-169.

10. Levis, A.G., V. Bianchi. 1982. In : Biological and Environmental Aspects of Chromium (S. Langård, ed.). Elsevier Biomedical Press, Amsterdam. New York. Oxford. pp. 171-208.

11. Tandon, S.K.. 1982. In : Biological and Environmental Aspects of Chromium (S. Langård, ed.). Elsevier Biomedical Press, Amsterdam. New York. Oxford. pp. 209-220.

12. Wallaeys, B., R. Cornelis, E. Sabbioni, F. Belpaire, N. Lameire. 1986. Bull.Soc.Chim.Belg. 95 (in press).

13. Connett, P.H., K.E. Wetterhahn. 1983. In : Structure and Bonding 54 - Inorganic Elements in Biochemistry (M.J. Clarke, ed.). Springer-Verlag, Berlin. Heidelberg. New York. Tokyo. pp. 93-124

14. Robson, M.D., A. Faivoseviz,, H. Malmoud. 1979. In : Continuous Ambulatory Peritoneal Dialysis - Proc.Int.Symp. - Paris (M. Legrain, ed.). Exerpta Medica. pp. 194-198.

15. Wallaeys, B., R. Cornelis, E. Sabbioni. Sci.Tot.Environ. (in press).

16. Cochran, M., D. Patterson, J.H. Coates, P.T.H. Coates. 1984. In : Trace Element - Analytical Chemistry in Medicine and Biology, Vol. 3 (P. Brätter, P. Schramel, eds). Walter de Gruyter & Co., Berlin. New York. pp. 311-318.

17. Hopkins, L.L., K. Schwarz. 1964. Biochem.Biophys.Acta 90, 484-491.

18. Venezia, C., M.H. Karol. 1984. Toxicol. 30, 125-133.

ALUMINIUM TRANSFER DURING HEMODIALYSIS

A. Känzig, R. Knutti, Ch. Schlatter
Institute of Toxicology, Swiss Federal Institute of Technology,
CH-8610 Schwerzenbach, Switzerland

U. Vogelsang, G. Zàruba
Renal Unit, City Hospital Waid, CH-8037 Zurich, Switzerland

Introduction

Excessive uptake of aluminium (Al) in long-term dialysis patients, caused by $Al(OH)_3$ as phosphate binders or high Al-concentrations in the dialysis fluid, may be the probable cause of dialysis encephalopathy, vitamin D resistant osteomalacia and non-irondeficient, microcytic anemia [1]. The plasma-Al concentrations of such patients are in the order of 50-500 µg Al/l which vastly exceed those found in normal, healthy subjects (2.0±0.4 µg Al/l [2]).

Whereas it is agreed upon that high concentrations of Al in the dialysis fluid will increase the body burden of Al, published data concerning the transfer of Al during hemodialysis with a dialysate of low Al-concentration are controversial. Some investigators reported that Al can be removed from plasma [3][4][5], while others claimed that it cannot [6][7]. Their data mainly based on measurements of plasma-Al concentrations before and after hemodialysis sessions.

This preliminary study evaluates a method which permits a direct assessment of the Al-transfer across the dialyser membrane, concerning both direction and actually exchanged amount of aluminium.

Trace Element – Analytical Chemistry in Medicine and Biology, Vol. 4

Patients and Methods

Al-transfer across the dialyser membrane was determined during two independent hemodialysis sessions in each of 3 chronic hemodialysis patients with plasma-Al concentrations of 100-150 µg Al/l. Hollow fiber dialysers (Hemoflow F60, Fresenius AG, Bad Homburg) with a dialysate proportioning system (Fresenius MTS 2008) with water purified by reversed osmosis were used. The patients were on a schedule of 4 hours of maintainance hemodialysis 3 times a week.

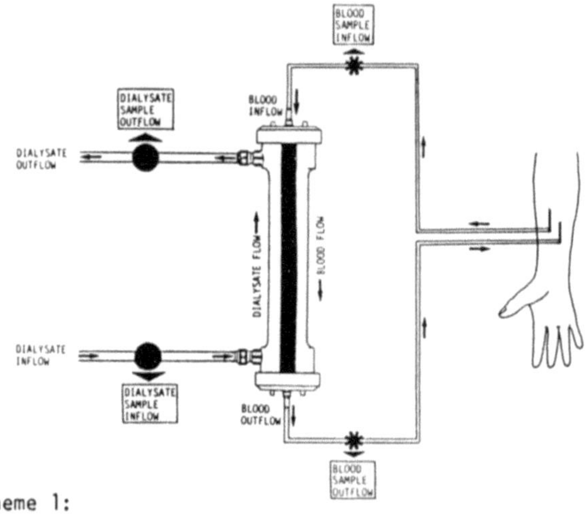

Scheme 1:

Hemodialysis equipment and sample collection sites during dialysis

The experimental set up is depicted in scheme 1. Ultrafiltrate of the blood (UF) was sampled from the membrane into acid washed polystyrene tubes before and after the dialysis session after removing the dialysate. 1-2 ml blood was drawn after 15 min and 240 min both at the inflow and outflow site of the dialyser into precleaned plastic syringes (Monovetten^R, Sarstedt AG, Sevelen), containing 2 mg Na_2EDTA and then centrifuged at 2000 rpm for 15 min. In each case an additional blood sample was taken for determination (3 times) of the packed cell volume by the capillary centrifugation methode ($Hct\%_{in}$ and $Hct\%_{out}$). Dialysate was sampled into acid washed polystyrene tubes both at the dialyser inflow and outflow site after 5, 15, 60, 120, 180 and 240 min. All samples were stored at 4°C until analysis.

Al was determined by electrothermal atomic absorption spectrophotometry (Perkin Elmer Model 5000, equipped with graphite furnace, Model HGA 500 and autosampler, Model AS-1). Standard graphite tubes were pretreated pyrolytically (200 sec. at 2000°C with 5% methane in argon). Analog absorption signals were recorded on a chart recorder and digital peak height values were transferred to a data station. The instrumental setting was: Al-hollow-cathode lamp at 15 mA, 309.3 nm, slit width 0.7 nm, tungsten-iodine background correction lamp, sample size 10 µl. Ultrafiltrate was diluted (1+1) with 0.01M HNO_3, plasma was diluted (1+4) with 0.4% ($^v/v$) Triton X-100 in 0.08M HNO_3 and dialysate was diluted (4+1) with 0.1% ($^v/v$) Nonidet P-40 in 0.17M HCl/0.01M HNO_3.

The temperature program is given in Table 1. The method of standard additions was used for calibration.

Table 1: Typical furnace temperature program for the determination of alumimium in ultrafiltrate, plasma and dialysate.

Step	Ramp (sec)	Hold (sec)	Temp. (°C)	Argon (ml/min)
Dry	5	15*	110	300
Ash	10	10	1500	300
	1	2	20	0
Atomise	0	4	2200	0
Clean	1	3	2500	300

* 25 sec for ultrafiltrate (no detergent added)

Because of loss of plasma fluid by ultrafiltration, as measured by an increase of the packed cell volume (Hct%), plasma-Al concentrations at the dialyser outflow site (AlP_{out}) have to be corrected in order to be comparable with

$$AlP_{out,corr} = AlP_{out} \cdot \left[\frac{Hct\%_{in}}{Hct\%_{out}} \cdot \frac{(100-Hct\%_{out})}{(100-Hct\%_{in})}\right] \quad (1)$$

$$EAl_P = QB_{in} \cdot \left[1-\frac{Hct\%_{in}}{100}\right] \cdot (AlP_{in} - AlP_{out,corr}) \quad (2)$$

$$EAl_D = AlD_{out} \cdot (QD_{in} + UFR) - AlD_{in} \cdot QD_{in} \quad (3)$$

plasma-Al at the inflow site (AlP_{in}). The corrected outflow concentration ($AlP_{out,corr}$) and the actual plasma-Al elimination rate (EAl_P) are derived from equations 1 and 2. The blood flow rate (QB_{in}) was 240 ml/min (patient A.S.) or 200 ml/min (patients K.B. and M.H.). For calculating the Al-elimination rate in the dialysate (EAl_D), a mean ultrafiltration rate (UFR) of 9 ml/min, as measured by an increase of packed cell volume along the dialyser ($Hct\%_{in}$ and $Hct\%_{out}$), was added to dialysate flow rate (QD_{in}) of 500 ml/min (equation 3).

Results and Discussion

From table 2 it can be seen that $AlP_{out,corr}$ was always lower than the simultaneously sampled AlP_{in}. Correspondingly, Al-concentrations in dialysate increased across the dialyser. Al-elimination rates in plasma (EAl_P) were in the range of 0.6-2.1 µg Al/min (tab. 2), resulting in 140-500 µg Al eliminated from plasma during 4 hours of dialysis (fig. 1). These results were in good agreement with the Al-elimination rates determined in dialysate (EAl_D) of 0.1-2.0 µg Al/min (tab. 2). The lower values determined in the dialysate (tab. 2/fig. 1) might be due to Al-adsorption effects at the dialyser membrane or at the protein gel layer of a "secondary membrane" forming on the surface of the dialyser membrane [8]. Two of the elimination rate values (table 2, note *) were relatively high and did not agree with the other results. Sample contamination is a probable explanation, because this is indeed a severe problem of Al-analysis.

The data in table 2 show, that plasma-Al concentrations AlP_{in} and AlP_{out} were, with one exception, distinctly higher at the end of hemodialysis sessions than at the beginning. This might imply an Al-transfer from dialysate into blood, in contrast to the fact that inflow dialysate-Al concentrations (AlD_{in}) of 0.2-3.2 µg Al/l were lower than the ultrafiltrable portion of plasma-Al (AlUF) of 12-22 µg Al/l (tab. 2). These data rather indicate that during hemodialysis considerable redistribution of Al takes place in body fluids and tissues.

The net Al-elimination is not paralleled by a corresponding decrease in the Al-concentration in plasma. Only the measurement of Al-concentrations at both sites of the dialyser permits a direct assessment of the Al-transfer during dialysis. Kaehny et al. [9] showed in a study with healthy persons that about 300 µg of Al are excreted in urine after an oral dose of 2.2 g of Al, a dose similar to that taken by dialysis patients. Therefore a dialysis session of four hours with an Al-elimination up to 500 µg can substitute the Al-output by a healthy kidneys in two days. We feel that this Al-transfer to the dialysate can play an important role in keeping the Al-burden low in patients with chronic renal disease.

In the artificial kidney the exchange of a substance between plasma and dialysate depends mainly on the free concentration plasma/dialysate gradient [10]; therefore dialysate-Al should be kept as low as possible. In our study only 17 ± 3 µg Al/l or about 14 ± 3 % of plasma-Al was ultrafiltrable (AlUF). We assume that dialysate-Al concentrations of 10-15 µg Al/l, proposed by the Commission of the European Communities [11] as an upper limit, and higher levels, will prevent Al-elimination during hemodialysis.

<u>Table 2</u>: Aluminium concentrations and elimination rates during 4 hours hemodialysis in 3 chronic renal failure patients (mean ± SEM). Blood flow rate 240 ml/min (patient A.S.) or 200 ml/min (K.B. and M.H.), dialysate flow rate 500 ml/min.

AlPin/AlPout, Hct%in/Hct%out: Plasma-Al concentration and packed cell volume of blood at dialyser inflow/outflow site. AlPout,corr: corrected outflow plasma-Al concentration. AlUF: blood ultrafiltrate Al-concentration. EAlP: Al-elimination rate from plasma. AlDin, AlDout): dialysate-Al concentration at the dialyser inflow/outflow site. EAlD: Al-elimination rate in dialysate.(* Contamination?)

time [min]	AlP_{in} [µg/l]	$Hct\%_{in}$ [%]	AlP_{out} [µg/l]	$Hct\%_{out}$ [%]	$AlP_{out,corr}$ [µg/l]	Al_{UF} [µg/l]	EAl_P [µg/min]	AlD_{in} [µg/l]	AlD_{out} [µg/l]	EAl_D [µg/min]
Hemodialysis 1			<u>Patient A.S.</u>							
5						19±1		2.2 ± 0.1	4.3 ± 0.2	1.1
15	141±1	21.0	138±1	22.0	130		2.1	1.6 ± 0.1	4.0 ± 0.1	1.2
60								1.8 ± 0.1	3.4 ± 0.1	0.8
120								1.7 ± 0.1	3.2 ± 0.1	0.8
180								1.7 ± 0.2	3.3 ± 0.1	0.8
240	134±1	22.7	133±2	24.0	124	13±1	1.9	1.8 ± 0.1	3.4 ± 0.1	0.8
Hemodialysis 2										
5						22±1		2.2 ± 0.1	5.2 ± 0.2	1.5
15	132±1	21.7	125±3	22.3	121		2.1	2.2 ± 0.2	5.4 ± 0.1	1.6
60								2.3 ± 0.1	4.5 ± 0.4	1.1
120								2.2 ± 0.1	4.5 ± 0.1	1.2
180								2.0 ± 0.2	5.3 ± 0.2	1.7
240	133±3	24.0	130±2	25.0	123	16±1	1.8	2.8 ± 0.2	5.6 ± 0.3	1.5

- continued -

Table 2: Continued

Hemodialysis 1			Patient K.B.							
5						14±1		2.4 ± 0.4	7.7 ± 0.4	2.7*
15	118±3	27.7	115±2	28.3	112		0.9	2.3 ± 0.1	2.9 ± 0.1	0.3
60								2.1 ± 0.1	2.7 ± 0.4	0.3
120								2.9 ± 0.1	3.2 ± 0.3	0.2
180								3.2 ± 0.1	3.3 ± 0.2	0.1
240	127±2	28.0	134±3	29.7	123	16±1	0.6	1.2 ± 0.1	2.6 ± 0.1	0.7
Hemodialysis 2										
5						12±1		2.4 ± 0.4	2.5 ± 0.1	0.1
15	93±1	25.0	99±1	27.0	89		0.6	2.4 ± 0.1	2.5 ± 0.1	0.1
60								2.4 ± 0.1	2.4 ± 0.4	<0.1
120								2.0 ± 0.1	2.6 ± 0.2	0.3
180								2.0 ± 0.2	2.4 ± 0.2	0.2
240	115±1	27.0	111±1	27.0	111	15±1	0.6	2.1 ± 0.2	3.0 ± 0.2	0.5
Hemodialysis 1			Patient M.H.							
5						17±1		1.4 ± 0.1	3.8 ± 0.3	1.2
15	106±1	22.0	104±1	23.0	98		1.2	1.4 ± 0.1	3.9 ± 1.0	1.3
60								1.5 ± 0.2	3.2 ± 0.2	0.9
120								1.3 ± 0.2	2.8 ± 0.3	0.8
180								1.1 ± 0.1	3.8 ± 0.3	1.4
240	140±1	23.3	135±1	24.0	130	19±1	1.5	1.7 ± 0.1	4.0 ± 0.2	1.2
Hemodialysis 2										
5						22±1		0.4 ± 0.1	4.3 ± 0.3	2.0
15	103±2	21.7	85±2	23.7	76		4.2*	0.4 ± 0.1	3.9 ± 0.3	1.8
60								0.6 ± 0.5	3.4 ± 0.1	1.4
120								0.2 ± 0.1	3.3 ± 0.2	1.6
180								0.6 ± 0.1	2.6 ± 0.1	1.0
240	110±1	22.0	109±1	22.7	105	15±1	0.8	0.7 ± 0.1	3.4 ± 0.6	1.4

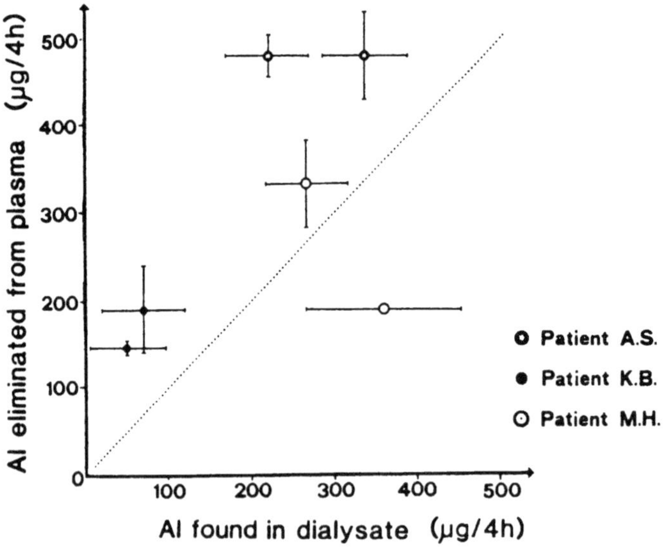

Figure 1: Amount of Al eliminated from plasma (µg Al/4h) and Al-quantity found in dialysate (µg Al/4h) during a four hours hemodialysis (mean ± SD).

Conclusion

I. In situations of Al-concentrations in plasma of 100-150 µg Al/l and in dialysate of < 3 µg Al/l a net transfer of aluminium from plasma to dialysate was found.

II. In most cases the plasma-Al concentration increased during hemodialysis, although Al was eliminated through the dialyser membrane.

III. A four hours dialysis session (dialysate-Al < 3 µg/l, ultrafiltration rate 9±5 ml/min, mean ± SD) removed from plasma up to 500 µg Al in patients with plasma-Al concentrations of 100-150 µg/l. These findings were in good agreement with simultaneously measured Al-content eliminated in dialysate.

IV. An dialysate-Al concentration of 10-15 µg Al/l and more most probably prevents the elimination of Al during hemodialysis.

References

1. Wills, M.R. and J. Savory. 1983. Lancet 2, 29-34.

2. D'Haese P. C., F.L. Van de Vyver, F.A. de Wolf, M.E. De Broe. 1985. Clin. Chem. 31, 24-29.

3. Graf, H., H.K. Stummvoll, V. Meisinger, J. Kovarik, A. Wolf, W.F. Pinggera. 1981. Kid. Int. 19, 587-592.

4. Hosokawa, S., T. Imai, T. Okumura, J. Kawamura, K. Sawanishi, O. Yoshida. 1984. Tr. Elem. Med. 1, 59-64.

5. Allain, P., H.-E. Thebaud, L. Dupouet, P. Coville, M. Pissant, J. Spiesser, P. Alquier. 1978. Nouv. Presse Méd. 7, 92-96.

6. Kaehny, W. D., A.C. Alfrey, R. E. Holman, W.J. Shore. 1977. Kid. Int. 12, 361-365.

7. King, S. W., J. Savory, M.R. Wills. 1982. Ann. Clin. Lab. Sci. 12, 143-149.

8. Sprenger K.B.G., W. Kratz, A.E. Lewis, U. Stadtmüller. 1983. Kidney Int. 24, 143-151.

9. Kaehny W.D., A.P. Hegg, A.C. Alfrey. 1977. N. Engl. J. Med. 296, 1389-1390

10. Wolf, A.V., D.G. Remp, J.E. Kiley, G.D. Currie. 1951. J. Clin. Invest. 30, 1062-70

11. Commission of the European Communities. 1982. Int. Workshop Luxenburg, 5-7 July 1982. J. Clin. Chem. Clin Biochem. 20, 837-839.

THE INFLUENCE OF ALUMINIUM HYDROXIDE AND SUCRALFATE ON SERUM ALUMINIUM LEVELS IN CHRONIC DIALYSIS PATIENTS

E.Kozak, V.Hudnik
Boris Kidrič Institute of Chemistry, Ljubljana, Yugoslavia

J.Drinovec, M.Benedik, A.Kandus, H.Ponikvar, A.Bren, R.Kveder, M.Močivnik
University Medical Center, Dialysis Center, Ljubljana, Yugoslavia

Introduction

The normal trace element levels in human tissues and body fluids can be altered by pathological conditions, occupational exposure as well as by therapeutic medical treatments. These trace element disturbances can have severe medical consequences. Aluminium toxicity has been found in the pathogenesis of a number of disorders in patients with chronic renal failure. Due to increased quantity of aluminium in human body can develop osteomalacia, dialysis encephalopathy and microcytic anemia (1,2,3). The application of various analytical techniques for aluminium determination in biological samples has been described (4,5,6). Wide variations in "normal" values for aluminium in human serum reported in the literature raise serious questions pertaining as to the methodologies used (7).

The present study was concentrated on the development of electrothermal atomic absorption spectrometry method (ETAAS) for the reliable determination of aluminium concentrations in serum and to investigate possible sources of contamination. The procedure was applied to the determination of aluminium in serum of chronic hemodialysis patients. Comparatively, two phosphate binders, aluminium hydroxide and sucralfate, were studied and their influence on the serum concentration levels of aluminium and phosphorus monitored in chronic dialysis patients.

Trace Element – Analytical Chemistry in Medicine and Biology, Vol. 4

Experimental

Patients and Methods

The comparative and prospective study was divided into 3 periods: A, B and C. In period A, lasting 4 weeks, the patients were given aluminium hydroxide 4.73 ± 1.64 g daily. During the period B, lasting 1 week (wash-out period) no phosphate binder was taken. In period C, lasting 4 weeks, sucralfate was substituted for aluminium hydroxide on the basis of 1 g of sucralfate for 475 mg of $Al(OH)_3$. The average daily dose of sucralfate in period C was 9.84 ± 3.64 g. During the study the patients took their normal diet; no antacids or Calcitriol were taken. The dialysis regime was kept constant. There were 20 patients at the beginning of the investigation, at the end there were 17 patients. Three patients were removed from the investigation due to undesirable side effects of sucralfate. The group of 17 patients comprised 10 women and 7 men with an average age of 49.5 ± 10.5 years. All patients cooperating in the investigation were stable dialysis patients that were on dialysis program for 70.5 ± 37.5 months. They were dialyzed for 14.1 ± 1.8 hours per week. 13 patients were on acetate - and 4 on bicarbonate dialysis. Several types of dialyzers were used: G.Plate (7 patients), Plivadial sce - 135 (4 patients), Hemomed 1.0 (3 patients), and Cordis 4000 (3 patients).

Sampling for Al Determination

Venous blood samples were taken by disposable needles into polyethylene vessels from patients several times in the morning before dialysis on an empty stomach. Serum was separated by centrifugation for 10 minutes at 3000 rpm immediately after collection. All samples were stored in polyethylene vessels at -20°C until analyzed.

Procedure

200 µl of serum was transferred into polyethylene tube and diluted with 200 µl of redistilled water. 10 µl of the solution was

injected into graphite tube furnace and the absorption was measured. Measurements were made with a Perkin Elmer 300 S spectrometer equipped with HGA 72 graphite tube furnace. Westinghouse hollow cathode lamp was used. The measurement conditions are given in Table 1.

Table 1: Measurement conditions

Wavelength	309.3 nm
Spectral bandwith	0.7 nm
Drying	150°C, 30 s
Ashing	1400°C, 50 s
Atomization	2700°C, 7 s
Purge gas	argon 50 ml min^{-1}
Sample volume	10 μl

All chemicals were of A.R. quality, acids were Merck "suprapur" quality. Redistilled water from Heraeus quartz apparatus was used.

All containers and pipette tips were cleaned with ultrapure nitric acid prepared in our laboratory by subboiling distillation in special apparatus made of quartz, and rinsed several times with redistilled water. After this procedure all vessels were tested for contamination. All sample treatments were carried out in clean bench.

For the determination of calcium, phosphorus, serum alkaline phosphatase, hemoglobin (Hb) and MCV standard methods used in our laboratory were performed. For the determination of serum PTH RIA method for the determination of C-terminal was used.

Results and Discussion

Aluminium is found in very low concentrations in body fluids and tissues in the normal population (7,8). Considerable deviation of serum aluminium concentrations can be found in chronic dialysis patients taking aluminium-containing phosphate binding agents. The main requirement for studying correlations between trace metal levels in human body and disorders are precise and accurate analytical procedures. In trace element analysis the accuracy can be affected to a large extent by contamination arising from sampling and analytical procedure. This is particularly important in determination of aluminium owing to its ubiquitous distribution. Exogenous contamination must be carefully taken into account and controlled during sample handling. Moreover, it should be emphasized that the risk of contamination increases with the number of manipulation steps in the analytical procedure. In our experiments we used ETAAS because of its simplicity and low detection limits. To study the accuracy of the ETAAS procedure the concentration of aluminium in serum samples was determined from an aqueous standard calibration curve and by a standard addition technique. As it can be seen from Figure 1 it can be concluded that under the experimental conditions

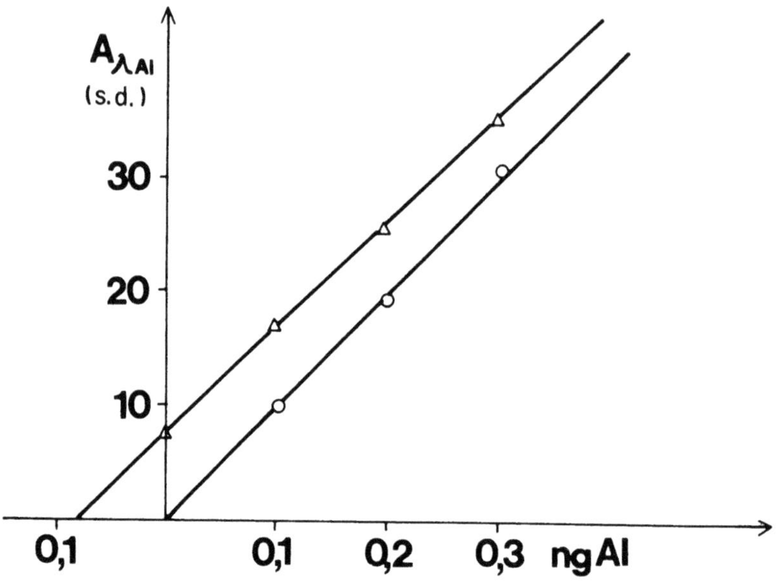

Figure 1: Calibration curve for aluminium in diluted (1+1) serum - Δ ;aqueous solution - O

used, other ions present in serum do not interfere with the determination of aluminium. The precision of the procedure for 9 determinations at 40 ng Al ml^{-1} in serum was 8 % (RSD); the statistical detection limit was 3 ng ml^{-1}.

To establish the possible source of contamination we tested the glass and polyethylene blood collection tubes. We filled the tubes with diluted nitric acid (0.001 $moll^{-1}$) and determined the concentration of aluminium over ten days period. The results of these experiments are shown in Figure 2.

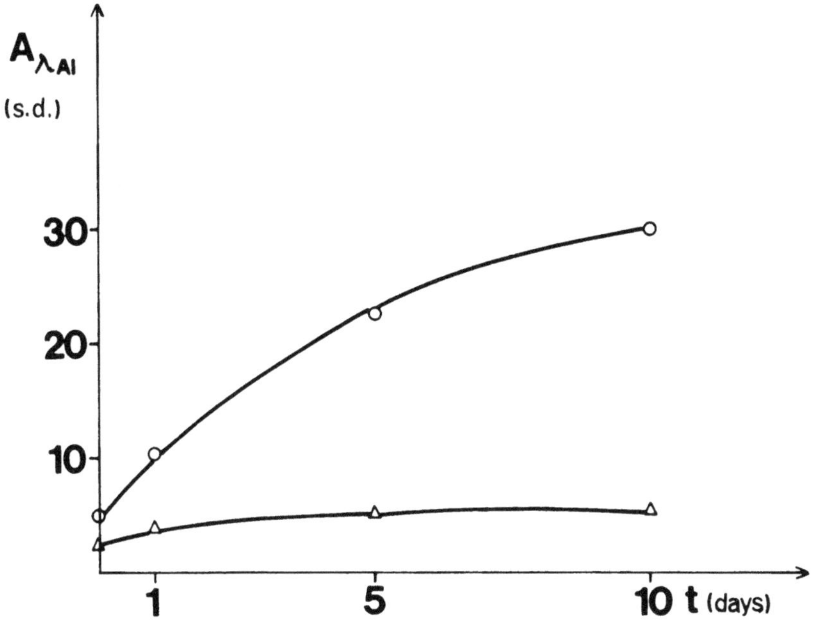

Figure 2: Contamination with aluminium in blood sampling devices: (O) glass, (Δ) polyethylene

It can be seen that an increase of aluminium concentration in glass vessels is observed, which can be ascribed to the leaching of aluminium from glass walls. Therefore the use of glass vessels must be avoided for the determination of aluminium in serum. We also tested disposable needles normally used in the hospital for blood sampling and no detectable amounts of aluminium were found neither in the plastic tube nor in the stainless steel needle.

The procedure described was applied to the determination of aluminium in serum of chronic hemodialysis patients. In our study the influence of two different phosphate binders: aluminium hydroxide and sucralfate on aluminium and phosphate serum levels was investigated. During the periods A and C the concentrations of aluminium, phosphorus, calcium, alkaline phosphatase, MCV were frequently determined. PTH concentration was determined only once. At the end of period B the concentration of aluminium, phosphorus, and calcium in serum were determined. The results for aluminium and phosphorus in serum of chronic dialysis patients treated in different ways are given in Table 2.

Table 2: Aluminium and phosphorus levels in patients treated with aluminium hydroxide (A) and sucralfate (C)

Period	Al ($\mu g\ l^{-1}$)	P ($mmol\ l^{-1}$)
A	69.71 ± 49.64^{x}	1.54 ± 0.28^{xx}
B	44.24 ± 26.62	2.09 ± 0.37
C	$90.88 \pm 79.81^{+,xx}$	$1.35 \pm 0.31^{++,xx}$
"Normal" values	< 3	0.8 ± 1.4

Statistical method: Student´s test:

A/B, C/B: $^{x}p < 0.02$ C/A: $^{+}p < 0.05$

$^{xx}p < 0.01$ $^{++}p < 0,01$

Serum aluminium concentrations during period A was $69.71 \pm 49.64\ \mu g\ l^{-1}$ and decreased to $44.24 \pm 26.62\ \mu g\ l^{-1}$ at the end of the period B (wash-out period, $p < 0.01$). The increase of serum aluminium concentration during period C to $90.88 \pm 79.81\ \mu g\ l^{-1}$ was significant ($p < 0.01$ versus period B an $p < 0.05$ versus period A). This significant decrease of aluminium level at the end of 1 week wash-out period shows that the intake of aluminium with aluminium hydroxide is initial for serum aluminium levels. This fact is corroborated by the repeated increase of the aluminium concentration during the period C when sucralfate was used.

The serum concentration of phosphorus during period A was 1.54 ± 0.28 mmol l^{-1}, and increased to 2.09 ± 0.37 mmol l^{-1} during period B ($p < 0.01$); during period C the serum concentration of phosphorus decreased to 1.35 ± 0.31 mmol l^{-1} ($p < 0.01$ versus period B and period A).

The concentration of phosphorus during sucralfate treatment was significantly lower compared to that during aluminium hydroxide treatment.

The results from Table 3 show that serum concentrations of calcium, Hb, PTH, and MCV during period A and C were not significantly different. The increase of concentration of alkaline phosphatase from 92.55±41.19 Ul^{-1} in period A to 103.44±49.5 Ul^{-1} in period C was significant ($p < 0,01$). It is well known that alkaline phosphatase level in blood is unspecific for diagnosis for uremic osteopathy.

Table 3: Serum levels of Ca, PTH, al.phosph., HB and MCV

Period	Ca (mmol l^{-1})	PTH ($\mu g\ l^{-1}$)	Alk.phosph. (Ul^{-1})	Hb (g l^{-1})	MCV (fl)
A	2.37±0.15	3.64±1.72	92.55±41.19	87.49±18.22	91.69±9.52
B	2.39±0.17	-	-	-	-
C	2.36±0.17	2.65±1.75	103.41±49.5	85.86±18.95	91.12±9.41

Statistical method: Student's test: C/A: $^{x}p < 0.01$

The aluminium concentrations in dialysis fluids were below 3 $\mu g\ l^{-1}$. The average pH value was 7.1±0.2. In our study the influence of aluminium in dialysis fluid on serum aluminium levels was found to be negligible.

The results of a comparative study of aluminium hydroxide and sucralfate show that equivalent doses of sucralfate decreased serum phosphate concentration more effectively than aluminium hydroxide. While using aluminium hydroxide and sucralfate the concentrations of aluminium in serum did not exceed the tolerance limit for chronic dialysis patients.

Acknowledgement

This work was supported by the Research Community of Slovenia, which is gratefully acknowledged.

References

1. Cartier, F., M.Chatel, P.Allain, 1981. In: Proceedings of 8th International Congress of Nephrology. Athens, p. 1022.

2. Savory, J., R.L. Bertholf, M.R. Wills. 1985. Clin. Endocrinol. Metab. 14, 681.

3. Gorsky, J.E., A.A. Dietz. 1981. Clin. Chem. 27/6, 932.

4. Kostyniak, P.J. 1983. J. Anal. Toxicol. 7, 20.

5. Giordano, R., S. Costantini, J. Vernillo, B. Casetta, F. Aldrighetti. 1984. Microchem. J. 30, 435.

6. Kayhan, G., A.J. Blotcky, E.P. Rack. 1978. Anal. Chem. 50, 144.

7. Versieck, J., R. Cornelis. 1980. Anal. Chim. Acta. 116, 217.

8. Berlyne, G.M., A.J. Adler. 1985. Am. J. Kidney Dis. 1985. VI/5, 288.

EXPOSURE OF INFANTS TO ALUMINIUM FROM MILK FORMULAE AND INTRAVENOUS FLUIDS

C.D. Hewitt, M. O'Hara and J.P. Day,
Department of Chemistry, University of Manchester, England.

N. Bishop,
Special Care Baby Unit, Hope Hospital, Manchester, England.

Introduction

Aluminium is a toxic element. It is also relatively abundant and of widespread occurrence, and the average adult may ingest about 5 mg/day from a normal diet (1). However, the element is poorly absorbed from the gut (ca. 0.3%) and efficiently excreted by the kidneys, leading to low retention and apparent absence of toxic effect in healthy adults (2).

Aluminium intoxication in patients with chronic renal failure is well established (3). Accumulation of the element in long term dialysis patients may cause encephalopathy, bone disease (osteomalacia), and microcytic anaemia. For such patients, concerted efforts are now made to control and reduce aluminium exposure (4). Preventive measures include the removal of aluminium from water supplies and pharmaceutical products used for dialysis or intravenous administration, reduction in the use of aluminium-containing antacids (frequently used for phosphate control), and regular screening of patients' aluminium status by serum-aluminium measurement.

The effects of aluminium accumulation and toxicity have also been seen in children and infants with reduced renal function, including premature infants in whom renal function may not be

Trace Element – Analytical Chemistry in Medicine and Biology, Vol. 4

fully developed (5). Elevated plasma, urine, and bone aluminium concentrations were found in premature infants who had received several weeks of intravenous therapy, and the source of exposure was ascribed to aluminium contamination of the intravenous solutions (6). Two fatal cases of aluminium toxicity in neonatal uraemia have been described (7), with elevated brain aluminium levels. These infants had been entirely orally fed, and had never received phosphate binders or intravenous fluids; the inferred source of exposure was the aluminium content of the powdered milk formulae, approximately 230 µg/l. In another case (8), severe bone aluminium toxicity in a uraemic baby resulted in respiratory failure due to thoracic wall instability; the suggested source of exposure was the use of an aluminium-containing phosphate binder. Presumably, the elevated gut absorption of the first few months, combined in some cases with immature or impaired renal function, predispose these infants to toxic levels of aluminium accumulation. Thus, the control of aluminium exposure for these patients would seem to be as important as for patients with chronic renal failure.

Recently, we have reported measurements of aluminium concentrations in milk formulae, sterile water and intravenous fluids commonly used in paediatric practice (9). In the present paper, we extend the scope of these data and calculate the consequential aluminium exposure in both orally fed infants and premature infants on total parenteral nutrition.

Analytical Methods.

Aluminium has been determined in the milk feeds and intravenous fluids used in our hospital. The selection covers various manufacturers and batches. Aluminium concentrations were determined by atomic absorption spectrometry, using graphite furnace atomisation and Zeeman effect background correction (Perkin Elmer Zeeman-3030 spectrometer). Samples were diluted appropriately with 0.1M nitric acid (for aqueous substrates) or, for fat-containing substrates, with 0.001M EDTA containing 0.5% aqueous Triton-X100 (a non-ionic detergent), and run against aluminium standards in appropriate matrices. The analytical response in the various sample types was tested by the addition of standard aliquots, and satisfactory aluminium recoveries (between 95 - 105%) were obtained.

Discussion: Milk Formulae.

Aluminium concentrations in milk formulae varied between 35 and 2080 µg/l (Table 1), similar to a previously reported range (6). Aluminium concentrations in the formulae intended for full term infants are generally lower than those in the specialised (pre-term and soy-based) formulae. The latter are also generally higher in calcium, potassium and phosphate, and it is probable that the aluminium contamination arises largely from the mineral additives.

These milk-aluminium levels result in infant aluminium intakes ranging from about 10 to 400 µg/kg body weight/day. Intakes at this level might not be regarded as excessive in adults, but although gut absorption factors for aluminium in children have not been established, higher gut absorption (10 to 100-fold) is normally found in infants compared to older children and adults. Additionally, with immature renal function, aluminium excretion will be reduced leading to excessive accumulation. Thus, Freundlich et al. (7) found elevated brain levels of aluminium in two infants after one and three months feeding on a milk formulae containing aluminium at 230 µg/l.

In hospital and home practice, powdered feeds are usually reconstituted with boiled tap water, hence the final level of

Table 1. Aluminium Content of Milk Formulae.

PRODUCT		ALUMINIUM - μg/l Range
Standard Formulae:		
Babymilk plus	(C & G)	88 - 107
Premium	(C & G)	107 - 191
Osterfeed	(Farley)	59 - 84
Ostermilk	(Farley)	191 - 370
Gold cap	(S.M.A.)	40 - 57
White cap	(S.M.A.)	35 - 63
Pre-term Formulae:		
Nenatal	(C & G)	499 - 660
Prematalac	(C & G)	168 - 192
Pregestimil	(Bristol Meyers)	1346 - 2080
Soya Protein Formulae:		
Formula S	(C & G)	419 - 1205
Wysoy	(Wyeth)	910 - 1330

aluminium in milk feed will depend also on the concentration of aluminium in the water supply. Whilst natural aluminium levels in potable water are commonly below 50 μg/l, in some areas much higher aluminium concentrations are common, and generally result from the use of aluminium sulphate in the water treatment process (10, 11). The EEC permitted maximum is currently 200 μg/l, but in the Manchester area over the past 8 years we have frequently found tap-water aluminium levels up to 400 μg/l, and on occasion up to 2000 μg/l (11). It is clear that if aluminium levels ca. 200 μg/l in milk formulae for pre-term infants are cause for concern (as seems to be the case), then consideration must also be given to the contribution from the mixing water. In this respect, the permitted maximum in tapwater (200 μg/l) seems excessive.

Discussion: Intravenous Solutions.

The aluminium content of a range of fluids used in our Units for parenteral nutrition is given in Table 2. Our results are similar to those reported by Sedman et al. (6) for comparable materials. As with the milk products, high aluminium concentrations appear to be associated particularly with the presence of calcium, potassium, and phosphate.

Table 2. Aluminium in Fluids for Parenteral Nutrition.

MATERIAL	ALUMINIUM CONCENTRATION - µg/l		
	Mean	Minimum	Maximum
Vamin 9 Glucose (KV)	86	44	127
Glucose 5%	2	1	3
Dextrose 5%	2	0	4
Dextrose 50% (Evans)	156	1	242
Addiphos (KV)	7172	7050	7293
Ped-el (KV)	1626	1336	1870
Solvito (KV)	29	9	69
Intralipid 10% (KV)	11	9	12
Vitlipid (KV)	307	307	307
Calcium gluconate 10% (PP)	4810	3430	6120
Potassium chloride 15% (PP)	212	164	290
Potassium phosphate 13% (McCarthy)	2130	1880	2390
Sodium chloride 0.9%	6	1	11
Water for injection	90	4	189

In Table 3, the aluminium intake (per kg body weight) which would result from the use of the intravenous solutions in Table 2 in a formulation for total parenteral nutrition (TPN) has been calculated. The example taken is typical of the formulation administered to premature infants of 1 to 2 kg body weight, and the resulting aluminium concentrations in TPN solutions range up to 150 µg/l, with an upper limit (worst case) concentration about 300 µg/l (cf. (6), 13 - 131 µg/l).

Table 3. Aluminium Intake (ug/kg/day) from Intravenous Feeds.

COMPONENT	DAILY INTAKE /ml		Al CONTENT /µg	
	Typical	Maximum	Typical	Maximum
Glucose (Vamin 9)	30.0	30.0	2.57	3.81
Dextrose 5%	117.0	140.0	0.20	0.56
Dextrose 50%	7.5	9.0	1.17	2.18
Addiphos	0.0	1.0	0.00	7.29
Ped-el	4.0	5.0	6.51	9.35
Solvito	0.5	0.7	0.01	0.05
Intralipid 10%	30.0	35.0	0.32	0.42
Vitlipid infant	1.0	2.0	0.31	0.61
Calcium gluconate	2.0	6.0	9.63	36.72
Potassium chloride	0.7	0.9	0.15	0.26
Potassium phosphate	1.7	2.2	3.63	5.25
Sodium chloride	16.0	20.0	0.10	0.22
Sterile water	0.0	20.0	0.0	3.78
TOTAL ALUMINIUM (µg/kg/day)			24.6	70.5

A major contribution to the total aluminium load arises from the use of calcium gluconate solution, which is used to supplement calcium. All samples of calcium gluconate we examined were relatively high in aluminium, but we found that the available samples of calcium chloride solution were low in aluminium (below 30 µg/l). We suggest that consideration could be given to the substitution of calcium gluconate, possibly by calcium chloride.

Another large contribution arises from the supplementation of phosphate, using potassium acid phosphate or 'Addiphos'. Although we did not find any paediatric potassium phosphate low in aluminium, samples of a more dilute preparation of this material for adult use contained aluminium at about 1% of the levels in the paediatric preparations. Clearly, manufacturing processes and quality control are an important factor in the elimination of aluminium contamination.

Aluminium intake from TPN is potentially more serious than from oral feeds, as retention may be much greater; a factor of 80% retention of intravenous aluminium has been quoted (6). Aluminium accumulation from contaminated parenteral nutrition solutions, and possibly consequential bone disease, has already been observed in adults with normal renal function (12); in premature infants the risk is likely to be significantly greater.

Conclusions.

There is significant aluminium contamination of infant milk formulae, particularly the specialised formulae used by premature and other infants, and commercial paediatric solutions for parenteral nutrition. Under particular circumstances, for example in pre-term infants, the aluminium content of these materials may lead to the absorption and retention of aluminium at toxic levels. We think that this presents a significant risk to children with impaired or immature renal function. This risk is surely avoidable. We re-iterate previous recommendations (9, 13) that manufacturers should make strenuous efforts to reduce the aluminium content of their products, and that specific recommendations for acceptable levels should be sought urgently.

Acknowledgement

We wish to thank the Kidney Unit Endowment Fund, Withington Hospital, Manchester, for their generous support of this research.

References

1. Greger, J.L., M.J. Baier. 1983. Food Chem. Toxicol. 21, 473.

2. Alfrey, A.C. 1983. Adv. Clin. Chem. 23, 69.

3. Kerr, D.N.S. and M.K. Ward. 1986. In: Aluminium and other Trace Elements in Renal Disease (A. Taylor, ed.). Bailliere Tindall, p. 1.

4. Commission of the European Communities. 1982. International Workshop on the Role of Biological Monitoring in the Prevention of Aluminium Toxicity in Man. Summary and conclusions in: Clin. Chem. Biochem. 20, 837.

5. Andreoli, S.P., J.M. Bergstein, D.J. Sherrard. 1984. New Eng. J. Med. 310, 1079.

6. Sedman, A.B., G.L. Klein, R.J. Merritt, N.L. Miller, K.O. Weber, W.L. Gill, H. Anand, A.C. Alfrey. 1985. New Eng. J. Med. 312, 1337.

7. Freundlich, M., G. Zilleruelo, C. Abitbol, J. Strauss. 1985. Lancet ii, 527.

8. Koch, H., H. Reich, D. Franke. 1985. Lancet ii, 831.

9. McGraw M., N. Bishop, R. Jameson, M.J. Robinson, M. O'Hara, C.D. Hewitt, J.P. Day. 1986. Lancet. i, 157.

10. Platts, M.M., G. Owen, S. Smith. 1984. Br. Med. J. 288, 969.

11. Day, J.P., M. O'Hara. Survey of Aluminium Concentrations in Water used for Home Dialysis by 200 Patients in the Manchester Area, 1978 - 1986. Reported annually to Hospital Dialysis Units but otherwise unpublished.

12. Ott, S.M., N.A. Maloney, G.L. Klein. 1983. Ann. Intern. Med. 98, 910.

13. Fell, G.S., A. Shenkin, D.J. Halls. 1986. Lancet, i, 380.

INTRAVENOUS INFUSIONS CAN CAUSE ALUMINIUM TOXICITY

B. Sampson[*], E.R. Maher[**], and J.R. Curtis[**]
Departments of Chemical Pathology[*] and Medicine[**], Charing Cross Hospital, Fulham Palace Road, London, W6 8RF, U.K.

Introduction

Aluminium toxicity in patients with chronic renal failure (crf) is now a well known problem (1). It is commonly due to raised aluminium concentrations in the dialysate due to the use of water with a high aluminium content (2) or to the use of orally administered aluminium hydroxide drugs to control phosphate absorption from the gut (3). We have previously described the more unusual problem of aluminium toxicity during the use of a sorbent regenerated ('REDY') dialysis system (4). Raised plasma aluminium concentrations in crf patients being treated with haemoperfusion and in patients with normal renal function receiving total parenteral nutrition (tpn) have also been described (5,6). We have recently reported a crf patient on haemodialysis receiving high volume albumin infusions who developed potentially toxic plasma aluminium concentrations (7). In this instance clinical toxicity did not result, but this may have been due to the early recognition of the source of the aluminium as the albumin infusions. We have now carried out a study of the effect of such infusions on the plasma aluminium concentration in haemodialysis patients.

Methods

Stable haemodialysis patients all with an initial plasma aluminium concentration of less than 2μmol/L were given an infusion of 200ml of 20% human albumin (Blood Products Laboratory, Elstree, Herts, U.K.) during a dialysis session. As a control an equivalent volume of 0.9% saline was infused during a separate

Trace Element – Analytical Chemistry in Medicine and Biology, Vol. 4

dialysis session. Blood was taken for biochemical estimations at the beginning and end of each session. A sample of each batch of infusate was stored for aluminium analysis. Aluminium was determined by atomic absorption spectroscopy with electrothermal atomisation (Perkin-Elmer 3030 with HGA500 furnace and AS40 autosampler). Plasma samples were diluted 1:4 in water, albumin solutions and some salt solutions required a higher dilution; in some instances ammonium nitrate was added as a matrix modifer. The HGA 500 programme used for aluminium assays is given in table1.

Table 1 Aluminium Assay Conditions

Step	Temperature (°C)	Ramp (sec)	Hold (sec)
1[a]	90	5	10
2	120	20	10
3	750	20	10
4	1200	10	10
5[b]	2450	0	3
6	2600	1	5
7	20	1	5

a 10 ul sample injected into pyrolytic coated tube, internal gas at 300ml/min except in step 5.

b Max power heating, gas stop, read peak area

Ultrafiltration of albumin solutions performed using Sartorius membrane filters (SM 13200). Membranes were freed of endogenous aluminium before use by extensive soaking in 20mM EDTA and 10mM nitric acid followed by a water rinse until no aluminium could be detected in the ultrafiltrate. A 5ml aliquot of sample was placed in the cone and the first 0.5-1ml of ultrafiltrate was discarded. The following 2ml of ultrafiltrate was collected as separate fractions and analysed for aluminium content. Albumin was not detected in the ultrafiltrate.

Results

The effect of the infusion of albumin solutions on the plasma albumin and aluminium concentrations of the haemodialysis patients is shown in table 2. Although the absolute changes in aluminium and albumin concentration seen do not differ significantly between the two groups due to the wide spread of initial concentrations, if the net change is considered then the albumin infused patients show a significant increase in both albumin and aluminium. The increase in plasma aluminium seen in these patients does not represent the total amount of aluminium infused. As can be seen from the date in table 3, summarising the aluminium content of some blood products, the aluminium content of 200ml of 20% albumin is 20-25µmol. If this were wholly retained in the plasma compartment the expected increase in plasma aluminium would be 3-5µmol/L.

Table 2 Changes In Plasma Aluminium And Albumin After Intravenous Infusions

		Aluminium µmol/L	Albumin g/L
Saline	Pre	0.71 ± 0.40	35.9 ± 3.6
(n=8)	Post	0.78 ± 0.38	36.9 ± 5.5
	Change	0.07 ± 0.05	0.9 ± 2.9
Albumin	Pre	0.84 ± 0.55	33.9 ± 4.4
(n=8)	Post	1.09 ± 0.46	39.6 ± 5.7
	Change	0.25 ± 0.18[a]	5.8 ± 2.1[b]

a) $p < 0.02$ b) $p < 0.01$ (paired t test)

Table 3 Aluminium Content Of Some Blood Products

Sample	Aluminium (µmol/L)	
4.5% Human Albumin	10.5 ± 2.6	(n=13)
20% Human Albumin	114.7 ± 25.4	(n=4)
Packed Red Cells	1.2 ± 0.5	(n=3)
Whole Blood	1.5	(n=1)

In order to identify a possible cause of the lower than expected increase in plasma aluminium the proportion of ultrafiltrable aluminium in the 4.5% albumin was determined. This was found to be 90-95% of the total aluminium. The 20% albumin used in this study was considered too viscous for ultrafiltration, but can be expected to contain a similar proportion of ultrafiltrable aluminium.

The aluminium content of other solutions intended for intravenous administration is shown in table 4. Many of the solutions have a relatively low aluminium content but some, especially the phosphate salts and calcium gluconate, have a very high aluminium content and will account for much of the aluminium in the tpn solutions. It is also of interest to note that there is not an inconsiderable aluminium content in the heparin administered to the haemodialysis patients regularly and in the desferrioxamine used to treat aluminium toxicity.

Table 4 Aluminium Content Of Solutions For Intravenous Administration

Sample	Aluminium (μmol/L)
5% dextrose	<0.1
0.9% saline	<0.1
water for injection BP	0.63
6% sodium chloride	0.80
12.5% sodium chloride	0.58
30% sodium chloride	0.46
15% potassium chloride	1.06
8.7% sodium phosphate	21.0
17.4% sodium phosphate	82.0
17.4% potassium phosphate	38.1
10% calcium gluconate	172.5
10% magnesium sulphate	0.82
'Solvito'	0.63
'Multibionta'	2.75
Trace element supplement	7.8
TPN solution	1.4
Heparin 1000IU/ml	2.20
Desferrioxamine 6g/L	5.04

Discussion

The increment in plasma aluminium seen in these patients is, as mentioned previously, not as high as would be expected were the total amount of aluminium infused retained in the plasma. This may be due to the high proportion of ultrafiltrable aluminium in the albumin infusate which may lead to a rapid clearance of aluminium during the dialysis.

There may also be a rapid tissue uptake of aluminium, possibly by binding to transferrin, the major serum binding protein for aluminium (8,9). It is possible that some patients' serum may exhibit an increased affinity for aluminium due to the acute phase response, which has amongst its biochemical features a decrease in the saturation of transferrin by iron (10,11). We have previously found that up to 50% of all dialysis patients have biochemical evidence of the acute phase response as measured by serum C-reactive protein concentrations (12), although the levels in haemodialysis patients are generally not high.

The high concentrations of aluminium found in the albumin preparations and in other solutions intended for parenteral use is in agreement with the results reported by other workers (13,15). However the load of aluminium administered even to crf patients by the occassional use of the solutions is not high in most instances. Patients with normal renal functions are able to excrete high loads of aluminium with little retention (16,17).

The amount of aluminium is undoubtedly high in many of these preparations and some manufacturers are now taking steps to reduce this (personal communication, Blood Products Laboratory). The origin of the aluminium has not been established in all cases, but in one instance the filters used in purification have been implicated (18).

Aluminium is not alone in being a contaminant of such fluids: nickel (19), manganese (20) and chromium (Sampson, Barlow and Wilkinson, unpublished observations) have also been shown to be present at high concentration in blood products and tpn solutions. The same may be true for other trace elements.

References

1. Alfrey, A.c., G.R. LeGendre, W.D. Kaehny. 1986. N.Engl.J.Med. 264, 184.

2. Platts, M.M., G.C. Goode, J.S. Hislop. 1977. Br.Med.J. 2, 657.

3. Fleming, L.W., W.F. Stuart, G.S. Fell, D.J. Halls. 1982. Clin.Nephrol. 17, 222.

4. Sampson, B., E.A. Brown, J.R. Curtis, M.E. Phillips, N.R.B. Cary, A.C.E. McCartney. 1985. In:Trace Elements in Man and Animals, Vol 5 (C.F. Mills, I. Bremner and J.K. Chesters, eds) Commonwealth Agricultural Bureau, p. 830.

5. Elliot, H.L., A.I. MacDougall, G.Haase, R.L.C. Cumming, P.H.E. Gardiner, G.S. Fell. 1978. Lancet 2, 940.

6. Klein, G.L., A.C. Alfrey, N.L. Miller, D.J. Sherrard, T.K. Haylett, M.E. Ament, J.W. Coburn. 1982. Am.J.Clin.Nut 35, 1425.

7. Maher, E.R., E.A. Brown, J.R. Curtis, M.E. Phillips, B. Sampson. 1986. Br.Med.J. 1, 306.

8. Trapp, G.A. 1983. Life Sci. 33, 311.

9. Rahman, H., A.W. Skillen, S.M. Channon, M.K. Ward, D.N.S. Kerr. 1985. Clin.Chem. 31, 1969.

10. Dinarrello, C.A. 1984. Rev.Infect.Diseases 6, 51.

11. Myers, M., A. Fleck, B. Sampson, C.M. Colley, J. Bent, G. Hall. 1984. J.Clin.Path 37, 862.

12. Sampson, B., J.R. Curtis, B.R. Muller, G.D. Carter, J. Alaghband-Zadeh. 1986. In:Aluminium and other Trace Elements in Renal Disease (A. Taylor, ed.) Bailliere-Tindall, p. 348.

13. Fuchs, C., V.W. Armstrong, E. Quellhorst, F. Scheller. 1984. Contr.Nephrol. 38, 81.

14. Sedman, A.B., G.L. Klein, R.J. Merritt, N.L. Miller, K.O. Weber, W.L. Gill, H. Anand, A.C. Alfrey. 1985. N.Engl.J.Med. 312, 1337.

15. McGraw, M., N. Bishop, R. Jameson, M.J. Robinson, M. O'Hara, C.D. Hewitt, J.P. Day. 1986. Lancet 1, 158.

16. Kovalchick, M.T., W.D. Kaehny, A.P. Hegg, J.T. Jackson, A.C. Alfrey. 1978. J.Lab.Clin.Med. 22, 712.

17. Henry, D.A., W.G. Goodman, R.K. Nudleman, N.C. DiDomenico, A.C. Alfrey, E. Slatopolsky, T.M. Stanley, J.W. Coburn. 1984. Kidney. Int. 25, 362.

18. Milliner, D.S., F. Feldman, J.H. Shinaberger, J.W. Coburn. 1985. N.Engl.J.Med. 312, 1390.

19. Leach, C.N. Jr., F.W. Sunderman. 1985. N.Engl.J.Med. 313, 1232.

20. Sampson, B., G.B. Barlow, A.W. Wilkinson. 1983. Pediatr.Res. 17, 263.

SPECIATION OF TRACE ELEMENTS IN HUMAN TISSUES: ROLE OF METALLOTHIONEIN

Herbert E. Heilmaier(1), Peter Schramel(2), Gustav A. Drasch(3), Elisabeth Kretschmer(3) and Karl H. Summer(1)

(1) Institut für Toxikologie und (2) Institut für Ökologische Chemie der Gesellschaft für Strahlen- und Umweltforschung, D-8042 Neuherberg; (3) Institut für Rechtsmedizin, Universität München, D-8000 München 2

Introduction

In mammalian organisms the vast majority of trace elements like essential Zn and Cu and toxic Cd are bound to biological ligands. The biological activity and toxicity of trace metals therefore is determined by both their concentrations in the tissues and their actual binding forms (speciation). A bioligand of particular high affinity for Zn, Cu and Cd is metallothionein (MT) an unique protein of low molecular weight and high cysteine content [1,2]. In mammals MT has been detected in all tissues investigated and high induction of the protein can be achieved by metals like Zn, Cu and Cd [3]. There is a considerable lack of information on MT levels and MT-inducing capacities of human tissues. In the present investigation we report on the content of MT and the levels of Zn, Cu and Cd in 10 tissues of man. The results are compared with data previously reported for the rat [4].

Materials and methods

Samples of human tissues were obtained from 10 male individuals (20-50 years old, mean age 43 ± 9, all smokers) necropsied usually within 2 days after sudden death. None of the individuals had intensive medical treatment before death. Tissue samples were stored at -80°C until being processed further.
Human and Wistar-rat (males, 180-200 g) tissues were homogenized

Trace Element – Analytical Chemistry in Medicine and Biology, Vol. 4

(Potter-Elvehjem or Ultra-turrax) in 4 or 9 volumes of 30 mM TRIS-HCl buffer, pH 7.4 and subsequently centrifuged at 18,000 g. MT was determined in the tissue supernatants by a Cd saturation assay as described previously [5].
Zn, Cu and Cd were detected in tissue homogenates by atomic emission spectroscopy with inductively coupled plasma excitation (ICP-AES) [6] or atomic absorption spectroscopy (AAS) after wet ashing of the samples at 140°C with nitric acid [7]. Cd content of all rat tissues was lower than the detection limit of 20 ng Cd/g tissue wet wt.

Results and discussion

MT was detected in all human tissues investigated (Fig. 1). The mean tissue content of MT varied between 4 µg/g wet wt. in lung and spleen and 495 µg/g wet wt. in kidney cortex. In all tissues except small intestine and brain higher levels of MT were observed in tissues of man than in rat.
High MT levels as compared to the rat particularly in human kidney cortex and liver may reflect a different transcriptional control of the protein but the high exposure of man to environmental metals may contribute as well. Particularly kidneys and liver play a crucial role in the sequestration of trace elements like Cd [8] and these organs exhibit high MT synthesizing capacities [9].
The levels of Zn, Cu and Cd in the tissues investigated are shown in Fig. 2. Human tissues had Zn levels between 10 and 52 µg/g tissue wet wt. being highest in liver. Mean levels of Cu varied between 1 and 6.2 µg/g tissue wet wt. again being highest in liver. Pronounced variations were seen in the Cd levels variing from 0.1 µg/g in brain and heart to 1.0 µg/g in liver and 30 µg/g tissue wet wt. in kidney cortex.
As compared to rats, Zn and Cu levels of human tissues except kidney were approximately similar within a factor of three. Both metals showed higher levels in human liver, whereas Cu in human kidney cortex approximated 25 % of the respective rat value only. Due to low Cd levels in rat tissues, the Cd contents of all human tissues were remarkable higher.

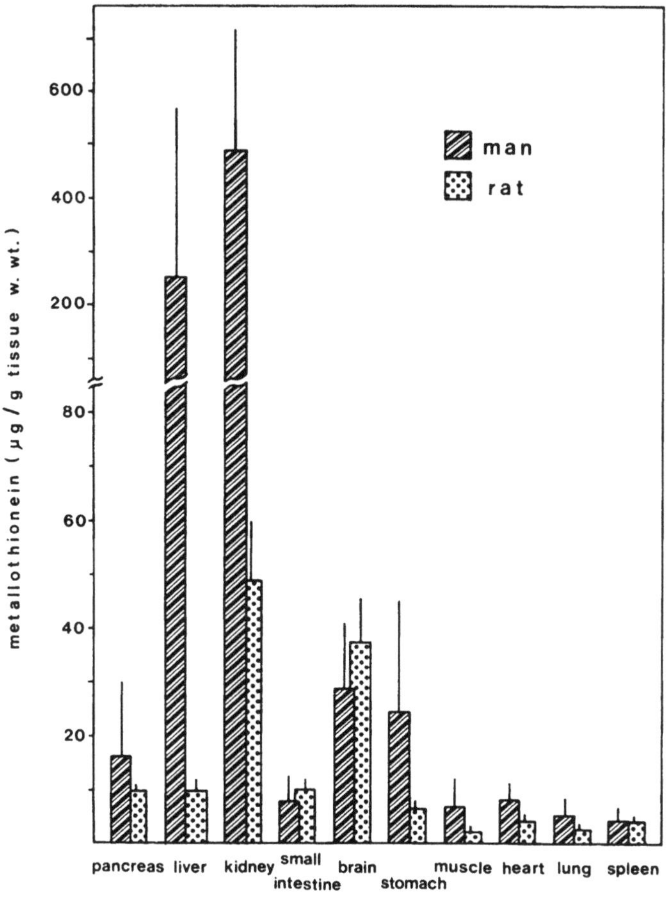

Figure 1

Metallothionein content in 10 tissues of man and rat. Samples were obtained from male individuals autopsied (n=10) and male Wistar rats (n=6). Metallothionein was determined by the Cd saturation assay as previously described [4]. Values are expressed as means and S.D.

Particularly in human liver and kidney cortex high levels of Cd are associated with high levels of MT. This may suggest that Cd promoted MT induction in these tissues and confirms earlier findings of a linearly increased content of renal cortex MT with increasing tissue Cd due to smoking [10].

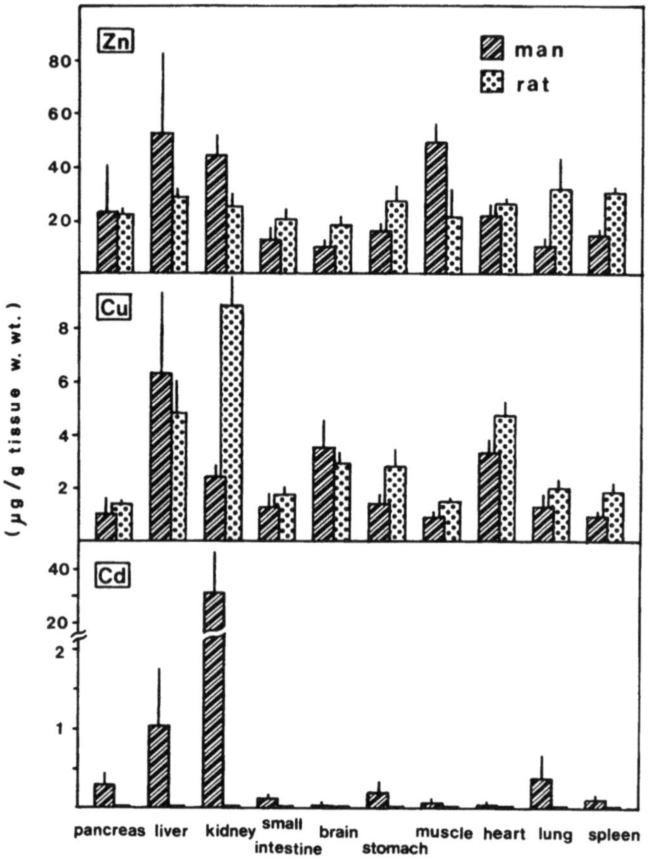

Figure 2

Zn, Cu and Cd content of 10 tissues of man and rat. Samples were obtained from male individuals autopsied (n=10) and male Wistar-rats (n=6). Metals were determined by ICP-AAS or AAS after wet ashing of the samples with nitric acid. The Cd content of all rat tissues was lower than the detection limit of 20 ng Cd/g tissue wet wt. Values are expressed as means and S.D.

MT is always saturated with metals. The ratio of metals in the protein is subject to the tissue concentration of the metals and their relative affinity to the protein. MT in human liver and kidney cortex at most may bind 58% and 80% of Zn and more than 100% of Cu and Cd present in the respective homogenates. The possibility that MT may bind total Cd present in liver and kidney cortex supports the presumptive protective role of MT in the detoxication of Cd. However according to our previous findings [10], MT in the renal cortex of smokers does not contain total tissue Cd. Non-MT Cd as a consequence may significantly contribute to the toxicity of the metal.

In conclusion, these results demonstrate that, similar to rat, MT is present in various tissues of man. Furthermore, high levels of both MT and Cd in liver and kidney cortex of man suggest that MT may be inducible by Cd also in human tissues. Increased MT formation as a consequence will enlarge intracellular metal binding sites but also may cause adverse effects by disturbing the homeostasis of endogenous and toxic metals.

Acknowledgement

The authors wish to thank Mrs. Marianne Ederer, Mrs. Gisela Lill and Mr. Josef Lichtmannegger for expert technical assistance.

References

1. Kägi, J.H.R., M. Nordberg (eds.). 1979. Metallothionein. Birkhäuser, Basel.

2. Foulkes, E.C. (ed.). 1982. Biological Roles of Metallothionein. Elsevier/North-Holland, New York.

3. Eaton, D.L., N.H. Stacey, K.L. Wong, C.D. Klaassen. 1980. Dose-response effects of various metal ions on rat liver metallothionein, glutathione, heme oxygenase, and cytochrome P-450. Toxicol. Appl. Pharmacol. 55, 393-402.

4. Heilmaier, H.E., K.H. Summer. 1985. Metallothionein content and zinc status in various tissues of rats treated with iodoacetic acid and zinc. Arch. Toxicol. 56, 247-251.

5. Heilmaier, H.E., P. Schramel, K.H. Summer. 1984. In: Trace Element-Analytical Chemistry in Medicine and Biology, Vol. 3 (P. Brätter, P. Schramel, eds.) Walter de Gruyter, Berlin, New York. pp. 669-676.

6. Schramel, P., B.J. Klose, S. Hasse. 1982. Die Leistungsfähigkeit der ICP-Emissionsspektroskopie zum Nachweis von Spurenelementen in biologisch-medizinischen und in Umweltproben. Fresenius Z. Anal. Chem. 310, 209-216.

7. Schramel, P., A. Wolf, R. Seif, B.J. Klose. 1980. Eine neue Apparatur zur Druckveraschung von biologischem Material. Fresenius Z. Anal. Chem. 302, 62-64.

8. Waku, K. 1984. The chemical form of cadmium in subcellular fractions following cadmium exposure. Environ. Health Perspect. 54, 37-44.

9. Onosaka, S., M.G. Cherian. 1981. The induced synthesis of metallothionein in various tissues of rat in response to metals. I. effect of repeated injection of cadmium salts. Toxicology 22, 91-101.

10. Summer, K.H., G.A. Drasch, H.E. Heilmaier. 1986. Metallothionein and cadmium in human kidney cortex: Influence of smoking. Human Toxicol. 5, 27-33.

SPECIATION OF CADMIUM AND COPPER IN LETTUCE LEAVES

Wolfgang Henze and Fritz Umland

Institute of Inorganic Chemistry
University of Münster, FRG

Summary

Cadmium and copper-binding compounds were isolated from commercial lettuce by gel filtration of the cytosol on Sephadex G-75. The metals were analyzed by electrothermal atomic absorption spectroscopy (ET-AAS).
The majority of cadmium is bound to a low molecular weight protein of about 3,200 dalton and to a high molecular weight protein of more than 75,000 dalton. The first one is heatstable and has a low content of aromatic amino acids. Ion exchange chromatography of this compound on DEAE-Sephadex A-25 leads to one single protein containing cadmium and copper.

Introduction

The analysis of cadmium and other trace elements in food, including plants, has been elucidated over the past 25 years. Concerning the bioavailability, however, not only the total content, but also the species of a certain trace element is of special interest. Speciation is therefore required for both, toxic and essential elements. Since the first characterization of metallothionein by Kägi and Vallee in 1960 (3), this protein that binds a number of heavy metals has been in the centre of speciation research. It has been

Trace Element – Analytical Chemistry in Medicine and Biology, Vol. 4

detected in various creatures, even in plants which have been cultured in heavy metal enriched solutions (1,2,4,5).
In this report, a separation procedure is presented that allows the isolation and quantification of cadmium and copper-species in commercially available native lettuce (Lattuca sativa).

Materials and Methods

Chemicals: All chemicals were of high analytical grade. Nitric acid, hydrochloric acid, potassium chloride and sodium azide were from E. Merck, Darmstadt, FRG, Tris-(hydroxymethyl)-aminomethane and cadmium and copper standard solutions were from Riedel-de Haën, Seelze, FRG, Sephadex G-75, DEAE-Sephadex A-25, and a protein calibration kit with proteins of specified molecular weights were purchased from Pharmacia Fine Chemicals, Uppsala, Sweden.
To prevent metal contamination, all buffers and dilutions were prepared with double distilled water.

Analytical methods: Metal analysis was performed with an atomic absorption spectrophotometer (Perkin Elmer 400 equipped with graphite furnace HGA 72, Überlingen, FRG) after sample digestion with nitric acid and hydrogenperoxide (6) using the standard addition method. Chromatographic fractions were measured directly against calibration curve. It was prooved that there was reasonable agreement with wetashed samples. UV-spectra were obtained using a double beam spectrophotometer (Perkin Elmer 320). Protein contents were estimated according to their 280 nm absorption. Ultrafiltration was performed in an Amicon ultrafiltration device, containing a membrane Diaflo YM 2 with an exclusion limit of 1,000 dalton (Amicon, Danvers, USA). Molecular weights of the metal binding proteins were determined on a Sephadex G-75 superfine column 930 mm x 16 mm Ø in Tris-HCl buffer pH 8.0, 20 mmol. The column was calibrated with Ribonuclease A, mol. wt. 13,700, Chymotrypsinogen A, mol. wt. 25,000, Ovalbumin, mol. wt. 43,000 and Bovine Serum Albumin, mol. wt. 67,000.

Isolation procedure: Fresh commercial lettuce was homogenized in Tris-HCl buffer pH 8.0, 20 mmol, sodium azide 1 mmol (1 : 2 w/v), using a Warring Blender (Braun, Frankfurt/M., FRG) and an Ultra Turrax (IKAA-Werk Stauffen, FRG), each 4 minutes. The homogenate was centrifuged for 90 minutes at 100,000 g (Beckmann ultracentrifuge Spinco L2 65, USA). The cytosol S 100 was separated by gel filtration on a Sephadex G-75 column 730 mm x 30 mm Ø at a flow rate of 17.5 ml/h. Afterwards, the low molecular weight protein was chromatographed on a DEAE-Sephadex A-25 column 200 mm x 30 mm Ø. After sample application, the column was eluted with buffer at a flow rate of 12.0 ml/h, then a linear KCl-gradient 0 - 0.6 mol/l was introduced. Elution profiles were detected at 254 and 280 nm, using a dual path UV monitor (UV-2, Pharmacia Fine Chemicals). After each clean-up step, the cadmium and copper content was determined. The total isolation procedure is demonstrated in figure 1.

Results and Discussion

The Sephadex G-75 chromatography yielded two main cadmium and copper-binding proteins (see fig. 2). The high molecular weight protein (>75,000 dalton) was not further examined. Besides, there was some copper bound to proteins of intermediate size. But its concentration was very low so that nonspecific adsorption of copper cannot be excluded. In some samples, there was also cadmium eluting at 600 ml, the elution volume of free metal ions. The molecular weight of the smaller one, containing the majority of cadmium and copper, was determined to about 3,200 dalton. Although the copper containing protein (Cu-BP) eluted earlier than the cadmium species (Cd-BP), ion exchange chromatography led to one single protein containing cadmium and copper (see fig. 3). This protein eluted at a chloride concentration of 0.22 mol/l. The cadmium analysis demanded a desalting of the sample by ultrafiltration. The different elution behaviour on Sephadex G-75 is probably caused by small changes in the protein structure depending on the bound metal.
Thermal denaturation experiments showed that this protein is heat-stable. Electron spectra show a relative low absorption at 280 nm, indicating a low content of aromatic amino acids (see fig. 4). The UV-spectrum resembles that of <u>Agrostis gigantea</u> metallothionein,

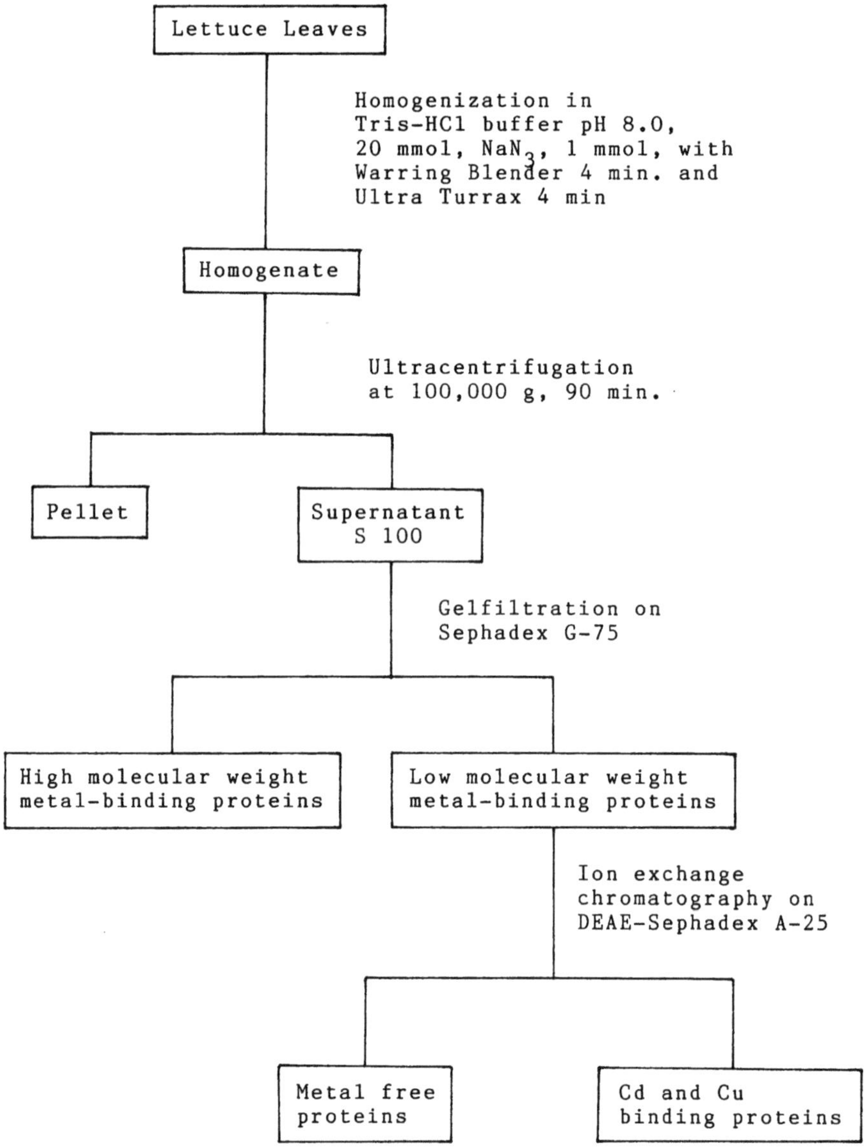

Figure 1: Isolation procedure of cadmium and copper-binding compounds from lettuce leaves

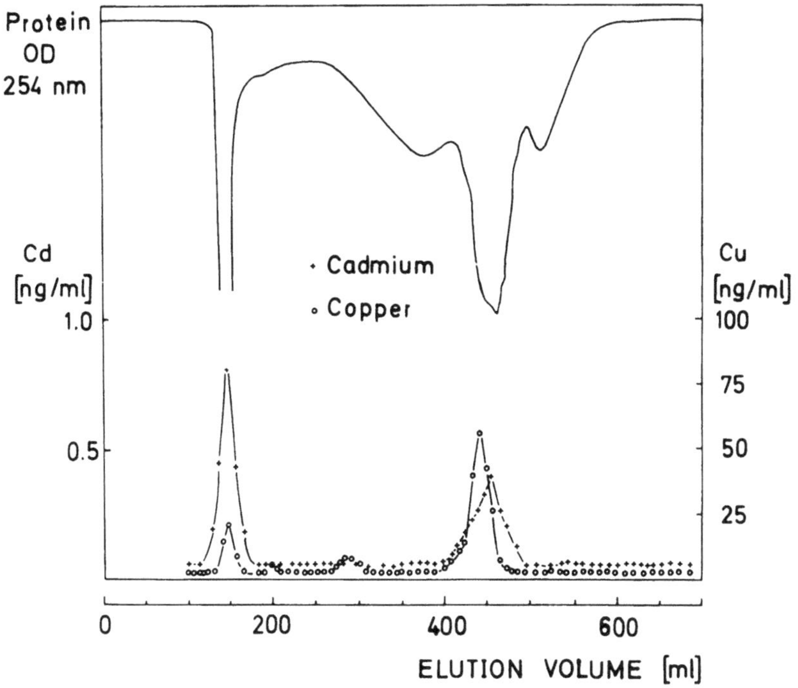

Figure 2: Sephadex G-75 chromatography of lettuce cytosol

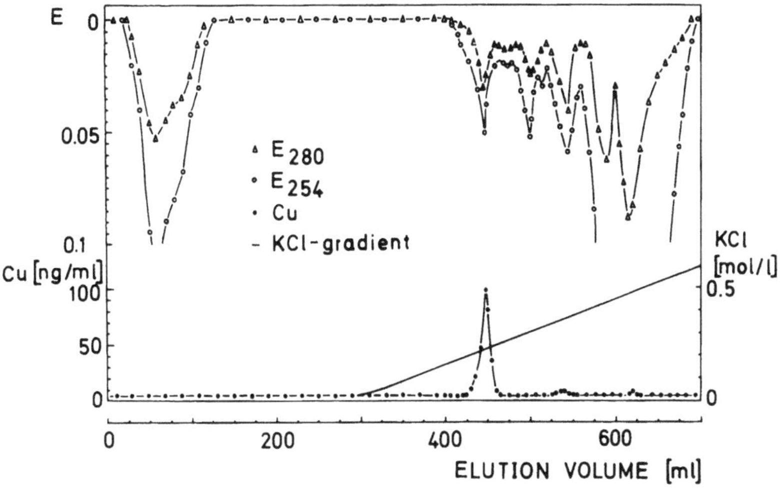

Figure 3: DEAE-Sephadex A-25 elution pattern of the low molecular weight Cd-BP and Cu-BP after Sephadex G-75 chromatography

fully loaded with copper (2). These facts lead to the supposition that the isolated protein is a phytometallothionein.
Quantitative estimations of the metal-binding protein content and the determination of the metal/protein stoichiometry failed, because it was impossible to quantify the protein content by either biuret or dye binding. Besides, the content of such proteins in native lettuce is rather low.

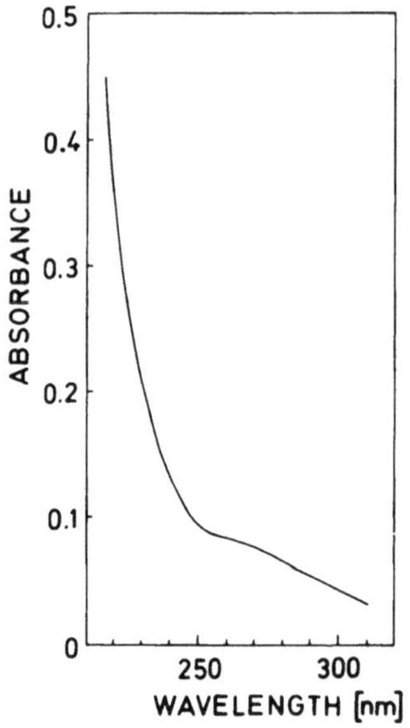

Figure 4: UV absorption spectrum of the Cd-BP and Cu-BP

Acknowledgement

The support of this study by the German Research Society DFG is kindly remarked.

References

(1) Bartolf, M., Brennan, E., and Rice, C. A. (1980) Plant Physiol. 66, 438 - 441

(2) Casterline, J. L., Barnett, N. M. (1982) Plant Physiol. 69, 1004 - 1007

(3) Kägi, J. H. R., Vallee, B. L. (1960) J. Biol. Chem. 235, 3460 - 3464

(4) Rauser, W. E., Curvetto, N. R. (1980) Nature 287, 563 - 564

(5) Rauser, W. E., Hartmann, H.-J., and Weser, U. (1983) FEBS Lett. 164, 102 - 104

(6) Windemann, H., Müller, U. (1975) Mitt. Geb. Lebensmittelunters. Hyg. 66 (1), 64 -68

THE IN-VITRO AVAILABILITY OF COPPER, ZINC AND SELENIUM AND THE SPECIATION OF SELENIUM IN FISHERY PRODUCTS

J.B. Luten, W. Bouquet, M.M. Burggraaf, J. Rus

TNO-CIVO Food Technology Institute, Department Fish Technology Institute, P.O.Box 183, 1970 AD Ymuiden, the Netherlands

Introduction

Evaluation of diets for adequate trace element intake requires knowledge of both the amount, the chemical form (speciation) and availability of trace elements in foods.

While information on the trace element content of foods is reasonably adequate, knowledge of the availability and speciation is still incomplete.

Measuring the availability of trace elements by human balance studies is tedious and susceptible to errors. The measurement of a trace element in plasma after an oral dose is a rather simple technique for estimating the availability. However the amount entering the blood does not necessarily reflect the amount retained and often non-physiological amounts of the trace element are tested. Absorption studies using radionuclides can be useful however the administration of radionuclides to humans is not always possible. This disadvantage of radionuclides can be overcome using stable isotopes.

An interesting alternative for measuring the availability is the use of in-vitro techniques. In-vitro methods have been used in particular for estimating the availability of the essential element iron (1-3).

Crews et al (4) reported a detailed enzymolysis and acid digestion procedure for measuring the solubility of lead, cadmium, zinc, iron and copper in different groups of food. The results showed that the amount of soluble analyte varies from element to element and depends on the type of food. The solubility of the elements added as their inorganic salts was often different from that of the endogenous element.

In two recent detailed studies Crews et al (5, 6) determined the solubility of the elements in some individual food items and in combinations of certain food items.

However detailed information about the in-vitro availability of essential elements like zinc, copper and selenium in important fishery products is still incomplete. In this paper the results of an in-vitro availability study of zinc, copper and selenium in mussels, shrimps and cod, are presented.

Trace Element – Analytical Chemistry in Medicine and Biology, Vol. 4

As already mentioned the chemical form of the trace element is important. Therefore the need for speciation analysis of trace elements in foods is still growing. Combination of fast separation techniques like high performance liquid chromatography with sensitive elemental detection methods gives nowadays the opportunity to gather more knowledge in the field of trace element speciation.

It is known that selenium occurs naturally in some foods in organic compounds, the primary forms of which are selenomethionine, Se-methylselenomethionine, selenocystine and selenocysteine (7, 8). Speciation of selenium in fishery products has been studied by Cappon and Smith (9, 10) and Braddon (11). Cappon and Smith (9, 10) studied the distribution of selenium in some specific protein fractions of fish and other marine animal organisms (molluscs, crusteceans and pods). For all fish samples 55-60% of the total selenium content was water-extractable while for other marine species only 35-40% was water-extractable. Braddon (11) isolated and purified selenoproteins from the livers of black tip shark and tuna. It was shown that in the shark liver selenium was present in two proteins with a molecular weight of 26,000 and 80,000 Dalton. Selenium in tuna liver was present in one protein fraction with a molecular weight of approx. 2,500 Dalton. More information about the selenium speciation in fishery products is needed. In this paper the preliminary results of a speciation study of selenium in some fishery products are presented.

Experimental

<u>Materials</u>. Cod (Gadus morhua), herring (Clupea harengus), plaice (Pleuronectes platessa), mackerel (Scomber scombrus), mussels (Mytilus edulis) and shrimps (Crangon crangon) were purchased from local retailers. Approx. 2.5 kg of the edible part was collected and homogenized. All homogenized samples were stored at -20°C in polyethylene containers until required.

From Sigma Chemicals pepsine (Porcine; cat.no. P7000), pancreatine (Porcine; cat.no. P1750), amylase (Porcine; cat.no. A6880) and bile salts (cat.no. B8756) were purchased. All other chemicals were from Merck (Suprapur) and Baker (Baker Analyzed).

Simulated gastric juice was prepared by dissolving pepsine (10 g/l) and sodium chloride (8.8 g/l) in bidistilled water and adjusting the pH to 1.8 with hydrochloric acid (36%).

Simulated intestinal juice was prepared by adding equal volumes of a saline solution (8.8 g NaCl/l) of pancreatin (30 g/l) and amylase (10 g/l) to a saline solution (8.8 g NaCl/l) of bile salts (1.5 g/l).

In-vitro digestion procedure

All determinations were carried out 4-6 times on mussels, shrimps and cod. The in-vitro digestion procedure was performed according to the method of Crews et al (5, 6).
Fifty grams of the samples were weighed into stomacher bags. To each sample 100 ml gastric juice was added. The digests were pummeled for one minute and then gently shaken during 4 hours at 37°C. After one hour the pH was measured and if necessary adjusted to pH 3.0. At the end of the gastric juice digestion 20 ml of the digest was transferred to polycarbonate centrifuge tubes (50 ml) and centrifuged at 1000 g for one hour. To the rest of the digest a saturated sodiumbicarbonate solution was added until a pH of 6,8 was reached. Then 100 ml intestinal juice was added, the digests were pummeled for one minute and incubated at 37°C for a further four hours. During the incubation period the digests were shaken gently.
At the end of the intestinal juice digestion 20 ml of the digest was centrifuged at 1000 g for one hour.
Prior to the determination of zinc, copper and selenium five ml of the supernatants of both enzymatic digestion steps were preconcentrated to approx. 1 ml by evaporation.

Extraction procedure for the determination of the selenium distribution in various protein fractions

The extraction procedure for the measurement of the selenium distribution in various protein fractions is based on the method of Cappon and Smith (10).
Ten grams of the homogenized sample (plaice, mackerel, herring, mussels and shrimps) was thoroughly mixed with 20 ml of bidistilled water with an Ultra Turrax. Then 20 ml of bidistilled water was added and the mixture was shaken intensively for one hour. This mixture was centrifuged at 0°C during 15 minutes at 30,000 g. To the aqueous extract 8 ml of a 5% trichloroacetic acid solution was added and thoroughly mixed during 30 minutes. This homogenate was centrifuged for 15 minutes at 30,000 g. The three fractions (solid residue, the final supernatant and the TCA precipitate) were analysed for selenium.

High performance gel filtration chromatography

The high performance gel filtration chromatography (HPGFC) procedure for separation of proteins on the basis of molecular weight was carried out according to the method of Suzuki (12) and Pfannkoch et al (13) with a high pressure liquid chromatograph (Perkin Elmer Series 4) equipped with a gel filtration column (Biosil TSK-400, 600 x 7.5 mm) with a precolumn (100 x 7.5 mm).
Proteins were detected with a variable UV-wave length detector (Perkin Elmer LC-85B) at 280 nm. The pH of the aqueous TCA-supernatant of herring and mussels was adjusted to approx. 7.0. Then the supernatant was injected with a 400 µl loop and eluted with a 50 mM tris-HCl buffer (pH = 6.8) containing 0.2 M KH_2PO_4 and 0.1% NaN_3 with a flow of 1.0 ml/min.
Eluate fractions of 1 ml were collected from 5 runs of each sample and then pooled. These pooled eluate fractions were preconcentrated to 2-3 ml by evaporation and then analysed for selenium.
The gel filtration column was calibrated for molecular weight estimation with a protein mixture of bovine γ-globuline (MW = 160,000), ovalbumine (MW = 45,000), bovine myoglobuline (MW = 17,000) and cyanobalamine (MW = 1,355).

Analysis of zinc, copper and selenium

Zinc and copper were determined by flame atomic absorption spectrometry respectively furnace atomic absorption spectrometry.
To an appropriate amount of the sample 5 ml of 70% nitric acid was added and the destruction was carried out in a teflon-lined stainless steel decomposition vessel for at least 8 hours at 150°C. After the destruction the nitric acid was evaporated completely. Then 0.5 ml 37% hydrochloric acid and 1 ml bidistilled water were added. The solution was made up to 25 ml with bidistilled water.
The concentrations of zinc and copper were determined by standard addition with an Instrumentation Laboratory atomic absorption spectrophotometer (IL 551).
Selenium was determined fluorimetrically (14). After a destruction similar to zinc and copper a post treatment with perchloric acid and hydrochloric acid was carried out in order to complete the destruction and convert all selenium to the quadrivalent state. After coupling to 2,3 diaminonaphtalene the resulting benzopiazselenol was extracted into cyclohexane and measured fluorimetrically (excitation 378 nm, emission 520 nm).

Results and discussion

In-vitro availability of zinc, copper and selenium. It was shown that after one hour of gastric juice digestion the pH increases from 3.0 to approx. 4.5.
So it is necessary to check and adjust the pH during this digestion procedure. Approx. 10% of the mussels, shrimps and cod was not digested at the end of the digestion procedure.
In table 1 the content of zinc, copper and selenium in mussels, shrimps and cod is given. The content of copper in cod was too low to measure accurately the copper content in the supernatants of the digests.

Table 1. The Content of Zinc, Copper and Selenium in Mussels, Shrimps and Cod

	Zinc (mg/kg)	Copper (mg/kg)	Selenium (mg/kg)
Mussels	19.4 (1.0)*	1.93 (0.38)	0.41 (0.01)
Shrimps	25.8 (2.3)	11.3 (1.3)	0.36 (0.02)
Cod	5.2 (0.2)	0.23 (0.04)	0.35 (0.03)

*) standard deviation, n = 6

The results of the average in-vitro availability of the elements from the fishery products after the two enzymatic digestion steps are presented in figure 1 - 3. In order to establish significant differences in the in-vitro availability of zinc, copper and selenium in relation to the digestion stage and type of fishery product the individual results were tested according to the rank sign test of Wilcoxon. Under gastric juice conditions, 75 - 91% of the zinc present in the investigated fishery products was soluble. The in-vitro availability of zinc from shrimps under gastric juice conditions seems to be somewhat lower than from mussels and cod. At the neutral pH of the intestinal juice conditions the in-vitro availability of zinc in shrimps and cod decreases significantly to 63% respectively 30%. The solubility of zinc from mussels under gastric juice conditions (85%) does not differ significantly from the solubility under intestinal juice conditions. Crews et al (5) reported a decrease in the solubility for zinc present in crabmeat from 90% at gastric juice condition to 25% at neutral pH. This decrease in solubility of zinc might be due to precipitation of phosphates or to hydroxide formation. In mussels however soluble complexes seem to be formed between zinc and amino

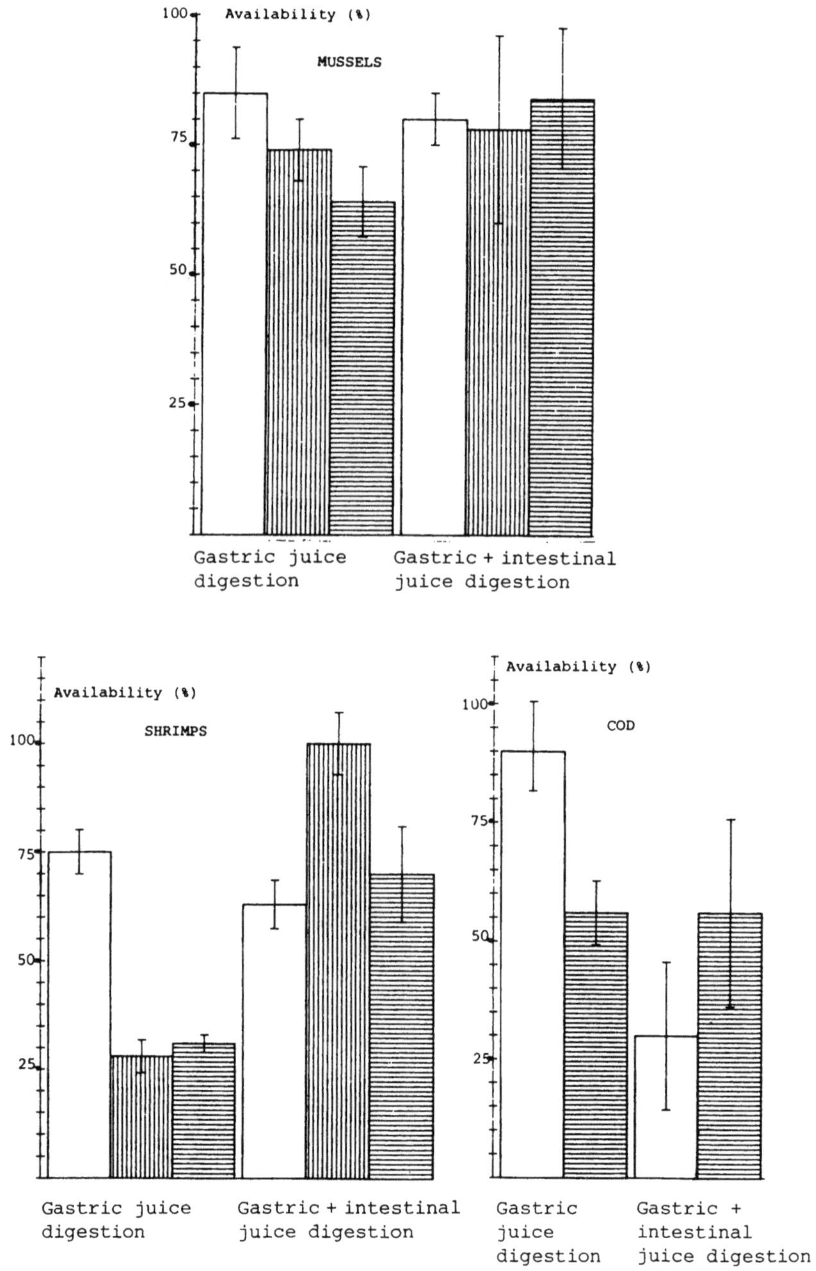

Figures 1-3. In-vitro availability of zinc, copper and selenium from mussels, shrimps and cod. The results are given as the mean ± SEM:

Zn; Cu; Se

acids or peptides during the intestinal juice digestion.
The in-vitro availability of copper from mussels is approx. 75% in both stages of digestion. Copper in shrimps is slightly soluble (28%) under gastric juice conditions and completely soluble during intestinal juice digestion. The solubility of copper from crabmeat, reported by Crews et al (5), is of the same magnitude as observed in this study for mussels. The solubility of selenium from mussels and shrimps after gastric juice digestion (64% respectively 31%) is significantly lower than after intestinal juice digestion (84% respectively 70%. No difference in the in-vitro availability of selenium from cod between the two digestion stages was observed.

Speciation of selenium. The distribution of selenium in the three fractions after the described extraction procedure for fishery products is given in table 2.

Table 2. Selenium Distribution in various Protein Fractions

	Percentage of total selenium content		
	Residue	Water-extractable	TCA-precipitate
Mussels	45.4%	27.4%	4.9%
Shrimps	79.9%	19.8%	5.6%
Cod	40.0%	30.9%	5.6%
Herring	45.1%	21.5%	9.2%
Mackerel	60.0%	18.2%	8.2%

The aqueous extract, containing amino acids, peptides and polar proteins, contained 18-40% of the total selenium content of the fishery product.
Cappon and Smith (10) observed that 25% of the selenium in mussels and shrimps was water-extractable. Our results are in agreement with this observation.
The water extractability of selenium from mackerel (18%) is significantly lower than found by Cappon and Smith (10). Only 5-8% of the total selenium content of the fishery products was present in the TCA-residue. The major part of the selenium content in the investigated samples (40-80%) was under the extraction conditions not soluble in the aqueous phase. In the case of cod, mussels and

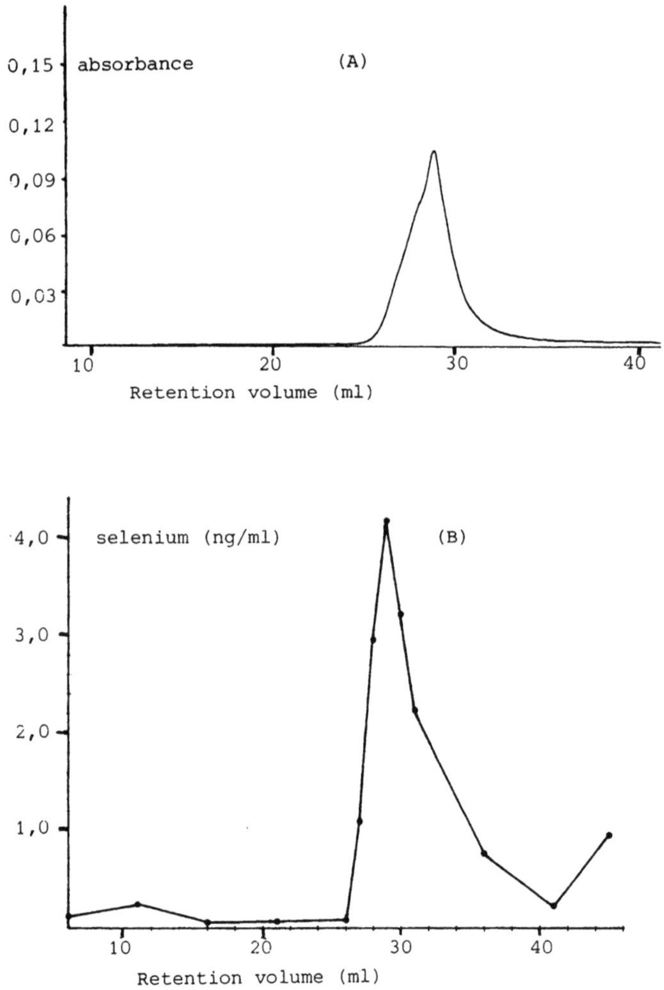

Fig. 4. HPGFC of the aqueous supernatant of herring (A) at 280 nm; (B) selenium

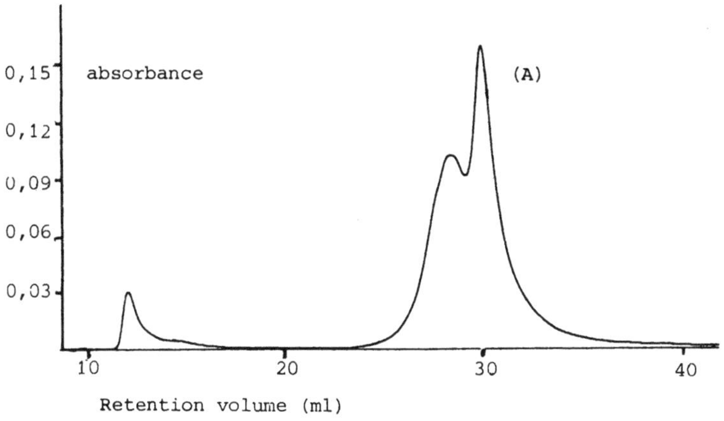

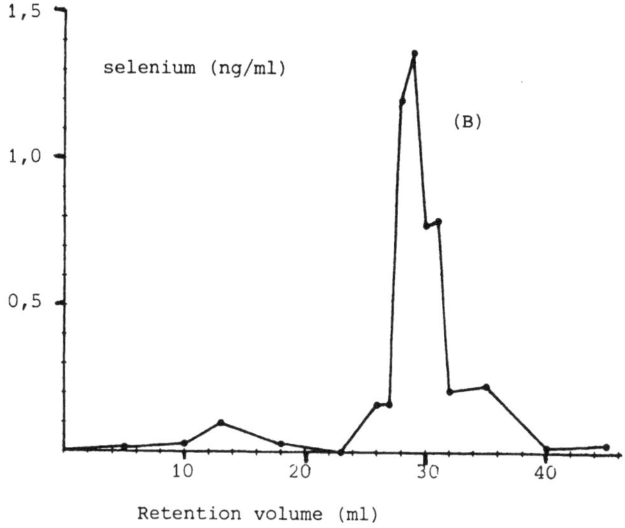

Fig. 5. HPGFC of the aqueous supernatant of mussels (A) at 280 nm; (B) selenium

herring about 25% of the selenium was not recovered during the extraction procedure. The results of HPGFC of the aqueous supernatant of herring and mussels are given in figure 4 and 5. Only a few peaks were observed with UV-detection at 280 nm during the elution of the supernatants of mussels and herring.

Selenium was eluted in one main peak corresponding with a peak detected at 280 nm. Some tailing of the selenium peak occurs. The maximum selenium content was present in the eluate after a retention volume of 28-29 ml. Molecular weight estimations, based on the retention volume of several proteins with known molecular weight, showed that selenium in mussels and herring is associated with a protein with a molecular weight of approx. 4,000 Dalton. Braddon (11) reported that in the liver of tuna the major part of selenium was present in a low molecular weight protein (MW $\approx$2,500 Dalton).

The significance of this low molecular selenoprotein in fishery products is rather speculative. It is known that the percentages of bio-availability of selenium for elevation the selenium content in the red blood cell population or in liver of rats were similar when rats were fed with kidney, wheat and tuna. However percentages of bio-availability of selenium for induction of glutathionperoxidase activity in red blood cell population or in liver were about 70-100% for kidney and wheat but less than 60% for tuna (15). Selenium in fishery products is probably present in a chemical form which is less potent to induce glutathion peroxidase activity than selenium in wheat or kidney. Therefore it is interesting to focus on a further detailed speciation of this low molecular weight selenoprotein in fishery products.

Conclusions

On the basis of the foregoing results the following conclusions are made.

The in-vitro availability of zinc, copper and selenium in fishery products depends on the type of fishery product and digestion conditions.

High performance gel filtration chromatography showed that at least a small amount of the selenium present in mussels and herring is associated with a low molecular weight protein fraction.

References

1. Lock, S., A.E. Bender. 1980. Br.J.Nutr. 43, 413.
2. Narasinga Rao, B.S., T. Prabhavathi. 1978. Am.J.Clin.Nutr. 31, 169.
3. Miller, D.D., B.R. Schricker, R.R. Rasmussen, D. van Campen. 1981. Am.J.Clin. Nutr. 34, 2248.
4. Crews, H.M., J.A. Burrell, D.J. McWeeny. 1983. J.Sci.Food Agric. 34, 997.
5. Crews, H.M., J.A. Burrell, D.J. McWeeny. 1985. Z.Lebensm.Unters.Forsch. 180, 221.
6. Crews, H.M., J.A. Burrell, D.J. McWeeny. 1985. Z.Lebensm.Unters.Forsch. 180, 405.
7. Olson, O.E., E.J. Novacek, E.I. Whitehead, I.S. Palmer. 1970. Phytochemistry 9, 1181.
8. Schrift, A. 1969. Ann.Prev.Plant Physiol. 20, 475.
9. Cappon, C.J., J.C. Smith. 1981. Arch.Environ.Contam.Toxicol. 10, 305.
10. Cappon, C.J., J.C. Smith. 1982. J.Anal.Toxicol. 6, 10.
11. Braddon, S.A. 1981. In: Selenium in Biology and Medicine (J.E. Spallholz, J.L. Martin, H.E. Ganther, Eds.) Westport AVI, p. 56.
12. Suzuki, K.T. 1980. Anal.Biochem. 102, 31.
13. Pfannkoch, E., K.C. Lu, F.E. Regnier, H.G. Barth. 1980. J.Chrom.Science 18, 430.
14. Koh, T. T.H. Benson. 1983. J.AOAC 66, (4), 918.
15. Douglass, J.S. V.C. Morris, J.H. Soares jr, O.A. Levander. 1981. J.Nutr. 111, 2180.

HEAVY METAL BINDING TO PROTEINS EXTRACTED FROM RATS

R. Tykva, I. Votruba, J. Veselý

Institute of Organic Chemistry and Biochemistry, Czechoslovak Academy of Science, CS-166 10 Prague 6

W.E. Kollmer, D. Berg

Abt. für Nuklearbiologie, Gesellschaft für Strahlenund Umweltforschung m.b.H. München, Ingolstädter Landstraße 1, D-8042 Neuherberg

Introduction

The environmental and/or industrial exposure of living organisms to heavy metals have been investigated in an increasing extent (1, 2, 3). The use of radioisotopes makes it easy to study the metal binding (4). If hair mineral analysis is evaluated in connection with internal body burden the knowledge of metal binding could be helpful in estimating different changes of metal concentrations due to washing procedures (5). Moreover, it also allows to assess possible correlations of metal levels in hair and organs (or serum) (6).
In the present study the binding of mercury to soluble proteins from rat hair, skin, blood plasma, liver and kidney was examined following intraperitoneal (Hg 203-chloride) or intravenous (Methyl-Hg 203-chloride) administration. Part of the results were compared with data obtained after application of Cd 109-chloride or Pb 210-nitrate.

Materials and Methods

Male adult Wistar rats kept in metabolic cages were given drinking water and food ad lib. They were epilated by plucking in the mid-dorsal region (4 x 4 cm) 10 days before the

Trace Element – Analytical Chemistry in Medicine and Biology, Vol. 4

administration of 7.4 MBq of Hg 203-Cl_2 i.p. (32.93 MBq mg^{-1} Hg) or 1.48 MBq of Methyl-Hg 203-Cl i.v. (0.65 MBq mg^{-1} Hg) and new hair was harvested at the time of sacrifice (7.8) or at day 11 in case of Methyl-Hg treated animals. The animals were sacrificed 7 ($HgCl_2$), or 11 (Methyl-Hg) days following the administration of radioactive mercury. In another group of animals 3.7 MBq of Cd 109-chloride was administered intravenously. The radioactivity of organs under study was determined as described previously (7, 8).

The protein fractions from liver and kidney were prepared by sonicating the tissues (100 mg portions) for 3 min in 0.02 M Tris-HCl buffer (pH 7.4) and by spinnning down the sonicate at 30 000 g. The activity of the mercury in plasma proteins was followed directly. The soluble protein fractions from skin (36 mg from 1 g of wet weight) and hair (390 mg from 1 g) were obtained as described previously (4).

The "total" and "bound" radioactivity in the protein extract was determined by liquid scintillation spectrometer using solubilizing the respective aliquots in 1 ml of Tissue-Solubilizer (Beckman, BTS-450) with 10 ml of Ready-Solv NA (Beckman) and following the precipitation with 5% TCA onto the membrane filters (Millipore, HAWP, 0.45 um) measured in toluene scintillator.

Polyacrylamide gel electrophoresis of protein extracts treated with mercaptoethanol was carried out as described (4).

Results and Discussion

At day 11 after the administration of radioactive Methyl-Hg the body burden was 78.9 $\pm$ 1.5 % of the administered radioactivity. The radioactivities in the tissues of Methyl-Hg treated animals are depicted in Table 1. The highest concentration of radioactivity was found in hair followed by kidney and liver. Comparing the data in Tables 1 and 2 it was calculated that approxi-

Table 1: The radioactivity distribution ($x \pm s_x$) in tissues of Methyl^{203}Hg treated animals

	% administered activity per g fresh weight	
Liver	0.487 ± 0.026	(N=3)
Kidney	1.860 ± 0.040	(N=3)
Plasma	0.019 ± 0.002	(N=3)
Skin	0.118 ± 0.011	(N=3)
Hair	10.370 ± 0.480	(N=9)

Table 2: Distribution of ^{203}Hg in soluble tissue proteins. A: $^{203}HgCl_2$; B: Methyl^{203}HgCl; --: not measured. The standard deviation was within the limits of ±3%.

	10^{-2}% of appl. activity per mg proteins					
	Total after extr.		Protein-bound			
					Mercaptoethanol added	
	A	B	A	B	A	B
Liver	0.072	0.197	0.070	0.078	--	--
Kidney	11.200	2.694	6.600	0.817	0.02	0.005
Plasma	0.028	0.094	0.023	--	--	--
Skin	0.219	--	0.172	--	--	--
Hair	0.023	0.784	--	--	0.0003	0.005

Table 3: ^{109}Cd-activity in induced hair. Standard deviation was within the limits of ±20%.

10^{-2}% of appl. activity per g hair		
Total hair	Extr. sol. proteins	Protein - bound
28.0	3.0	2.1

mately 30 % of the radioactivity was extracted from hair. Comparing specific radioactivity of extracted proteins (Table 2) it is evident that a large amount of mercury is tightly bound to proteins regardless whether Hg 203-Cl_2 given i.p. or Methyl-Hg 203-Cl with considerably different specific activity given i.v. had been administered; in both cases the protein-bound Hg 203 coprecipitated in 5 % TCA.

In the liver following Hg 203-Cl_2 administration on the average 97 % of radioactivity was retained by the filter, in the case of kidney, plasma and skin 59 %, 82 % and 79 %, respectively. Following the administration of Methyl-Hg 203-Cl the corresponding percentage was only 40 % in the liver and 30 % in the kidney. In the hair proteins which were solubilized in the presence of a high concentration of mercaptoethanol (4) the filter retained less than 1 % of the extracted radioactivity (Table 2). Similar results were obtained when mercaptoethanol (10 ul per 60 ug of proteins) was added to extracted proteins from kidney 2 min prior to TCA precipitation at room temperature. Analogously, no radioactive component was found after polyacrylamide gel electrophoresis of the analysed protein extraction from the kidney using mercaptoethanol. These findings correspond to the covalently bound mercury to proteins via sulfhydryl groups (9); the respective linkage is stable in 5 % TCA whereas in the presence of mercaptoethanol it breaks down. This contention is further strenghtened by the observation that the radioactivity in the examined hair subjected to washing procedure (5, 10) remained unchanged. Moreover, the relatively high Hg 203 activity in hair (Table 1) is explained by the high concentration of cystine (11).

The relatively low transfer of Cd 109 to hair and the correspondingly low radioactivity found in extracted proteins (Table 3) makes possible only to prove the protein-bound radioactivity after TCA precipitation. The corresponding specific activity is less than 10^{-4} % of the applied activity (3.7 MBq) per mg of extracted soluble proteins. This level eliminated application of the electrophoretic analysis using the PAGE-SDS procedure

(4). Such a separation in combination with semiconductor topographic evaluation of the radioactivity position (12) and computer-controlled smoothing (13) enables for somewhat higher activities a non-destructive and automated determination of the radiometal binding. In a previous investigation (4) using the PAGE-SDS procedure to binding of Pb 210 a fraction with molecular weight of about 16 000 was detected in rat skin.

Acknowledgement

This work was performed in the frame of IAEA Co-ordinated Research Programme (Research Contract 3673/RB and Research Agreement 3674/CF).

References

1. United Nations Environment Programme Report. 1984. No. 2.

2. Schwermetalle-Biologische Wirkung und Expositionsquellen. 1984. Gesellschaft für Strahlen- und Umweltforschung, München. Rep. A 6-84.

3. Health-Related Monitoring of Trace Element Pollutions Using Nuclear Techniques. 1985. IAEA Vienna. Techn. Doc. 330.

4. Votruba, I., Veselý, J., Tykva, R. 1985. The putative lead-binding carrier protein in rat epidermis using. Pb. J. Radioanal. Nucl. Chem., Letters 96, 557.

5. Das, H.A., M. Dejkumhang, R.F.M. Herber, D. Hoede, H.A. van der Sloot. 1981. Instrumental neutron activation analysis of human hair and related radiotracer experiments on washing and leaching. Netherland Energy Research Foundation. Rep. ECN-107.

6. Gibson, R.S., I.L. Gibson. 1984. The interpretation of human hair trace element concentrations. Sci. Total Environ. 39, 93.

7. Berg, D., W.E. Kollmer. 1983. The deposition of inorganic mercury in newly grown hair of the rat. In: Trace Element-Analytical Chemistry in Medicine and Biology. Vol. 2 (P. Brätter, P. Schramel, eds.). Walter de Gryter, Berlin-New York, p. 571.

8. Kollmer, W.E., D. Berg. 1979. Interrelations between hair content, intravenous dose and retention of Cd in different rat tissues. J. Radioanal. Chem. 52, 189.

9. West, E.S., W.R. Todd, H.S. Mason, J.T. van Bruggen. 1966. Textbook of Biochemistry, 4. ed. Macmillan Comp., New York. p. 852.

10. Kollmer, W.E., D. Berg. unpublished.

11. Wolfram, J.L. 1981. In: Hair Research - Status and Future Aspects. (C.E. Orfanos, W. Montagna and G. Stüttgen, eds.). Springer, Berlin. p. 479.

12. Tykva, R., F. Franěk . 1977. Nondestructive and quantitative evaluation of radioactive spots on two-dimensional peptide maps by an automated procedure. Anal. Biochem. 78, 572.

13. Tykva, R., R. Jisl . Correction of the continual scanning record of radioactivity distribution. I. Smoothing. Nucl. Instrum. Methods Phys. Res. (in press).

PIXE-MICROANALYSIS OF PSORIATIC HUMAN SKIN*

W. Bischof and B. Gonsior
Institut für Experimentalphysik III der Ruhr-Universität Bochum, D 4630 Bochum 1, F.R. Germany

K. Kurz and G.K. Steigleder
Universitäts-Hautklinik Köln, D 5000 Köln 41, F.R. Germany

Introduction

Psoriasis vulgaris is a chronic inflammatory skin disease with hyper- and parakeratosis. Since the elemental distribution varies with the functional state of the keratinocytes, investigations have been focussed on electrolytes because of their influence on cell metabolism and cell proliferation (1) as well as on trace elements such as iron, copper and zinc because of their crucial role in a great number of enzymes (2). The role of zinc in psoriatic lesions was first estimated by histochemical methods (3). Later on, trace elements were determined by neutron activation analysis in mechanically separated psoriatic epidermis, which revealed increased levels of zinc (4) and iron (5). Recently, the elemental composition of psoriatic skin was investigated by means of STEM (Scanning Transmission Electron Microscope) X-ray microanalysis revealing higher levels of phosphorus and calcium in the stratum corneum (6). Applying the same technique, increased concentrations of Mg, K, and P were found in involved psoriatic epidermis as a typical pattern of highly proliferative, non-neoplastic cells (7).

Elemental analysis in human skin using PIXE in combination with a proton microprobe has already been reviewed (8,9). In comparison with the STEM-method PIXE achieves lower detection limits, which allow the determination of elements down to levels of 1 ppm.

* Supported in part by Minister für Wissenschaft und Forschung des Landes Nordrhein-Westfalen, D 4000 Düsseldorf, F.R. Germany

In our present investigation we quantified elemental distributions, including trace elements in pin point lesions, old psoriatic plaques, uninvolved psoriatic skin, and skin from healthy control persons. This should help us to estimate the role of the different elements in the development of a psoriatic lesion.

Materials and Methods

Biopsies were taken from the lumbal region of 5 psoriatics and 4 controls. None of the patients was on systematic treatment. Except petrolatum, no local treatment was applied two weeks before taking the biopsies. After anaesthesia with 1 ml Xylocain 1 % s.c., biopsies were obtained from an early psoriatic lesion with a maximum diameter of 3 mm (pin point lesion), an old psoriatic plaque, and uninvolved psoriatic skin, at least 5 cm from a psoriatic lesion. The biopsies were frozen and stored in liquid nitrogen at -196 $^{\circ}$C. Cryosections of 10 µm thickness were cut perpendicular to the skin surface at a temperature of -20 $^{\circ}$C. The sections were mounted on thin (< 0.5 µm) Formvar foils without any supplement, smoothed and adhered to the foils by short-term thawing and then completely dried at a temperature of -20 $^{\circ}$C.

For ion-movement in the thawed sections may equalize the concentrations this method was compared to completely dry mounting of freeze-dried sections between two Formvar foils as well as to drying the completely thawed cryosections at +20 $^{\circ}$C. The comparison was accomplished by analyzing neighbouring parallel sections, treated in the different way as described. In all three cases the elemental distributions did not differ markedly in regions larger than the cell dimension. The drying process seems to be much faster than diffusion of ions through the cell membranes.

In addition we noticed that the mass distribution profiles measured in human skin sections by Forslind et al. (8) are similar to most of our results though all there samples were freeze-dried and completely dry mounted.

For the determination of element distributions the PIXE-method in combination with the Bochum Proton Microprobe (9) was used.

The microprobe is equipped with a magnetic scanner and a PDP 11/44 /CAMAC data handling system (10). The analyses described in this paper were accomplished using a lateral resolution of 5 x 5 μm^2, a beam current of 1 nA and a proton energy of 3 MeV. In each of the samples the distribution of the elements P, S, Cl, K, Fe, Cu, and Zn as well as the sample mass per unit area were determined quantitatively taking a line-scan perpendicular to the skin surface. Calibration was done by measuring the elastic scattering of the protons at the target nuclei. The matrix mass per unit area was determined by evaluation of the secondary electron bremsstrahlung background in two energy ranges between the X-ray energies 3.05 and 3.95 keV. Normally this range is free from pile-up effects. A Kapton foil (830 $\mu g/cm^2$) was used for calibration.

The significance levels of differences between samples from psoriatics and healthy controls were calculated using Student's t-test. For the comparison of involved and uninvolved skin a paired t-test was used.

The reproducibility attainable with Micro-PIXE analysis depends in a crucial way on the condition, that losses of elements due to evaporating sample material are negligible. Normally it is asserted, that the microprobe operation is non-destructive. Nevertheless in case of the analysis of organic samples one finds black spots on the target caused by a pyrometamorphism of the organic matrix in consequence of the heat in the proton beam's focus. So material losses can not be excluded.

To get an idea of the analysis conditions for those material losses are possible, we vaporized a layer of 10 $\mu g/cm^2$ Se upon a 10 μm thick Nuclepore polycarbonat foil. Selenium is a very volatile element whose vapor pressure reaches the pressure in the target chamber (10^{-6} mb) at a temperature of 125 ^{o}C. Selenium dioxide possibly formed during the storage of the targets in open air sublimes even at a temperature of 37 ^{o}C. Polycarbonat was used as backing foil to simulate a non conductive organic matrix.

The Selenium mass per unit area versus irradiation time was measured with a fixed proton beam of 8 µm diameter at different beam currents. The result is shown in Fig. 1.

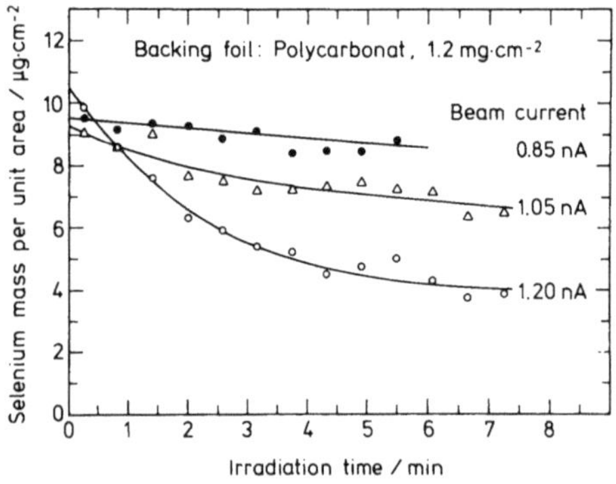

Fig. 1: Selenium mass per unit area versus irradiation time at three different beam current values.

Limiting the beam current to 1 nA only slow evaporation processes will take place so that material losses are detectable by repeating the measurements. This was practiced with two skin samples by six repetition analyses of each during the whole irradiation time lasting five hours. Table 1 shows the result of this measurements. Significant material losses are not detectable.

Table 1: Standard Deviation of the Results of Repetition Measurements Using Human Skin Samples as Target. (Details in the text).

Element	Mass per unit area ± SDV / %	Concentration ± SDV / %
P	2.3	5.5
S	0.8	3.7
Cl	0.5	3.9
K	0.8	4.3
Fe	8.8	10.8
Cu	15.7	17.6
Zn	10.6	9.6
Matrix	3.4	--

Results

Elemental distribution in freeze-dried skin sections of 5 psoriatics and 4 controls were determined. In order to localize the stratum corneum and the basal layer, a photograph was taken from each skin section measured (e.g. Fig. 2). The irradiated area

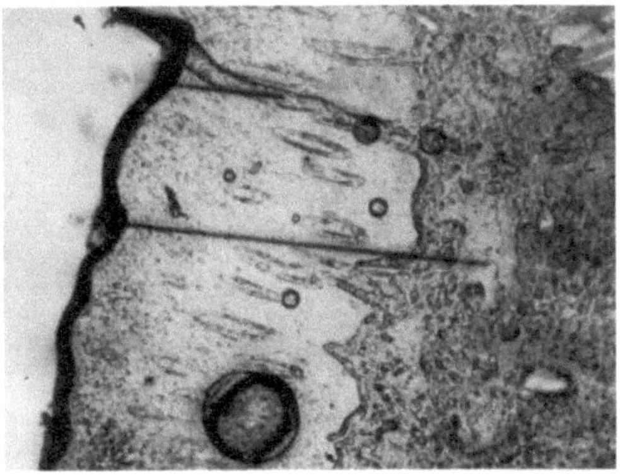

Fig. 2: Micrograph of a dried cryosection of an old psoriatic plaque after irradiation with the proton microprobe. The irradiated area is visible as a dark line.

is visible as a dark line. In the epidermis and upper dermis concentration profiles of the elements P, S, Cl, K, Fe, Cu, and Zn could be obtained.

Each concentration profile was plotted in a diagram (e.g. Fig. 3); mean values in the epidermis and upper dermis were calculated. The results are summarized in Tab. 2.

Compared to the controls we found significantly decreased dermal Cl concentrations in involved and uninvolved psoriatic skin. In the epidermis the Cl distribution was unaffected, showing a linear increase from the stratum corneum down to the basal layer. All other elements did not show significant differences in dermal concentrations. However, it should be noticed, that Xylocain used for anaesthesis contains about 0.5 % Cl, but this should not lead to differences in the dermal Cl content in skin of psoriatics and controls, for all samples were taken under identical conditions.

Within the epidermis, all striking differences were found between involved and uninvolved psoriatic skin. The involved psoriatic epidermis showed increased concentrations of K, P, and Zn, as well as decreased levels of Fe. We could demonstrate a linear corre-

Table 2: Elemental Concentration of Normal and Psoriatic Skin.

Values are given in µg/g dry weight for elemental concentration in the epidermis and in 1 case in the dermis (Cl), as means ± SDV in 4 healthy control persons and 5 psoriatic patients.

a,b Significantly different from uninvolved psoriatic skin ([a] $p < 0.05$; [b] $p < 0.01$).

c,d Significantly different from control ([c] $p < 0.02$; [d] $p < 0.01$).

	P	S	Cl	K	Fe	Cu	Zn	Cl (Dermis)
Control	12500 ± 4300	11400 ± 900	10200 ± 3900	8900 ± 2200	240 ± 225	8.3 ± 0.6	77 ± 22	20400 ± 2000
Uninvolved psoriatic skin	12000 ± 3700	10700 ± 2100	9300 ± 2500	6800 ± 1900	98 ± 54	12.7 ± 6.5	64 ± 20	14400 ± 3300[c]
Pin point lesion	18400 ± 2700[b]	11500 ± 2100	8600 ± 2800	8800 ± 4000[a]	55 ± 23	9.9 ± 4.5	85 ± 11[b]	15100 ± 1900[d]
Old psoriatic plaque	16600 ± 1900[b]	11000 ± 1800	9300 ± 2500	11000 ± 3800[b]	59 ± 15[b]	14.9 ± 8.2	71 ± 7	14100 ± 2500[d]

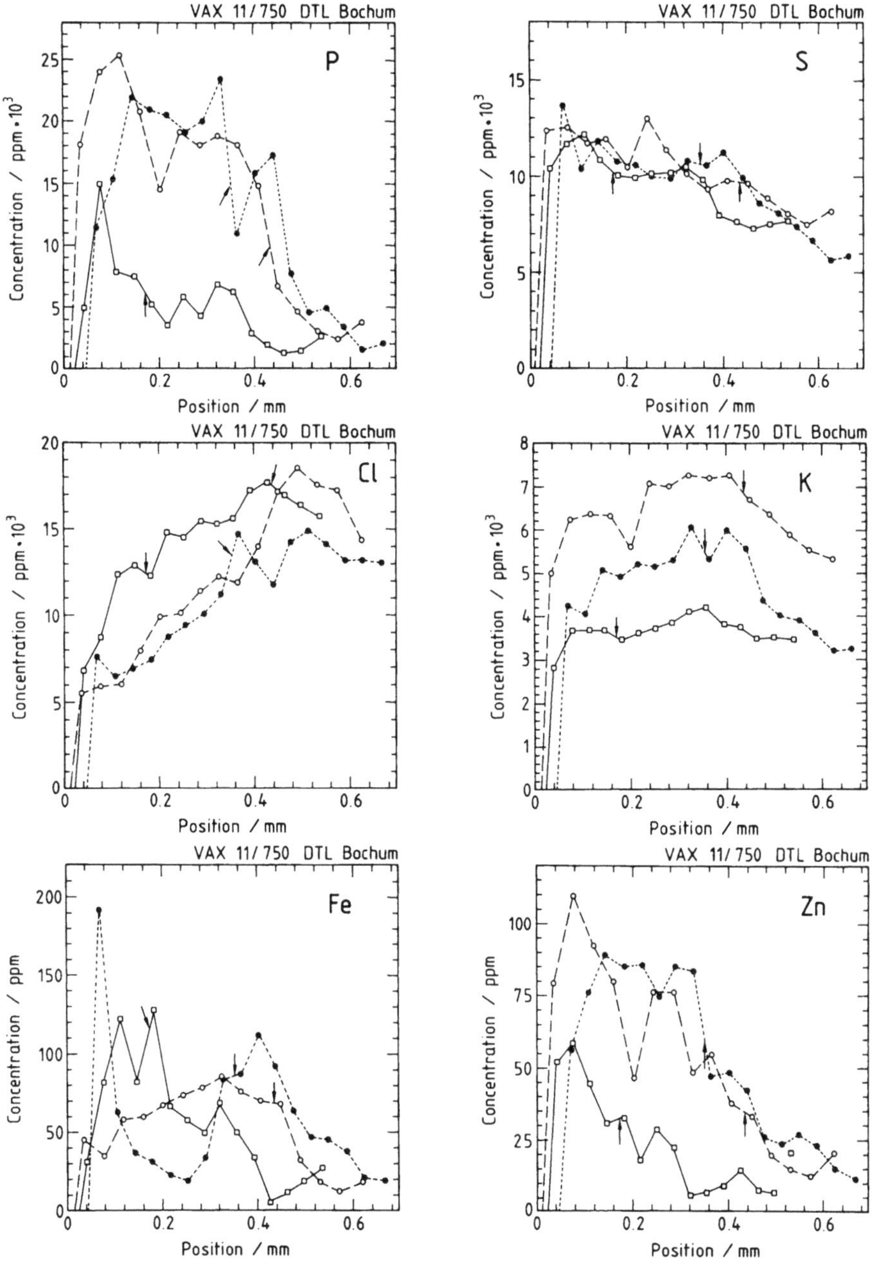

Fig. 3: Elemental concentrations in cryosections of one representative patient: The stratum corneum starts approximately at position 0 mm, the arrows indicate the dermoepidermal junction. o—–o old psoriatic plaque, ●-----● pin point lesion, □——□ uninvolved psoriatic skin. All concentrations are given in µg/g (ppm) dry weight.

lation between the elevated K levels and the stage of the psoriatic lesion (Fig. 3). Compared to the controls, K concentration seemed to be lower in uninvolved psoriatic skin, but the difference was statistically not significant. We found markedly increased Zn levels in pin point lesions, whereas in old psoriatic plaques this increase was no more significant. The concentration profiles of Zn and P were remarkably correlated in all psoriatic skin samples. The levels of S in psoriatics resembled those in the controls, sometimes showing slight concentration maxima in the upper epidermis. The iron levels were significantly decreased in old psoriatic plaques, whereas they varied considerably in the controls. The concentrations of Cu were close to the detection limit, showing varying concentration profiles.

Discussion

In general, most differences regarding the elemental distribution were found between involved and uninvolved psoriatic epidermis, and not between psoriatics and controls. This may be due to the relatively high interindividual variability observed in psoriatics and controls. Any localization factors could be excluded as all measured biopsies were taken from the lumbal region.

It has been suggested that measured K concentrations in correspondence with the stage of the psoriatic lesion may be the result of hyper- and parakeratosis (11). The markedly decreased Cl concentrations we found in involved and uninvolved psoriatic dermis have not been described yet. Further measurements without the use of Xylocain are still required to interpret these interesting findings. In addition, psoriatics and controls did not show any significant differences concerning serum levels of Na, K, Ca, and Cl.

The increased P levels in lesional psoriatic epidermis could be explained by higher concentrations of phospholipids (12) or higher contents of nucleic acids (13). The latter possibility seems to be more likely in view of the remarkable similarity of the Zn and P concentration profiles in psoriatic skin. As Zn levels were significantly increased in pin point lesions, we assume that DNA- and RNA-polymerases dependant on Zn were elevated, increasing the metabolic turnover and the proliferation rate of the keratinocytes in the early stage of a psoriatic lesion (14).

In contrast to former investigations showing elevated Fe levels in psoriatic skin (5), we found decreased iron levels in the old psoriatic plaques. Those authors discussed a marked loss of iron by epidermal desquamation in psoriatic patients. Our results do not support this theory. Since neutron activation analysis was applied in mechanically separated epidermis, the difference may be caused in the different analytical methods.

Our results are in good agreement with the data obtained by Grundin et al. (7), who investigated psoriatic epidermis using the electron probe X-ray microanalysis. They additionally found elevated Mg levels and unaffected concentrations of Na, but caused by higher detection limits they were not able to determine trace element concentrations. The present investigation shows, that Micro-PIXE is a valuable quantitative technique in addition to electron probe X-ray analysis and morphological methods. For better understanding of the measured results, however, further investigations applying the PIXE technique in other skin diseases are required.

References

1. Hennings, H., K.A. Holbrook and S.H. Yuspa. 1983. Potassium mediation of Calcium-induced terminal differentiation of epidermal cells in culture. J. Invest. Dermatol. 81, 505-553.

2. Gordon, M. and W.C. Johnson. 1967. Histopathology and histochemistry of psoriasis. I. The active lesion and clinically normal skin. Arch. Dermatol. 95, 402-407.

3. Braun-Falco, O. and B. Rathjens. 1956. Histochemische Untersuchungen über das Verhalten von Zink in der Haut bei Psoriasis und anderen Hauterkrankungen. Derm. Wschr. 134, 837-841.

4. Molokhia, M.M. and B. Portnoy. 1970. Neutron activation analysis of trace elements in skin. V. Copper and zinc in psoriasis. Br. J. Dermatol. 83, 376-381.

5. Molin, L. and P.O. Wester. 1973. Iron content in normal and psoriatic epidermis. Acta Dermatovener. (Stockholm) 53, 473-476.

6. Burkhart, C.G. and J.C. Burnham. 1983. Elevated phosphorus in psoriatic skin determined by energy-dispersive X-ray microanalysis. J. of Cutaneous Pathology 10, 171-177.

7. Grundin, T.G., G.M. Roomans, B. Forslind, M. Lindberg and Y. Werner. 1985. X-ray microanalysis of psoriatic skin. J. Invest. Dermatol. 85, 378-380.

8. Forslind, B., G.M. Roomans, L.E. Carlsson, K.G. Malmqvist and K.R. Akselsson. 1984. Elemental analysis on freeze-dried sections of human skin: Studies by electron microprobe and particle induced X-ray analysis. Scan. Electron Microsc. (part two), 755-759.

9. Bischof, W., M. Höfert, B. Raith, H.R. Wilde, B. Gonsior and K. Enderer. 1984. In: Trace Element Analytical Chemistry in Medicine and Biology, Vol. 2 (P. Brätter and P. Schramel, eds.). Walter de Gruyter, Berlin New York. pp. 1053-1061.

10. Höfert, M., W. Bischof, A. Stratmann, B. Raith and B. Gonsior. 1984. Determination of lateral trace element distributions with the Bochum Proton Microprobe. Nucl. Instr. and Meth. B3, 572-579.

11. Herrmann, W.P., K.G. Ippen, K. Schäfer and G. Stüttgen. 1973. Biochemie der Haut. Thieme Verlag Stuttgart. pp. 118-121.

12. Yardley, H.J. and R. Summerly. 1981. Lipid composition and metabolism in normal and diseased epidermis. Pharmac. Therapy 13, 357-383.

13. Zs Nagy, I. 1983. Energy dispersive X-ray microanalysis of biological bulk specimens: a review on the method and its applications to experimental gerontology and cancer research. Scan. Electron Microsc. (part 3), 1255-1268.

14. Norris, D. 1985. Zinc and cutaneous inflammation. Arch. Dermatol. 121, 985-989.

CORRELATION OF SERUM TRACE ELEMENTS WITH OTHER BIOLOGICAL PARAMETERS IN HEMOPROLIFERATIVE DISORDERS

Y. Beguin *, J. Bury, G. Fillet,
Department of Hematology.

J. M. Delbrouck, G. Robaye, I. Roelandts, G. Weber **,
Department of Experimental Nuclear Physics, University of Liège, Sart-Tilman B 15, B-4000 Liège, Belgium.

Abstract

Serum trace elements were measured by PIXE in 88 patients with hemoproliferative disorders (Lymphoma, Hodgkin's disease, chronic lymphocytic leukemia, multiple myeloma, polycythemia vera and myeloproliferative disorders). Calcium, copper, zinc, selenium and bromine were correlated with other biological parameters, including hematocrit, WBC and platelet counts, ESR, fibrinogen, transferrin, ceruloplasmin, haptoglobin, orosomucoid, antitrypsin, serum protein, albumin, alpha 2-globulin, alkaline phosphatase, SGPT, LDH, gamma-GT, cholesterol, triglycerides, glycemia, urea, uric acid, creatinine, beta 2-microglobulin, CEA and C1q assay for circulating immune complexes. Copper was correlated with the different inflammatory tests (all P values < 0.01) and with the value of the C1q assay ($P < 0.001$). Zinc was inversely correlated with the inflammatory tests. The Cu/Zn ratio was thus significantly related to the importance of inflammation (all P values < 0.01). Calcium and zinc were correlated with albumin ($P < 0.05$). Selenium was positively correlated with gamma-GT ($P < 0.001$) and alkaline phosphatase ($P < 0.001$), and also with glycemia ($P < 0.001$). Bromine was not significantly associated with any other biological parameter. The significance of these associations is discussed.

---

* Research Assistant of the National Fund for Scientific Research
** Research Associate of the National Fund for Scientific Research (Belgium)

Trace Element – Analytical Chemistry in Medicine and Biology, Vol. 4

Introduction

There are few reports on serum trace element (STE) levels in hemopro-liferative disorders (HPD). Usually, serum copper levels are elevated and serum zinc levels are lowered in patients with active desease (1 - 10). Cu levels and Cu/Zn ratio might be useful parameters to monitor disease activity in patients treated for Hodgkin's disease (1, 2) or lymphoma (3, 4).

The present study was undertaken to measure serum calcium, copper. zinc, selenium and bromine in chronic HPD and to investigate the significance of the observed abnormalities in 88 patients. The STE values were correlated with a number of biological parameters to find out whether disease activity, inflammation, bone marrow, renal and hepatic functions or metabolic factors could significantly alter their levels in patients with HPD.

Patients and methods

88 patients with hemoproliferative disorders (HPD) were studied. There were 23 patients with multiple myeloma (MM), 13 with Hodgkin's disease (HD) or non-Hodgkin's lymphoma (NHL), 18 with chronic lymphocytic leukemia (CLL), 14 with polycythemia vera (PV) and 20 with myeloproliferative disorders (MPD). There were 56 males and32 females. Age ranged from 25 to 82 years. The patients were compared with a group of 100 healthy subjects.

Serum trace elements (STE) were measured by the PIXE method (Particle-Induced X-Ray Emission) (11), allowing for a simultaneous determination of calcium (Ca), copper (Cu), zinc (Zn), selenium (Se) and bromine (Br). The copper-to-zinc ratio (CZR) was calculated from the actual Cu and Zn levels.

At the same time as for STE determination, blood was drawn for the measurement of other biological parameters, including hematocrit, white blood cell (WBC) and platelet counts, erythrocyte sedimentation rate (ESR), fibrinogen, transferrin, ceruloplasmin, orosomucoid, haptoglobin, antitrypsin, serum protein and protein electrophoresis, alkaline phosphatase, serum glutamic pyruvic transaminase (SGPT), gamma-glutamyl transferase (gamma-GT), total cholesterol, HDL cholesterol, triglycerides, glycemia, urea, uric acid, creatinine, beta 2-microglobulin, carcinoembryonic antigen (CEA) and C 1 q assay for circulating immune complexes.

Results

Mean STE levels for the 88 patients and for each diagnostic group are shown in Table 1. All the groups were characterized by decreased serum Ca ($P < 0.001$) and Zn ($P < 0.001$) levels and by increased Cu levels ($P < 0.001$), as compared to normal controls. Therefore, the CZR was significantly raised in patients ($P < 0.001$). Serum Se and Br levels in patients did not differ significantly from those found in controls.

Table 1
Mean STE levels in hemoproliferative disorders

Diagnosis	N	Ca	Se	Br	Cu	Zn	CZR
MM	23	96.2	0.095	4.63	1.23	0.85	1.48
NHL-HD	13	94.5	0.079	4.23	1.34	0.80	1.72
CLL	18	95.8	0.111	5.61	1.34	1.00	1.39
PV	14	94.4	0.072	5.39	1.26	0.99	1.32
MPD	20	95.8	0.099	4.91	1.30	0.96	1.42
TOTAL	88	95.5	0.094	4.96	1.29	0.92	1.45
CONTROLS	100	102.5	0.097	5.36	1.10	1.10	1.03

The values are given in µg/ml except for the CZR.
MM = Multiple Myeloma. NHL = Non-Hodgkin's Lymphoma.
HD = Hodgkin's Disease. CLL = Chronic Lymphocytic Leukemia.
PV = Polycythemia Vera. MPD=Myeloproliferative Disorders.

Table 2 displays the R correlation coefficients between STE and the other biological parameters. Only the statistically significant associations are shown. Ca ($P < 0.001$) and Zn ($P < 0.05$) were positively correlated with albumin. Se showed a direct relationship with gamma-GT ($P < 0.001$) and alkaline phosphatase ($P < 0.001$), but not with SGPT. Se was also correlated with glycemia ($P < 0.001$), but not with triglycerides nor cholesterol. Cu was significantly associated with the various inflammatory tests except with alpha 2-globulin.

Table 2

R correlation coefficients between STE and other biological parameters

	Ca	Se	Br	Cu	Zn	Cu/Zn
Hematocrit				-0.237°		-0.355*
Platelets	0.227°					
C1q assay			0.251°	0.379*		0.278°
ESR				0.265°	-0.283+	0.442*
Fibrinogen				0.329+	-0.221°	0.447*
æ1 antitrypsin				0.330+		0.307+
Ceruloplasmin				0.550*	-0.276+	0.628*
Orosomucoid				0.279+	-0.307+	0.422*
æ2-globulins					-0.255°	0.310+
Albumin	0.374*		0.230°		0.268°	
Alkalinephosphatase		0.387*		0.227°		0.224°
Gamma-GT		0.580*				
SGPT						0.244°
ß2 microglobulin					-0.249°	
Uric acid					-0.240°	
Glycemia		0.402*				
Triglycerides			-0.226°			
CEA		0.269°				
Bromine.	0.250°					

R correlation coefficients not reaching statistical signifiance are not shown.
P values : ° : $P < 0.05$ + : $P < 0.01$ * : $P < 0.001$

The strongest association was found to be the one with ceruloplasmin ($R = 0.550$, $P < 0.001$). Cu was also correlated with the value of the C1q assay ($P < 0.001$). Zn was inversely correlated with most of the inflammatory tests. The opposite effects of Cu and Zn accounted for the strong relationship between inflammation and the CZR. The CZR also bore an inverse relationship with the hematocrit ($P < 0.001$).

Discussion

The results of this study urge us to stress some problems of methodology in trace element research. The first problem would be the use of several samples taken from the same patient at different times, in order to increase sample size (1, 9). Repeated extrem values in a same patient may create some "highly significant" correlations (data not shown) and may considerably alter mean values. Another problem arises from weak but significant correlation coefficients between two parameters (Table 1). Figures with marginally significant P values should be interpreted very cautiously and be further investigated in larger studies before definitive conclusions can be drawn. Although HPD are diseases sharing many characteristics in common, there are important differences that may change the significance of some parameters from one disease to another. Therefore, some significant associations of STE with other biological parameters could be missed when amalgaming patients with different diagnosis. However, in a previous study (12), we found similar results among the different groups of patients with various HPD. The findings were confirmed in the present study (Table 1). Each group bore a similar pattern, i. e. decreased Zn and increased Cu levels, both being moderately but highly significantly altered. The elevation of the CZR was more impressive, perhaps to a larger extent in HD and NHL than in the other groups. Similar results have been reported by others for MM (6, 8, 9), NHL (3, 4, 6 - 9), HD (1, 2, 5- 1o) and CLL (6, 10). However, the significance of these findings are difficult to assess and the mechanism may be multiple. Decreased serum Zn in HPD can be accounted for by anorexia and decreased Zn intake (13), tissue catabolism (14), antimetabolite drugs (13), and intercurrent infections

(13, 14). The role of the so-called leucocytic endogenous mediator (LEM) remains unclear (15). The increased Cu levels in cancer patients have often been considered as a consequence of ceruloplasmin production as part of a nonspecific acute phase reaction (3), but this assumption has been questionned by others (16).

Because we found a similar pattern of STE levels in the different HPD studied, we examined the significance of the observed anomalies in the entire group of 88 patients.

Decreased Zn and increased Cu levels represent, to a certain degree, disease activity in HPD. The inverse relationship between the CZR and the hematocrit also supports this idea. The elevation of serum Cu seemed to be largely due to an aspecific inflammatory process (3, 16). In this prospect, it was not surprising to find a strong association between ceruloplasmin and Cu. However, part if this inflammation could be caused by clinical or subclinical infectious episodes (13, 14).

Decreased Ca and Zn levels probably resulted from impaired balance between increased tissue catabolism and depressed food intake (13, 14). Serum Ca and Zn are normally mainly bound to albumin (17), and both were significantly correlated with serum albumin (Table 2). Zn deficiency may be responsible for the relative immunosuppression of patients with HPD (18 - 20). A number of studies have demonstrated that Zn deficiency impaired lymphocyte function (20,21) and that Zn supplementation could reverse immunosuppression induced by Zn deficiency (20) or by age (22).

The correlation between serum Se on one hand, and gamma-GT and alkaline phosphatase on the other hand, may be due to elevated VLDL levels in patients presenting some degree of cholestasis (17). However, we were enable to show any correlation between triglycerides and Se. This relationship seems to be on contradiction with reported Se deficiency in alcoholic subjects with or without liver disease (23). However, one cannot anticipate the result of liver damage in patients with presumable normal Se stores, from the situation of chronic selenium deficient alcoholics.

Although we found no siginificant association between triglycerides and Se, the relationship between Se and glycemia is to be compared to a recent report on elevated serum Se levels in diabetic children (24). This observation warrants further investigation.

Acknowledgements

We are indepted to the Institut Interuniversitaire des Sciences Nucléaires for financial support. We wish to express our thanks to the Centre de recherche du Cyclotron for its technical assistance.

References

1. Hrgovcic M., Tessmer C. F., Thomas F. B., Fuller L. M., Gamble J. F., Shullenberger C. C. Significance of serum copper levels in adult patients with Hodgkin's disease. Cancer 31, 1337-1345 (1973).

2. Thorling E. B., Thorling K. The clinical usefulness of serum copper determinations in Hodgkin's disease. A retrospective study of 241 patients from 1963 - 1973. Cancer 38, 225 - 231 (1976).

3. Shah I., Lewkow L. M., Khilanani U. Correlation of hypercupremia with other acute phase reactants in malignant lymphoma. Cancer 51, 851 - 854 (1983).

4. Cohen, Y., Epelbaum R., Haim N., McShan D., Zinder, O. The value of serum copper levels in non-Hodgkin's lymphoma. Cancer 33, 296 - 300 (1984).

5. Cavdar A. O., Babacan E., Arcasoy A., Erten J., Ertem U. Zinc deficiency in Hodgekin's disease. Eur. J. Cancer 16, 317 - 321 (19807.

6. Auerbach S. Zinc content of plasma, blood and erythrocytes in normal subjects and in patients with Hodgkin's disease and various hematolic disorders. J. Lab. Clin. Med. 65, 628 - 637 (1965).

7. Koch H. J., Smith E. R., McNeely J. Analysis of trace elements in human tissues. II. The Lymphomatous disease. Cancer 10, 151 - 160 (1957).

8. Herring W. B., Leavell B. S., Paixao L. M., Yoe H. J., Trace metals in human plasma and red blood cells. A study of magnesium, chromium, nickel, copper and zinc. II. Observations of patients with some hematologic disease. Amer. J. Clin. Nutr. 8, 855 - 863 (1960).

9. Pizzolo G., Savarin T., Molino A. M., Ambrosetti A., Todeschini G., Vettore L. The diagnostic value of serum copper levels and other hematochemical parameters in malignancies. Tumori 64, 55 - 61 (1978).

10. Lahey M. E., Gubler C. J., Cartwright G. E., Wintrobe M. M. Studies on copper metabolism. VII. Blood copper in pregnancy and various pathologic states. J. Clin. Invest. 32, 329 - 339 (1953).

11. Johansson T. B., Akselsson R., Johansson S. A. E. X-Ray analysis: elemental trace analysis at the 10 - 12 /ug level. Nucl. Instrum. Methods 84, 141 - 143 (1970).

12. Beguin Y., Delbrouck J. M., Roelandts I., Weber G. Serum trace elements in hemoproliferative diseases. In: "Trace element-analytical chemistry in medicine and biology", Vol. 3, P. Brätter and P. Schramel Eds, Walter deGruyter, Berlin-New York. pp. 383 - 389 (1984).

13. Sandstead H. H., Vo-Khactu K. P., Solomons N. Conditioned zinc deficiencies. In: "Trace elements in human health and disease", Vol. 1, A. S. Prasad Ed., Academic Press, New York, pp. 33 - 49 (1976).

14. Cuthbertson D. P., Fell G. S., Smith C. M., Tolstone W. J. Metabolism after injury. I. Effect of severity, nutrition and environmental temperature on protein, potassium, zinc and dreatine. Brit. J. Surg. 59, 925 - 931 (1972).

15. Beisel W. R., Pelarek R. S., Wannemacher R. W. Homeostatic mechanisms affecting plasma zinc levels in acute stress. In: "Trace elements in human health and disease", Vol. 1, A. S. Prasad Ed., Academic Press, New York. pp 87 - 106 (1976).

16. "Copper and lymphomas" Hrgovcic M. J. and Shullenberger C. C. Eds, CRC Press, Boca Raton, Florida (1984).

17. "Trace elements in human health and disease", Vol. 1, A. S. Prasad Ed, Academic Press, New York (1976).

18. Acar S., Ersey F., Sanal O., Bechel I., Gebri N. Cell-mediated immunity, serum and lymphocyte zinc levels in Hodgkin's disease Prog. Clin. Biol. Res. 129, 255 - 260 (1983).

19. Cavdar A. O., Babacan E., Ertem U., Gozdasoglu S., Arcasoy A. Zinc status and cellular immunity in pediatric Hodgkin's disease. Prog. Clin. Biol. Res. 129, 207 - 220 (1983).

20. Allen J. I., Bell E., Boosalis M. G., Oken M. M., McClain C.J. Levine A. S., Morley J. E. Association between urinary zinc excretion and lymphocyte dysfunction in patients with lung cancer. Amer. J. Med. 79, 209 - 215 (1985).

21. Pelarek R. S., Sandstead H. H., Jacob R. A., Barcome D. F. Abnormal cellular immune responses during acquired zinc deficiency. Amer. J. Clin. Nutr. 32, 1466 - 1471 (1979).

22. Duchateau J., Delepesse G., Vrijens R., Collet H. Beneficial effects of oral zinc supplementation on the immune response of old people. Amer. J. Med. 70, 1001 - 1004 (1981).

23. Dwork B., Rosenthal W. S., Jankowski R. H., Gordon G. G., Haldea D. Low blood selenium levels in alcoholics with and without advanced liver disease. Correlations with clinical and nutritional status. Dig. Dis. Sci. 30, 838 - 844 (1985).

24. Gebre-Medhin M., Ewald U., Plantin L. O., Tuvemo T. Elevated serum selenium in diabetic children. Acta Paediatr. Scand. 73, 109 - 114 (1984).

EVALUATION OF NORMAL LEVELS OF MAJOR AND TRACE ELEMENTS IN HUMAN LUNG TISSUE

R. Cornelis*, and C. Vanoeteren

Laboratory for Analytical Chemistry, Rijksuniversiteit Gent
Proeftuinstraat 86, B-9000 Gent, Belgium

E. Sabbioni

Commission of the European Communities, Radiochemistry Division,
Joint Research Centre, 21020 Ispra (Varese) - Italy

*Senior Research Associate of the National Fund for Scientific Research

Introduction

Over the past decades research on the major and trace element composition of human lung tissue has produced an impressive number of data, scattered all through the medical and analytical chemistry literature. A state of the art description may be a useful document in establishing baseline concentrations for future biological monitoring with epidemiological and occupational objectives. Lungs are the more relevant as they constitute the important entry into the body for various air-borne substances.

All the values are expressed in $\mu g.g^{-1}$ wet tissue. If necessary the published data have been recalculated either on the basis of conversion factors given by the authors or on the assumption of a 20% dry weight or a 1% ash content.

Material

As a rule the lung concentrations only relate to persons not occupationally exposed to any kind of heavy atmospheric contamination, and exclude patients who died as a result of a lung disease or suffered one. In the more than 100 papers reviewed the number of subjects studied varied between 1 and 500. The majority of the authors analysed only a single, quite often a very small part of the lung. Moreover, it is rare to find an

Trace Element – Analytical Chemistry in Medicine and Biology, Vol. 4

explicit description of its particular anatomical site within the lung. A serious consideration in tissue sampling is whether the sites analysed yield values that are representative of the organ as a whole. In the case of lungs, there is evidence that the distribution of a majority of elements is very inhomogeneous.

The intrinsic value of the trace element levels in human lung raises significantly if information on the donors is available. Data with respect to age, sex and living environment are rather frequent, be they often vague and equivocal. A full record of the donors' backgrounds is missing in most cases.

Sample Preparation

Sample preparation procedures appear to be not uniform at all and the precautions taken to avoid contamination vary between nihil and quite substantial. Stainless steel implements are not to be used as they may introduce uncontrollable blanks of Co, Cr, Fe, V,... . The best strategy consists without any doubt in selecting a high purity material for cutting that will produce no consequential contamination. For example, the use of a quartz knife, respectively a Ti knife when there is no analytical interest in Si or Ti, but in other elements which are absent from the knife.

Washing procedures of the lung tissue e.g. with distilled H_2O, have disappeared from the more recent papers. In the majority of the investigations, tissue samples are stored frozen before the analysis. Very often there is a drying procedure included (lyophilization, dry ashing), occasionally a homogenization. A few researchers resorted to preservatives such as formalin to store their lung samples but this carries the risk of introducing blanks and of leaching out some elements. Such practises should not be applied any more.

As stated, previously all the results are compared on a wet weight basis.

The published data bear much more credibility if the investigators have indicated the steps taken to ensure both the precision and the accuracy of their analyses results.

Post-mortem Changes

Nearly all of the determinations were done on autopsy specimens. Post-mortem changes such as cell swelling, tissue dehydration, imbibition, putrefaction and autolysis occur during the time lapse between death and sample collection. This may affect the elemental composition profiles of the lung tissue through weight changes as demonstrated by Iyengar (1) and Pietra et al. (2). The impact of possible post-mortem changes on the published data is, however, unknown.

Major and Trace Element Concentrations

The literature survey resulted in a compilation of concentrations for 67 elements in human lung tissue. These data, together with extensive comments, can be read in a special CEC-report (3).

The following paragraphs briefly report on the effect of different parameters liable to influence the elemental composition of lung tissue.

Anatomical distribution of the elements over the lung

A small number of investigators analysed the different anatomical segments of one lung pair in order to learn the elemental distribution pattern (4-12).

Vanoeteren et al. (9-11) observed that the elements Br, Cs, Cu, K, Na, Rb, Se and Zn are relatively homogeneously spread over one lung pair and also show little interindividual concentration differences. On the other hand Cd, Co, Cr, Pb, Sb, Sc and V are very inhomogeneously distributed. Those are the elements which occur at the $ng.g^{-1}$ levels, sometimes up to a few hundred $ng.g^{-1}$. The other researchers yielded data which imply similar conclusions.

Some authors (4-7) established that certain elements (Co, Cr, Ni, Sb, Sr, Ti,...) exhibit a positive concentration gradient from the lung bases to the lung tops (apexes), whereas Cd appears to be often higher in the middle lobe of the right lung (5). Vanoeteren et al. (9-11) found a similar pattern for Co, Ce, Sb, Sc, V in their oldest donors (> 70 years) only.

Age factor

A certain effect of age on the trace element content of lung tissue may be anticipated as a result of a cummulative contribution of substances of exogenous origin with time. The outcome of sixteen studies on this aspect implies, however, that there exists no unanimous proof for a general relationship between age and element levels in lung. Al is most often mentioned as being positively related with age (13, 14, 15). The same applies to Pb, be it to a lesser degree (16,17). Cu and Zn appear independent of aging. The situation of Cd is more obscure. Detailed studies seem to reveal a peak in the Cd levels at maturity (18,19). For the other elements inconsistent conclusions were drawn. The evaluation of the different studies would be highly facilitated with a more systematic and uniform planning of the age groups. Age, however, is not the only factor influencing the lung concentrations.

Sex

As men's and women's life-style still differs, it may be worthwhile to look for variations in the trace element composition of lung tissue between both sexes. There appears, however, from fourteen studies dealing with this parameter, no overall proof that particular elements occur on different levels in the lungs of males and females. The higher concentrations found in the lungs of men for Cr, Fe, Mn, Pb, Ti (13), for Cd (20) and for Pb (16,17) may be the result of fortuitous exposures to higher concentrations in the working environment or of more intensive cigarette smoking.

Geographical environment

It is most plausible that marine, montain, tropical, desert, urban, industrial, etc. locations influence the trace element composition of the lungs through difference in inhaled air. A few research centres undertook to study this parameters.

A major contribution has been delivered in the early sixties by Tipton et al. who analysed tissues of persons from the United States of America (13), the African continent, the Near East, the Far East and Switzerland (21). Some of these data mentioned may, however, lack the necessary accuracy, attributable to insufficient sensitivity of the analytical method and to some deficiencies in sample handling and storing.

Other researchers compared lung concentrations of patients who lived their entire lives in an industrial area to those of patients of more rural habitats (15,16,22). The various studies cited point out discrete influences of the geographical environment on the concentration of some trace elements in human lung tissue.

Ambient air

A few research groups tried to put into evidence a causal relationship between the trace elements present in lung and the ambient air composition (15,23,24). The problem is very complex since many factors concerning inhalation, deposition and clearance influence the lung-air concentration relationship. Only exceptional exposures allow an irrefutable link between cause and effect. It may be interesting to cite the study by Bernstein et al. (24) who analysed right lungs of New York City residents for Cd, Cr, Cu, Fe, Mn, Ni, Pb and Zn. There appeared to be a direct correlation between the trace metal content of New York City air and that found in the lungs and lymph nodes of the subjects.

Distribution of the lung concentrations over a population

Some researchers undertook to describe the trace element concentration pattern in specific groups of individuals.

Sweet et al. (25) performed a statistical analysis of their data on lung tissue of 100 urban residents. They concluded that the elements Al, Cd, Co, Cr, Cu, Ge, I, Mg, Mn, Mo, Pb, Se, Ti, V and Zn were log normally distributed over their particular group. Tipton et al. (13,21,22) came to a similar conclusion for the elements they studied. Along the same lines Stein et al. (22) described a log normal distribution for Hg in the lungs of 2 groups from different geographical areas.

Historical evaluation of lung levels

Kevorkian et al. (26) analysed many preserved human autopsy tissues, including lung, for total Hg and methyl Hg. The tissues had been sampled from 1910 till 1970 from persons who had lived in Michigan. They demonstrated that the Hg content in almost all organs, certainly in lungs, declined sharply in the early decades of this century, levelling off recently. The reason for the rather logarithmic decline could only be guessed at. One logical factor could be the decline and disappearance of coal, an important source of Hg for heating individual homes. This study is the only one to compare tissue concentrations over such a long period of time. It suffers, however, from the important setback that the tissues were all fixed in a formalin solution and preserved in ethyl alcohol, some of them for almost 60 years. The blank value of the preservative is a big unknown parameter, seriously jeopardizing the validity of Kevorkian et al.'s conclusion.

Dust content of the lungs

Dust accumulated in lung tissue largely contributes to its trace element content. Several procedures are described in the literature attempting to isolate the dust fraction from the lung tissue as a whole (4,9,25,27). The enzymatic procedure (9,25,27) is a more elegant and safe way to destroy the organic material than high temperature ashing (4).

Sweet et al. (25) report a geometric mean of 15.92 $mg.g^{-1}$ wet lung tissue, a value surprisingly similar for their urban residents and their West-Virginia coal miners.

The mean total dust reported by Vanoeteren et al. (27) reached 3 $mg.g^{-1}$ wet tissue only. The dust content was higher in the lungs of older persons. A major part of the Co, Cr, Sb and Sc present in the tissue was recovered in the dust residues but only small fractions of the Cs, Rb and Zn content.

Concluding Remarks

An attempt was made to interpret the vast amount of literature data on major and trace elements in normal human tissue and to put forward indicative baseline data. These are summarized in table 1. The reader be cautioned against the high degree of speculation built into these figures. This very unscientific approach arises from a suspicion about the intrinsic quality of some lung samples and/or about the lack of accuracy and sensitivity of some methodologies (sample handling and detection method). For many elements these doubts are also substantiated by the modest number of studies and the very limited number of lungs analysed.

The proposed baseline data for trace elements present in lung tissue at the $ng.g^{-1}$ level more often than not cover a very broad range (several orders of magnitude). It may seem an easy disguise for the uncertainty of some data, yet it is meant to reflect the outspoken inhomogeneous distribution over the lungs and in the population as a result of a cumulative effect over the lifespan. The fraction of each of these trace elements controlled by homeostatic mechanisms undoubtedly covers a relatively small spread. However, as the respiratory system is the gateway for a continuous flow of atmospheric contaminants, the resulting trace element pattern will be governed by each individual's own processing mechanisms and his lifestyle. Such a multivariate system can yet hardly be unravelled by lack of a statistically sufficient number of data and a complete description of all possible parameters. Only for Cd and Pb the situation appears to be slightly better documented.

The essential elements Cu and Zn, who do exceed the $\mu g.g^{-1}$ level and cover a narrow range, are less liable to be influenced by relatively smaller amounts of exogenous material. On top of that when this is soluble, it will readily be eliminated by the bloodstream. On the contrary the non-essential element Al, also occurring at the $\mu g.g^{-1}$ level, is supposed to originate mainly from poorly soluble Al compounds, exerting a cumulative effect.

Although valuable work has already been done on the trace element composition of normal human lung tissue, a closer look at the contents of table 1 invites for many more investigations. This bears especially on polluting elements such as As, Cr, Co, F, Hg, Ni, Sn, Ti, Tl, V and even Cd and Pb, to cite just a dozen. There is a great need for establishing reference ranges as narrow as possible. This imposes the separate knowledge of the variations due to the many biological and analytical parameters involved. Such an elaborate endeavour would greatly benifit from the co-ordinated effort of several research groups.

Table 1 : Indicative Baseline Data for Major and Trace Elements in Normal Human Lung Tissue, Derived from the Compilation of Literature Data Summarized in ref. (3). The Values between Brackets are More Speculative than the Others.

Element	Concentration µg.g^{-1} wet lung tissue	Element	Concentration µg.g^{-1} wet lung tissue
Actinides :			
Thorium	(0.001 - 0.02)*	Europium	(0.0001 - 0.004)*
Uranium	(0.005)	Gadolinium	(<0.01)*
Aluminium	1 - 50	Holmium	(<0.001)*
Antimony	(0.002 - 0.1)*	Lanthanum	(0.001 - 0.2)*
Arsenic	0.001 - 0.1	Lutetium	(0.0001 - 0.01)*
Barium	(0.03 - > 1.1)*	Neodymium	(0.006 - 0.09)*
Beryllium	(0.0001 - 0.03)*	Samarium	(0.00007- 0.006)*
Bismuth	(0.002 - 0.2)*	Terbium	(0.0004 - 0.003)*
Boron	no value recommended	Ytterbium	(0.0002 - 0.006)*
Bromine	0.4 - 24	Lead	0.02 - 0.5
Cadmium	0.01 - 1.0	Lithium	no value recommended
Caesium	0.003 - 0.3	Magnesium	20 - 250
Calcium	50 - 300	Manganese	(0.01 - 3)*
Chlorine	1000 - 2200	Mercury	(<1)*
Chromium	0.002 - 0.5	Molybdenum	0.001 - 1.0
Cobalt	0.002 - 0.1	Nickel	(0.01 - 1.0)*
Copper	1 - 5	Niobium	no value recommended
Fluorine	(0.01 - 0.1)*	Phosphorus	500 -2000
Gallium	(0.0004)*	Platinum	(0.5)*
Germanium	(0.02 - 1.0)*	Potassium	500 -2000
Gold	(0.0001 - 0.01)*	Rubidium	0.5 - 10
Hafnium	(0.002)*	Scandium	(0.0001 - 0.007)*
Indium	no value recommended	Selenium	0.05 - 0.5
Iodine	no value recommended	Silicium	(30 - 100)*
Iridium	(0.00003- 0.0003)*	Silver	0.001 - 0.01
Iron	40 - 500	Sodium	1000 -3000
Lanthanides :		Strontium	(0.010 - 1.0)*
Cerium	(0.001 - 0.4)*	Sulfur	no value recommended
Dysprosium	(<0.01)*	Tantalum	(0.004 - 0.05)*
Erbium	(<0.01)*	Tellurium	no value recommended

Thallium	(<0.0002- 0.004)*	Vanadium	0.0005 - 0.5
Tin	(0.1 - 4)*	Yttrium	no value recommended
Titanium	(0.1 -20)*	Zinc	1 - 30
Tungsten	(<0.07 - 0.6)*	Zirconium	(0.05 - 0.2)*

( ) * speculative data

References

1. Iyengar, V.: 1981. Autopsy sampling and elemental analysis : errors arising from post-mortem changes. J.Pathol. 134, 173-180.
2. Pietra, R., Sabbioni, E., Ubertalli, L., Orvini, E., Vocaturo, G., Colombo, F., Zanioni, M.: 1985. Trace elements in tissue of a worker affected by rare earths pneumoconiosis. A study carried out by neutron activation analysis. J.Radioanal.Nucl.Chem. 92, 247-259.
3. Vanoeteren, C., Cornelis, R., Sabbioni, E.: 1986. Critical Evaluation of Normal Levels of Major and Trace Elements in Human Lung Tissue, EUR-10440-EN, CEC-Luxembourg.
4. Gibbs, G.W., Bogdanovic, E.: 1974. In: Recent Advances in the Assessment of the Health Effects of Environmental Pollution, Paris. Proc. Symp., Paris pp. 2271-2283.
5. Molokhia, M.M., Smith, H.: 1967. Trace elements in the lung. Arch.Environ.Hlth. 15, 745-750.
6. Bartsch, P., Weber, G., Roelandts, J., Collignon, A., Delbrouck, J.M., Robaye, G.: 1981. In: Medical Applications of Cyclotrons, Turku (1980) Proc. Second Symp., Ann.Univ. Turkuensis D, Ed. Nöntö, V., Suolinna, E.M.,: 13, 168-173.
7. Bartsch, P., Collignon, A., Weber, G., Robaye, G., Delbrouck, J.M., Roelandts, I., Yujie, J.: 1982. Distribution of metals in human lung : analysis by particle induced X-ray emission. Arch.Environ.Hlth. 37, 111-117.
8. Schicha, H., Kasperek, K., Feinendegen, L.E., Siller, V., Klein, H.J.: 1972. Inhomogeneous but partly parallel distribution for cobalt, iron, selenium, zinc and antimony in different regions of liver, lung, kidney, heart and aorta, measured by neutron activation analysis. Beitr.Path.Bd. 146, 55-62.
9. Vanoeteren, C., Cornelis, R., Versieck, J., Hoste, J., De Roose, J.: 1982. Trace element patterns in human lung tissues. J. Radioanal.Chem. 70, 219-238.
10. Vanoeteren, C., Cornelis, R., Hoste, J., Vander Haeghen, L., Versieck, J.: 1983. In: Trace Element Analytical Chemistry in Medicine and Biology, Vol.2, (Brätter, P. Schramel, P., eds) Walter de Gruyter & Co., Berlin; New York, pp. 315-332.

11. Vanoeteren, C., Cornelis, R., Versieck, J.: 1986. Evaluation of trace elements in human lung tissue. Part 1. Sci.Tot.Environ. in press.

12. Mylius, E.A., Ophus, E.M.: 1977. Pulmonary distributions of lead in human subjects. Bull.Environ.Contam.Toxicol. 17, 302-310.

13. Tipton, I.H., Cook, M.J.: 1963. Trace elements in human tissue. Part II. Adult subjects from the United States. Health Phys. 9, 103-145.

14. Stitch, S.R.: 1957. Trace elements in human tissue. 1. A semi-quantitative spectrographic survey. Biochem.J. 67, 97-109.

15. Persigehl, M., Kasperek, K., Klein, M.J., Feinendegen, L.E.: 1976. Influence of industrialization on trace element concentrations in human lungs. Beitr. Path.Bd. 157, 260-268.

16. Ophus, E.M., Mylius, E.A.: 1977. Lead levels in human lungs. Bull.environm.Contam.Toxicol. 18, 734-741.

17. Stringer, C.A. Jr., Zingaro, R.A., Creech, B., Kolar, F.L.: 1974. Lead concentrations in human lung samples. Arch.Environ.Hlth. 29, 268-270.

18. Casey, C.E., Guthrie, B.E., McKenzie, J.M.: 1981. New Zealand Workshop on Trace Elements in New Zealand, Dunedin, Proc. Workshop, Univ. Otago, 210-217.

19. Vuori, E., Huunan-Seppällä, A., Kilpiö, J.O., Salmela, S.S.: 1979. Biologically active metals in human tissues. II. The effect of age on the concentration of cadmium in aorta, heart, kidney, liver, lung, pancreas and skeletal muscle. Scand.J.Work.Environ.Hlth. 5, 16-22.

20. McKenzie, J.M.: 1974. Tissue concentration of cadmium, zinc and copper from autopsy samples. NZ. Med.J. 12, 1016-1019.

21. Tipton, I.H., Schroëder, H.A., Perry, H.H., Cook, M.J.: 1965. Trace elements in human tissue. Part III. Subjects from Africa, the near and far East and Europe. Health Phys. 11, 403-451.

22. Stein, P.C., Campbell, E.E., Moss, W.D., Trujillo, P.: 1974. Mercury in man. Arch.Environ.Hlth. 29, 25-27.

23. Brandone, A., Borroni, P.A., Genova, N.: 1983. Determination of arsenic, cadmium and mercury in biological samples by neutron activation analysis. Radiochem.Radioanal.Lett. 57, 83-94.

24. Bernstein, D.M., Kneip, T.J., Kleinman, M.T., Riddick, R., Eisenbud, M.: 1974. Trace Substances in Environmental Health VIII, Proc.Symp., University Missouri, Columbia, Ed. Hemphill D.D., 329-337.

25. Sweet, D.V., Crouse, W.E., Crable, J.V.: 1978. Chemical and statistical studies of contaminants in urban lungs. Am.Ind.Hyg.Ass.J. 39, 515-526.

26. Kevorkian, J., Cento, D.P., Hyland, J.R., Bagozzi, W.M., Van Hollebeke, E.: 1972. Mercury content of human tissues during the twentieth century. Am.J. Public Hlth. 62, 504-513.

27. Vanoeteren, C., Cornelis, R., Dams, R.: 1986. Recovery and trace element analysis of inhaled particulates from human lungs. Part 2. Sci. Tot.Environ. In press.

SERUM SELENIUM LEVELS AND GLUTATHIONE PEROXIDASE ACTIVITIES IN ERYTHROCYTES OF RHEUMATIC AND DIABETIC CHILDREN AND THEIR FAMILIES LIVING IN WESTERN FINLAND

A.-L. Mäkelä, M. Tasanne, E. Lazowska
Department of Paediatrics, University Central Hospital, Turku, Finland

K. Pulkki, H. Hyörä, V. Näntö
Department of Clinical Chemistry, University Central Hospital, Turku, Finland

The determination of the activity of erythrocyte glutathione peroxidase (E-GPx) is a relatively simple and sensitive method for the assessment of long-term selenium depletion in man (1). Incorporation of selenium in the enzyme occurs during erythropoesis and cell maturation. E-GPx activity is thus not rapidly affected by temporary changes in selenium status, whereas prolonged selenium depletion in nutrition is well reflected in decreased enzyme activity (2)

Chronically ill patients, especially those living on selenium-depleted areas, may easily get a negative selenium balance. Low serum selenium values and decreased activities of GPx in plasma, blood and erythrocytes have been reported in rheumatic patients living in Scandinavia (3,4,5,6). Selenium supplementation does not cure the rheumatic disease in itself (7). Adequate selenium intake increases, however, the activity of GPx towards the normal in rheumatic patients also (8). The recognition of the selenium depleted state and adequate supplementation may benefit the patient clinically by preventing the development of fatal cardiomyopathy and by relieving otherwise unexplained muscle pains.

Children are more susceptible to the development of selenium depletion than adults. In the Keshan disease area, the most susceptib-

le groups were children below 10 years old (9). The selenium concentration in serum has been noted to be age-dependent (10,11,12, 13). The lowest values occur during the first half year of life. This may, however, at least partly, depend on the low selenium content of milk formulas or human milk. The highest concentrations are noted in adults in the age of 20-40 years. After that, a gradual decline occurs with age.
E-GPx activities have also been reported to be lower in infants than in normal adults (11). Later on in the childhood, however, no significant differences were found between the E-GPx activities of toddlers and school children. If, however, the selenium intake is very low, and the erythrocytes do not have sufficiently selenium available for enzyme synthesis, the marginal selenium deficiency state is reflected in decreased activities of E-GPx. At low erythrocyte selenium concentrations the E-GPx activity is known to become strongly selenodependent (2).
The purpose of our study was to detect the frequency of possible marginal selenium deficiency state in rheumatic and diabetic children in Finland. The levels of E-GPx activities and serum selenium concentrations (S-Se) were measured from the patients and their parents and siblings in our study.

Material and methods

The activities of E-GPx were determined by a modification (14,15) of the method of Paglia and Valentine (16) in 90 rheumatic and 77 diabetic children and 177 members of their families. 104 healthy persons (37 children under the age of 15 years) served as controls. Corresponding serum selenium levels were determined by flameless atomic absorption spectrophotometry by the modified method of Alfthan and Kumpulainen (17).
Intra-assay precision studies for the assays gave 5.1 % for a mean erythrocyte GPx activity of 74.5 (3.8) U/gHb (n=18) and 4.2 % for a mean serum selenium concentration of 0.94 (0.04) umol/l (n=10). Inter-assay precision gave 7.5 % for a mean erythrocyte GPx activity of 54.8 (4.1)U/gHb (n=25) and 5.2 % for a mean serum selenium concentration of 1.15 (0.06) umol/l (n=78).

Results

As a consequence of the poor selenium intake of the Finnish population, in general, the S-Se concentrations were relatively low in all patients as well as in controls (Table 1).

Table 1. Serum Selenium Levels (S-Se) umol/l in Patients, in the Members of Their Families and in Controls.

	Rheumatic	Diabetic	Controls
			Children
Patients No	85	68	33
Age: x (SD)	9.7 (3.7)	10.3 (3.6)	9.0 (4.1)
range	2-15	2-15	2-15
S-Se x (SD)	0.87 (0.16)	0.99 (0.14)	0.81 (0.14)
Siblings No	46	37	
Age: x (SD)	8.7 (3.7)	8.9 (3.8)	
range	2-15	2-15	
S-Se x (SD)	0.86 (0.12)	0.90 (0.14)	
			Adults
Parents No	57	30	34
Age: x (SD)	37.3 (5.5)	36.7 (4.7)	37.9 (5.0)
range	29-49	29-49	29-49
S-Se x (SD)	1.01 (0.14)	1.03 (0.17)	0.94 (0.12)

In this material, too, the S-Se concentrations of children were significantly lower than those of adults ($p<0.001$), except for the diabetic patients, by whom S-Se concentrations corresponded to those found in adults. The S-Se concentrations of rheumatic children did not statistically differ from those of age-matched controls.

No age-dependency was observed in S-Se concentrations or E-GPx activities in children under the age of 15.

Young individuals, in general, had slightly lower E-GPx activities than older persons (Table 2).

In S-Se concentrations and E-GPx activities there were no significant differencies between males and females.

The E-GPx activities of rheumatic children and their siblings were significantly lower than those of age-matched controls ($p<0.001$). Diabetic children also had lower E-GPx activities than age-matched controls ($p<0.005$). Between rheumatic and diabetic children or

Table 2. The Activities of E-GPx (U/gHb) in Patients, in the Members of Their Families and in Controls.

	Rheumatic	Diabetic	Controls
			Children
Patients No	90	77	37
Age: x (SD)	9.9 (3.8)	10.4 (3.6)	9.4 (3.6)
range	2-15	2-15	2-15
E-GPx x (SD)	57.3 (13.9)	60.8 (16.5)	70.2 (15.4)
Siblings No	47	43	
Age: x (SD)	8.7 (3.6)	8.8 (3.6)	
range	2-15	2-15	
E-GPx x (SD)	58.5 (14.0)	63.8 (14.2)	
			Adults
Parents No	57	30	67
Age: x (SD)	37.3 (5.5)	36.7 (4.7)	37.9 (5.6)
range	29-49	29-49	29-49
E-GPx x (SD)	70.8 (16.6)	67.4 (14.3)	77.2 (15.4)

their siblings there were no significant differences in E-GPx levels. No correlation was observed between S-Se concentrations and corresponding E-GPx activities (Fig. 1).

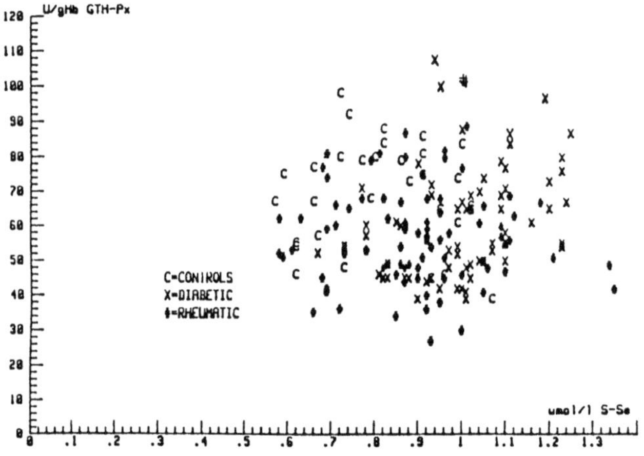

Figure 1. Lack of correlation between S-Se concentrations and E-GPx activities in patients and in control children.

Discussion

The purpose of this study was to assess the selenium status of diabetic and rheumatic children and their families.
Rheumatoid arthritis is often associated with a secondary anemia. It has been shown that iron-deficiency anemia is accompanied by decreased E-GPx activity, and supplementation with iron results in a rapid increase in the activity of E-GPx (18,19,20). Studies with experimental animals have shown also, that the decrease in E-GPx activity is manifest in iron-deficiency anemia even when the activity is expressed in terms of per unit of hemoglobin (21). This may partly explain the occurence of low E-GPx activities in rheumatoid arthritis, as observed in this study, too. An other theoretical explanation is a primary inability to utilize dietary selenium for E-GPx synthesis. On the other hand the poor appetite associated with chronic inflammatory diseases eg. rheumatoid arthritis may give some explanation.
In diabetic patients serum selenium levels have been reported to be normal (22) or even higher than in healthy age-matched controls (23). This may be explained by the specific diet recommended for diabetic patients. In our study also, diabetic children had higher S-Se concentrations than their siblings, children with rheumatoid arthritis or controls.

Conclusion

The results of our study suggest, that the selenium intake of Finnish population seems not always to be sufficient for GPx synthesis. Children with chronic diseases are apparently more prone to develop marginal selenium deficiency states than children with corresponding diseases on selenium-rich areas.
Selenium content of erythrocytes may give better information of the nutritional selenium status in individual cases and the correlation with E-GPx activities may be better.

References:

1. Van Rij, A.M., C.D. Thomson, J.M. McKenzie, M.F. Robinson. 1979. Selenium deficiency in total parenteral nutrition. Am.-J.Clin.Nutr. 32, 2076-2085.

2. Nève, J., F. Vertongen, N. Thonnart, Y.A. Carpentier. 1984. In: Trace Element Analytical Chemistry in Medicine and Biology, Vol. 3 (P. Brätter and P. Schramel, eds.). Walter de Gruyter, Berlin, New York. pp. 139-149.

3. Aaseth, J., E. Munthe, Ø. Førre, E. Steinnes. 1978. Trace elements in serum and urine of patients with rheumatoid arthritis. Scand J.Rheumatology 7, 237-240

4. Åkesson, B., U. Johansson, P.-A. Öckerman, S. Portinsson, A. Åkesson, H. Svantesson. 1984. In: Trace Element Analytical Chemistry in Medicine and Biology, Vol. 3 (P. Brätter and P. Schramel, eds.). Walter de Gruyter, Berlin, New York. pp. 167-174.

5. Mäkelä, A.-L., H. Hyörä, K. Vuorinen, V. Näntö. 1984. Trace elements (Fe, Zn, Cu and Se) in serum of rheumatic children living in western Finland. Scand.J.Rheumatology 53 (Suppl.) 94.

6. Tarp, U., K. Overvad, J.C. Hansen, E.B. Thorling. 1985. Low selenium level in severe rheumatoid arthritis. Scand. J. Rheumatology, 14, 97-101.

7. Tarp, U., K. Overvad, E.B. Thorling, H. Graudal, J.C. Hansen. 1985. Selenium treatment in rheumatoid arthritis. Scand.J. Rheumatology 14, 364-368.

8. Munthe, E., J. Aaseth. 1984. Treatment of rheumatoid arthritis with selenium and vitamin E. Scand.J.Rheumatology 53 (Suppl.) 103.

9. Keshan Disease Research Group of Chinese Academy of Medical Sciences. 1982. Keshan Disease and Selenium in China. Curr. Concepts Nutr. 11, 127-138.

10. Kasperek, K., H. Shicha, V. Siller, L.E. Feinendegen. 1972. Normalwerte von Spurenelementen im menschlichen Serum und Korrelation zum Lebensalter und zur Serum-Eiweiss-Konzentration. Strahlentherapie 143, 468-472.

11. Lombeck, I., K. Kasperek, H.D. Harbisch, L.E. Feinendegen, H.J. Bremer. 1977. The selenium state of healthy children. I. Serum selenium concentration at different ages; Activity of glutathione peroxidase of erythrocytes at different ages; Selenium content of food of infants. Europ.J.Pediat. 125, 81-88.

12. Westermarck, T., P. Raunu, M. Kirjarinta, L. Lappalainen. 1977. Selenium content of whole blood and serum in adults and children of different ages from different parts of Finland. Acta pharmacol. et toxicol. 40, 465-475.

13. Hatano, S., Y. Nishi, T. Usui. 1984. Plasma selenium concentration in healthy Japanese children and adults determined by flameless atomic absorption spectrophotometry. J Pediatr.Gastroenterol.Nutr. 3, 426-431.

14. Gunzler, W.A., H. Kremers, L. Flohe. 1974. An improved coupled test procedure for glutathione peroxidase (EC 1.11.1.9) in blood. Z.Klin.Chem.Klin.Biochem. 12, 444-448.

15. Pleban, P.A., A. Munyani, J. Beachum. 1982. Determination of selenium concentration and glutathione peroxidase activity in plasma and erythrocytes. Clin.Chem. 22, 311-316.

16. Paglia, D.E., W.N. Valentine. 1967. Studies on the quantitative and qualitative characterization of erythrocyte glutathione peroxidase. J.Lab.Clin.Med. 70, 158-169.

17. Alfthan, G., J. Kumpulainen. 1982. Determination of selenium in small volumes of blood plasma and serum by electrothermal atomic absorption spectrometry. Anal.Chim. Acta 140, 221-227.

18. MacDougall, L.G. 1972. Red cell metabolism in iron deficiency anemia. III. The relationship between glutathione peroxidase, catalase, serum vitamin E and susceptibility of iron-deficient red cells to oxidative hemolysis. J.Pediatr. 80, 775-782.

19. Hopkins, J., G.R. Tudhope. 1973. Glutathione peroxidase in human red cells in health and disease. Brit.J.Haematol. 25, 563-575.

20. Perona, G., R. Cellerino, G.C. Guidi, G. Moschini, B.M. Stievano, C. Tregnaghi. 1977. Erythrocytic glutathione peroxidase: its relationship to plasma selenium in man. Scand.J. Haematol. 19, 116-120.

21. Rodvien, R., A. Gillum, L.R. Weintraub. 1974. Decreased glutathione peroxidase activity secondary to severe iron deficiency: A possible mechanism responsible for the shortened life span of the iron-deficient cell. Blood. 43, 281-289.

22. Burk, R.F. Selenium in man. In: Trace Elements in Human Health and Disease. Vol. 2, (A.S. Prasad, ed.). Academic Press, New York. pp. 105-133.

23. Gebre-Medhin, M., U. Ewald, L.-O. Plantin, T. Tuvemo. 1984. Elevated serum selenium in diabetic children. Acta Paediatr. Scand. 73, 109-114.

TRACE ELEMENT LEVELS IN CHILDREN WITH JUVENILE RHEUMATOID ARTHRITIS

H. Mussalo-Rauhamaa, J. Lehto

Department of Public Health, University of Helsinki
SF-00290 Helsinki, Finland

P. Pelkonen

Children´s Hospital, University of Helsinki
SF-00290 Helsinki, Finland

V. Honkanen

The Rheumatism Foundation Hospital
SF-18120 Heinola, Finland

T. Westermarck

The Helsinki Central Institution for the Mentally Retarded
SF-02400 Kirkkonummi, Finland

Introduction

It has been reported that the homeostasis of many trace elements in patients with rheumatoid arthritis (RA) is disturbed, a phenomenon which is considered to be linked to the body´s inflammatory response to the disease. Selenium levels have been found to be significantly lower in adult RA patients than in control subjects (1,2,3,4,5), while copper levels in the serum and synovial fluid have been found to increase (6,7). Contradictory findings have been reported regarding the level of zinc in the serum of RA patients; some investigators report lowered concentrations (8,9), while others find no difference in the levels of subjects with RA and their controls (3,10,11). It should be noted that one study which found no difference in zinc levels (10,12) did note a decreased zinc level in RA patients once they began anti-rheumatic treatment (ASA or Indomethacin).

Trace Element – Analytical Chemistry in Medicine and Biology, Vol. 4

Despite the volume of research which has been done in this field, only a few recent studies have dealt with patients suffering from juvenile rheumatoid arthritis (JRA) (13,14,15). To gain further information in this area, selenium, copper and zinc concentrations in the serum of JRA patients and the glutathione peroxidase and (Zn,Cu)superoxide dismutase enzymes in the serum of some of these subjects were investigated.

Patients and methods

The group studied consisted of 105 children with juvenile rheumatoid arthritis (JRA), 80 girls and 25 boys. The mean age of the patients at the time of the study was 8.3 years (range 2.4 to 17.5 years); 8.6 years for the girls and 7.4 years for the boys. The mean duration of affliction was 4.0 years (range 0.1 to 15.8 years). The onset type was systemic in 19% of the patients, and 10% had systemic symptoms at the time of the study. The present state of the disease was oligoarticular in 45% and polyarticular in 55% of the patients. Eight per cent of the patients were positive for rheumatoid factor, 41% had antinuclear antibodies. Seventy per cent of the patients were receiving nonsteroidal antiinflammatory drugs, 44% had been on prednisolone during the 3-month period prior to the study, and 28% had received intra-articular steroids during this period. Most patients were on second-line antirheumatic treatment. The clinical activity of the disease at the time of the study was recorded as absent or minor in 38% and moderate or severe in 62% of the patients.

The patients and their parents were informed about the study in advance, and the patients were advised to abstain from taking any extra medications in form of vitamins, iron and other trace element-containing preparations during two days prior to the sampling. The parents filled in a detailed questionnaire concerning the use of the above-mentioned preparations during the previous month. Only patients without any trace element supplementation were accepted to this present study.

Blood samples were drawn using stainless steel needles (Venoject^R, Terumo, Belgium) following an overnight fast. The serum and cells were separated by centrifugation, and the cells were washed with 0.9% NaCl solution to be analyzed later for glutathione peroxidase and (Zn,Cu)superoxide dismutase enzymes. The methods of analysis assay and the results of these studies will be reported in detail elsewhere at a later date. Serum and erythrocyte samples were stored in polyethylene and glass tubes, respectively, and kept at -20°C until analysis.

Zinc and copper concentrations were determined by flame ionization atomic absorption spectrometry (Perkin Elmer 300 AAS) using a method developed in this laboratory (16). Flameless AAS (Perkin Elmer 5000, HGA-400) was used to analyze the selenium concentrations, utilizing the method described by Alfthan and Kumpulainen (17).

Accuracy was corroborated by comparison with the commercial standard reference material Seronorm^R, batch 157 (Nyegaard and Co, Oslo), and Moni-Trol^R (Dade, Miami, Florida).

Reference values for trace element concentrations in the serum of healthy children have been published earlier by another group (18), utilizing data from 3,480 subjects (Table II). Reference values for trace selenium concentrations were obtained by analyzing serum samples from 920 of these subjects (Finnish multicentre study, Mussalo-Rauhamaa et al., unpublished results).

The statistical analyses were carried out by Student´s t-test.

Results and discussion

The results of the analyses are given in Table I. When compared to reference values (Table II), significant differences can be noted. In children with JRA the mean serum concentrations of zinc and selenium were significantly lower ($p<0.01$) than in healthy subjects. This result was obtained for both boys and girls and in most age groups. The mean concentrations of selenium in the serum of healthy subjects were 63, 73 and 74 µg/l for 3, 9 and 15 year olds (not shown in the Table II); while levels were 56, 57 and 51 µg/l for the roughly same age group of JRA children. One JRA girl had a very high serum Se-value (187 µg/l) but she denied selenium supplementation. Our results support earlier findings by Mäkelä et al. (13) which reported low selenium and zinc levels in subjects with juvenile chronic arthritis.

The serum copper concentrations of JRA patients can be seen to be significantly ($p<0.01$) higher than those reported for healthy subjects. Also, contrary to copper levels in healthy subjects, levels in subjects with JRA are not subject to a significant decrease with age. For female subjects, the phenomenon of a decrease in zinc concentration with an increase in age which was observed in healthy subjects was also found to hold true for subjects with JRA. The increase in zinc concentrations with increased age which was noted in healthy male subjects was not evident, however, in male subjects with JRA.

The ratios of trace elements in the serum of JRA patients were quite steady with respect to the different age groups (Table III). For the healthy ones these ratios increased with age; e.g. for girls the Zn/Cu-ratio varied from 0.69 to 0.81 and for boys from 0.67 to 0.93.

Table I. The zinc, copper and selenium concentrations in the serum of JRA children

Age (mean) (years)	No of cases	Concentration (SD) Cu (µmol/l)	Zn (µmol/l)	Se (µg/l)
Girls				
1.5 - 4.5 (3.5)	12	25.3 (5.5)	13.0 (2.1)	57 (14)
4.5 - 7.5 (6.3)	18	26.2 (4.4)	13.9 (2.1)	53 (10)
7.5 -10.5 (9.0)	24	23.9 (6.7)	12.6 (2.0)	58 (10)
10.5-13.5 (11.7)	18	22.2 (4.0)	11.9 (1.5)	56 (12) a)
13.5-16.5 (14.9)	6	25.1 (3.2)	11.0 (1.2)	52 (14) a)
16.5- (17.3)	1	23.3	15.3	53
Boys				
1.5 - 4.5 (3.3)	6	24.7 (5.2)	13.8 (1.2)	54 (10)
4.5 - 7.5 (5.9)	8	22.4 (5.7)	13.0 (1.4)	59 (11)
7.5 -10.5 (8.2)	6	24.5 (4.6)	12.6 (1.9)	51 (13)
10.5-13.5 (11.6)	3	24.5 (4.2)	14.9 (1.6)	63 (9)
13.5-16.5 (16.4)	1	29.1	12.0	44
16.5- (16.9)	1	17.6	12.0	48

a) one girl with a level of 187 µg/l was not included in the mean value

Table II. The serum copper and zinc concentrations in the healthy Finnish cildren according to Vuori et al. (18)

Age (years)	No of cases	Concentration (SD) µmol/l Cu	Zn
Girls			
3	255	21.3 (4.2)	14.6 (1.9)
6	287	19.8 (3.4)	14.6 (1.7)
9	316	19.1 (3.1)	14.5 (1.8)
12	328	16.6 (2.9)	14.5 (1.6)
15	308	15.9 (3.1)	14.2 (1.7)
18	276	17.3 (5.0)	14.1 (1.8)
Boys			
3	270	21.1 (4.2)	14.3 (1.7)
6	273	20.1 (3.5)	14.6 (1.7)
9	315	19.1 (3.1)	14.7 (1.6)
12	316	17.8 (3.5)	14.8 (1.6)
15	282	15.0 (2.5)	14.8 (1.7)
18	253	15.1 (2.7)	15.0 (1.8)

Table III. The ratios of selenium, copper and zinc in serum samples of JRA children. Mean values (SD) are given

Age (years)	Zn/Cu	Se/Zn	Se/Cu
Girls			
1.5 - 4.5	0.54(0.13)	3.8(0.7)	2.1(0.6)
4.5 - 7.5	0.55(0.18)	4.5(1.2)	2.4(1.2)
7.5 -10.5	0.56(0.15)	4.7(1.0)	2.5(0.7)
10.5-13.5	0.56(0.13)	4.7(1.2)	2.6(0.7)
13.5-16.5	0.44(0.07)	4.9(1.9)	2.1(0.5)
16.5-	0.66	3.4	2.3
Boys			
1.5 - 4.5	0.58(0.12)	3.9(0.7)	2.3(0.7)
4.5 - 7.5	0.62(0.18)	4.6(1.1)	2.7(0.7)
7.5 -10.5	0.52(0.09)	4.0(0.7)	2.1(0.5)
10.5-13.5	0.61(0.07)	4.3(1.1)	2.7(0.8)
13.5-16.5	0.41	3.7	1.5
16.5-	0.68	4.0	2.7

Selenium is part of the enzyme glutathione peroxidase that catabolyes peroxides which are suggested to be actively involved in inflammation. In JRA children (35 cases) the mean E-GSH-Px concentration was 20.3 U/g Hb (SD 5.1); for the control population of 93 subjects a value of 21.3 U/g Hb (SD 5.3) was obtained. No difference in results with patients of RA and controls was reported by Åkesson et al. (15), either.

Superoxide dismutases are known to act in cells as protection agents againts toxic products of oxygen. The (Zn,Cu) superoxide dismutase concentrations (E-SOD) of JRA children was, in average, 3658 ± 549 U/g Hb (SD). In controls of 10 persons the average E-SOD concentration was 3614 ± 305 U/g Hb.

This study confirms that the disturbance of trace elements reported in adult RA patients are to be recorded also in children with juvenile chronic arthritis. Further analysis of our results will reveal how the type and duration of the disease and the various forms of therapy affect the trace element concentrations in children with JRA.

Acknowledgement

Ms. Laila Saarvo and Ulla-Riitta Nordberg, M.Sc., are thanked for skillful technical assistance.

References

1. Möttönen, T., P. Hannonen, P. Seppälä, G. Alfthan and M. Oka. 1985. Clin. Rheumatol. 3, 195.

2. Tarp, U., K. Overvad, J.C. Hansen, E.B. Thorling, H. Graudal. 1985. Scand. J. Rheumatol. 14, 97.

3. Aaseth, J., E. Munthe, Ø. Førre and E. Steinnes. 1978. Scand. J. Rheumatol. 7, 237.

4. Aaseth, J. and E. Munthe. 1983. Scand. J. Rheumatol. 49 (suppl.), 35.

5. Bikow, K. and B. Salbu. 1983. Scand. J. Rheumatol. 49 (suppl.), 35.

6. Youssef, A.A.R., B. Wood and D.N. Baron. 1983. J. Clin. Pathol. 36,14.

7. Banford, J.C., D.H. Brown, R.A. Hazelton, C.J. McNeil, R.D. Sturrock and W.E. Smith. 1982. Ann. Rheum. Dis. 41, 458.

8. Kenedy, A.C., G.S. Fell, P.J. Rooney, W.H. Stevens, W.C. Dick and W.W. Buchanan. 1975. Scand. J. Rheumatol. 4, 243.

9. Balogh, Z., A.F. El-Ghobarey, G.S. Fell, D.H. Brown, J. Dunlop and W.C. Dick. 1980. Ann. Rheum. Dis. 39, 329.

10. Ambanelli, U., G.F. Ferraccioli, W. Troise and G. Serventi. 1977. Zeit. Rheumatol. 36, 345.

11. Bonebroke, R.A., J.T. McCall, G.G. Hunder and H.F. Polley. 1972. Mayo Clin. Proc. 47, 746.

12. Ambanelli, U., G.F. Ferraccioli, G. Serventi and G.L. Vaona. 1982. Scand. J. Rheumatol. 11, 63.

13. Mäkelä, A-L., H. Hyörä, K. Vuorinen and V. Näntö. 1984. Scand. J. Rheumatol. 53 (suppl.), 94.

14. Svantesson, H., U. Johansson, S. Portinsson, A. Åkesson, B. Åkesson and P.A. Öckerman. 1984. Scand. J. Rheumatol. 53 (suppl.), 94.

15. Åkesson, B., U. Johansson, S. Portinsson, A. Åkesson, H. Svantesson and P.A. Öckerman. 1984. In: Proc. coll. of Int. Workshop Trace Element Analytical Chemistry in Medicine and Biology, April 16-18, Neuherberg, FRG, abstract no. 10.

16. Salmela, S.S. and E. Vuori. 1984. At. Spectrosc. 5, 146.

17. Alfthan, G., J. Kumpulainen. 1982. Anal. Chim. Acta 140, 221.

18. Vuori, E., S.S. Salmela, H. Åkerblom, J. Viikari, M. Uhari, P. Suoninen, M. Pietikäinen, E. Pesonen, P-L. Lähde, M. Dahl. 1985. Acta Pediatr. Scand. 318 (suppl.), 205.

SELENIUM DEFICIENCY IN DILATED CARDIOMYOPATHY

M.A.J. Weber, J. Schuhbeck, C. Hötte, K. Theisen

Medizinische Klinik Innenstadt der Universität München
8000 München 2, FR Germany

P. Schramel

Gesellschaft für Strahlen- und Umweltforschung mbH,
8042 Neuherberg, FR Germany

Introduction

Selenium is an essential trace element for man. A selenoenzyme, glutathione peroxidase, plays an important role in preventing tissues from peroxidative cell damage by destroying hydrogen peroxids. Selenium deficiency has been associated with Keshan disease, a form of dilated cardiomyopathy, which appears in selenium deficient areas in China; it has been treated with prophylactic selenium substitution.
Cases of patients who, after a longer period of total parenteral nutrition, developed selenium deficiency and a dilated CMP, also indicate a possible connection between selenium deficiency and CMP. We therefore determined selenium plasma levels in our patients with dilated cardiomyopathy.

Methods

The patients of our study were classified into four groups according to the following criteria.

1) Dilated cardiomyopathy
   n=8, 3 women, age 47±9 years
   Biplane LV-ejection fraction ≤45%, with or without elevated enddiastolic pressure, normal coronary angiogram.

Trace Element – Analytical Chemistry in Medicine and Biology, Vol. 4

2) "Minor LV-dysfunction"
   n=30, 10 women, age 47±9 years
   Biplane LV-ejection fraction 45-55%, or isolated elevated enddiastolic pressure.

3) Congestive heart failure
   n=15, 6 women, age 60±14 years
   CAD 8, valvular heart disease 6, cor pulmonale 1

4) Healthy controls
   n=9, 4 women, age 36±9

The selenium levels were determined by instrumental neutron activation analysis respectively by electrothermal atomic absorption spectroscopy using Zeeman background correction.

Statistics

Results are reported as mean ± SD,
Selenium plasma levels were compared by Krushal-Wallis-H-test, by Dumms-test I, and by Student's t-test.

Results

The selenium plasma levels of patients with dilated cardiomyopathy were significantly lower than those of
1) the control group
2) the patients with minor LV-dysfunction
3) the patients with congestive heart failure of different origin.

Only the selenium plasma levels of the patients with CMP differed significantly from those of the control group (v. fig. 1).
The selenium plasma levels of patients with dilated CMP in a group with alcohol abuse were not significantly lower compared with the group without alcohol abuse (v. fig. 2).

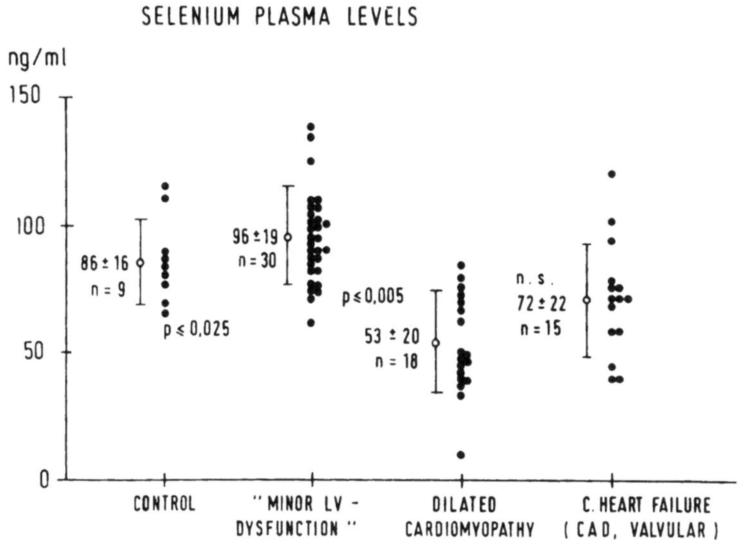

Fig. 1

Selenium plasma levels, p values refer to the comparison between selenium plasma levels of dilated cardiomyopathy to the other groups.

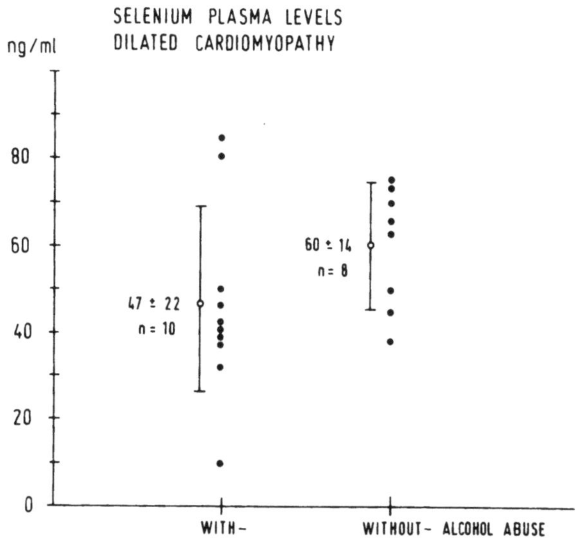

Fig. 2

Selenium levels in patients with dilated cardiomyopathy with or without alcohol abuse.

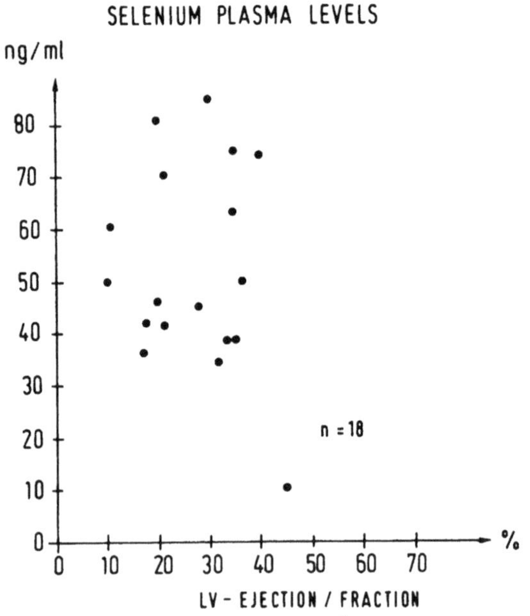

Fig. 3
Correlation between ejection fraction and selenium plasma levels in dilated cardiomyopathy.

The left ventricular ejection fraction did not correlate with the selenium levels (v. fig. 3).

Discussion

The results show that in Germany, too, patients with dilated cardiomyopathy may have a selenium deficiency. This confirms Oster's reports who likewise discovered a significantly reduced selenium plasma level in patients suffering from dilated cardiomyopathy (47,8 $\pm$ 16,2 ng/ml) compared with healthy controls (80,1 $\pm$ 13,2 ng/ml) (1). The suspicion that selenium deficiency does not appear in connection with CMP but either in connection with the resulting congestive heart failure with malabsorption and treatment with diuretics could be defeated. We didn't find correspondingly reduced selenium plasma levels among patients with heart failure of different origin or among patients with diuretic treatment for hypertension.

It is reported that alcoholics with and without liver cirrhosis have a selenium deficiency as a consequence of inadequate diet (2, 3). The lowest selenium levels were indeed found among patients with alcohol abuse, but the selenium average values of these patients were not significantly lower than those of patients without alcohol abuse (cf. fig. 2). The number of patients in the two groups was, however, so small that it is possible that an existing difference was not recognized.

The pathomechanism through which selenium deficiency could cause cardiomyopathy is unclear (4). Reduced glutathion peroxydase activity because of selenium deficiency causes liver cell necrosis in animal tests. Vitamin E which also plays a role in preventing cellular oxydation can compensate selenium deficiency to some degree. Some species of animal develop cardiac muscle defects, if a selenium deficiency appears in combination with Vitamin E deficiency. A reduced resistance, caused by selenium deficiency, fosters the formation or the further development of a myocarditis and is therefore discussed to be also a possible connecting link. The succes of a prophylactic selenium substitution in the treatment of Keshan disease would fit into this concept. Myocard biopsies, possible means for evaluating the rate of cases of myocarditis, do not exist from the patients in our study. In those cases of patients receiving total parenteral nutrition where a CMP appears along with selenium deficiency, an improvement of the already existing CMP after selenium substitution is described (5).

Unlike Oster we did not find a correlation between the reduction of the ejection fraction as a sign of left ventricle dysfunction and the selenium plasma levels. However, the ejection fraction showed a high variance due to atrial fibrilation in several patients.

Since selenium plasma levels depend to a high extent on the quantity and composition of diet, it is problematic to diagnose selenium deficiency merely by analyzing the plasma. Therefore an additional analysis of intracellular selenium levels (e.g in erythrocytes) would be desirable.

Conclusion

Patients with dilated cardiomyopathy showed significantly lower Selenium plasma levels but the role of this deficiency in the pathogenesis is not known.
Patients with congestive heart failure also showed lowered Selenium plasma levels, but to a lesser degree.

References

1. Oster O., Prellwitz W., Kasper W., Meinertz Th.
   Congestive cardiomyopathie and Selenium content of serum
   Clin. Chim. Acta 128:125-132 (1983)

2. Aaseth J., Thomassen Y., Alexander J., Nordheim G.
   Decreased serum Selenium in alcoholic cirrhosis
   N. England J. Med. 303:944-945 (1981)

3. Dworkin B.D., Rosenthal W.S.
   Selenium and the alcoholic
   Lancet I:1015 (1985)

4. Westermarck T.W.
   Consequences of low Selenium intake for man
   in: Trace element analytical chemistry in medicine and biology
   Eds.: P. Brätter, P. Schramel, Walter de Gruyter Verlag
   Berlin, New York 1984 S.:49-70

5. Johnson R.A., Baker S.S., Fallon J.T., Maynard E.P.,
   Ruskin J.N., Wen Z., Ge K., Cohen H.J.
   An Occidental Case of Cardiomyopathy and Selenium Deficiency
   N. Engl. J. Med. 304:1210-1212 (1981)

THE RELATIONSHIP BETWEEN SELENIUM AND CORONARY HEART DISEASE

O.Oster, W.Prellwitz, C. Luley
Department of Clinical Chemistry, University Hospital Mainz, Germany

T. Meinertz, A. Geibel, W. Kasper
II. Department of Internal Medicine, University Hospital Mainz
present address: Department of InternalMedicine, University Hospital Freiburg, Germany

Introduction

The importance of selenium for the function of the heart of men has been proven by the chinese medicine. The incidence rates of Keshan disease, an endemic cardiomyopathy of unknown cause in the peoples Republic of China that occours most frequently in children under 15 years of age and women of child-bearing age, was reduced by treating the children with sodium selenite (1). Keshan disease is a congestive cardiomyopathy that affects persons living in rural areas of a selenium deficient zone in China (2.3).
Our own investigations showed that patients with congestive heart failure also in the western hemisphere may have low serum selenium values (4,5). We found for patients with congestive cardiomyopathies a significant correlation of the serum selenium content with the left ventricular ejection fraction (5).
From epidemiological studies done in Eastern Finland it is found that serum selenium concentrations of less than 45 µg/l are associated with a relative risk of coronary and cardiovascular heart disease and of a risk of fatal and non-fatal myocardial infarction (6). The low dietary intake of selenium in Finland may be the reason for the increased mortality in Finland from coronary and cardiovascular heart disease.
Germany has in regard of the serum selenium concentration some similarities with Finland. As it is shown in Table 1 Germany is as Finland in an international comparison of the serum selenium concentrations below the average, especially when the serum selenium concentrations of the industrialized countries are compared.

Trace Element – Analytical Chemistry in Medicine and Biology, Vol. 4

Table 1: International comparison of the serum selenium concentrations

Serum Selenium [µg/l]		Serum Selenium [µg/l]	
Germany		Finland	
66,4 ± 12,9[1]		55,3 ± 14,7	(6)
100,0 ± 20,0	(8)	83,1 ± 15,7	(17)
80,6 ± 13,8[1]	(7)	108,0 ± 33,0	(18)
102,0 (86-125)	(9)	42,0 ± 13,0	(18)
109,0 ± 21,0	(10)	New Zealand	
75,0	(11)	49,0 ± 12,0	(19)
67,0 ± 12,0[2]	(12)	47,0 ± 11,0	(20)
75,0 ± 13,3[3]	(12)	Canada	
USA		142,9 ± 16,1	(21)
96,3 ± 14,2	(13)	144,0 ± 28,6	(22)
80,0 ± 10,0	(14)	Belgium	
136,0 ± 2,0	(15)	88,0 - 185	(23,24)
Norway		Japan	
92,0 - 140	(16)	120,0 (98-139)	(25)
Italy		Great Britain	
79,0 (43-114)	(26)	116,0 ± 15,9	(27)
		120,4 ± 4,9	(28)

1 Mainz - 2 Giessen - 3 Heidelberg

In Eastern Finland in the 1970s the world highest mortality from coronary heart disease and atherosclerotic cardiovasvular disease for men was recorded (6). Finland is today on the fourth position with regard to the coronary heart disease mortality in the world (29). For Germany up to 1975 an increase in the mortality from arterial diseases for men and women was reported (30). After the midth of the 1970s in many industrialized countries declining

trends of the coronary heart disease mortality was reported especially for the USA (31 - 33) but also for Europe (34). For Germany this declining trend is not observed.
This report is written because of several reasons:

1. From the Finnish epidemiological investigations it is to expect that a relative high percentage of patients with coronary heart disease may have lower serum selenium concentrations than an healthy control group. It is the aim of this study to prove this.
2. To investigate whether there are correlations of the serum selenium concentrations of the patients with coronary heart disease with the risk factors for coronary heart disease.
3. To investigate whether there is a relationship between the serum selenium concentration and the severity of the coronary heart disease.
4. To confirm the positive significant correlation between the serum selenium concentration and the left ventricular ejection fraction as we found for patients with congestive cardiomyopathy (5).

Materials and methods

59 male patients with an average age of 51,1 $\pm$ 6,8 years were included in the study. Of the 59 patients with coronary heart disease 50 had myocardial infarction and 9 were without myocardial infarction.
The control group were 41 male middle aged employees of the University of Mainz.
The ejection fraction of 47 patients was determined by standard routine left ventricular contrast angiography using the right anterior oblique view.
Electrocardiograms were abtained in all patients and were abnormal in each. The most common abnormalities of the electrocardiographic findings were left bundle branch block (9 patients) and left ventricular hypertrophy (5 patients), 37 patients showed infarction signs. Blood was collected before the angiography was done.
The serum selenium was determined by flameless AAS as described by us (7).

Results and discussion

The serum selenium content of 59 middle aged male (average age 51,1 ± 6,8 years) patients not older than 60 years with coronary heart disease was compared with that of an healthy male control-group (average age 51,8 ± 6,5 years). The serum selenium content of the patients with coronary heart was determined with 55,02 ± 15 µg Se/l and is significant different from that of the control group (78,8 ± 11,5 µg Se/l SD) as it is shown in Table 1 and Fig. 1. If the 95 percentile (mean value - 2 SD) of the serum selenium content of the healthy control group is calculated the lower margin of the serum selenium content is determined with 55,8 µg Se/l. From the 59 patients with coronary heart disease 33 have a serum selenium content lower than 55,8 µg Se/l, that is 55,9 %.

There is no significant difference between those patients having had myocardial infarction (mean serum selenium 56,1 ± 16,2 µg Se/l) and those without myocardial infarction (mean serum selenium 49 ± 4,3 µg Se/l). There is also no significant difference in the serum selenium content of the patients having had myocardial infarction

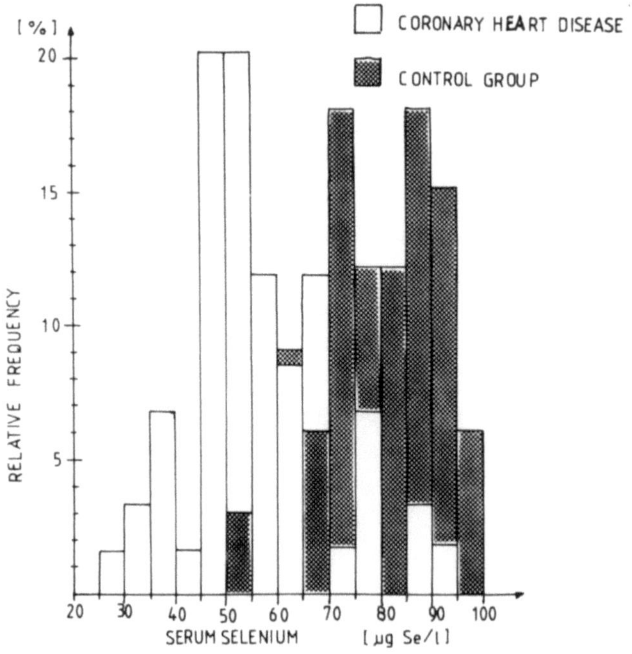

Fig.1: Histogram showing the frequency distribution of a healthy male age matched control group and patients with coronary heart disease

Table 2: Summary of the serum selenium content of the patients with coronary heart disease under the consideration of the risk factors cholesterol (chol), hyperuricemia, hyperlipidemia, adipositas, smoking, hypertension, diabetes and myocardial infarction (MI)

Patients/control group	N	serum selenium [µg Se/l] [1] (range)	Student's t
Healthy control group	41	78,8 ± 11,5 (50-95)	7,9
Whole patient group	59	55,0 ± 15,0 (27-92)	p < 0,0001
Patients with MI	50	56,1 ± 16,2 (27-86)	1,3
Patients without MI	9	49,0 ± 4,3 (43-56)	p > 0,05
Patients with MI before less than 12 months	11	60,1 ± 13,2 (38-85)	0,94
Patients with MI before more than 12 months	39	55,0 ± 16,7 (27-92)	p > 0,05
Hypertensive patients	16	59,8 ± 15,7 (38-86)	1,56
Non hypertensive patients	43	53,0 ± 14,6 (27-92)	p > 0,05
Patients with adipositas	25	55,5 ± 19,4 (27-92)	0,29
Patients without adipositas	34	54,3 ± 11,1 (38-85)	p > 0,05
Diabetic patients	6	52,5 ± 8,9 (43-67)	0,43
Non diabetic patients	53	55,3 ± 15,6 (27-92)	p > 0,05
Hyperuricemic patients	7	62,3 ± 17,0 (28-78)	1,38
Non hyperuricemic patients	52	54,0 ± 19,6 (27-92)	p > 0,05
Patients with chol < 2600 mg/l	34	55,3 ± 14,2 (27-92)	0,31
Patients with chol > 2600 mg/l	25	56,4 ± 13,0 (35-86)	p > 0,05
Hyperlipidemic patients	12	56,0 ± 12,4 (27-92)	0,25
Non hyperlipidemic patients	47	55,1 ± 16,6 (34-86)	p > 0,05
Smokers of the patients	29	55,4 ± 17,7 (27-92)	0,20
Non smokers	30	54,6 ± 12,1 (34-85)	p > 0,05

1 = mean value and SD

before less than twelve months (mean serum selenium concent 60,1 ± 13,2 µg Se/l and those before more than twelve months (55 ± 16,7 µg Se/l). No significant difference in the serum selenium content is found if the serum selenium content of the patients with the risk factors high cholesterol, hyperlipidemia, hyperuricemia, adipositas, diabetes, hypertension and smoking were compared with the serum selenium content of those patients without these risk factors (see Table 1). It has to be mentioned that from the patient group heavy smokers with more than 15 cigarettes a day and alcoholics are excluded.

From our own not published results we know that heavy smokers (more than 20 cigarettes a day) and drinkers have lower serum selenium than non smokers and moderate alcoholic consumers. The connection between lower serum selenium content and alcohol consumption as well as smoking was also found by Lloyd et al (35). It has however also to be considered that heavy smokers are very often heavy drinkers and vice versa. As we excluded the heavy smokers and alcoholics from our patient group the lower serum selenium of the patients with coronary heart disease is most probably not due to excessive alcohol consumption and cigarette smoking.

Our results are in agreement with epidemiological investigations done by Salonen et al. (6) in Finland who found in an epidemiological study which was conducted to investigate the association between serum selenium and risk of death from acute coronary heart disease as well as risk of fatal and non-fatal myocardial infarction that low serum selenium was associated with a relative risk of cardiovascular disease. According to the results of Salonen et al. a serum selenium content of less than 45 µg/l was associated with an adjusted relative risk of coronary heart disease death, a relative risk of cardiovascular disease death and a relative risk of fatal and non-fatal myocardial infarction. From the results of Salonen et al. one can assume that in a patient group with coronary heart disease a lower serum selenium content is found compared to a healthy control group, this is what we found. In our patient group 18 patients have a lower serum selenium content than 45 µg/l that is 22 %. The combination of the epidemiological results of Salonen et al. with our results let us suggest that it has to be considered that low serum selenium might be a risk factor for coronary heart disease. This is supported also by our results that patients with acute myocardial infarction have also low serum selenium concentrations and a similar distribution pattern of the serum selenium than patients with coronary heart disease (37).

If the serum selenium content of the patients with coronary heart was correlated with the severity of the coronary heart disease

Table 3: Summary of the statistic concerning the severity of the coronary heart disease taking the number of vessels included and the extent of stenoses as criterion for the severity

Patient group	N	serum selenium µg Se/l [1] (range)	Student's t
A One and two vessel disease stenosed $<75\%$	13	$51,3\pm12,5$ (34-77)	0,84 compared with D $p>0,05$
B One vessel disease stenosed $>75\%$	33	$56,3\pm16,2$ (28-92)	1,00 compared with A $p>0,05$
C Two vessel disease stenosed $>75\%$	8	$52,8\pm15,6$ (27-86)	0,21 compared with A $p>0,05$ 0.50 compared with B $p>0,05$
D One and two vessel disease stenosed $>75\%$	41	$55,6\pm17,1$ (27-92)	

1 = mean value and SD

taking the number of vessels included and the extent of stenoses as criterion no significant differences are found. Patients with one and two vessel disease and an extent of stenoses < 75 % have no significant lower serum selenium content than patients with an extent of stenoses >75 %. Patients with one vessel disease and an extent of stenoses >75 % show no significant difference in the serum content when compared with those having a two vessel disease and an extent of stenoses >75 %. In Table 2 the statistic is summarized. Moore et al. (36) reported that patients with coronary atherosclerosis have a significant inverse correlation between the plasma selenium and the severity of the coronary atherosclerosis. Moore et al. compared four patient groups (patients with zero-, one-, two- and three-vessel disease) with different severity of coronary atherosclerosis and found that patients with two- and three-vessel disease have significant lower plasma selenium content than patients with a zero-vessel disease. Our results do not confirm a correlation of the serum selenium content with the severity of the coronary heart disease. As severity criteria for the coronary heart disease we have taken the number of vessels stenosed in combination with the extent of stenoses. It has however to be considered that the severity criteria of Moore et al. differed from ours and a relationship between the severity of the coronary heart disease and the serum selenium concentration cannot be excluded. Our results are therefore only partially comparable with the results of Moore et al. Also differences between the serum selenium concentrations of Moore and us are found. The plasma selenium content of the american patients with coronary atherosclerosis reported by Moore et al. (136,3 µg/l) for the zero-vessel disease and 105,2 µg/l for the three-vessel-disease) is more than twice higher than that found by us (55,02 µg Se/l) for German patients with coronary heart disease.

From 47 patients with coronary heart disease besides the serum selenium also the ejection fraction of the left ventricle was determined. If the serum selenium concentration of the patients is correlated with the ejection fraction of the left ventricle than a positive linear correlation is found which is significant (N = 47, correlation coefficient r = 0,40). In Fig. 2 the result is summarized as a graphical presentation. The correlation is significant on the level $p > 0,01$. This is in agreement with previously published results by us (5), where we found for adult patients with congestive cardiomyopathies a similar positive correlation. The reason for

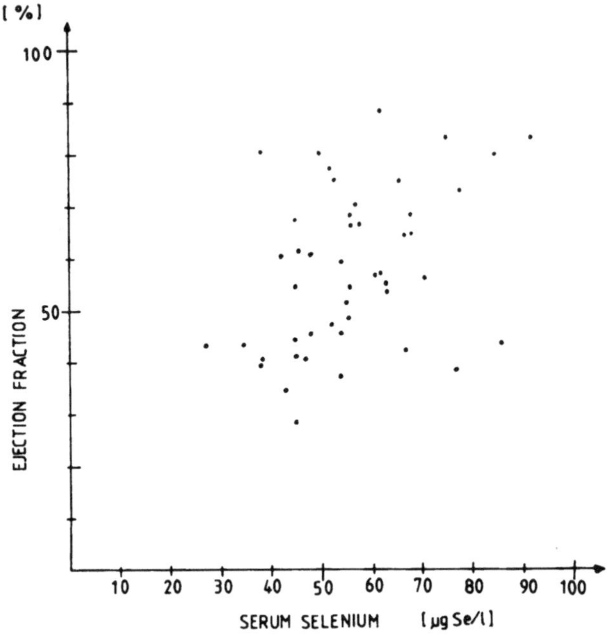

Fig. 2: Correlation of the serum selenium concentration with the left centricular ejection fraction

the positive correlation of the serum selenium concentration with the left ventricular ejection fraction is not known but it has to be considered that we found this positive correlation in two heart diseases (congestive cardiomyopathy and coronary heart disease) most probably of different origin. From this result it can be concluded that this might be a general phenomenon independent of the kind of heart disease.

References

1. Keshan Disease Research Group of The Chinese Academy of Medical Sciences, Beijing. 1979. Chin. Med. J. 92, 477.
2. Xiaoshu Chen, Guangqi Yang, Junshi Chen, Xuecun Chen, Zhimei Wen, Keyon Ge. 1980. Biological Trace Element Research 2, 91.
3. Keshan Disease Research Group of The Chinese Academy of Medical Sciences, Beijing. 1979. Chin. Med. J. 92, 471.
4. Oster, O., W. Prellwitz. 1981. XIth International Congress of Clinical Chemistry and IVth European Congress of Clinical Chemistry, Vienna.
5. Oster, O., W. Prellwitz, W. Kasper, T. Meinertz. 1983. Clin. Chim. Acta 128, 125.
6. Salonen, J.T., G. Alfthan, J. Pikkarainen. J.K. Huttunen, P. Puska. 1982. The Lancet 175.
7. Oster, O., W. Prellwitz. 1982. Clin. Chim. Acta 124, 272.
8. Lombeck, I., K. Kasperek, L.E. Feinendegen, H.J. Bremer. 1975. Clin. Chim. Acta 64, 57.
9. Behne, D., H. Jürgensen. 1978. J. Radioanal. Chem. 42, 447.
10. Kasperek, K., G.B. Iyengar, J. Kiem, H. Borberg, L.E. Feinendegen. 1979. Clin. Chem. 25, 711.
11. Schaller, K.H. 1985. 3. Colloquium Atomspektrometrische Spurenanalytik Konstanz.
12. Thorling, E. 1985. Selenium symposium, Helsinki
13. Pleban, P.A., A. Munyani, J. Beachum. 1982. Clin. Chem. 28, 311.
14. Lane, H.W., D.C. Warren, F. Martin, J. MC Lowan. 1983. Nutr. Res. 3, 805.
15. Willet, W.C., J.S. Morris, S. Pressel, J.O. Taylor, B.F. Polk, M.J. Stampfer, B. Rosner, K. Schneider, C.G. Hames. 1983. The Lancet 130.
16. Saed, K., Y. Thomassen, F.J. Langmyr. 1979. Anal. Chim. Acta 110, 285.
17. Välimäki, M.J., K.J. Harju, R.H. Ylikahoi. 1983. Clin. Chim. Acta 13, 291.

18. Westermark, T., P. Raunu, M. Kirjarinta, L. Cappalainen. 1977. Act. Pharmacol. et Toxicol. 40, 465.

19. Robinson, M.F., D.R. Campbell, W.H.F. Sutherland, G.P. Herrison, J.M. Paulin, F.O. Simpson. 1983. New Zealand J. Med. 96, 755.

20. Thompson, C.C., H.M. Rea, M.F. Robinson, F.O. Simpson. 1978. Proc. University Otago Med. School 56, 31.

21. Lalonde, L., Y. Jean, K.D. Roberts, A. Chapdelaine, G. Bleau. 1982. Clin. Chem. 28, 172.

22. Dickson, R.C., R.H. Tomlinson. 1967. Clin. Chim. Acta 16, 311.

23. Cornelis, R., C. Mees, S. Ringoir, J. Hoste. 1979. Min. Electrol. Metab. 2, 88.

24. Versieck, J., R. Cornelis. 1980. Anal Chim. Acta 116, 217.

25. Kurahashi, K., S. Inone, S. Yonehava, Y, Shimshi, K. Toei. 1980. Analyst 105, 690.

26. Calautti, P., G. Moschine, B.M. Stievano, L. Tomio, F. Calzarava, G. Perona. 1980. Scand. J. Haematology 23, 63.

27. Lloyd, B., P. Holt, H.T. Delves. 1983. In Trace Element Analytical Chemistry in Medicine and Biology, Volume 2 page 1129 P. Editor Brätter, P. Schramel, de Gruyter Berlin

28. Lloyd, B., A.T. Delves. 1982. Analyst 107, 927.

29. Huttunen, J.K. 1985. An. Clin. Res. in press

30. Ioannidis, P.J., G. Efthymiopoulou. 1982. Am. J. Epidemiology 115, 278.

31. Epstein, F.J. 1982. Sixth International Symposium on atherosclerosis, Berlin in: Atherosclerosis VI, page 723, G. Editors Schettler, A.M. Gotto, G. Middelhoff, A.J.R. Habenicht, K.R. Jurrutka.

32. Walter, W.J. 1974. J. Am. Med. Ass. 227, 1045.

33. Gordon, T., T. Thom. 1975. Prevent. Med. 4, 115.

34. Epstein, F.H. 1980. Nutr. Metab. 24 (Suppl. 1), 45.

35. Lloyd, B., R.S. Lloyd, B.E. Clayton. 1983. J. Epidemiology Community Health 37, 213.

36. Moore, J.A., N. Robert, J.C. Wells. 1984. Clin. Chem. 30, 1171.

37. Oster, O., M. Drexler, J. Schenk, T. Meinertz, W. Kasper, C.J. Schuster, W. Prellwitz. 1986. Ann. Clin. Research 18, 36.

STRAIN DEPENDENT DIFFERENCES IN SENSITIVITY TO COPPER INTOXICATION IN THE RAT: THE RELATION BETWEEN INTRAHEPATIC CU DISTRIBUTION AND CU-INDUCED LIVER DAMAGE

H.Nederbragt, Marion Gijbels, A.J.Lagerwerf.

Department of Veterinary Pathology, Faculty of Veterinary Science, State University, Utrecht, the Netherlands.

Introduction

Biliary excretion of copper (Cu) and its incorporation into metallothionein (MT) are assumed to prevent Cu-induced liver damage in conditions of chronic or acute Cu overload. Failure of these mechanisms may be the underlying cause of such (hereditary) Cu accumulation diseases as Wilson's disease in man and Cu toxicosis in Bedlington terriers and varying breed susceptibility to chronic Cu toxicity in sheep (1). However, neither these mechanisms nor the mechanism by which Cu causes cell death are clearly understood.
In the study presented here we describe two inbred rat strains that differ both in the metabolism of Cu after intravenous (i.v.) or oral administration and in the sensitivity to Cu intoxication. We suggest that these findings may be causally related and that these strains offer a suitable model for a further study of the mechanism of Cu intoxication.

Methods

Inbred male Fischer and Brown Norway rats were purchased from Dutch laboratory animal breeders and housed under standardized conditions; commercial food and tap water were given ad libitum. The initial weight of the rats was 200 g in experiments 1 and 2 and 60 g in experiment 3.
In exp. 1 bile canulation and bile sampling was performed as described before (2) except that body temperature was maintained by placing the rats on a heating pad. Cu was injected i.v. in doses of 10, 30 and 50 µg/100 g body weight. In exp. 2 rats were injected i.v. under light ether anesthesia with doses of 50, 100 and 200 µg Cu/100 g body weight and killed by exsanguination 1, 2 and 4 h later. The liver was excised, a part used for total Cu determination and the remaining part for centrifugal fractionation. Only the Cu concentration of the

Trace Element – Analytical Chemistry in Medicine and Biology, Vol. 4

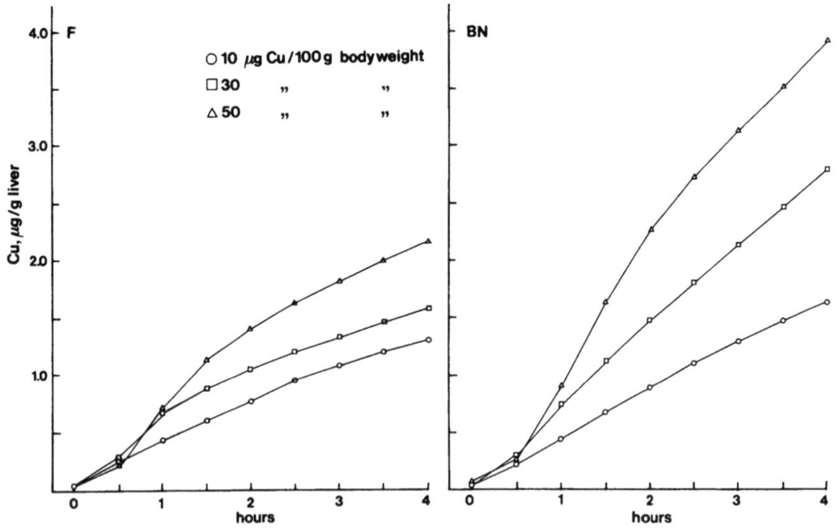

Fig.1. Cumulative biliary Cu excretion after i.v. injection of 10 (○), 30 (□) and 50 (Δ) μg Cu/100 g body weight in Fischer (F) and Brown Norway (BN) rats.

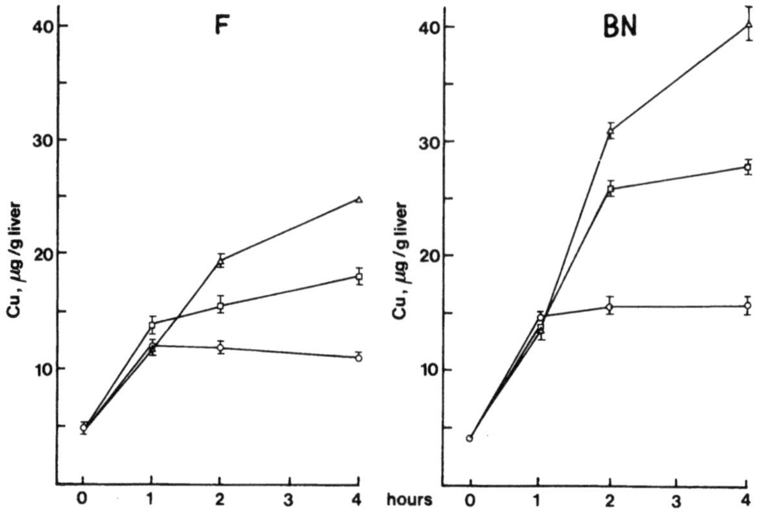

Fig.2. Cu concentration of the liver of Fischer (F) and Brown Norway (BN) rats after i.v. injection of 50 (○), 100 (□) and 200 (Δ) μg Cu/100 g body weight.

supernatant (30 min.,75.000 g) is reported here. A portion of the supernatant was applied to a Sephadex G-75 column (2.6 x 90 cm) and eluted at room temperature with a Tris-acetate buffer (0.01 M, pH 7.4). Fractions of 6 ml were collected; elution rate was 20 ml/h. In exp. 3 Cu was administered to the rats by adding Cu to the diet in concentrations of 500, 1500 and 2500 mg/kg. Rats were killed at intervals of 3 weeks. Livers were processed as described for exp. 2, except that a piece of liver was fixated in 4% buffered formaldehyde for microscopical evaluation. Staining with hematoxilin-eosine and with rubeanic acid for localization of Cu were performed routinely.
For Cu determination bile samples were diluted 10x with 6% butanol; liver samples and samples of fractions were digested with concentrated nitric acid under high pressure (2); supernatants were digested with nitric acid in Cu-free closed glass tubes during 1 h at 80°C; gel filtration fractions were used directly. Cu was determined by atomic absorption spectrophotometry, using an air-acetylene flame.

Results

Biliary Cu excretion after i.v. administration was studied in Fischer (F) and Brown Norway (BN) rats; the results are shown in fig. 1 and indicate that biliary Cu excretion in BN rats is far more efficient than in F rats, particularly after higher doses.
Cu distribution in the liver after high i.v. Cu doses to F and BN was compared in exp. 2. Although total Cu concentration of the liver was similar 1 h after injection of each of the three doses, BN rats accumulated more Cu than F rats at later timepoints (fig. 2). After gel filtration of the supernatants of those livers three Cu-containing protein peaks were found, designated peak 1 (Ve/Vo = 1.00), peak 2 (Ve/Vo = 1.45) and peak 3 (Ve/Vo = 1.90) respectively. The third peak is assumed to represent metallothionein (MT) because of its lack of absorbance at 280 nm (details not shown) and because of identification as MT of similar proteins after parenteral Cu administration to rats (3). Total supernatant Cu concentration and Cu concentration of the three peaks is shown in fig. 3. Although the supernatant Cu concentration varies with strain, Cu dose and time it is apparent that BN rats accumulate much more Cu in peak 3. This is particularly evident 1 h after injection of all three doses when total Cu concentration of the livers of rats of both strains was approximately similar: at this timepoint the Cu concentration of peak 3 was 2.5 times higher in BN rats than in F rats. Compared to F rats the Cu concentration of peak 1 was of minor importance

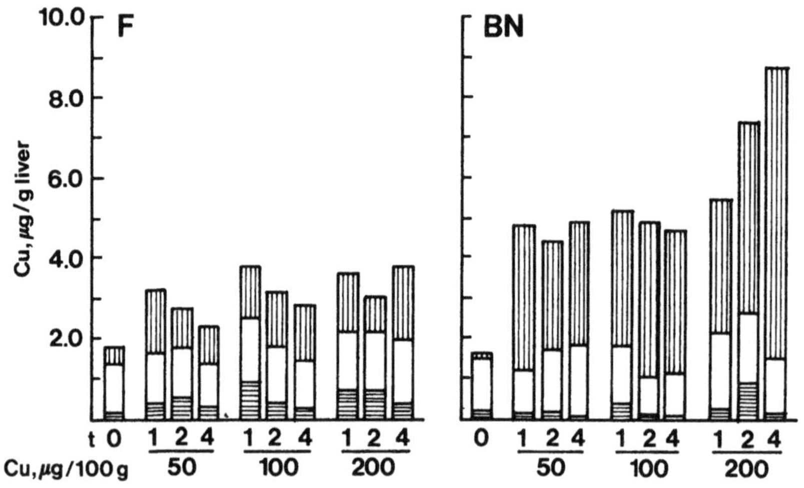

Fig.3. Cu concentration of the supernatant (full Column) and of protein peaks 1 (▤), 2 (□) and 3 (▥) of Fischer (F) and Brown Norway (BN) rats after indicated times and i.v. doses.

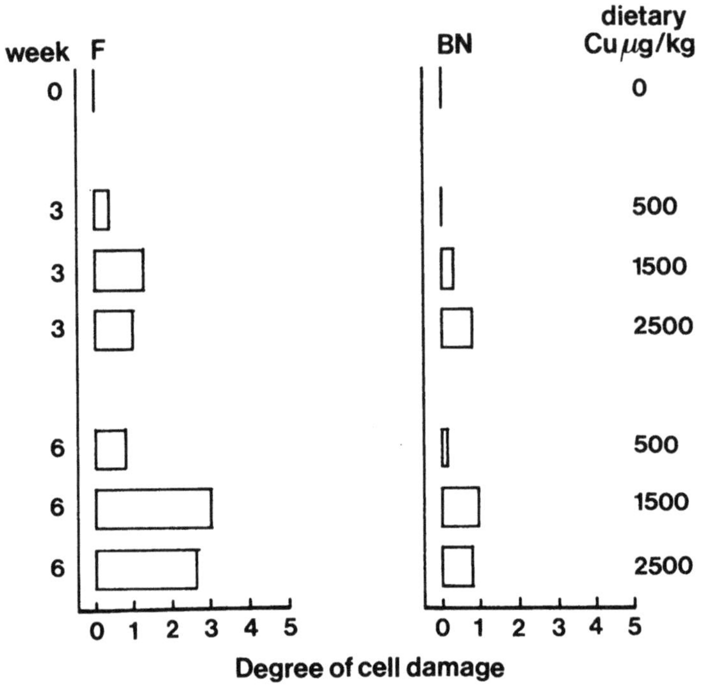

Fig.4. Score of liver damage in Fischer (F) and Brown Norway (BN) rats as observed by microscopical examination after indicated times and dietary Cu doses (mg Cu/kg). Criteria for degree of damage are given in the text.

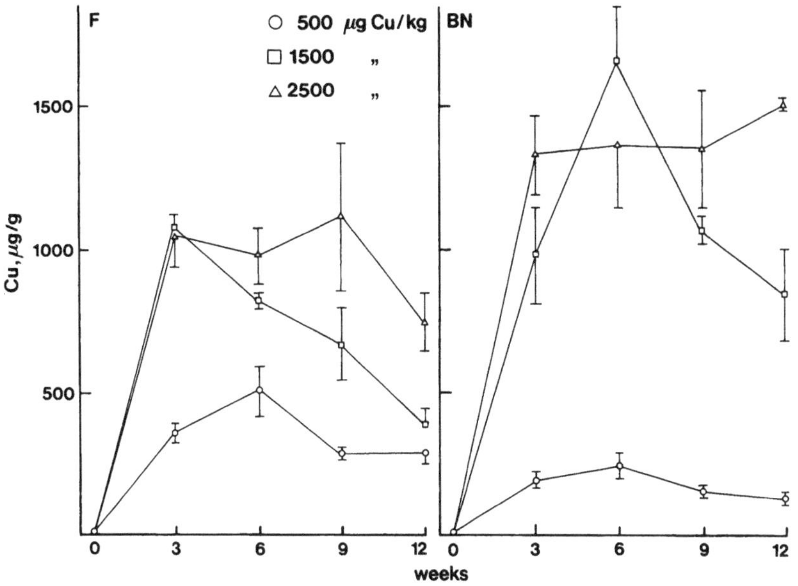

Fig.5. Cu concentration of the liver of Fischer (F) and Brown Norway (BN) rats after dietary administration of 500 (○), 1500 (□) and 2500 (△) mg Cu/kg.

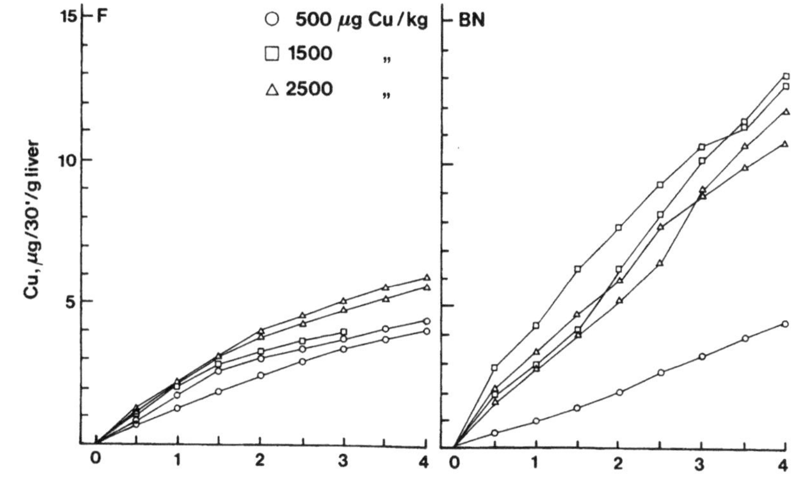

Fig.6. Cumulative biliary Cu excretion of Fischer (F) and Brown Norway (BN) rats after 5 weeks of dietary administration of 500 (○), 1500 (□) and 2500 (△) mg Cu/kg.

in BN rats.

Since F rats excrete less Cu in the bile and accumulate less Cu in peak 3 (which we assume to be MT) than BN rats they should be more sensitive to Cu intoxication. This hypothesis was tested in a feeding experiment in which 500, 1500 and 2500 mg Cu/kg diet was given to young rats of both strains for a prolonged period of time.

Liver damage was evaluated by microscopic examination. It was scored according to a scale of the following gradation. Grade 1: less than 10% of cells with increased eosinophilia and megakaryocytosis; grade 2: 10 - 25% of cells showing increased eosinophilia, megakaryocytosis and karyopyknosis; small dispersed nodules of necrosis are visible; grade 3: as grade 2 but 25 - 50% of the cells affected; grade 4: as grade 2 but 50 - 75% of the cells affected. The results based on the means of 3 livers per group are given in fig. 4 and show that liver damage is more serious in F rats and can be observed at a lower Cu dose and at earlier timepoints than in BN rats.

Despite a more serious liver damage in F rats they accumulated less Cu in the liver than BN rats (fig. 5), particularly after the higher Cu doses. Adaptation to these high Cu doses is reflected in a plateauing or a decrease of liver Cu levels between 3 and 12 weeks and in a decrease of the degree of liver damage between 6 and 12 weeks (not shown).This finding of adaptation confirms similar findings of Haywood (4). For this reason microscopic examination of liver morphology and determination of intrahepatic Cu distribution was performed at 3 and 6 weeks only.

In the liver most of the Cu was found in the insoluble particulate fraction: 81-89% in F rats and 90-95 % in BN rats (all three doses, 3 and 6 weeks). In control rats these values were 80 and 75% for F rats and BN rats respectively. This suggests that Cu accumulation in particulate fractions is more important in BN rats than in F rats.

Gel filtration of supernatants revealed three Cu containing protein peaks; their elution volumes corresponded to those found after i.v. Cu administration. Distribution of Cu over peaks 1 and 3 is presented in table 1.

It should be noted that relative to total liver Cu concentration the supernatant Cu concentration of F rats was twice as high as in BN rats. Nevertheless, it can be seen that on a percentage base Cu in BN rats accumulated preferentially in peak 3; however, the most important difference was found in the accumulation of Cu in peak 1 which was far more important in F rats.

Biliary Cu excretion was determined in some of the rats of both strains that had been on the different Cu diets for 5 weeks. The results (fig. 6) suggest that the rate of biliary Cu excretion was saturated at dietary Cu concentrations

Table 1. Cu Concentration of Supernatant and Peaks 1 and 3 in the Liver of Fischer and Brown Norway Rats after Chronic Oral Cu Administration

		FISCHER					BROWN NORWAY				
		Cu, μg/g liver					Cu, μg/g liver				
wk	Cu dose mg/kg	sup	pk 1	%	pk 3	%	sup	pk 1	%	pk 3	%
	0	1.1	0.1	10.3	0.2	14.0	1.0	0.5	51.5	---	---
3	500	27.7	14.0	50.7	4.4	15.9	7.9	0.7	8.6	3.6	45.1
3	1500	144.2	38.0	26.4	68.4	47.4	54.2	11.2	20.6	22.7	41.9
3	2500	161.2	32.1	19.9	72.8	44.8	79.2	6.4	8.1	45.7	57.6
6	500	52.5	16.0	30.4	22.8	43.3	5.5	0.8	14.8	3.0	54.5
6	1500	114.3	30.9	27.0	54.8	48.0	98.2	24.7	21.2	46.5	47.3
6	2500	97.3	28.1	28.9	45.8	47.1	101.5	9.9	9.8	76.8	75.7

below 500 mg/kg in F rats and between 500 and 1500 mg/kg in BN rats. Although BN rats accumulated more Cu in the liver than F rats (see fig. 5) this difference seems not sufficient to explain the difference in biliary Cu excretion.

Discussion

The current theory about the susceptibility of sheep to Cu intoxication is that in sheep liver insufficient MT synthesis takes place, leading to high toxic Cu concentrations (4) and that sheep are unable to excrete Cu in the bile in an efficient way (5) and thus can not maintain Cu homeostasis by biliary Cu excretion (6).

Therefore, the finding of reduced biliary Cu excretion and reduced accumulation of Cu in MT in F rats compared to BN rats in experiments 1 and 2 indicated that the approach of comparing different inbred strains of rats may provide a model for further investigation of the mechanism of Cu intoxication. Subsequently the value of this model seemed to be confirmed by the feeding experiment in which it was shown that F rats are more sensitive to Cu-induced liver damage than BN rats.

It is tempting to speculate that the increased sensitivity to Cu intoxication as observed in F rats after Cu feeding is related directly to reduced biliary Cu excretion and Cu accumulation in MT as it was observed after i.v. Cu administration. However, several objections can be made against this interpretation. First, although the Cu excretion in the bile after Cu feeding is more efficient in BN rats it did not lead to lower liver Cu concentrations. This may partly be due to differences in interorgan distribution since kidney Cu concentrations in BN rats are several times lower than in F rats (not shown). Secondly, the results of table 1 show that F rats accumulated similar or even higher amounts of Cu in peak 3 compared to BN rats; however, this did not protect them against liver damage. Furthermore we have to prove that after oral Cu administration peak 3 Cu really is MT bound. Thus, our experiments show that F rats are more sensitive to Cu intoxication but do not permit the conclusion that this is due to less efficient biliary Cu excretion and/or accumulation of Cu in MT.

An interesting finding is the important contribution of peak 1 Cu to total supernatant Cu in F rats: both absolute and relative Cu levels of this fraction are much higher than in BN rats. In fact, it is the only fraction observed thus far in which the Cu concentration is positively correlated with liver damage. Although it can not be concluded whether Cu binding to high molecular weight proteins of the supernatant is a cause or a consequence of cell damage, supposing there is a real causal relation, we still feel that an examination of the relation between Cu distribution and liver damage, more detailed than presented here, can be very useful.

References

1. Nederbragt, H., T.S.G.A.M. van den Ingh and P.Wensvoort. 1984. Pathobiology of copper toxicity. Vet.Quart. 6, 179-186.

2. Nederbragt, H. 1985. Strain- and sex-dependent differences in response to a single high dose of copper in the rat. Comp.Biochem.Physiol. 81C, 425-431.

3. Winge, D.R., B.L.Geller and J.Garvey. 1981. Isolation of copper thionein from rat liver. Archs.Biochem.Biophys. 208, 160-166.

4. Haywood, S. 1980. The effect of excess dietary copper on the liver and kidney of the male rat. J.Comp.Pathol. 90, 217-224.

5. Bremner, I. 1979. In: Metallothionein. (J.H.R.Kägi and M.Nordberg, eds). Birkhäuser, Basel. pp.273-280.

6. Soli, N.E. and J.P.Rambaek. 1978. Excretion of intravenously injected copper-64 in sheep. Acta Pharmacol.Toxicol. 43, 205-210.

7. Weber, K.M., R.C.Boston and D.D.Leaver. 1980. A kinetic model of copper metabolism in sheep. Aust.J.Agric.Res. 31, 773-790.

SELENIUM TREATMENT OF MYOTONIC DYSTROPHY

Gustav Örndahl, Ulla Sellden, Sverker Hallin, Hakon Wetterqvist
Department of Medicine, Östra Hospital, Göteborg, Sweden

Anders Rindby, Eva Selin
Department of Physics, Chalmers University of Technology, Göteborg, Sweden

Introduction

It is now well known that muscular dystrophy in cattle and sheep in New Zealand, Scandinavia and the US is a selenium (Se) deficincy disease which can be prevented by intake of Se (1). The soil is Se-deficient in these countries, and cereals, vegetables and hay therefore contain low levels of Se. The safe use of Se in veterinary medicine has been reported as a very important advance during the last quarter of this century (2). The question whether this observation in animals is applicable to human medicine is unsettled because little is known about the possible existence of a Se deficiency state in man. It would also be of interest to study if any muscular disease in man can be improved by Se.

Myotonic Dystrophy (MD) is the most common muscular disease among adults in Gothenburg and its surroundings. The aetiology of muscle diseases in man is unknown. The primary aim of this study was to determine the Se concentration in serum from patients with MD and secondly to study the clinical effect of Se-treatment in a patient with MD. As this patient improved, 5 more patients were included in the treatment-study.

Study Population

This study was based on 22 patients with MD and 18 sex- and age-matched healthy persons. All patients exhibited both active and mechanical myotonia, i.e. abnormal continuation of the contraction

Trace Element – Analytical Chemistry in Medicine and Biology, Vol. 4

of the voluntary muscles. The majority was investigated by electromyography (EMG) and showed typical findings. Muscular atrophy, weakness and other characteristic features in addition to myotonia were observed in all patients.

In order to document the severity of the disabling symptoms, we adopted an arbitrary classification of the functional state. Grade I patients (5 men and 3 women) could maintain their jobs. Grade II patients (3 men and 4 women) were unable to continue their jobs but could perform very light indoor work. Grade III patients (3 men and 4 women) were able to take care of their personal hygiene, but unable to perform any kind of useful physical or mental activity. Almost all patients had a family history of MD. Six of these patients have been selected for treatment with Se and vitamin E. All of these six patients (which are more thoroughly presented in tables 1 and 2) underwent opthalmological examinations and EMG registration before treatment and later once a year.

Furthermore ECG registration were performed for all patients. The exercise tests were carried out on an electrically braked bicycle ergometer. An ordinary direct-writing electrocardiograph (Mingograph, Siemens-Elema) was used, ECG heart rate, blood pressure (arm cuff) and respiratory frequency were recorded.

Table 1 Study Population in the Therapy Group

Patient (func. cap.)	Sex	Age at start of therapy	Approximate duration of symptoms in years	Profession
1 (I)	f	42	20 (15)*	Nurse assistant in a hospital Heavy job.
2 (I)	m	45	15 (10)*	Porter at a high school Heavy job.
3 (II)	f	46	30 (15)*	Wife working at home. Only light indoor work.
4 (I)	f	39	20 (9) *	Nurse assistant in a hospital Heavy job.
5 (II)	m	33	20 (15)*	Assistant in a workshop Light-heavy job.
6 (II)	m	39	25 (20)*	Unemployed Can only do light job.

* Duration of diagnosis myotonic dystrophy made by physicians.

Rate of perceived exertion according to Borg (3), dyspnoea or pain were also recorded during exercise. Each patient participated in two work tests, 21-27 months apart. Physical working capacity at a heart rate of 130/min (PWC) was calculated from each individual exercise test. A heart rate exceeding 130 was reached in every test.

Table 2 Observations before and during Se therapy

	Before treatment	Changes during treatment
Muscular findings:		
In the supine position on the couch -with maximal flexion of feet, you can pull him (her) by the feet ?	Yes(pat. 2,4)	Yes(pat. 1)
-can he raise his head and shoulder ?	Yes(pat. 2,4,5)	Yes(pat. 6)
-can he sit up without turning to one side and climbing with the arms ?	Yes(pat. 1,2,4,5,6)	
Lying on the floor -stand up without any help ?	Yes(pat. 1,2,4,5,6)	Yes(pat. 3)
Sit down on heels stand up ?	Yes(pat. 2,5)	Yes(pat. 4,6)
Stand or walk on toes ?	Yes(pat. 4)	Yes(pat. 2)
Stand or walk on heels ?	Yes(pat. 4)	
Normal walk (no cock's gait) ?	Yes(pat. 4)	Yes(pat. 1,2,5)
Carry bags in hand (and not in the crook of the arm or wrist) ?	-	Yes(pat. 1,2,4,5)
Normal features of face (no myopathic faces) ?	Yes(pat. 1,2)	Yes(pat. 4)
Normal position of upper eyelid	Yes(pat. 1,3)	Yes(pat. 2,6)
Other clinical findings:		
Eye lens opacities	Yes(pat. 2,4,5,6)	
Hair on the head (no premature baldness)	Yes(pat. 1,3,4,5)	Yes(pat. 6)
Normal testis (no significant atrophy)	Yes(pat. 2)	

Method

Selenium determination

Blood serum samples were analysid with a secondary target Energy Dispersive X-ray Fluorescence Spectrometer as described by Standzenieks et al. (4,5). The basic principle of the instrument is that primary radiation from an X-ray tube impinges on a secondary target (of molybdenum). The characteristic radiation from this target then excites characteristic radiation from the elements in the sample (4,5,6).

Examples of blood serum spectra are shown in figure 1. The peak heights for different elements are a measure of the concentrations of the element in question (7) but not directly proportional to them.

Preparation of serum samples

Blood samples of 5 ml were taken from each patient in the morning. The blood samples were centrifuged and 50 µl of serum were pipetted on a plastic foil of polyprophylene mounted on a plastic frame. Internal reference elements (yttrium and scandium) were added directly into the serum drop. The samples were dried in low temperature in order to remove excess water.

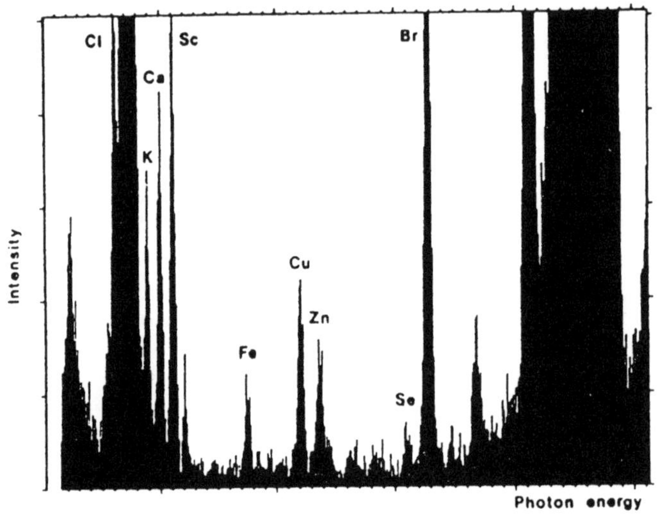

Fig. 1. Blood serum from a selenium-treated patient with myotonic dystrophy.

Treatment

The time point of initiation, the first daily dose of Se and vitamin E, the daily dose of Se and vitamin E in Dec. 1983 are given in Table 3. The drugs were only given orally two or three times a day. No side-effects were observed. None of these patients received any other treatment. They were not treated by any physiotherapist, nor had they increased their physical activity. The concentration of Se and other trace elements in blood serum and urine have been analysed during the treatment.

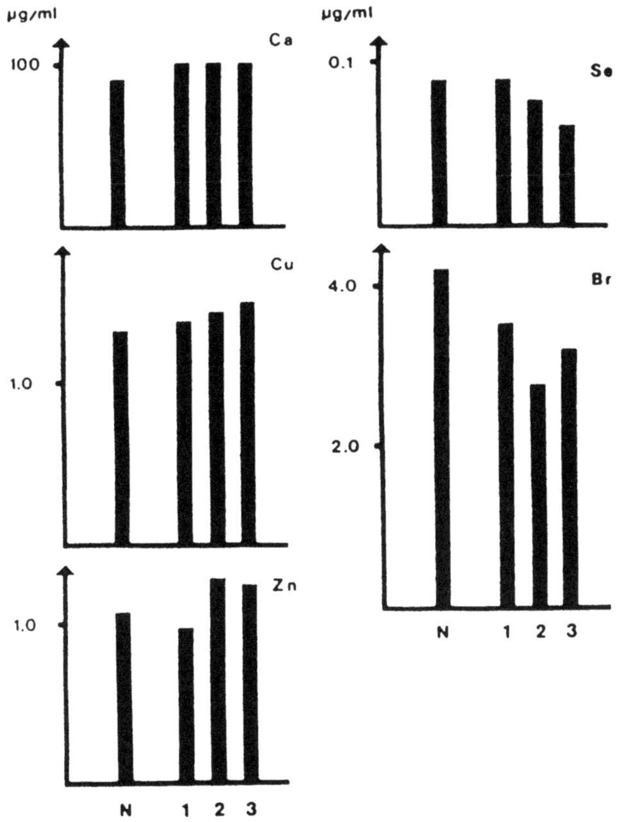

Fig. 2. Blood serum concentrations of some elements in the control group (N) and in the three functional groups (labelled 1, 2 and 3).

Results

Trace elements

The concentrations of Se and some other trace elements in serum of the primary group of MD patients are given in figure 2. Although there is a spread in the individual values, the decrease in the Se level is statistically significant for patients with functional capacity of grade II and III. The average values for the functional groups are 0.086 +- 0.01 µg/ml for group I, 0.076 +- 0.01 for group II and 0.063 +- 0.01 for group III. The mean values for all patients was 0.075 +- 0.01µg/ml. The control group had a Se concentration of 0.090 +- 0.01 µg/ml. (All errors are given as 95% confidence levels).

As Energy-Dispersive X-ray Fluorescence Analysis is a multi-element method, we have illustrated the values of some other elements in figure 2. Serum concentration of both copper and zinc tends to increase with progressing illness however no significant difference was determined. For calcium and bromine no obvious trend were observed.

For the first patient subjected to Se therapy, blood, liver, kidney and urine tests were carried out every 4th-6th week. All values were normal except S-creatine kinase, U-creatine and U-creatinine, which sometimes were slightly increased. These values as well as the serum Se-concentrations and the daily intake of Se and vitamin E are represented in figure 3. The serum Se-concentrations varied from 0.06 to 0.19 μg/ml during the period of treatment. At the beginning of the treatment there is a build-up of serum Se when the dose is suddenly increased, and a subsequent return to a lower values as the patient adapts to the increased dose. This pattern is less marked in the later stage of the treatment. The concentrations of other elements (calcium, iron, copper, zinc and bromine) measured varied somewhat during the treatment but with no obvious pattern except for zinc which seemed to decrease slowly with time. As can be seen from figure 2, the serum zinc concentrations were somewhat higher in the MD patients as compared with the control group.

The other patients in the treatment group showed a very similar respond-pattern concerning serum concentrations as the first patient represented in figure 3. The mean serum Se levels during period of constant intake are given in Table 3. Although the values fluctuated during the period, (SD about 15%) the overall values were not higher than those reported as normal in some other countries, e.g. the US (1). At the same time, the stable levels tend to increase slowly.

Table 3 Treatment, Duration and Dose.

Patient	Start of therapy	Daily dose at start Na SeO /vitamin E	Conc of serum Se at start of therapy	Daily dose in Dec. -83 Na SeO /vitamin E	Conc of serum-Se in Dec. -83
1	March 1980	0.5/10 mg	0.06 ug/ml	4.5/600 mg	0.23 ug/ml
2	April 1981	0.5/10 "	0.09 "	4.0/600 "	0.20 "
3	April 1981	0.5/10 "	0.06 "	4.0/500 "	0.22 "
4	April 1982	1.0/100 "	0.08 "	3.0/600 "	0.18 "
5	April 1982	0.5/50 "	0.10 "	4.0/600 "	0.15 "
6	March 1982	0.5/50 "	0.10 "	4.0/600 "	0.18 "

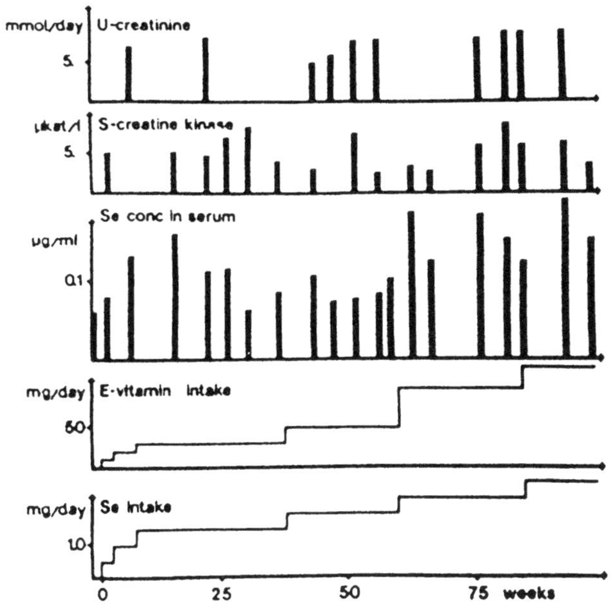

Fig. 3. Variation in U-creatinine, S-creatine kinase, serum Se and the intake of vitamin E and Se as a function of time. Every time when Se intake is increased, there is a sudden increase in serum Se concentration, followed by a return to a lower value. S-creatine kinase shows a similar tendency, but the effect is somewhat delayed as compared to serum Se. U-creatinine shows no obvious correlation to the therapy. However, the number of samples may be too small.

No significant correlations were recorded between Se and other trace elements.

Trace element concentrations in urine were also analysed by the same technique mentioned above. The total amount of Se in 24-hour urine samples from patients during treatment averaged 400 μg which is approximately 10 times higher than normal levels (1). The other trace elements recorded in urine (copper, zinc, bromine and rubidium) were within the normal range. No significant correlations were observed between Se and other trace elements in urine.

Clinical findings

Following the initiation of therapy there was a slow regression of the symptoms. During the first 6-12 months it was very difficult to decide whether or not there was any improvement, but later on both subjective and objective improvements have been observed. There were no significant muscle pains during the treatment.

Patients in an early stage of MD improved more markedly and earlier during treatment than patients in later stages. According to the ECG-examination almost all patients increased there physical capacity.

The mental tiredness of all patients in the therapy group regressed and 2 patients showed regression of ptosis.

Conclusion

From the analytical result of the MD patients we conclude that MD is a Se-dependent disease. In spite of the slow regression of symptoms we conclude that our six patients in the therapy group have benefited from the treatment with Se and vitamin E.

Acknowledgement

This study was supported by grants from the Göteborg Medical Society, the Swedish Natural Science Research Council and the Swedish Working-Environmental Fund.

References

1. Vokal-Borek, H.. 1979. Selenium. USIP Report 79-16, University of Stockholm.

2. Frost, D.V.. 1972. The two faces of selenium - can selenophobia be cured? CRC Crit Rev Toxicol 1, 467-480.

3. Borg, G.. 1962. Physical performance and perceived exertion. Lund: Glerup.

4. Standzenieks, P., A. Rindby and E. Selin. 1978. Development of a low power monoenergetic X-ray tube for trace element analysis. Nucl. Instrum. Methods 153, 269-276.

5. Standzenieks. P and E. Selin. 1979. Background reduction of X-ray fluorescence spectra in a secondary target energy dispersive spectrometer. Nucl. Instrum. Methods 165, 63-65.

6. Alenius, G., H. Riedl, A. Rindby, E. Selin and P. Standzenieks. 1977. Trace element analysis in thick organic specimens by photon excited X-ray fluorescence. Phys. Scripta 15, 220-224.

7. Rindby, A, P. Standzenieks and E. Selin. 1978. Report from the Swedish Work Environment Fund no. 134, 22.27.

LIST OF PARTICIPANTS

Abdulla, Mr. M.
Univ. Hospital
Dept. of Clin. Chem.
S-22185 Lund
Sweden

Allain, Mr. Pierre
Lab. de Pharm.
Gie, centre Hospitalier
Universitaire
F-49040 Angers
France

Alnor, Mr. C.C.
Kirkevejen 20
Teestrup
DK-4690 Haslev
Denmark

Alt, Mr. Friedrich
Inst. f. Spektrochemie
Bunsen-Kirchhoff-Straße 11
D-4600 Dortmund
FR Germany

Anke, Mr. M.
Karl-Marx-Univ.
Dornburgerstraße 24
DDR-6900 Jena
Germ. Dem. Rep.

Aras, Mr. Namik
Middle East Techn. Univ.
Dept. of Chem.
TR-Ankara
Turkey

Asche, Mr. Wolfgang
Chem. Rundschau, Presse
Pappelallee 3
D-6940 Weinheim
FR Germany

Aubree, Mr. Eric
Fa. BSN, Paris
7, Rue Teheran
F-75008 Paris
France

Backhaus, Mr. Gerd
Instruments S.A. GmbH
Hauptstraße 68
D-8025 Unterhaching
FR Germany

Baumgardt, Mrs. Barbara
Inst. f. Spektrochemie
Bunsen-Kirchhoff-Straße 11
D-4600 Dortmund 1
FR Germany

Behne, Mr. Dietrich
Hahn-Meitner-Institut
Glienicker Straße 100
D-1000-Berlin
FR Germany

Benes, Mr. Josef
Czech. Academie of Science
Inst. of Nuclear Biology
and Radiochemistry
CS-14220 Prague 4-KRC
Czechoslovakia

Berg, Mr. Dieter
GSF, Abt. f. Nuklearbiologie
Ingolstädter Landstraße 1
D-8042 Neuherberg
FR Germany

Beyerle-Pfuer, Mrs. Rut
GSF, Inst. f. Ök. Chemie
Ingolstädter Landstraße 1
D-8o42 Neuherberg
FR Germany

Bischof, Mr. Wolfgang
Ruhr Univ.
Inst. f. Experim. Physics
D-4630 Bochum 1
FR Germany

Brätter, Mr. Peter
Hahn-Meitner-Institut
Glienickerstraße 100
D-1000-Berlin 39
FR Germany

Bruhn, Mr. Carlos
Inst. f. Spektrochemie
P.O.Box 778
D-4600 Dortmund
FR Germany

Burow, Mrs. Mechthild
KFA Jülich
Inst. f. Chemie
Postfach 1913
D-5170 Jülich
FR Germany

Byrne, Mr. Anthony
"J. Stefan" Institute
Jamova 39
YU-61000 Ljubljana
Yugoslavia

Camara, Mrs. Carmen
Univ. Complutense, Facultad
de Dciencias Quimicas, Dept. de
Quimica Analitica
E-28040 Madrid
Espania

Chatt, Mr. A.
Trace Analysis Research
Center Dept. of Chemistry
Dalhousie Univ.
CDN-B3H4 Ji Halifax, N. Scotia
Canada

Chavez, Mr. Eduardo
Faculty of Agriculture
Mc Gill Univ.
Mac Donald College
CDN-Montreal
Canada

Christensen, Mr. Jytte Molin
Danish National Institute
of Occupational Health
Baunegaardsvej 73
DK-2900 Hellerup
Denmark

Christophe, Mr. C.
1. Quai des Bateliers
F-67000 Strasbourg
France

Cornelis, Mrs. Rita
Inst. for Nucl. Sciences
Dept. of Anal. Chemistry
Proeftuinstraat 86
B-9000 Ghent
Belgium

Cossak, Mr. Z.T.
Odense Univ.
Inst. of Comm. Health
J.B. Winslows Vej 19
DK-5000 Odense C
Denmark

Couzy, Mr. Francois
Div. de Chimie
Toxicologie
F-75731 Paris 15
France

Dammers, Mrs. Heide
Wierenweg 2
D-2000 Hamburg
FR Germany

Damsgaard, Mrs. Else
Isotope Div.
Riso Nat. Lab.
DK-4000 Roskilde
Denmark

Deschuytere, Mrs. Antoinette
Univ. of Antwerp
Universiteitsplein 1
B-2610 Wilryk
Belgium

Dirschl, Mr. Cyrill
Inst. f. Hygiene und
Technologie des Lebensmittel-
techn. Ursprungs LMU München
D-8000 München
FR Germany

Döring, Mrs. Angela
GSF, MEDIS-Institut
Ingolstädter Landstraße 1
D-8042 Neuherberg
FR Germany

Dornemann, Mr. A.
Bayer AG
GB Anorganica Forschung
(Untersuchungslabor)
D-4150 Krefeld 11
FR Germany

Edlund, Mr. Olof
ACO Laekemedel, Sweden
P.O.Box 3026
S-17103 Solna
Sweden

Eisenmann, Mr. Thomas
GSF, Inst. f. Ökolog. Chemie
Ingolstädter Landstraße 1
D-8042 Neuherberg
FR Germany

Erler, Mr. Wolfgang
Bodenseewerk
Perkin-Elmer & Co GmbH
Postfach 11 20
D-7770 Überlingen
FR Germany

Erzberger, Mrs. Angela
BGA, Strahlenhygiene
D-8042 Neuherberg
FR Germany

Everett, Mr. George
Dept. of Agriculture ARS
Ithaca, NY 14850
USA

Faassen van, Mrs. Arendina
P.O.Box 360
NL-Zeist 3700 AJ
The Netherlands

Fremmer, Mrs. Roswitha
GSF, Projektträgerschaft
Josephspitalstraße 15
D-8000 München 2
FR Germany

Frenzel, Mr. W.
Hahn-Meitner-Institut
Glienickerstraße 100
D-1000 Berlin 39
FR Germany

Fresenius, Mr. Wilhelm
Inst. Fresenius GmbH
Im Meisel
D-6204 Taunusstein 4
FR Germany

Furtmayer-Schuh, Mrs.
F A Z
D-8032 Gräfelfing
FR Germany

Galorini, Mr Mario
CNR National Research
Counsil. Center for Radiochem.
and Activation Analysis
Via Taranelli 12
I-27100 Pavia
Italy

Gardiner, Mr. Ph.
Hahn-Meitner-Institut
Glienicker Straße 100
D-1000 Berlin 39
FR Germany

Gatschke, Mr. W.
Hahn-Meitner-Institut
Glienicker Straße 100
D-1000 Berlin 39
FR Germany

Gawlik, Mr. D.
Hahn-Meitner-Institut
Glienicker Straße 100
D-1000 Berlin 39
FR Germany

Geldmacher-Malinckrodt von, Mrs. M.
Inst. f. Rechtsmedizin
Univ. Erlangen
Universitätstraße 22
D-8520 Erlangen
FR Germany

Gerken, Mr.
Hahn-Meitner-Institut
Glienicker Straße 100
D-1000 Berlin 39
FR Germany

Gessner, Mrs. H.
Hahn-Meitner-Institut
Glienicker Straße 100
D-1000 Berlin 39
FR Germany

Gharib, Mr. Ahmad
Atomic Energy Org. of Iran
IR-Teheran
Iran

Grill, Mr. Peter
GSF, Inst. f. Ökolog. Chemie
Ingolstädter Landstraße 1
D-8042 Neuherberg
FR Germany

Grobecker, Mr. Karlheinz
D-6336 Solms
FR Germany

Gutierrez, Mrs. Ana
Univ. Complutense de Madrid
Quimicas, Dept. de Quimica
E-28040 Madrid
Espania

Halbach, Mr. Stefan
GSF, Pharmakologie
Ingolstädter Landstraße 1
D-8042 Neuherberg
FR Germany

Hamann, Mr. Hans
GSF, IB
Ingolstädter Landstraße 1
D-8042 Neuherberg
FR Germany

Hamer van den, Mr. Cornelis
Interuniversity Reactor Inst.
Mekelweg 15
NL-2629 Delft
The Netherlands

Hansen, Mrs. Christine
GSF, Biophys. Strahlenforschung
Paul-Ehrlich-Straße 15
D-6000 Frankfurt
FR Germany

Harlozinska-Szmyrka, Mrs. Antonina
Zaklad Anatomii Patolog.
Ul. Marcinkowskiego 1
Pl-50368 Wroclaw
Poland

Harmuth-Hoene, Mrs. Anna-E
BFA f. Ernährung
Engesserstraße 20
D-7500 Karlsruhe
FR Germany

Hartfiel, Mr. Wilhelm
Inst. f. Tierernährung
Univ. Bonn
Endenicher Allee 15
D-5300 Bonn
FR Germany

Harto, Mr. Anders
Univ. Complutense Madrid
Facultad de Ciencias
Quimicas, Dept. de Quimica Anal.
E-28040 Madrid
Spain

Hasse, Mrs. Sylvia
GSF, Inst. f. Ökolog. Chemie
Ingolstädter Landstraße 1
D-8042 Neuherberg
FR Germany

Hecht, Mr. Hermann
BA f. Fleischforschung
D-8650 Kulmbach
FR Germany

Heck, Mr. Dieter
Kernforschungszentrum Karlsruhe
Inst. f.Kernphysik
Postfach 36 40
D-7500 Karlsruhe 1
FR Germany

Heilmaier, Mr. Herbert
GSF, Abt. f. Toxikologie
Ingolstädter Landstraße 1
D-8042 Neuherberg
FR-Germany

Heinitz, Mr. Matthias
Leopoldstraße 23
D-7570 Baden-Baden
FR Germany

Hense, Mr. Hans-Werner
GSF, MEDIS-Institut
Ingolstädter Landstraße 1
D-8042 Neuherberg
FR Germany

Henze, Mr. Wolfgang
Anorg. Chem. Institut
Münster
Correnstraße
D-4400 Münster
FR Germany

Herak, Mr. Mark
Faculty of Science
Univ. of Zagreb
Lab. of Analytical Chemistry
Strossmayerov Trg. 14
YU-41000 Zagreb
Yugoslavia

Herber, Mr. Robert
Coronel Lab. for Occupational and
Environm. Health
Univ. of Amsterdam
Meibergdreef 15
NL-1105 AZ Amsterdam
The Netherlands

Hewitt, Mrs. Caroline
Univ. of Manchester
Dept. of Chemistry
GB-Manchester MB9PL
Great Britain

Hofner, Mrs. Eva
GSF, Abt. f. ATS
Ingolstädter Landstraße 1
D-8042 Neuherberg
FR Germany

Hurell, Mr. Richard
Nestec Ltd. Research Dept.
Av. Nestle 55
CH-1800 Vevey
Switzerland

Ingrao, Mr. Giovanni
Enea-Cre Casaccia
Pas-Scamb C.P. 2400
I-00100 Roma
Italy

Iyengar, Mr. G.V.
Gaithersburg NBS, MD 20899
USA

Johansson, Mr. Erland
Gustav-Werner-Institute
Dept. of Physical Biology
S-75121 Uppsala
Sweden

Jorgensen, Mr. Poul
Klinisch Kemisch Abt.
Univ. Krankenhaus
DK-5000 Odense
Denmark

Kaenzig, Mr. A.
Inst. f. Toxikologie der
Univ. Zürich
Schorenstr. 16
Ch-8603 Scherzenbach
Switzerland

Keim, Mr. Alfred
Bundesforschungsanstalt
f. Ernährung
Engesserstraße 20
D-7500 Karlsruhe
FR Germany

Kerbl, Mr. Hannes
Leopold & Co
A-8010 Graz
Austria

Kiefer, Mr. Felix
Wander AG
Postfach 2747
Ch-3001 Bern
Switzerland

Kiem, Mr. Josef
KFA Jülich
Inst. f. Medizin
Postfach
D-5170 Jülich
FR Germany

Kirchgessner, Mr. M.
Inst. f. Ernährungsphys.
der TU München
D-8050 Freising-Weihenstephan
FR Germany

Klevay, Mr. Leslie
Res. Medical Officer
P.O.Box 7166, 2420, 2ND Av.
Grand Forks, N.D. 58202
USA

Klug, Mr. Jürgen
Wierenweg 2
D-2000 Hamburg
FR Germany

Knapp. Mr. Günter
Techn. Univ. Graz
Inst. f. Analyt. Chemie
Mikro- u. Radiochemie
Technikerstraße 4
A- 8010 Graz
Austria

Knutti, Mr. R.
Inst. f. Toxikologie
ETH Univ. Zürich
Schorenstraße 16
Ch-8603 Schwerzenbach
Switzerland

Köhler, Mr. G.
Hochstraße 14
D-6146 Alsbach
FR Germany

Kollmer, Mr. Willy
GSF, Abt. f. Nuklearbiologie
Ingolstädter Landstraße 1
D-8042 Neuherberg
FR Germany

Kozak, Mrs. Elizabeth
Boris Kidric Inst. of Chemistry
Haldrihova 19
YU 61000 Lijubliana
Yugoslavia

Kron, Mr. Thomas
GSF, Inst. für
Biophys. Strahlenforschung
Paul-Ehrlich-Straße 20
D-6000 Frankfurt
FR Germany

Kruse-Jarres, Mr. J.D.
Katharinenhospital
Klinisch-Chem. Institut
Kriegsbergstraße 60
D-7000 Stuttgart
FR Germany

Kuipers, Mr. Folkert
Dept. of Pediatrics
Univ. of Groningen
Bloemsingel 10
NL-9712 K2 Groningen
The Netherlands

Kumpulainen, Mrs. Jorma
Agric. Res. Centre
Central Laboratory
SF-31600 Jokioinen
Finland

Kuusniemi, Mr. Ossi
Huhtamaeke oy, Novamed
PL 415
SF-20101 Turku
Finland

Laakso, Juha
Mila 04
SF-00200 Helsinki
Finland

Ladefoged, Mrs. Karin
Rigshospitalet
9 Blegdamsvej
DK-2100 Kopenhagen
Denmark

Lamberty, Mr. Andree
CBNM-EURATOM
B-2440 Geel
Belgium

Ledersteger, Mr. Alfred
Karl-Franzens Univ. Graz
Inst. f. Analyt. Chemie
Universitätsplatz 1
A-8010 Graz
Austria

Lehto, Mr. Jari
Univ. of Helsinki
Dept. of Public Health
Haartmaninkatu 3
SF-00290 Helsinki
Finland

Leroy, Mr. Maurice
Ecole Nationale Sup. de Chemie
1, Rue B. Pascale
F-6700 Strasbourg
France

Lieser, Mr. Karl Heinrich
D-6100 Darmstadt
FR-Germany

Lill, Mrs. Gisela
GSF, Inst. f. Ökolog. Chemie
Ingolstädter Landstraße 1
D-8042 Neuherberg
FR Germany

Lörinci, Gabor
GSF, Inst. f. Ökolog. Chemie
Ingolstädter Landstraße 1
D-8042 Neuherberg
FR Germany

Luten, Mr. Joop
TNO-CIVO Food Techn. Inst.
P.O.Box 183
NL-1970 AD Ijmuiden
The Netherlands

Maekelea, Mrs. Anna-Liisa
Turku Universityhospital
Dept. of Paediatrics
SF-20520 Turku
Finland

Maenhaut, Mr. Willy
Inst. voor Nucleaire
Wetenschappen
Proeftuinstraat 86
B-9000 Ghent
Belgium

Maichin, Mrs. Brigitte
Inst. f. Analyt. Chemie
TU-Graz
Techniker Straße 4
A-8010 Graz
Austria

Maier, Mr. Eddie
Ecole Nationale Sup. de Chemie
1, Rue B. Pascale
F-67000 Strasbourg
France

Mansour, Mr. M.
GSF, Inst. f. Ökolog. Chemie
D-8051 Attaching
FR-Germany

Massol, Mr. Michel
Lab. de Chimie
Bioinorganique
Univ. Paul Sabatier
38, Rue D.36 Ponts
F-31400 Toulouse
France

Mattsson, Mr. Per
The Swedisch Nutrional
Food Administration
S-75126 Uppsala
Sweden

Mayer, Mr. Rolf
Varian GmbH
Arnulfstraße 44
D-8000 München 2
FR-Germany

Merian, Mr. Ernst
Chemosphere u.a. Wiss.
Zeitschriften
Im Kirschgarten 22
CH-4106 Therwil
Switzerland

Messerschmidt, Mr. Jürgen
Inst. f. Spektrochemie u.
Angew. Spektroskopie
Bunsen-Kirchhoff-Straße 11
D-4600 Dortmund
FR-Germany

Miha, Mr. Benedik
Univ. Clinical Centre
Dialysis Center, Ljubljana
YU-Ljubljana
Yugoslavia

Mills, Mr. C.F.
Dept. of Inorganic Chemistry
Greenborn Road
GB-AB2 Bucksburn, Aberdeen, 9 SB
Great Britain

Müller, Mr. Wilhelm
GSF, Ingolstädter Landstraße 1
D-8042 Neuherberg
FR Germany

Muntau, Mr. H.
JRC, Commission of the
European Communities
I-21020 Ispra, Varese
Italy

Nabarawy, Mrs. F.
GSF, Inst. f. Biologie
Ingolstädter Landstraße 1
D-8042 Neuherberg
FR Germany

Naentoe, Mr. Veikko
Univ. of Turku
Dept. of Clinical Chemistry
SF-20520 Turku
Finland

Nederbragt, Mr. H.
Dept. of Veterinary
Pathology
Biltstraat 172
NL-3572 Utrecht, BP
The Nederlands

Negretti de Brätter, Mrs. V.E.
Hahn-Meitner-Inst.
Glienicker Straße 100
D-1000 Berlin 39
FR Germany

Nepveu, Mrs. Francoise
Lab. d. Chimie Bioinorg.
Univ. Paul Sabatier
38. Rue d. Trente-Six-Ponts
F-31400 Toulouse
France

Neve, Mr. Jean
Univ. of Brussel
Institute of Pharmacy
Campus Plaine 205-1
B-1050 Brussels
Belgium

Ochs, Mr. Andreas
Med. Univ. Klinik Freiburg
Habichtweg 9
D-7800 Freiburg
FR Germany

Oerndahl, Mr. Gustav
Kampegaten 36
Molndal
Sweden

Omang, Sverre
Oslo Board of Health
Dept. of Food Control
St. Olavs Plass 5
N-0165 Oslo 1
Norway

Oster, Mr. Oskar
Abt. Klin. Chem. Univ. Mainz
D-6500 Mainz
FR Germany

Pais, Mr. Istvan
Dept. of Chemistry at the
University of Horticulture
XI. Villany ut 29/31
H-1114 Budapest
Hungary

Parr, Mr. Robert
International Atomic
Energy Agency
P.O.Box 100
A-1400 Vienna
Austria

Pedersen, Mr. Ejivind
Consult Pharma APS
Toftebakken 13
DK-3460 Birkerod
Denmark

Perzl, Mr. Franz
GSF, Wiss.-techn. Abt.
Ingolstädter Landstraße 1
D-8042 Neuherberg
FR Germany

Pfannhauser, Mr. Werner
Forschungsinst. f. Ernährung
Blaasstraße 29
A-1190 Vienna
Austria

Pohl, Mrs. Barbara
Varian GmbH
Arnulfstraße 44
D-8000 München 2
FR Germany

Prellwitz, Mr. W.
Univ. Mainz, II. Med. Klinik
Langenbeckstraße 1
D-6500 Mainz
FR Germany

Rastegar, Mr. F.
Ecole Nationale Sup. de Chimie
1, Rue B. Pascale
F-6700 Strasbourg
France

Ribas, Mr. Bartolome
Complutense Univ.
Consejo Sup. Invest. Cientificas
Faculty of Pharmacy
E-28040 Madrid
Espania

Rindby, Mr. Anders
Chalmers Univ. of Technol.
S-41296 Goeteborg
Sweden

Runkel, Mr. Karl-Heinz
GSF, Abt. f. Toxikologie
Ingolstädter Landstraße 1
D-8042 Neuherberg
FR Germany

Saeed, Mr. Khalid
Collett-Harwell Hauge
P.O.Box 205
N-1371 Asker
Norway

Sampson, Mr. Barry
Dept. Chemical Pathology
Charing Cross Hospital
Fulham
GB-W6RF London
Great Britain

Sandberg, Mr. Finn
Journalist
DK-4520 Svinninge
Denmark

Sansoni, Mr. Bruno
Kernforschungsanlage Jülich
ZCH
D-5170 Jülich
FR Germany

Sargentini, Mrs. Maria Laura
Ecole Nationale Sup. de Chimie
1, Rue B. Pascale
F-6700 Strasbourg
France

Schaupp, Mrs. Karin
Leopold & Co. GmbH
Hafnerstraße 36
A-8055 Graz
Austria

Schelenz, Mr. Rieder
International Atomic
Energy Agency
P.O. Box 100
A-1400 Vienna
Austria

Schenkel, Mr. Hans
Univ. Hohenheim
Inst. f. Tierernährung
D-7000 Stuttgart
FR Germany

Schlegel, Mr. Heinz
Heinz Schlegel AG
Industriestraße 3
Ch-4313 Möhlin
Switzerland

Schmidt, Mr. Peter-Fritz
Inst. f. Med. Physik
Hüfferstraße 68
D-4400 Münster
FR Germany

Schoor van, Mr. Omer
University of Antwerp
Universiteitsplein 1
B-2610 Wilryk
Belgium

Schothorst, Mr. Ronald
Univ. of Amsterdam
Lab. f. Occup. & Env. Health
Meibergdreef 15
NL-1105 Amsterdam A2
The Netherlands

Schramel, Mr. Peter
GSF, Inst. f. Ökolog. Chemie
Ingolstädter Landstraße 1
D-8042 Neuherberg
FR Germany

Schrauzer, Mr. G.N.
Univ. of California, San Diego
La Jolla, CA. 92093
USA

Shamberger, Mr. Raymond
Section of Enzymology
9500 Euclid Avenue
Cleveland, Ohio 44106
USA

Smit, Mr. Egbert
Dept. of Physics, State
Univ. of Groningen
Westersingel 34
NL-Groningen 9718 CM
The Netherlands

Smith, Mr. Norman
Health and Safety
Executive, Room 2/32
West Street, Sheffield S1 2GQ
403 Edgeware Road
GB-London NW 26LN
Great Britain

Srianujata, Mr. Songsak
Faculty of Medicine
Ramathibodi Hospital
Research Center
Rama Vi Road
T-10400 Bangkok
Thailand

Stahl, Mr. Martin
URSAPHARM
Schulze-Delitzsch-Straße 2
D-4790 Paderborn
FR Germany

Stahle, Mr. Inga
United Lab. Ltd.
P.O.Box 70
SF-00511 Helsinki
Finland

Stampfl, Mr. Andreas
GSF, Abt. f. Pharmakologie
Ingolstädter Landstraße 1
D-8042 Neuherberg
FR Germany

Stelte, Mr. Wilhelm
Inst. f. Ernährungswiss.
Justus-Liebig-Univ.
Goethestraße 55
D-6300 Giessen
FR-Germany

Stockhausen von, Mr. Hans-B.
Universität Würzburg
Dept. of Pediatrics
D-8700 Würzburg
FR Germany

Stoeppler, Mr. Markus
Kernforschungsanlage
Jülich, ICH 4
P.O.Box 19 13
D-5170 Jülich
FR Germany

Strehlke, Mr. A.F.P.
Neue Ärztl. Zeitung
Presse
Josephsburgerstraße 26
D-8000 München 80
FR Germany

Summer, Mr. K.H.
GSF, Abt. f. Toxikologie
Ingolstädter Landstraße 1
D-8042 Neuherberg
FR Germany

Teigen, Mr. S.W.
Collet-Marwell Hauge
Gruppen
P.O.Box 205
N-1371 Asker
Norway

Thomassen, Mr. Yngvar
Inst. of Occup. Health
P.O.Box 81 49
N-0033 Oslo 1
Norway

Tiggelen van, Mr. Cees
Consultant Geriatrician
Farmers Road
AUS-3956 Meeniyan, Victoria
Australia

Tobler, Mr. Leonhard
Eidg. Inst. f. Reaktor-
forschung Wuerenlingen
Ch-5303 Wuerenlingen
Switzerland

Vandenhaute, Mr. Joris
Inst. voor Nucleaire
Wetenschappen
Proeftuinstraat 86
B-9000 Ghent
Belgium

Verbueken, Mr. Armand
University of Antwerp
B-2610 Antwerp-Wilrisk
Belgium

Versieck, Mr. Jacques
Dept. of Intern. Med.
Div. of Gastroenterology
De Pintelaan 185
B-9000 Ghent
Belgium

Viswanathan, Mr. R.
GSF, Inst. f. Ökolog. Chemie
Ingolstädter Landstraße 1
D-8042 Neuherberg
FR Germany

Voget, Mr. Michael
GSF, Inst. f. Ökolog. Chemie
Ingolstädter Landstraße 1
D-8042 Neuherberg
FR Germany

Wallaeys, Mr. Bart
Lab. of Analyt. Chem.
Proeftuinstraat 86
B-9000 Ghent
Belgium

Weber, Mr. Georges
IPNE
Univ. of Liege
Sart-Tilman B 15
B-4000 Liege
Belgium

Weigert, Mr. Peter
BGA, (ZEBS)
Werner-Voss-Damm 62
D-1000 Berlin 42
FR Germany

Werner, Mr. Eckard
GSF, Inst. f. Biophys.
Strahlenforschung
Paul-Ehrlich-Straße 20
D-6000 Frankfurt
FR Germany

Wigfield, Mr. Donald
Dept. of Chemistry
Carleton Univ. Otawa
CDN-Ottawa, KIS 5B6
Canada

Wittl, Mrs. Christiane
GSF, Inst. f. Ökolog. Chemie
Ingolstädter Landstraße 1
D-8042 Neuherberg
FR Germany

Woittiez, Mr. Joost
NL-1755 ZG Petten
The Netherlands

Wolf, Mr. Wayne
United State
Dept. of Agriculture
Beltsville Maryland 20105
USA

Wyttenbach, Mr. Armin
Eidg. Inst. f. Reaktor-
forschung Wuerenlingen
CH-5303 Wuerenlingen
Switzerland

Zeisler, Mr. Rolf
Center for Analytical Chemistry
National Bureau of Standards
Bldg. 235, B-108
Gaithersburg, MD 20899
USA

Zwennis, Mr. W.C.M.
Medisch Biologisch Lab.
Lange Kleiweg 139
NL-2200 Ryswik
The Netherlands

AUTHOR INDEX

SUBJECT INDEX

A

D

E

F

G

H

I

P

Q

R

S

T

Bei Fragen zur Produktsicherheit wenden Sie sich bitte an:
If you have any questions regarding product safety,
please contact:

Walter de Gruyter GmbH
Genthiner Straße 13
10785 Berlin
productsafety@degruyterbrill.com